Albrecht Dold was born on August 5, 1928 in Triberg (Black Forest), Germany. He studied mathematics and physics at the University of Heidelberg, then worked for some years at the Institute for Advanced Study in Princeton, at Columbia University, New York and at the University of Zürich. In 1963 he returned to Heidelberg where he has stayed since, declining several offers to attractive positions elsewhere.

A. Dold's seminal work in algebraic topology has turned him into a figure of recognition beyond the world of mathematics itself. In particular, his work on fixed-point theory has made his a household name in economics, and this book a standard reference among economists as well as mathematicians.

Albrecht Dold was born on August 5, 1928 in Triberg (Black Forest), Germany. He studied mathematics and physics at the University of Heidelberg, then worked for some years at the Institute for Advanced Study in Princeton, at Columbia University, New York and at the University of Zürich. In 1963 he returned to Heidelberg where he has stayed since, declining several offers to attractive positions elsewhere.

A. Dold's seminal work in algebraic topology has brought him international recognition beyond the world of mathematics itself. In particular, his work on fixed-point theory has made his a household name in economics, and this book a standard reference among economists as well as mathematicians.

Classics in Mathematics

Albrecht Dold   Lectures on Algebraic Topology

Albrecht Dold

Lectures on Algebraic Topology

Reprint of the 1972 Edition

Springer

Albrecht Dold
Mathematisches Institut der Universität Heidelberg
Im Neuenheimer Feld 288
69120 Heidelberg
Germany

Originally published as Vol. 200 of the
Grundlehren der mathematischen Wissenschaften

Mathematics Subject Classification (1991):
Primary 57N65, 55N10, 55N45, 55U15-25, 54EXX
Secondary 55M20-25

ISBN 3-540-58660-1 Springer-Verlag Berlin Heidelberg New York

Photograph by kind permission of Alfred Hofmann

CIP data applied for

Springer-Verlag is a part of Springer Science+Business Media

springeronline.com

SPIN 11326144 41/3111 - 5 4 3 2 – Printed on acid-free paper

A. Dold

Lectures on Algebraic Topology

With 10 Figures

Springer-Verlag
Berlin Heidelberg New York 1972

Albrecht Dold
Mathematisches Institut der Universität Heidelberg

AMS Subject Classifications (1970)
Primary 57A65, 55B10, 55B45, 55J15-25, 54E60
Secondary 55C20-25

ISBN 3-540-05777-3 Springer-Verlag Berlin Heidelberg New York

ISBN 0-387-05777-3 Springer-Verlag New York Heidelberg Berlin

 Library of Congress Catalog Card Number 72-79062
Monophoto typesetting, offset printing and binding: Universitätsdruckerei H. Stürtz AG, Würzburg

Foreword

This is essentially a book on singular homology and cohomology with special emphasis on products and manifolds. It does not treat homotopy theory except for some basic notions, some examples, and some applications of (co-)homology to homotopy. Nor does it deal with general(-ised) homology, but many formulations and arguments on singular homology are so chosen that they also apply to general homology. Because of these absences I have also omitted spectral sequences, their main applications in topology being to homotopy and general (co-)homology theory. Čech-cohomology is treated in a simple ad hoc fashion for locally compact subsets of manifolds; a short systematic treatment for arbitrary spaces, emphasizing the universal property of the Čech-procedure, is contained in an appendix.

The book grew out of a one-year's course on algebraic topology, and it can serve as a text for such a course. For a shorter basic course, say of half a year, one might use chapters II, III, IV (§§ 1–4), V (§§ 1–5, 7, 8), VI (§§ 3, 7, 9, 11, 12). As prerequisites the student should know the elementary parts of general topology, abelian group theory, and the language of categories – although our chapter I provides a little help with the latter two. For pedagogical reasons, I have treated *integral* homology only up to chapter VI; if a reader or teacher prefers to have general coefficients from the beginning he needs to make only minor adaptions.

As to the outlay of the book, there are eight chapters, I–VIII, and an appendix, A; each of these is subdivided into several sections, § 1, 2, ... Definitions, propositions, remarks, formulas etc. are consecutively numbered in each §, each number preceded by the §-number. A reference like III, 7.6 points to chap. III, § 7, no. 6 (written 7.6) – which may be a definition, a proposition, a formula, or something else. If the chapter number is omitted the reference is to the chapter at hand. References to the bibliography are given by the author's name, e.g. Seifert-Threlfall; or Steenrod 1951, if the bibliography lists more than one publication by the same author.

The exercises are meant to provide practice of the concepts in the main text as well as to point out further results and developments. An exercise or its solution may be needed for later exercises but not for the main text. Unusually demanding exercises are marked by a star, $*$.

I have given several courses on the subject of this book and have profited from many comments by colleagues and students. I am particularly indebted to W. Bos and D.B.A. Epstein for reading most of the manuscript and for their helpful suggestions.

Heidelberg, Spring 1972 ALBRECHT DOLD

Contents

To Z.

Chapter I

Preliminaries on Categories, Abelian Groups and Homotopy

The purpose of this chapter is to provide the reader of the book with quick references to the subjects of the title. The content is motivated by the needs of later chapters, and not by intrinsic considerations. The reader should have some elementary knowledge of categories and abelian groups; otherwise he might find the treatment too concise. But even with very little knowledge he should probably start the reading with Chapter II, and refer to Chapter I only when necessary. He may then find the reference in I too short, insufficient (some proofs are omitted), or too ad-hoc; in that case he should consult the relevant literature, samples of which are listed at the end of §1 and §2.

The customary language and notation of set theory (such as $\cup$, $\cap$, $\subset$, $\in$, $\emptyset$, $X \times Y$, $f\colon X \to Y$, $x \mapsto y$, $\{x \in X \mid x \text{ has property } P\}$, etc.) are used without comment. Similarly, the reader is assumed to know the elementary parts of general topology.

Some basic sets and spaces are denoted by special symbols which are fixed throughout the book. For instance,

$\mathbb{N}$ = set of natural numbers,
$\mathbb{Z}$ = ring of integers,
$\mathbb{Q}, \mathbb{R}, \mathbb{C}$ = field of *rational* numbers, *real* numbers, *complex* numbers, with the usual topology,
$\mathbb{R}^n = \mathbb{R} \times \mathbb{R} \times \cdots \times \mathbb{R}$, $\mathbb{C}^n = \mathbb{C} \times \mathbb{C} \times \cdots \times \mathbb{C}$, ($n$ factors),
$\mathbb{B}^n = \{x \in \mathbb{R}^n \mid \|x\| \leq 1\}$, where $\|x\|^2 = \sum_{i=1}^n x_i^2$,
$\mathbb{S}^{n-1} = \{x \in \mathbb{R}^n \mid \|x\| = 1\} = (n-1)$-sphere,
$[0,1] = \{t \in \mathbb{R} \mid 0 \leq t \leq 1\}$ = unit interval.

1. Categories and Functors

1.1 Definition. A *category* $\mathscr{C}$ consists of

(i) a class of *objects*, denoted by $\mathrm{Ob}(\mathscr{C})$. When there is no danger of confusion we also write $\mathscr{C}$ instead of $\mathrm{Ob}(\mathscr{C})$.

(ii) For every pair X, Y of objects, a set of *morphisms from X to Y*, denoted by $\mathscr{C}(X, Y)$ or $[X, Y]$. If $\alpha \in \mathscr{C}(X, Y)$ then X is called the *domain* of α and Y the *range* of α; one also writes $\alpha: X \to Y$, or $X \xrightarrow{\alpha} Y$, or simply $X \to Y$ to denote morphisms from X to Y.

(iii) For every ordered triple of objects X, Y, Z a map from $\mathscr{C}(X, Y) \times \mathscr{C}(Y, Z)$ to $\mathscr{C}(X, Z)$, called *composition*; the image of (α, β) is denoted by $\beta \circ \alpha$ or $\beta\alpha$, and is called the *composite* of α and β.

These data have to satisfy the following two axioms

(iv) $\gamma \circ (\beta \circ \alpha) = (\gamma \circ \beta) \circ \alpha$ (*associativity*) whenever $X \xrightarrow{\alpha} Y \xrightarrow{\beta} Z \xrightarrow{\gamma} W$.

(v) There exists an *identity* morphism $\mathrm{id} = \mathrm{id}_X: X \to X$, for every object X, such that

$$\alpha \circ \mathrm{id}_X = \alpha, \quad \mathrm{id}_Y \circ \alpha = \alpha$$

whenever $\alpha: X \to Y$. These identities are easily seen to be unique ($\mathrm{id}_X^1 = \mathrm{id}_X^1 \circ \mathrm{id}_X^2 = \mathrm{id}_X^2$).

1.2 Examples. (i) The category of sets, $\mathscr{C} = \mathscr{S}ets$. The objects of this category are arbitrary sets ($\mathrm{Ob}(\mathscr{S}ets)$ = class of all sets), morphisms are maps ($[X, Y]$ = set of all maps from X to Y), and composition has the usual meaning.

(ii) The category of abelian groups, $\mathscr{C} = \mathscr{A}\mathscr{G}$. Here, $\mathrm{Ob}(\mathscr{A}\mathscr{G})$ is the class of all abelian groups, $[X, Y] = \mathrm{Hom}(X, Y)$ is the set of all homomorphisms from X to Y, and composition has the usual meaning.

(iii) The category of topological spaces, $\mathscr{C} = \mathscr{T}op$. Here, $\mathrm{Ob}(\mathscr{T}op)$ is the class of all topological spaces, $[X, Y]$ is the set of all continuous maps from X to Y, and composition has the usual meaning.

(iv) The homotopy category, $\mathscr{C} = \mathscr{H}tp$, as defined in I.3, has the same objects as $\mathscr{T}op$, but the morphisms are not mappings in the usual sense.

(v) Every quasi ordered set C can be viewed as a category $\mathscr{C}$ as follows: $\mathrm{Ob}(\mathscr{C}) = C$, $\mathscr{C}(X, Y) = \emptyset$ for elements $X, Y \in C$ which are not comparable, and $\mathscr{C}(X, Y)$ consists of a single element (X, Y) if $X \leq Y$. Conversely, if $\mathscr{C}$ is a category such that no $\mathscr{C}(X, Y)$ has more than one element and if $\mathrm{Ob}(\mathscr{C})$ is a set then $\mathrm{Ob}(\mathscr{C})$ is quasi-ordered by putting $X \leq Y \Leftrightarrow \mathscr{C}(X, Y) \neq \emptyset$.

(vi) Every group G gives rise to a category $\mathscr{C}$ with a single object e, $\mathrm{Ob}(\mathscr{C}) = \{e\}$, with $\mathscr{C}(e, e) = G$, and composition defined by multiplication.

(vii) If $\mathscr{C}$ is a category then the *dual* or *opposite* category $\mathscr{C}^{\mathrm{op}}$ is defined as follows: $\mathrm{Ob}(\mathscr{C}^{\mathrm{op}}) = \mathrm{Ob}(\mathscr{C})$, $\mathscr{C}^{\mathrm{op}}(X, Y) = \mathscr{C}(Y, X)$, $\beta * \alpha = \alpha \circ \beta$ where $*$ denotes composition in $\mathscr{C}^{\mathrm{op}}$.

(viii) If $\mathscr{C}_1$ and $\mathscr{C}_2$ are categories, then the product category $\mathscr{C}=\mathscr{C}_1\times\mathscr{C}_2$ is defined as follows. $\mathrm{Ob}(\mathscr{C})=\mathrm{Ob}(\mathscr{C}_1)\times\mathrm{Ob}(\mathscr{C}_2)=$ class of all pairs (X_1, X_2) where $X_i\in\mathrm{Ob}(\mathscr{C}_i)$; $\mathscr{C}((X_1, X_2), (Y_1, Y_2))=\mathscr{C}_1(X_1, Y_1)\times\mathscr{C}_2(X_2, Y_2)$; $(\beta_1, \beta_2)\circ(\alpha_1, \alpha_2)=(\beta_1\circ\alpha_1, \beta_2\circ\alpha_2)$.

1.3 Definition. If $\mathscr{C}'$, $\mathscr{C}$ are categories then $\mathscr{C}'$ is called a *subcategory* of $\mathscr{C}$ provided

(i) $\mathrm{Ob}(\mathscr{C}')\subset\mathrm{Ob}(\mathscr{C})$,

(ii) $\mathscr{C}'(X, Y)\subset\mathscr{C}(X, Y)$ for all $X', Y'\in\mathrm{Ob}(\mathscr{C}')$,

(iii) the composites of $\alpha\in\mathscr{C}'(X', Y')$, $\beta\in\mathscr{C}'(Y', Z')$ in $\mathscr{C}'$ and $\mathscr{C}$ coincide,

(iv) the identity morphisms of $X\in\mathrm{Ob}(\mathscr{C}')$ in $\mathscr{C}'$ and $\mathscr{C}$ coincide.

If, furthermore, $\mathscr{C}'(X, Y)=\mathscr{C}(X, Y)$ for all $X', Y'\in\mathrm{Ob}(\mathscr{C}')$ then $\mathscr{C}'$ is called a *full* subcategory. A full subcategory $\mathscr{C}'$ of $\mathscr{C}$ is therefore completely determined by the class $\mathrm{Ob}(\mathscr{C}')$. For instance, the category of finite sets and (all maps) is a full subcategory of $\mathscr{S}ets$. Non-full subcategories of 1.2 (i), (ii) or (iii) are obtained by taking for $\mathscr{C}'(X, Y)$ the set of all injective (or all surjective) morphisms, and $\mathrm{Ob}(\mathscr{C}')=\mathrm{Ob}(\mathscr{C})$.

1.4 Definition. If $\alpha\colon X\to Y$, $\beta\colon Y\to X$ are morphisms (in a category $\mathscr{C}$) such that $\beta\alpha=\mathrm{id}$ then β is called a *left inverse* of α, and α a *right inverse* of β. If α admits a left inverse β_l and also a right inverse β_r then $\beta_l=\beta_l(\alpha\beta_r)=(\beta_l\alpha)\beta_r=\beta_r$; in this case, α is called an *equivalence*, or *isomorphism*, and the *inverse* (or *inverse isomorphism*) $\beta_r=\beta_l$ is denoted by α^{-1}. Two objects X, Y are said to be *equivalent* or *isomorphic*, in symbols $X\sim Y$, if an isomorphism $\alpha\in\mathscr{C}(X, Y)$ exists. For instance, an equivalence in $\mathscr{C}=\mathscr{S}ets$ is a bijective map, an equivalence in $\mathscr{C}=\mathscr{T}op$ is a homeomorphism, an equivalence in $\mathscr{C}=\mathscr{A}\mathscr{G}$ is an isomorphism in the usual sense.

1.5 Definition. Let $\mathscr{C}$ and $\mathscr{D}$ be categories. A (covariant) *functor* T from $\mathscr{C}$ to $\mathscr{D}$, in symbols $T\colon\mathscr{C}\to\mathscr{D}$, consists of

(i) a map $T\colon\mathrm{Ob}(\mathscr{C})\to\mathrm{Ob}(\mathscr{D})$, and

(ii) maps $T=T_{XY}\colon\mathscr{C}(X, Y)\to\mathscr{D}(TX, TY)$, for every $X, Y\in\mathrm{Ob}(\mathscr{C})$, which preserve composition and identities, i.e. such that

(iii) $T(\beta\circ\alpha)=(T\beta)\circ(T\alpha)$, for all morphisms $X\xrightarrow{\alpha}Y\xrightarrow{\beta}Z$ in $\mathscr{C}$,

(iv) $T(\mathrm{id}_X)=\mathrm{id}_{TX}$, for all $X\in\mathrm{Ob}(\mathscr{C})$.

A *cofunctor* (or contravariant functor) from $\mathscr{C}$ to $\mathscr{D}$ is, by definition, a functor from $\mathscr{C}$ to the dual category $\mathscr{D}^{\mathrm{op}}$. Its explicit definition is as above with (ii) replaced by $T\colon\mathscr{C}(X, Y)\to\mathscr{D}(TY, TX)$, and (iii) replaced by $T(\beta\circ\alpha)=(T\alpha)\circ(T\beta)$. Equivalently, a cofunctor from $\mathscr{C}$ to $\mathscr{D}$ is a functor from $\mathscr{C}^{op}$ to $\mathscr{D}$. A functor $\mathscr{C}_1 \times \mathscr{C}_2 \to \mathscr{D}$, where $\mathscr{C}_1 \times \mathscr{C}_2$ is a product category (1.2 (viii)) is called a *functor of two variables* (with values in $\mathscr{D}$).

1.6 Examples of Functors. (i) The *identity functor* $ID\colon \mathscr{C} \to \mathscr{C}$ which is given by $ID(X)=X$, $ID(\alpha)=\alpha$, for all objects X and morphisms α.

(ii) If $T\colon \mathscr{C} \to \mathscr{D}$ and $U\colon \mathscr{D} \to \mathscr{E}$ are functors then so is the composite $UT\colon \mathscr{C} \to \mathscr{E}$, defined by $(UT)X = U(TX)$, $(UT)(\alpha) = U(T\alpha)$.

(iii) For any fixed $D \in \mathrm{Ob}(\mathscr{D})$ we have the *constant functor* $T\colon \mathscr{C} \to \mathscr{D}$ such that $TX = D$, $T\alpha = \mathrm{id}_D$ for all X and α.

(iv) For any fixed $A \in \mathrm{Ob}(\mathscr{C})$ we have the *morphism functors* $\mathscr{C}_A\colon \mathscr{C} \to \mathscr{S}ets$, $\mathscr{C}^A\colon \mathscr{C} \to \mathscr{S}ets^{op}$, defined as follows. $\mathscr{C}_A(X) = \mathscr{C}(A,X)$, $\mathscr{C}^A(X) = \mathscr{C}(X,A)$ for all $X \in \mathrm{Ob}(\mathscr{C})$, $\mathscr{C}_A(\xi) = \xi\,\circ =$ composition with ξ on the left, $\mathscr{C}^A(\xi) = \circ\,\xi$ composition with ξ on the right, for all $\xi \in \mathscr{C}(X,Y)$. Thus,

$$(1.7)\qquad \begin{aligned} &\mathscr{C}_A(\xi)\colon \mathscr{C}(A,X) \to \mathscr{C}(A,Y), && \alpha \mapsto \xi \circ \alpha \\ &\mathscr{C}^A(\xi)\colon \mathscr{C}(Y,A) \to \mathscr{C}(X,A), && \beta \mapsto \beta \circ \xi. \end{aligned}$$

(v) If we view the groups G, H as categories, as in 1.2 (vi), then functors correspond to homomorphisms $G \to H$, and cofunctors to antihomomorphisms.

1.8 Proposition. *Let $T\colon \mathscr{C} \to \mathscr{D}$ be a (co-)functor. If $\alpha \in \mathscr{C}(X,Y)$ is an isomorphism then so is $T\alpha$, and $(T\alpha)^{-1} = T(\alpha^{-1})$.*

Indeed, $\alpha\alpha^{-1} = \mathrm{id} \Rightarrow T(\alpha)\,T(\alpha^{-1}) = T(\alpha\alpha^{-1}) = T(\mathrm{id}) = \mathrm{id}$. ∎

1.9 Definition. Let $S, T\colon \mathscr{C} \to \mathscr{D}$ be functors. A *natural transformation* Φ from S to T, in symbols $\Phi\colon S \to T$, consists of a system of morphisms $\Phi_X \in \mathscr{D}(SX, TX)$, one for each $X \in \mathrm{Ob}(\mathscr{C})$, such that all diagrams

$$(1.10)\qquad \begin{array}{ccc} SX & \xrightarrow{S\alpha} & SY \\ \downarrow{\scriptstyle \Phi_X} & & \downarrow{\scriptstyle \Phi_Y} \\ TX & \xrightarrow{T\alpha} & TY \end{array}$$

(for all $\alpha \in \mathscr{C}(X,Y)$) are commutative; in formulas, $\Phi_Y \circ (S\alpha) = (T\alpha) \circ \Phi_X$. If every Φ_X is an equivalence then Φ is called a *natural equivalence*. In this case, $\Psi_X = \Phi_X^{-1}$ *is also a natural equivalence* (just reverse the vertical arrows in (1.10)), and it is called the *inverse* natural equivalence.

1.11 Examples of Natural Transformations. (i) For every functor $T\colon \mathscr{C} \to \mathscr{D}$ the identity morphisms $\Phi_X = \mathrm{id}_{TX}\colon TX \to TX$ constitute a natural equivalence.

(ii) If $S, T, U\colon \mathscr{C} \to \mathscr{D}$ are functors, and $\Phi\colon S \to T$, $\Psi\colon T \to U$ are natural transformations then so is the *composite transformation* $\Psi \circ \Phi\colon S \to U$, where $(\Psi \circ \Phi)_X = \Psi_X \circ \Phi_X$.

(iii) Let $S = \mathscr{C}_A\colon \mathscr{C} \to \mathscr{S}ets$ a morphism functor as in (1.6 (iv)) where A is a fixed object of $\mathscr{C}$. Let $T\colon \mathscr{C} \to \mathscr{S}ets$ an arbitrary functor and let $a \in TA$ denote a fixed element in the set TA. Define $\Phi^a\colon S \to T$ as follows.

$$\Phi_X^a\colon SX = \mathscr{C}(A, X) \to TX, \qquad \Phi_X^a(\xi) = (T\xi)\,a.$$

We verify that 1.10 commutes:

$$\begin{aligned}(\Phi_Y^a \circ (S\alpha))(\xi) &= \Phi_Y^a((S\alpha)(\xi)) = \Phi_Y^a(\alpha\xi)\\ &= T(\alpha\xi)\,a = (T\alpha)(T\xi)\,a = ((T\alpha) \circ \Phi_X^a)(\xi).\end{aligned}$$

Similarly for cofunctors $T\colon \mathscr{C} \to \mathscr{S}ets$; i.e. if $A \in \mathrm{Ob}(\mathscr{C})$ and $a \in TA$ then $\Phi_X^a\colon \mathscr{C}^A(X) = \mathscr{C}(X, A) \to TX$, $\Phi_X^a(\xi) = (T\xi)\,a$, defines a natural transformation $\Phi^a\colon \mathscr{C}^A \to T$. These transformations Ψ^a are in fact the only transformations of morphism functors. More formally,

1.12 Proposition (Yoneda-Lemma). *If $T\colon \mathscr{C} \to \mathscr{S}ets$ is a functor and $\Phi\colon \mathscr{C}_A \to T$ is a natural transformation ($A \in \mathrm{Ob}(\mathscr{C})$) then there is a unique element $a \in TA$ such that $\Phi = \Phi^a$, namely $a = \Phi_A(\mathrm{id}_A)$.*

Thus, natural transformations $\mathscr{C}_A \to T$ are completely determined by their value on $\mathrm{id}_A \in \mathscr{C}_A(A)$, and this value $\Phi_A(\mathrm{id}_A)$ can be arbitrarily chosen in TA. Similarly for cofunctors $\mathscr{C} \to \mathscr{S}ets$.

Proof. If $\Phi\colon \mathscr{C}_A \to T$ is a natural transformation then the diagram

$$\begin{array}{ccc} \mathscr{C}_A(A) & \xrightarrow{\mathscr{C}_A(\xi)} & \mathscr{C}_A(X) \\ \downarrow{\scriptstyle \Phi_A} & & \downarrow{\scriptstyle \Phi_X} \\ TA & \xrightarrow{T\xi} & TX \end{array}$$

must commute for every $\xi \in \mathscr{C}_A(X) = \mathscr{C}(A, X)$. In particular, $\Phi_X(\mathscr{C}_A(\xi)(\mathrm{id}_A)) = (T\xi)(\Phi_A(\mathrm{id}_A))$. But $\mathscr{C}_A(\xi)(\mathrm{id}_A) = \xi \circ \mathrm{id}_A = \xi$, hence $\Phi_X(\xi) = (T\xi)\,a = \Phi_X^a(\xi)$, where $a = \Phi_A(\mathrm{id}_A)$. ∎

1.13 Definition. If $T\colon \mathscr{C} \to \mathscr{S}ets$ is a (co-)functor, and $A \in \mathrm{Ob}(\mathscr{C})$ then $u \in TA$ is said to be *universal* (for T) if $\Phi^u\colon \mathscr{C}_A \to T$ is a natural equivalence. Not every (co-)functor $T\colon \mathscr{C} \to \mathscr{S}ets$ admits a universal element. If it does then T is said to be *representable*, and the object A resp. the pair (A, u) are said to *represent the (co-)functor* T. Up to equivalence the pair (A, u) is uniquely determined, as follows.

1.14 Proposition. *Let* $T\colon \mathscr{C} \to \mathscr{S}ets$ *be a representable functor, with universal element* $u \in TA$. *If* C *is an object in* $\mathscr{C}$ *and* $c \in TC$ *then there is a unique morphism* $\gamma\colon A \to C$ *such that* $(T\gamma)u = c$ (by universality of u). *If* c *is also universal then* γ *is an equivalence. Similarly, for cofunctors.*

Proof. If c also universal then there is $\beta\colon C \to A$ with $(T\beta)c = u$, hence $T(\beta\gamma)u = (T\beta)(T\gamma)u = u$, hence $\beta\gamma = \mathrm{id}$ by universality of u; similarly $\gamma\beta = \mathrm{id}$. ∎

1.15 One can therefore use (co-)functors $T\colon \mathscr{C} \to \mathscr{S}ets$ to define objects in $\mathscr{C}$ (up to equivalence). This method of "definition by universal properties" is very common and very important in many branches of mathematics. As an example we consider the product of two morphism functors, say

$$T = \mathscr{C}_B \times \mathscr{C}_C\colon \mathscr{C} \to \mathscr{S}ets,$$

$$TX = \mathscr{C}(B, X) \times \mathscr{C}(C, X), \qquad T\alpha = (\mathscr{C}_B\alpha) \times (\mathscr{C}_C\alpha) = (\alpha\circ) \times (\alpha\circ).$$

If T is representable then the representing object is called the *coproduct* of B and C, and is denoted by $B \sqcup C$. The universal element $u \in T(B \sqcup C) = \mathscr{C}(B, B \sqcup C) \times \mathscr{C}(C, B \sqcup C)$ is a pair of morphisms $u_B\colon B \to B \sqcup C$, $u_C\colon C \to B \sqcup C$, called the *injections* (of the cofactors). By definition, for every pair of morphisms $\alpha_B\colon B \to X$, $\alpha_C\colon C \to X$ there is a unique morphism $\alpha\colon B \sqcup C \to X$ such that $\alpha u_B = \alpha_B$, $\alpha u_C = \alpha_C$. It is customary to write $\alpha = (\alpha_B, \alpha_C)$. — Similarly, one can define the coproduct of any family of objects $\{B_\gamma\}_{\lambda \in \Lambda}$; it is denoted by $\bigsqcup_{\lambda \in \Lambda} B_\lambda$, and it is characterised by the natural equivalence $\mathscr{C}(\bigsqcup_\lambda B_\lambda, X) \approx \prod_\lambda \mathscr{C}(B_\lambda, X)$, for $X \in \mathrm{Ob}(\mathscr{C})$.

Dually, the *product* $B \sqcap C$ of two objects $B, C \in \mathrm{Ob}(\mathscr{C})$ is defined (if it exists) by the natural equivalence $\mathscr{C}(X, B \sqcap C) \approx \mathscr{C}(X, B) \times \mathscr{C}(X, C)$, i.e. $B \sqcap C$ is that object of $\mathscr{C}$ which represents the cofunctor $T = \mathscr{C}^B \times \mathscr{C}^C$. The universal element $u \in T(B \sqcap C) = \mathscr{C}(B \sqcap C, B) \times \mathscr{C}(B \sqcap C, C)$ is a pair of morphisms $u_B\colon B \sqcap C \to B$, $u_C\colon B \sqcap C \to C$, called the *projections* onto the factors. If $\alpha_B\colon X \to B$, $\alpha_C\colon X \to C$ is any pair of morphisms then there is a unique morphism $\alpha\colon X \to B \sqcap C$ such that $\alpha_B = u_B\alpha$, $\alpha_C = u_C\alpha$. It is customary to write $\alpha = (\alpha_B, \alpha_C)$. — Similarly, the product $\sqcap_\lambda B_\lambda$ of an arbitrary family of objects is defined by (if it exists) the natural equivalence $\mathscr{C}(X, \sqcap_\lambda B_\lambda) \approx \prod_\lambda \mathscr{C}(X, B_\lambda)$.

In concrete categories such as $\mathscr{S}ets$, $\mathscr{T}op$, $\mathscr{AG}$ etc., other (ad hoc) notations are in use for products $\sqcap$ and coproducts $\sqcup$. For instance, the coproduct $\sqcup$ is called "disjoint union", "topological sum", "direct sum" in $\mathcal{S}ets$, $\mathcal{T}op$, $\mathcal{AG}$, and is denoted by $\cup, \oplus, \oplus$. Products $B \sqcap C$ resp. $\sqcap_\lambda B_\lambda$ are denoted by $B \times X$ resp. $\prod_\lambda B_\lambda$ in these categories; furthermore, $B \times C = B \oplus C$ in $\mathcal{AG}$.

MAC LANE, S.: Categories. Berlin-Heidelberg-New York: Springer 1972.
MITCHELL, B.: Theory of categories. New York: Academic Press 1965.
SCHUBERT, H.: Kategorien, 2 vols. Berlin-Heidelberg-New York: Springer 1970.

2. Abelian Groups (Exactness, Direct Sums, Free Abelian Groups)

Abelian groups and their homomorphisms form a category which we denote by $\mathcal{AG}$. If $\alpha: A \to B$ is a homomorphism between abelian groups, $\alpha \in \mathcal{AG}(A, B)$, then one defines

(2.1) kernel of $\alpha = \ker(\alpha) = \{a \in A \mid \alpha(a) = 0\}$,

(2.2) image of $\alpha = \operatorname{im}(\alpha) = \alpha A = \{b \in B \mid \exists\, a \in A \text{ with } \alpha(a) = b\}$.

These are subgroups of A resp. B. The corresponding quotients are

(2.3) coimage of $\alpha = \operatorname{coim}(\alpha) = A/\ker(\alpha)$,

(2.4) cokernel of $\alpha = \operatorname{coker}(\alpha) = A/\operatorname{im}(\alpha)$.

We say α is *monomorphic* if $\ker(\alpha) = \{0\}$, *epimorphic* if $\operatorname{coker}(\alpha) = \{0\}$.

A monomorphism is then the same as an injective homomorphism, an epimorphism is the same as a surjective homomorphism. And α is *isomorphic*, in symbols $\alpha: A \cong B$, if and only if it is both monomorphic and epimorphic. The homomorphism theorem asserts that

(2.5) $\operatorname{im}(\alpha) \cong A/\ker(\alpha) = \operatorname{coim}(\alpha)$.

Because of this, the coimage will play a minor role only.

2.6 Definition. A sequence $A \xrightarrow{\alpha} B \xrightarrow{\beta} C$ of homomorphisms is said to be *exact* if $\ker(\beta) = \operatorname{im}(\alpha)$. A longer sequence like $\cdots \to A_{-2} \to A_{-1} \to A_0 \to A_1 \to A_2 \to \cdots$ is exact if any two consecutive arrows form an exact sequence. An exact sequence of the form

(2.7) $0 \to A' \xrightarrow{\alpha'} A \xrightarrow{\alpha''} A'' \to 0$

is called a *short exact sequence*. For instance, if B is a subgroup of A then

$$0 \to B \xrightarrow{\iota} A \xrightarrow{\pi} A/B \to 0$$

is a short exact sequence where ι = inclusion, π = projection. Conversely, if 2.7 is exact then $B = \mathrm{im}(\alpha') = \ker(\alpha'')$ is a subgroup of A, and $B \cong A'$, $A/B \cong A''$ by 2.5.

2.8 Proposition. *If* $\cdots \xrightarrow{\alpha^-} A \xrightarrow{\alpha} B \xrightarrow{\alpha^+} \cdots$ *is an exact sequence then* α *is monomorphic if and only if* $\alpha^- = 0$, α *is epimorphic if and only if* $\alpha^+ = 0$. *Therefore,* α *is isomorphic if and only if both* $\alpha^- = 0$ *and* $\alpha^+ = 0$. ∎

This (rather obvious) fact will be used many times. Another useful result is the following (less obvious)

2.9 Five Lemma. *If*

$$\begin{array}{ccccccccc}
A_1 & \xrightarrow{\alpha_1} & A_2 & \xrightarrow{\alpha_2} & A_3 & \xrightarrow{\alpha_3} & A_4 & \xrightarrow{\alpha_4} & A_5 \\
\downarrow{\scriptstyle\varphi_1} & & \downarrow{\scriptstyle\varphi_2} & & \downarrow{\scriptstyle\varphi_3} & & \downarrow{\scriptstyle\varphi_4} & & \downarrow{\scriptstyle\varphi_5} \\
B_1 & \xrightarrow[\beta_1]{} & B_2 & \xrightarrow[\beta_2]{} & B_3 & \xrightarrow[\beta_3]{} & B_4 & \xrightarrow[\beta_4]{} & B_5
\end{array}$$

is a commutative diagram with exact rows, and if $\varphi_1, \varphi_2, \varphi_4, \varphi_5$ *are isomorphic then so is* φ_3.

Proof. Passing to quotients and subgroups the diagram induces the following commutative diagram with exact rows.

$$(2.10)\qquad \begin{array}{ccccccccc}
0 \to & \mathrm{coker}(\alpha_1) & \xrightarrow{\alpha_2'} & A_3 & \xrightarrow{\alpha_3'} & \ker(\alpha_4) & \to 0 \\
& \cong\downarrow{\scriptstyle\varphi_2'} & & \downarrow{\scriptstyle\varphi_3} & & \cong\downarrow{\scriptstyle\varphi_4'} & \\
0 \to & \mathrm{coker}(\beta_1) & \xrightarrow[\beta_2']{} & B_3 & \xrightarrow[\beta_3']{} & \ker(\beta_4) & \to 0
\end{array}$$

This reduces the problem to a special (easier) case. Now

$$\ker(\varphi_3) \subset \ker(\beta_3' \varphi_3) = \ker(\varphi_4' \alpha_3') = \ker(\alpha_3') = \mathrm{im}(\alpha_2'),$$

hence $\ker(\varphi_3) \cong \ker(\varphi_3 \alpha_2') = \ker(\beta_2' \varphi_2') = \{0\}$, i.e. φ_3 is monomorphic. Dually, $\beta_3' \varphi_3 = \varphi_4' \alpha_3'$ is epimorphic, hence $B_3 = \mathrm{im}(\varphi_3) + \ker(\beta_3')$; but $\ker(\beta_3') = \mathrm{im}(\beta_2') = \mathrm{im}(\beta_2' \varphi_2') = \mathrm{im}(\varphi_3 \alpha_2') \subset \mathrm{im}(\varphi_3)$; hence $B_3 = \mathrm{im}(\varphi_3)$, i.e. φ_3 is epimorphic. ∎

As an exercise, the reader might prove the 5-lemma directly, without using the reduction 2.10.

2.11 Proposition and Definition. A short exact sequence 2.7 is said to *split* if one of *the following equivalent conditions* holds

(i) α' has a left inverse $\beta'\colon A \to A'$, $\beta' \alpha' = \mathrm{id}_{A'}$,
(ii) α'' has a right inverse $\beta''\colon A'' \to A$, $\alpha'' \beta'' = \mathrm{id}_{A''}$.

In fact, the equation

$$\alpha' \beta' + \beta'' \alpha'' = \mathrm{id}_A \tag{2.12}$$

establishes a one-one correspondence between left inverses β' of α' and right inverses β'' of α''. Moreover, $\beta' \beta'' = 0$.

Proof. If β'' is a right inverse of α'' then $\alpha''(\mathrm{id}_A - \beta''\alpha'') = \alpha'' - (\alpha''\beta'')\alpha'' = 0$, hence $\mathrm{im}(\mathrm{id}_A - \beta''\alpha'') \subset \ker(\alpha'') = \mathrm{im}(\alpha')$, and we can define β' by $\alpha'\beta' = \mathrm{id}_A - \beta''\alpha''$, i.e. by 2.12; since α' is monomorphic this defines β' uniquely. Moreover, if we compose this equation (or 2.12) with α' on the right, and use $\alpha''\alpha' = 0$, we get $\alpha'(\beta'\alpha') = \alpha'$, hence $\beta'\alpha' = \mathrm{id}$ because α' is monomorphic. This proves that every right inverse β'' of α'' determines a unique left inverse β' of α' such that 2.12 holds.

If β' is any left inverse of α' then $(\mathrm{id}_A - \alpha'\beta')\alpha' = \alpha' - \alpha'(\beta'\alpha') = 0$, hence $(\mathrm{id}_A - \alpha'\beta')$ vanishes on $\mathrm{im}(\alpha') = \ker(\alpha'')$; since α'' is epimorphic there is a unique $\beta''\colon A'' \to A$ such that $\beta''\alpha'' = (\mathrm{id}_A - \alpha'\beta')$, i.e. such that 2.12 holds. Moreover, if we compose this equation with α'' on the left we find $(\alpha''\beta'')\alpha'' = \alpha''$, hence $\alpha''\beta'' = \mathrm{id}$ because α'' is epimorphic. Finally, we compose 2.12 with β' on the left, and get $\beta' + (\beta'\beta'')\alpha'' = \beta'$, hence $(\beta'\beta'')\alpha'' = 0$, hence $\beta'\beta'' = 0$. ∎

2.13 Definition. Let $\{A_\lambda\}_{\lambda\in\Lambda}$ be a family of abelian groups. Consider the set of all functions a on Λ such that $a(\lambda) \in A_\lambda$ for all $\lambda \in \Lambda$. Under addition of values these functions form an abelian group, called the *direct product* of $\{A_\lambda\}_{\lambda\in\Lambda}$, and denoted by $\prod_{\lambda\in\Lambda} A_\lambda$. The elements $a_\lambda = a(\lambda)$ are called the *components* of $a = \{a_\lambda\} \in \prod_\lambda A_\lambda$. The homomorphism $\pi_\nu\colon \prod_\lambda A_\lambda \to A_\nu$ which assigns to each $a \in \prod_\lambda A_\lambda$ its ν-th component, $\pi_\nu a = a_\nu$, is called the *projection onto the factor* A_ν.

The *direct sum* of $\{A_\lambda\}_{\lambda\in\Lambda}$ is the subgroup $\oplus_{\lambda\in\Lambda} A_\lambda$ of $\prod_{\lambda\in\Lambda} A_\lambda$ which consists of all functions a of finite support, i.e.

$$\oplus_\lambda A_\lambda = \{a \in \textstyle\prod_\lambda A_\lambda \mid a_\lambda = 0 \text{ for almost all } \lambda \in \Lambda\}.$$

Clearly, $\oplus_\lambda A_\lambda = \prod_\lambda A_\lambda$ if Λ is finite. The homomorphism $\iota_\nu\colon A_\nu \to \oplus_\lambda A_\lambda$ such that $\pi_\nu \iota_\nu = \mathrm{id}$, $\pi_\lambda \iota_\nu = 0$ for $\lambda \neq \nu$, is called the *inclusion of the summand* A_ν; by definition, if $x \in A_\nu$ then all components of $\iota_\nu x$ vanish except the ν-th, and $(\iota_\nu x)_\nu = x$.

2.14 Proposition and Definition. (i) If $X \in \mathrm{Ob}(\mathscr{AG})$, and $\{\varphi_\lambda\colon X \to A_\lambda\}$, $\lambda \in \Lambda$, *is a family of homomorphisms then there exists a unique homomorphism* $\varphi\colon X \to \prod_\lambda A_\lambda$ *such that* $\varphi x = \{\varphi_\lambda x\}_{\lambda\in\Lambda}$, for all $x \in X$. We write $\varphi = \{\varphi_\lambda\}$, and call these $\varphi_\lambda = \pi_\lambda \varphi$ the *components* of φ.

(ii) *If* $X \in \mathrm{Ob}(\mathscr{AG})$, *and* $\{\psi_\lambda\colon A_\lambda \to X\}$, $\lambda \in \Lambda$, *is a family of homomorphisms then there exists a unique homomorphism* $\psi\colon \oplus_\lambda A_\lambda \to X$ *such that*

$\psi a = \sum_{\lambda\in\Lambda} \psi_\lambda a_\lambda$ (n.b. this sum is finite!). We write $\psi = \{\psi_\lambda\}$, and call these $\psi_\lambda = \psi\,\iota_\lambda$ the *components* of ψ.

In other words, $\prod_\lambda A_\lambda$ is the categorical product $\sqcap_\lambda A_\lambda$ in the sense of 1.15, and $\oplus_\lambda A_\lambda$ is the categorical coproduct $\sqcup_\lambda A_\lambda$; the family of projections $\{\pi_\lambda\}$ resp. inclusions $\{\iota_\lambda\}$ is the universal element for the corresponding functors $\prod_\lambda \mathscr{AG}(X, A_\lambda)$ resp. $\prod_\lambda \mathscr{AG}(A_\lambda, X)$.—Both parts of the proposition follow easily from the definitions 2.13. ▮

2.15 Definition. Let $\{A_\lambda\}_{\lambda\in\Lambda}$ and A denote abelian groups. A family of homomorphisms $\{p_\lambda\colon A\to A_\lambda\}_{\lambda\in\Lambda}$ resp. $\{i_\lambda\colon A_\lambda\to A\}_{\lambda\in\Lambda}$ is called a *direct product representation* resp. *direct sum representation* if $\{p_\lambda\}\colon A\to\prod_\lambda A_\lambda$ resp. $\{i_\lambda\}\colon \oplus_\lambda A_\lambda\to A$ is an isomorphism.

2.16 Proposition. *If Λ is finite and if* $\{p_\lambda\colon A\to A_\lambda\}$ resp. $\{i_\lambda\colon A_\lambda\to A\}$, $\lambda\in\Lambda$, *are families of homomorphisms such that*

$$(2.17)\qquad p_\lambda i_\lambda = \mathrm{id}_{A_\lambda},\quad p_\lambda i_\mu = 0 \quad \text{for } \mu\neq\lambda,\quad \textstyle\sum_\lambda i_\lambda p_\lambda = \mathrm{id}_A,$$

then $\{p_\lambda\}$ is a direct product representation and $\{i_\lambda\}$ is a direct sum representation.

Conversely, if $p=\{p_\lambda\colon A\to A_\lambda\}_{\lambda\in\Lambda}$ is a direct product representation then there is a unique family $\{i_\lambda\colon A_\lambda\to A\}$ which satisfies 2.17; *similarly, for direct sum representations.*

In particular (cf. 2.11), a short exact sequence $0\to A'\xrightarrow{\alpha'} A\xrightarrow{\alpha''} A''\to 0$ splits if and only if α' (resp. α'') is one component of a direct sum (resp. product) representation $A'\oplus A''\cong A$.

Proof. We first have to show that $i=\{i_\lambda\}\colon \oplus_\lambda A_\lambda\to A$ and $p=\{p_\lambda\}\colon A\to\prod_\lambda A_\lambda$ are isomorphic. But $\oplus_\lambda = \prod_\lambda$ because Λ is finite,

$$(i\,p)\,a = i\{p_\lambda a\} = \left(\textstyle\sum_\lambda i_\lambda p_\lambda\right) a = a,$$

and

$$(p\,i)\,a = p\left(\textstyle\sum_\mu i_\mu a_\mu\right) = \{p_\lambda\left(\textstyle\sum_\mu i_\mu a_\mu\right)\}_{\lambda\in\Lambda} = \{\textstyle\sum_\mu (p_\lambda i_\mu)\,a_\mu\}_{\lambda\in\Lambda} = \{a_\lambda\}_{\lambda\in\Lambda} = a,$$

hence p, i are reciprocal isomorphisms. For the converse, we can assume $A=\prod_\lambda A_\lambda = \oplus_\lambda A_\lambda$, and $p_\lambda = \pi_\lambda$ (because $p\colon A\cong\prod_\lambda A_\lambda$, $\pi_\lambda p = p_\lambda$). The first two equations 2.17 then show that $i_\lambda = \iota_\lambda$ (as defined in 2.13) so that only $\sum_\lambda \iota_\lambda \pi_\lambda = \mathrm{id}$ remains to be checked; this is easy, and left to the reader.—Similarly for direct sum representations. ▮

2.18 If A is an abelian group, and $A_1, A_2\subset A$ are subgroups then *we say A is the direct sum of A_1 and A_2* if the inclusion homomorphisms form a direct sum representation (i_1, i_2): $A_1 \oplus A_2 \cong A$. One easily proves that this is the case if and only if

(i) $A_1 \cup A_2$ generates A, and (ii) $A_1 \cap A_2 = \{0\}$.

A subgroup $A_1 \subset A$ is called a *direct summand* (of A) if A is the direct sum of A_1 and some $A_2 \subset A$. For instance, if $0 \to A' \xrightarrow{\alpha} A \to A'' \to 0$ is a short exact sequence then $\operatorname{im}(\alpha)$ is a direct summand of A if and only if the sequence splits (cf. remark after 2.16). Applying this to

$$0 \to A_1 \xrightarrow{i} A \to A/A_1 \to 0$$

we see that the subgroup $A_1 \subset A$ is a direct summand if and only if the inclusion map i has a left inverse $r: A \to A_1$, $r\,i = \mathrm{id}$.

If $\{A_\lambda\}_{\lambda \in \Lambda}$ is any family of subgroups of A such that the inclusion homomorphisms constitute a direct sum representation, $\{i_\lambda\}$: $\bigoplus_{\lambda \in \Lambda} A_\lambda \cong A$, then we also say that A is the direct sum of $\{A_\lambda\}$.

2.19 Definition. If A is an abelian group and $a \in A$ we define $i_a: \mathbb{Z} \to A$, $i_a n = n \cdot a$, for all integers $n \in \mathbb{Z}$; thus i_a is the unique homomorphism $\mathbb{Z} \to A$ such that $1 \mapsto a$. A subset B of A is said to be a *base of* A if the family $\{i_b\}_{b \in B}$ is a direct sum representation, $\{i_b\}$: $\bigoplus_{b \in B} \mathbb{Z} \cong A$. *Every element* $x \in A$ *then has a unique representation as a finite linear combination of base elements with integral coefficients* $x = \sum_{b \in B} x_b \cdot b$, $x_b \in \mathbb{Z}$, almost all $x_b = 0$. Not every abelian group has a base; if it does it is said to be *free*. Thus, an abelian group is free if and only if it is isomorphic to a direct sum of groups $\mathbb{Z}$. From 2.14ii we get

2.20 Proposition (Universal property of a base). *If B is a base of A, if X is an arbitrary abelian group and $\{x_b \in X\}_{b \in B}$ an arbitrary family of elements then there is a unique homomorphism $\xi: A \to X$ such that $\xi b = x_b$, for all $b \in B$.* I.e., the homomorphisms of a free group are determined by their values on a base, and these values can be chosen arbitrarily. ∎

2.21 Definition. For every set Λ we can form the direct sum $\bigoplus_{\lambda \in \Lambda} \mathbb{Z}$. This group is called *the free abelian group generated by* Λ; it is often denoted by $\mathbb{Z}\Lambda$. Its elements are functions $a: \Lambda \to \mathbb{Z}$ which vanish almost everywhere. If we identify $\lambda \in \Lambda$ with the function $\Lambda \to \mathbb{Z}$ such that $\lambda \mapsto 1$, $\nu \mapsto 0$ for $\nu \neq \lambda$, then Λ becomes a subset of $\mathbb{Z}\Lambda$, and *this subset* Λ *is a base of* $\mathbb{Z}\Lambda$. Thus, every $a \in \mathbb{Z}\Lambda$ has a unique expression $a = \sum_{\lambda \in \Lambda} a_\lambda \cdot \lambda$, $a_\lambda \in \mathbb{Z}$, almost all $a_\lambda = 0$; the group $\mathbb{Z}\Lambda$ consists of all finite linear combinations of elements $\lambda \in \Lambda$ with integral coefficients.

2.22 *Every abelian group A is isomorphic to a quotient of a free abelian group.* Indeed, if Λ is any subset of A which generates A then (by 2.20) there is a (unique) homomorphism ξ: $\mathbb{Z}\Lambda \to A$ such that $\xi(\lambda) = \lambda$.

This ξ is epimorphic because Λ generates A, hence $A \cong \mathbb{Z}\Lambda/\ker(\xi)$. Moreover, $\ker(\xi)$ is also free because

2.23 Proposition. *Every subgroup of a free abelian group is free* [Kurosh, § 19]. ∎

If a *quotient* group is free then it is a direct summand, i.e.

2.24 Proposition. *If F is a free abelian group then every short exact sequence $0 \to A' \to A \xrightarrow{\alpha} F \to 0$ splits (hence $A \cong A' \oplus F$).*

Proof. Take a base B of F, and choose elements $\{a_b \in A\}_{b \in B}$ such that $\alpha(a_b) = b$, for all $b \in B$. Define $\beta: F \to A$ by $\beta(b) = a_b$, as in 2.20; then $\alpha\beta(b) = b$, hence $\alpha\beta = \mathrm{id}$ by the uniqueness part of 2.20. ∎

For finitely generated groups 2.23 refines as follows.

2.25 Proposition. *If F is a finitely generated free abelian group, and $G \subset F$ is a subgroup then one can find bases $\{b_1, \ldots, b_m\}$ of F and $\{c_1, \ldots, c_n\}$ of G such that $n \le m$, $c_j = \mu_j b_j$ with $\mu_j \in \mathbb{Z}$ for $j \le n$, and μ_j divides μ_{j+1} for $j < n$.*—For a proof cf. [Kurosh, § 20]. ∎

The quotient group F/G is easily seen to be the direct sum of the cyclic subgroups C_j, where C_j is generated by the coset of b_j; the order of this subgroup is μ_j if $j \le n$, and is ∞ if $j > n$. Since every finitely generated abelian group A is of the form F/G, by 2.22, we have the

2.26 Corollary. *Every finitely generated abelian group A is a finite direct sum of cyclic subgroups $\{C_j \subset A\}$,*

(2.27) $\quad A = \oplus_{j=1}^{k} C_j, \quad C_j \cong \mathbb{Z}/\nu_j\mathbb{Z}, \quad \nu_j \in \mathbb{Z}, \quad \nu_j \ge 0.$ ∎

The partial sum $T = \oplus_{\nu_j > 0} C_j = \oplus_{\nu_j > 1} C_j$ is called the *torsion subgroup of A; it is a finite group and consists of all elements of A of finite order.* The quotient $A/T \cong \oplus_{\nu_j = 0} \mathbb{Z}$ is called the *free part of A*. The number of summands $\mathbb{Z}$ in A/T is called the *rank of A*. It does not depend on the particular direct sum decomposition 2.27; in fact, *rank (A) is the maximal number of linearly independent elements in A.*

The numbers $\nu_j > 1$ which occur in 2.27 are not unique. However, *they can be chosen as powers of prime numbers, $\nu_j = p_j^{\rho_j}$, p_j* prime, *$\rho_j > 0$, and then they are unique* (independent of the decomposition 2.27) *up to permutation* [Kurosh, § 20]. These $\{\nu_j\}$ are called the *torsion coefficients* of A. *Two finitely generated abelian groups are isomorphic if and only if they have the same rank and the same system of torsion coefficients.*

2.28 Proposition. *If A is a finitely generated abelian group, and $A' \subset A$ is a subgroup then A' and A/A' are also finitely generated, and* rank$(A)=$ rank$(A')+$rank(A/A'). Using 2.25, this is easy to prove; cf. [Kurosh, §19]. ∎

2.29 For *arbitrary* abelian groups G one can define a rank as follows: If G is free, rank(G) is the cardinality of a base; otherwise, rank(G) is the supremum of {rank(F)} where F ranges over all free subgroups of G. With this definition, rank$(G)=$rank$(G')+$rank(G/G'), for all $G' \subset G$.

FUCHS, L.: Abelian groups. Hung. Acad. Sci. Budapest 1954. New York: Pergamon Press 1960.
KUROSH, A.G.: The theory of groups, vol. I. New York: Chelsea Publ. Co. 1955.
V. D. WAERDEN, B.L.: Algebra, Bd. II. Berlin-Heidelberg-New York: Springer 1967.

3. Homotopy

Let X, Y denote topological spaces, and $f: X \to Y$ a continuous map. If we modify (disturb) f by a small amount then we might expect that its properties also change by small amounts only. Whether this is the case or not depends of course, on the property which we consider and, perhaps, on f. Many important properties, however, do behave in this way. If, in particular, such a property can only change in jumps (e.g. if it is expressed by an integer) then it will not change at all under slight modifications of f. It will then also be unchanged under large modifications provided the large modification can be decomposed into small steps, i.e. if the modification is the result of a continuous process. This, intuitively speaking, is the principle of homotopy invariance; the homotopy notion which we now discuss makes precise what is meant by a, "continuous process".

3.1 Definition. If X, Y are topological spaces and $[0,1]$ denotes the unit intervall then a *homotopy* or *deformation* (of X into Y) is a continuous map $\Theta: X \times [0,1] \to Y$. For every $t \in [0,1]$ we have

(3.2) $$\Theta_t: X \to Y, \quad \Theta_t(x) = \Theta(x,t),$$

a continuous map. Clearly, Θ is determined by the "one-parameter family" $\{\Theta_t\}_{0 \leq t \leq 1}$, and vice versa. Therefore $\{\Theta_t\}_{0 \leq t \in 1}$ is also called a *homotopy* or *deformation.*—The one-parameter-family notation $\{\Theta_t\}$ is more intuitive and sometimes more convenient, however, in order to properly express the continuity property of a homotopy it is preferable to write $\Theta\colon X\times[0,1]\to Y$. With $x\in X$ fixed and $t\in[0,1]$ variable we can also think of $\Theta(x,t)$ as the trajectory which x describes in Y during the time unit $[0,1]$; the deformation Θ is then a family of such trajectories in Y, indexed by the parameter $x\in X$.

3.3 Definition. Two continuous maps $f_0, f_1\colon X\to Y$ are said to be *homotopic* if a deformation $\{\Theta_t\colon X\to Y\}_{0\le t\le 1}$ exists such that $f_0=\Theta_0$, $f_1=\Theta_1$. We write $\Theta\colon f_0\simeq f_1$, or simply $f_0\simeq f_1$, and we say Θ *is a deformation of* f_0 *into* f_1.—If $A\subset X$ then $\Theta\colon X\times[0,1]\to Y$ is said to be a *homotopy rel. A* provided $\Theta_t|A=\Theta_0|A$ for all t; we write $\Theta\colon f_0\simeq f_1$ *rel. A*.—A homotopy Θ such that Θ_1 is a constant map is sometimes called a *nullhomotopy*, and $f=\Theta_0$ is said to be *nullhomotopic*.

3.4 Proposition and Definition. *The homotopy relation* $\simeq$ *is an equivalence relation.* The equivalence class (under $\simeq$) of f is denoted by $[f]$, and is called the *homotopy class of f*.

Proof. The constant homotopy $\{\Theta_t=f\}_{0\le t\le 1}$ is a deformation $f\simeq f$ (*reflexivity*). If $\{\Theta_t\}\colon f_0\simeq f_1$ then $\{\Theta_{1-t}\}\colon f_1\simeq f_0$ (*symmetry*). If $\Theta'\colon f_0\simeq f_1$, and $\Theta''\colon f_1\simeq f_2$, then $\Theta\colon f_0\simeq f_2$, where $\Theta_t=\Theta'_{2t}$ for $2t\le 1$, $\Theta_t=\Theta'_{2t-1}$ for $2t\ge 1$ (*transitivity*). ∎

3.5 Proposition and Definition. *The homotopy relation is compatible with composition*, i.e. *if* $f_0, f_1\colon X\to Y$, $g_0, g_1\colon Y\to Z$ *are maps such that* $f_0\simeq f_1$, $g_0\simeq g_1$, *then* $g_0 f_0\simeq g_1 f_1$. Indeed, if $\Theta'\colon f_0\simeq f_1$, and $\Theta''\colon g_0\simeq g_1$, then $\Theta\colon g_0 f_0\simeq g_1 f_1$, where $\Theta_t=\Theta''_t\,\Theta'_t$. ∎

We can therefore define *composition of homotopy classes* by $[g]\circ[f]=[g\circ f]$. This defines a new category $\mathcal{H}tp$: Its objects are topological spaces as in $\mathcal{T}op$, $\mathrm{Ob}(\mathcal{H}tp)=\mathrm{Ob}(\mathcal{T}op)$; the *morphisms*, however, *are homotopy classes* of continuous maps, $\mathcal{H}tp(X,Y)=\{[f]\,|\,f\in\mathcal{T}op(X,Y)\}$. If we assign to every continuous map $f\colon X\to Y$ its homotopy class $[f]$ we obtain a functor

$$(3.6)\qquad \pi\colon \mathcal{T}op\to\mathcal{H}tp,\quad \pi X=X\quad \text{for } X\in\mathrm{Ob}(\mathcal{T}op),\quad \pi f=[f].$$

3.7 Some of the main tools in algebraic topology are functors $t\colon\mathcal{T}op\to\mathcal{A}$ where $\mathcal{A}$ is some algebraic category (groups, rings, ...). In most cases these functors are homotopy-invariant, i.e., $f_0\simeq f_1\Rightarrow tf_0=tf_1$. Equivalently, t *factors through* π, i.e. $t=t'\circ\pi$ where $\mathcal{T}op\xrightarrow{\pi}\mathcal{H}tp\xrightarrow{t'}\mathcal{A}$. Thus, t looses all informations on $\mathcal{T}op$ which is lost by π. Due to this fact, algebraic topologists are often more interested in the category $\mathcal{H}tp$ than in $\mathcal{T}op$. In particular, they often do not distinguish between spaces X, Y if they are equivalent in $\mathcal{H}tp$. This means that mappings $f\colon X\to Y$,

$g: Y \to X$ exist such that $fg \simeq \mathrm{id}_Y$, $gf \simeq \mathrm{id}_Y$. Such mappings are called (reciprocal) *homotopy equivalences*, and X, Y are called *homotopy equivalent*, in symbols $X \simeq Y$. Functors t as above take the same value on homotopy equivalent spaces, in fact, they transform homotopy equivalences $f: X \simeq Y$ into equivalences $tf: tX \cong tY$.

3.8 The preceding notions and results generalize to *pairs of spaces*. By definition, a pair (X, A) of topological spaces consists of a space X and a subspace A. If (X, A), (Y, B) are pairs of spaces then a *map of pairs* $f: (X, A) \to (Y, B)$ is, by definition, a (continuous) map f of X into Y such that $fA \subset B$. Pairs and their maps constitute a new category (under ordinary composition) which we denote by $\mathcal{Top}^{(2)}$. If we assign to each space X the pair $(X, \emptyset)$ and to each map $X \to Y$ the corresponding map of pairs $(X, \emptyset) \to (Y, \emptyset)$ we obtain a functor $\mathcal{Top} \to \mathcal{Top}^{(2)}$. We use this functor to identify $\mathcal{Top}$ with a (full) subcategory of $\mathcal{Top}^{(2)}$, i.e. we shall write $X = (X, \emptyset)$.

If X is the disjoint union of a family $\{X_\lambda\}$, $\lambda \in \Lambda$, of open subsets, i.e. if $X = \biguplus_\lambda X_\lambda$ is the topological sum of the X_λ, and if $A_\lambda = A \cap X_\lambda$ then we write $(X, A) = \oplus_\lambda (X_\lambda, A_\lambda) =$ *topological sum* of $\{(X_\lambda, A_\lambda)\}$. It is easily seen that this agrees with the categorical *coproduct* in $\mathcal{Top}^{(2)}$, as defined in 1.15, i.e. $\oplus_\lambda = \bigsqcup_\lambda$. The categorical product is $(X, A) \sqcup (Y, B) = (X \times Y, A \times B)$ but this is not much in use. Instead we shall often encounter the following *product of pairs*, $(X, A) \times (Y, B) = (X \times Y, X \times B \cup A \times Y)$; this notation is misleading but generally accepted.

Occasionally, we shall also consider *triples* (X, A, B) consisting of spaces such that $X \supset A \supset B$, and *triads* $(X; X_1, X_2)$ consisting of spaces $X \supset X_1$, $X \supset X_2$ (no inclusion between X_1, X_2 required). Both notions give rise to categories which contain $\mathcal{Top}^{(2)}$, and also to obvious homotopy notions and -categories (as below).

3.9 *A homotopy between maps* $f_0, f_1: (X, A) \to (Y, B)$ is, by definition, a one-parameter family $\Theta_t: (X, A) \to (Y, B)$, $0 \le t \le 1$, as in 3.1–3.3, with $\Theta_0 = f_0$, $\Theta_1 = f_1$. We write $f_0 \simeq f_1$; then $\simeq$ is an equivalence relation (as in 3.4) which is compatible with composition (as in 3.5). Identifying homotopic maps defines the homotopy category $\mathcal{Htp}^{(2)}$, and a functor $\pi: \mathcal{Top}^{(2)} \to \mathcal{Htp}^{(2)}$ with $\pi(X, A) = (X, A)$, $\pi f = [f] =$ homotopy class of f.

Chapter II

Homology of Complexes

1. Complexes

1.1 Definition. A *complex* K is a sequence

$$\cdots \leftarrow K_{n-1} \xleftarrow{\partial_n} K_n \xleftarrow{\partial_{n+1}} K_{n+1} \leftarrow \cdots$$

of abelian groups K_n and homomorphisms ∂_n, called *boundary operators*, such that $\partial_n \partial_{n+1} = 0$ for all integers n.

We call *n-chains* the elements of K_n, *n-cycles* the elements of $Z_n K = \ker(\partial_n) = \partial_n^{-1}(0)$, and *n-boundaries* the elements of $B_n K = \operatorname{im}(\partial_{n+1}) = \partial_{n+1}(K_{n+1})$. The condition $\partial_n \partial_{n+1} = 0$ means $B_n K \subset Z_n K$. We can therefore form the quotient $H_n K = Z_n K / B_n K$, called *n-th homology group* of K; its elements are called *n-dimensional homology classes.* By definition, homology classes are equivalence classes of cycles; two cycles z_n, $z'_n \in Z_n K$ being equivalent, or "*homologous*", if and only if their difference is a boundary, $z - z' \in B_n K$. The homology class of a cycle z is denoted by $[z]$.

Given complexes K, K', we define a *chain map* $f\colon K' \to K$ to be a sequence of homomorphisms $f_n\colon K'_n \to K_n$ such that $\partial_n f_n = f_{n-1} \partial'_n$ for all $n \in \mathbb{Z}$. The composite $ff'\colon K'' \to K$ of two chain maps $K'' \xrightarrow{f'} K' \xrightarrow{f} K$ is defined by $(ff')_n = f_n f'_n$; it is again a chain map. Chain complexes and chain maps then form a category, which we denote by $\partial\mathscr{A}\mathscr{G}$. It follows immediately that a chain map f is an isomorphism (in $\partial\mathscr{A}\mathscr{G}$) if and only if every f_n is an isomorphism (in $\mathscr{A}\mathscr{G}$).

The relation $\partial_n f_n = f_{n-1} \partial'_n$ implies $f_n(Z_n K') \subset Z_n K$ and $f_n(B_n K') \subset B_n K$. Passing to quotients, f_n therefore induces a homomorphism

$$H_n f\colon H_n K' \to H_n K, \quad (H_n f)[z'] = [f z'],$$

and one easily checks that

(1.2) $$H_n(ff') = (H_n f)(H_n f'), \quad H_n(\mathrm{id}_K) = \mathrm{id}_{H_n K},$$

i.e., homology is a functor,

$$H_n\colon \partial\mathscr{AG} \to \mathscr{AG}.$$

We shall often omit indices when there is no danger of confusion; e.g. we shall write $\partial x, fx$ instead of $\partial_n x, f_n x$. We also abbreviate $H_n f = f_*$; the functor relation 1.2 thus becomes $(ff')_* = f_* f'_*$, $\mathrm{id}_* = \mathrm{id}$.

1.3 Examples. *1.* A complex $\cdots \leftarrow K_{n-1} \xleftarrow{\partial_n} K_n \xleftarrow{\partial_{n+1}} K_{n+1} \leftarrow \cdots$ is *exact* if and only if $\ker(\partial_n) = \mathrm{im}(\partial_{n+1})$ for all n, i.e. if and only if $H_n K = 0$ for all n. Homology then can be viewed as a *measure for the lack of exactness*. An exact complex is often called *acyclic* (it has no cycles besides boundaries).

2. A sequence $G = \{G_n\}_{n\in\mathbb{Z}}$ of (abelian) groups is called a *graded* (abelian) *group*. For instance, the cycles $ZK = \{Z_n K\}$, the boundaries $BK = \{B_n K\}$, or the homology $HK = \{H_n K\}$ of a complex are graded abelian groups. In fact, Z, B, H are covariant functors of the category $\partial\mathscr{AG}$ into the category $\mathscr{GAG}$ of graded abelian groups; the morphisms $\varphi\colon G \to G'$ of this category are sequences $\varphi_n\colon G_n \to G'_n$ of ordinary homomorphisms.

A complex K is a graded abelian group together with some extra structure given by the boundary operator ∂.

Every graded abelian group G can be made a complex by taking $\partial = 0$. This defines an embedding $\mathscr{GAG} \subset \partial\mathscr{AG}$; in particular, we can always view ZK, BK, HK as complexes (with vanishing boundary operator). If $G \in \mathscr{GAG}$ then $ZG = G$, $BG = 0$, $HG = G$.

If A is an abelian group and $k \in \mathbb{Z}$ we denote by (A, k) the following graded group: $(A, k)_n$ is A if $n = k$, and is zero for $n \neq k$; i.e. (A, k) is concentrated in dimension k, and equals A there. This defines embeddings $\mathscr{AG} \subset \mathscr{GAG}$.

3. If $\{K^\lambda\}_{\lambda\in\Lambda}$ is a family of complexes we define their direct sum $\oplus_\lambda K^\lambda \in \partial\mathscr{AG}$ by

(1.4) $$[\oplus_\lambda K^\lambda]_n = \oplus_\lambda (K^\lambda_n), \qquad \partial\{c^\lambda\} = \{\partial c^\lambda\},$$

i.e. we take the direct sum in each dimension and let the boundary $\oplus_\lambda K^\lambda_n \to \oplus_\lambda K^\lambda_{n-1}$ act componentwise. It follows easily that

(1.5) $$Z(\oplus_\lambda K^\lambda) = \oplus_\lambda ZK^\lambda, \quad B(\oplus_\lambda K^\lambda) = \oplus_\lambda BK^\lambda, \quad H(\oplus_\lambda K^\lambda) \cong \oplus_\lambda HK^\lambda.$$

Similarly for the direct product $\prod$.

In general, we shall translate notions from abelian groups $\mathscr{AG}$ to complexes $\partial\mathscr{AG}$ by applying them dimension-wise. Other examples are kernel, cokernel, quotient, monomorphism, exact sequence etc. Usually the translation will be quite obvious.

4. The mapping cone. This is a useful technical notion. If $f\colon K\to L$ is a chain map we define a new complex Cf, the mapping cone, as follows:

$$(1.6)\qquad (Cf)_n=L_n\oplus K_{n-1},\qquad \partial^{Cf}(y,x)=(\partial^L y+fx,\,-\partial^K x).$$

We verify that $\partial^{Cf}\partial^{Cf}=0$:

$$\partial\partial(y,x)=\partial(\partial y+fx,\,-\partial x)=(\partial\partial y+\partial fx-f\partial x,\,\partial\partial x)=(0,0).$$

If $L=0$, hence $f=0$, then $K^+=Cf$ is called the *suspension* of K. It is given by $(K^+)_n=K_{n-1}$, $\partial^{K^+}=-\partial^K$. Clearly $H_nK^+=H_{n-1}K$, in fact $H(K^+)=(HK)^+$.

We have a short exact sequence

$$(1.7)\qquad 0\to L\xrightarrow{\iota} Cf\xrightarrow{\pi} K^+\to 0$$

of chain maps given by $\iota y=(y,0)$, $\kappa(y,x)=x$. It splits in every dimension (obviously) but in general there will be no splitting chain map (e.g., take $K=L=(\mathbb{Z},0)$, and $f=\mathrm{id}$).

The mapping cone of $\mathrm{id}\colon K\to K$ is called *cone of* K, and is denoted by CK. The sequence 1.7 becomes

$$(1.8)\qquad 0\to K\xrightarrow{\iota} CK\xrightarrow{\kappa} K^+\to 0.$$

1.9 Exercise. If K, L are complexes define a new complex $\mathrm{Hom}(K,L)$ as follows

$$[\mathrm{Hom}(K,L)]_n=\textstyle\prod_{\nu\in\mathbb{Z}}\mathrm{Hom}(K_\nu,L_{n+\nu}),$$

i.e. an element of $\mathrm{Hom}(K,L)_n$ is a sequence

$$f=\{f_\nu\colon K_\nu\to L_{n+\nu}\}_{\nu\in\mathbb{Z}}$$

of homomorphism. Define

$$\partial(f)=\{\partial\circ f_\nu-(-1)^n f_{\nu-1}\circ\partial\}_{\nu\in\mathbb{Z}}$$

and verify that $\partial(\partial(f))=0$. Show that $Z_0\mathrm{Hom}(K,L)$ consists precisely of all chain maps $K\to L$. More generally, $Z_{-k}\mathrm{Hom}(K,L)$ consists of all chain maps of K into the k-fold suspension of L; these are often called chain maps of degree $-k$. Show that if $g\colon L\to L'$ is a chain map then so is

$$\mathrm{Hom}(K,f)\colon\ \mathrm{Hom}(K,L)\to\mathrm{Hom}(K,L'),\qquad \{f_\nu\}\mapsto\{gf_\nu\},$$

and its mapping cone $C\,\mathrm{Hom}(K,g)\cong\mathrm{Hom}(K,Cg)$. Similarly for chain maps $K'\to K$.

2. Connecting Homomorphism, Exact Homology Sequence

2.1 Definition. If K is a complex, and $K'_n \subset K_n$, $n \in \mathbb{Z}$, a sequence of subgroups such that $\partial(K'_n) \subset K'_{n-1}$ for all n then

$$\cdots \xleftarrow{\partial'} K'_n \xleftarrow{\partial'} K'_{n+1} \xleftarrow{\partial'} \cdots, \qquad \partial' = \partial | K'$$

is itself a complex, and the inclusion map $i: K' \to K$ is a chain map (by definition of ∂'). Such a K' is called *subcomplex* of K. Passing to quotients, ∂_n induces a homomorphism

$$\bar{\partial}_n: K_n/K'_n \to K_{n-1}/K'_{n-1},$$

and $\bar{\partial}_n \bar{\partial}_{n+1} = 0$. The resulting complex $K/K' = \{K_n/K'_n, \bar{\partial}_n\}$ is called *quotient complex* (of K by K'). The natural projection $p: K \to K/K'$ (which assigns to each $x \in K$ its coset in K/K') is a chain map (by definition of $\bar{\partial}$).

2.2 Examples. The kernel, $\ker(f)$, and the image, $\mathrm{im}(f)$, of a chain map $f: K \to L$ are subcomplexes (of K resp. L), defined by $(\ker(f))_n = \ker(f_n)$, $(\mathrm{im}(f))_n = \mathrm{im}(f_n)$. By the homomorphism Theorem I, 2.5 we have $K/\ker(f) \cong \mathrm{im}(f)$.

2.3 The sequence

$$0 \to K' \xrightarrow{i} K \xrightarrow{p} K/K' \to 0$$

of chain maps of Section 2.1 is *exact*, meaning that

(2.4) $$0 \to K'_n \to K_n \to (K/K')_n \to 0$$

is exact for every n. Conversely, if

(2.5) $$0 \to K' \xrightarrow{i} K \xrightarrow{p} K'' \to 0$$

is a short exact (in every dimension) sequence of chain maps then $K' \cong i(K)$ and $K'' \cong K/i(K)$ by 2.2, i.e. up to isomorphism every short exact sequence 2.5 is of the form 2.4.

2.6 Proposition. *If* $0 \to K' \xrightarrow{i} K \xrightarrow{p} K'' \to 0$ *is an exact sequence of chain maps then the sequence*

$$HK' \xrightarrow{i_*} HK \xrightarrow{p_*} HK''$$

is also exact (H is a half-exact functor; cf. VI, 2.10).

However, i_* is in general not monomorphic and p_* is not epimorphic.

Proof. We have to show $\operatorname{im}(i_*)=\ker(p_*)$. Since $p\,i=0$ we have $p_*\,i_*=(p\,i)_*=0_*=0$, hence $\operatorname{im}(i_*)\subset\ker(p_*)$. Conversely, let $[z]\in\ker(p_*)$, i.e. $p\,z=\partial''\,x''$ for some $x''\in K''$. Pick $x\in p^{-1}(x'')$. Then $p(z-\partial x)=\partial''\,x''-\partial''\,p\,x=0$, hence $z-\partial x=i\,z'$ for some $z'\in K'$. Further, $i\,\partial'\,z'=\partial\,i\,z'=\partial(z-\partial x)=0$, hence $\partial'\,z'=0$ because i is monomorphic. Thus z' is a cycle, and $i_*[z']=[i\,z']=[z-\partial x]=[z]$; in particular, $[z]\in\operatorname{im}(i_*)$. ∎

In general, i_* is not monomorphic and p_* is not epimorphic (H is neither right- nor left-exact). An example is provided by the sequence

$$0\to(\mathbb{Z},0)\xrightarrow{i=\iota}C(\mathbb{Z},0)\xrightarrow{p=\kappa}(\mathbb{Z},1)\to 0$$

of 1.8. One finds $HC(\mathbb{Z},0)=0$, $\ker(i_*)=(\mathbb{Z},0)$, $H(\mathbb{Z},1)=(\mathbb{Z},1)\neq\operatorname{im}(p_*)$.

We now propose to "measure" how much p_* (resp. i_*) differs from being epimorphic (resp. monomorphic). More precisely, we shall associate, in a natural way, with every $y''\in H_n K''$ an element $\partial_*\,y''\in H_{n-1}K'$ which is "the obstruction" for lifting y'' to $H_n K$; i.e., $y''\in\operatorname{im}(p_*)\Leftrightarrow\partial_*\,y''=0$. One can prove that these properties essentially characterize ∂_* (cf. exerc. 2).

2.7 Definition of $\partial_*\colon H_n K''\to H_{n-1}K'$. As before let

$$0\to K'\xrightarrow{i}K\xrightarrow{p}K''\to 0 \tag{2.8}$$

be an exact sequence of chain maps. Consider the homomorphisms

$$H_{n-1}K'\xleftarrow{\bar\partial}p^{-1}(Z_n K'')\xrightarrow{\bar p}H_n K''$$

where $\bar p\,x=[p\,x]$ (note that $p\,x\in Z_n K''$) and $\bar\partial x=[i^{-1}\,\partial x]$; the definition of $\bar\partial$ makes sense because $p\,\partial x=\partial''\,p\,x=0$, hence $\partial x\in\operatorname{im}(i)$, and $\partial'(i^{-1}\,\partial x)=i^{-1}\,\partial\,\partial x=0$. Clearly $\bar p=[\;]\circ p$ is epimorphic. We shall see that $\bar\partial\,|\ker(\bar p)=0$; therefore passage to the quotient yields a unique homomorphism

$$\partial_*=\bar\partial\,\bar p^{-1}\colon H_n K''\to H_{n-1}K',\qquad \partial_*[p\,x]=[i^{-1}\,\partial x],$$

called *connecting homomorphism* of the sequence (2.8).

We now show $\bar p\,x=0\Rightarrow\bar\partial\,x=0$. The assumption $\bar p\,x=0$ means $p\,x=\partial''\,p\,y=p\,\partial y$ for some $y\in K$. Because $\ker(p)=\operatorname{im}(i)$ this implies $x-\partial y=i\,y'$ for some $y'\in K'$, hence $i^{-1}\,\partial x=i^{-1}\,\partial i\,y'=\partial'\,i^{-1}\,i\,y'=\partial'\,y'$, hence $[i^{-1}\,\partial x]=0$. ∎

The main properties of ∂_* are as follows.

2.9 Proposition.

a) *Naturality: If*

$$\begin{array}{ccccccccc} 0 & \to & K' & \xrightarrow{i} & K & \xrightarrow{p} & K'' & \to & 0 \\ & & \downarrow{\scriptstyle f'} & & \downarrow{\scriptstyle f} & & \downarrow{\scriptstyle f''} & & \\ 0 & \to & L' & \xrightarrow[j]{} & L & \xrightarrow[q]{} & L'' & \to & 0 \end{array}$$

is a commutative diagram of chain maps with exact rows then

$$\begin{array}{ccc} H_n K'' & \xrightarrow{\partial_*} & H_{n-1} K' \\ \downarrow{\scriptstyle f''_*} & & \downarrow{\scriptstyle f'_*} \\ H_n L'' & \xrightarrow{\partial_*} & H_{n-1} L' \end{array}$$

is also commutative, i.e. $\partial_* f''_* = f'_* \partial_*$.

b) *Exactness: The sequence*

$$\cdots \xrightarrow{\partial_*} H_n K' \xrightarrow{i_*} H_n K \xrightarrow{p_*} H_n K'' \xrightarrow{\partial_*} H_{n-1} K' \xrightarrow{i_*} H_{n-1} K \xrightarrow{p_*} \cdots,$$

called homology sequence of 2.8, *is exact.*

Proof. (a) follows because all steps involved in the definition of ∂_* are natural. In detail:

$$f'_* \partial_* [px] = f'_* [i^{-1} \partial x] = [f' i^{-1} \partial x] = [j^{-1} f \partial x] = [j^{-1} \partial f x]$$
$$= \partial_* [q f x] = \partial_* [f'' p x] = \partial_* f''_* [p x].$$

(b) By Proposition 2.6, it remains to show exactness at HK' and at HK''. This is the assertion of the following 4 inclusions.

$\mathrm{im}(\partial_*) \subset \ker(i_*)$: Let $[px] \in HK''$. Then $i_* \partial_* [px] = i_* [i^{-1} \partial x] = [i i^{-1} \partial x] = [\partial x] = 0$.

$\ker(i_*) \subset \mathrm{im}(\partial_*)$: Let $[z'] \in HK'$ and $i_*[z'] = 0$. Then $iz' = \partial x$ for some $x \in K$, and $\partial'' px = p \partial x = p i z' = 0$. Hence $[z'] = [i^{-1} \partial x] = \partial_* [px]$.

$\mathrm{im}(p_*) \subset \ker(\partial_*)$: If $[z] \in HK$ then $\partial_* p_* [z] = \partial_* [pz] = [i^{-1} \partial z] = 0$ because $\partial z = 0$.

$\ker(\partial_*) \subset \mathrm{im}(p_*)$: Let $[px] \in HK''$ and $0 = \partial_* [px] = [i^{-1} \partial x]$. Then $i^{-1} \partial x = \partial' x'$ for some $x' \in K'$, hence $\partial(x - ix') = \partial x - i \partial' x' = 0$, and $p_*[x - ix'] = [px]$. ∎

2.10 Corollary. *If*

$$\begin{array}{ccccccccc} 0 & \to & K' & \longrightarrow & K & \longrightarrow & K'' & \to & 0 \\ & & \downarrow & & \downarrow & & \downarrow & & \\ 0 & \to & L' & \longrightarrow & L & \longrightarrow & L'' & \to & 0 \end{array}$$

is a commutative diagram of chain maps with exact rows and if two of the vertical arrows induce homology isomorphisms then so does the third.

Proof. The vertical arrows induce a map of exact homology sequences. Two out of three terms are mapped isomorphically; therefore the third maps isomorphically by the five Lemma I, 2.9. ∎

2.11 Definition. An exact sequence $0 \to K' \xrightarrow{i} K \xrightarrow{p} K'' \to 0$ of chain maps is said to be *direct* if it splits in every dimension. This means (I, 2.11) that mappings $K'_n \xleftarrow{j_n} K_n \xleftarrow{q_n} K''_n$, $n \in \mathbb{Z}$, exist such that $ji = \mathrm{id}$, $pq = \mathrm{id}$, $ij + qp = \mathrm{id}$. The connecting homomorphism $HK'' \to HK'$ then has a convenient description as follows.

2.12 Proposition. *The sequence of mappings $d_n = j_{n-1}\partial q_n\colon K''_n \to K'_{n-1} = (K')^+_n$ is a chain map $d\colon K'' \to (K')^+$, and the induced homomorphism $d_*\colon H_n K'' \to H_n(K')^+ = H_{n-1}K'$ coincides with the connecting homomorphism.*

Proof. *We have*

$$\begin{aligned} i(\partial' d) &= (i\partial')j\partial q = \partial(ij)\partial q = \partial(id - qp)\partial q = -\partial q(p\partial)q = -\partial q\partial''(pq) \\ &= -\partial q\partial'' = -(ij+qp)\partial q\partial'' = -i(j\partial q)\partial'' - q\partial''(pq)\partial'' \\ &= i(-d\partial''), \end{aligned}$$

hence $\partial' d = -d\partial''$ because i is monomorphic, hence $d\colon K'' \to (K')^+$ is a chain map. If $z'' \in ZK''$ then $\partial_*[z''] = [i^{-1}\partial q z''] = [j\partial q z''] = [dz''] = d_*[z'']$. ∎

2.13 Corollary. *If $f\colon K \to L$ is a chain map then the connecting homomorphism of the exact sequence* 1.7, $0 \to L \to Cf \to K^+ \to 0$, *coincides with $Hf\colon HK \to HL$.*

Indeed, the sequence is split in every dimension by $qx = (0, x)$, $j(y, x) = y$, and we have $j\partial q = f$. ∎

2.14 Corollary. *If $f\colon K \to L$ is a chain map then $Hf\colon HK \to HL$ is isomorphic if and only if the mapping cone Cf is acyclic, $H(Cf) = 0$.*

This follows from the exact homology sequence 2.9 b because of 2.13. ▮

2.15 Example. If K is a complex then $0 \to ZK \xrightarrow{i} K \xrightarrow{\partial} (BK)^+ \to 0$ can be viewed as exact sequence of chain maps (i = inclusion). The connecting homomorphism is given by $i^{-1} \circ \partial \circ \partial^{-1}$, that is by the inclusion map j: $BK \subset ZK$. The exact homology sequence therefore has the form

$$H_{n+1} K \xrightarrow{0} B_n K \xrightarrow{j_n} Z_n K \xrightarrow{[\ ]} H_n K \xrightarrow{0} B_{n-1} K,$$

i.e. essentially it coincides with the exact sequence

$$0 \to BK \xrightarrow{\subset} ZK \xrightarrow{[\ ]} HK \to 0.$$

2.16 Exercises. 1. The cone CK of every complex K is acyclic, $HCK = 0$.

2. Prove: The connecting homomorphism $\partial_*: H_{n+1} K'' \to H_n K'$ is determined up to sign ± 1 by the Properties 2.9 a), b).

Hint: Consider the exact sequence

$$(E) \qquad 0 \to (\mathbb{Z}, n) \to C(\mathbb{Z}, n) \to (\mathbb{Z}, n+1) \to 0$$

first. Then prove. For every $z'' \in Z_{n+1} K''$ there exists a map of the sequence (E) into $0 \to K' \to K \to K'' \to 0$ such that $1 \mapsto z''$. Apply 2.9 a).

3. *Chain-Homotopy*

According to exercise 1.9, chain maps f: $K \to L$ can be viewed as zero-cycles of $\operatorname{Hom}(K, L)$. What does it mean then for two chain maps f, g: $K \to L$ to be homologous in $Z_0 \operatorname{Hom}(K, L)$? It means that $s \in \operatorname{Hom}(K, L)_1$ exists such that $\partial(s) = f - g$. This notion, usually called *chain homotopy*, is of great importance.

3.1 Definition. Let f, g: $K \to K'$ be chain maps. *A homotopy s between f and g*, in symbols s: $f \simeq g$, is a sequence of homomorphisms, s_n: $K_n \to K'_{n+1}$, such that

$$\partial'_{n+1} s_n + s_{n-1} \partial_n = f_n - g_n \quad \text{for all } n \in \mathbb{Z}.$$

We write $f \simeq g$ and say *f and g are homotopic* if such an s exists.

3.2 Proposition. *The homotopy relation $\simeq$ is an equivalence relation.* The equivalence class of f: $K \to K'$ is denoted by $[f]$, and is called *homotopy class* of f.

Proof. Reflexivity 0: $f \simeq f$.
Symmetry s: $f \simeq g \Rightarrow -s$: $g \simeq f$.
Transitivity s: $f \simeq g$, t: $g \simeq h \Rightarrow s + t$: $f \simeq h$. ▮

3.3 Proposition and Definition. *The homotopy relation is compatible with composition, i.e. if* $f \simeq g\colon K \to K'$ *and* $f' \simeq g'\colon K' \to K''$ *then* $f'f \simeq g'g$.

We can therefore define a composition law for homotopy classes by $[f'] \circ [f] = [f' \circ f]$. This defines a new category $\mathcal{H}\partial\mathcal{G}$. Its objects are complexes as in $\partial\mathcal{A}\mathcal{G}$, the morphisms, however, are *homotopy classes* of chain maps. If we assign to each chain map $f\colon K \to K'$ its homotopy class $[f]$ we get a covariant functor $\pi\colon \partial\mathcal{A}\mathcal{G} \to \mathcal{H}\partial\mathcal{G}$.

A chain map $f\colon K \to K'$ whose class $[f]$ is an equivalence in $\mathcal{H}\partial\mathcal{G}$ is called *homotopy equivalence*, and K, K' are called *homotopy equivalent* if such an f exists; we write $K \simeq K'$. Explicitely this means that chain maps $K \xrightarrow{f} K' \xrightarrow{f^-} K$ exist such that $f^- f \simeq \mathrm{id}_K$, $ff^- \simeq \mathrm{id}_{K'}$. The map f^- is called a *homotopy inverse* of f.

Proof of 3.3. If $s\colon f \simeq g$ then $f's\colon f'f \simeq f'g$ because $\partial''(f's) + (f's)\partial = f'(\partial's + s\partial) = f'(f - g) = f'f - f'g$. Similarly, $s'\colon f' \simeq g' \Rightarrow s'g\colon f'g \simeq g'g$, hence by transitivity, $f'f \simeq g'g$. ∎

3.4 Proposition. *If* $f \simeq g\colon K \to K'$ *then* $f_* = g_*\colon HK \to HK'$, *i.e. homotopic chain maps induce the same homology-homomorphism.*

Proof. $f_*[z] - g_*[z] = [fz - gz] = [\partial sz + s(\partial z)] = [\partial(sz)] = 0$. ∎

3.5 Corollary. *If* $f\colon K \to K'$ *is a homotopy equivalence then* $f_*\colon HK \to HK'$ *is an isomorphism.*

Proof. $ff^- \simeq \mathrm{id}$, $f^- f \simeq \mathrm{id}$ imply $f_* f_*^- = (ff^-)_* = \mathrm{id}_* = \mathrm{id}$, and $f_*^- f_* = \mathrm{id}$. ∎

Clearly, Proposition 3.4 can also be formulated as follows: *The homology functor H factors through* $\mathcal{H}\partial\mathcal{G}$ i.e. there is a commutative diagram of functors

$$\begin{array}{ccc} \partial\mathcal{A}\mathcal{G} & \xrightarrow{\quad H \quad} & \mathcal{G}\mathcal{A}\mathcal{G} \\ & {\scriptstyle\pi}\searrow \quad \nearrow{\scriptstyle H'} & \\ & \mathcal{H}\partial\mathcal{G} & \end{array}$$

The corollary then simply states that the functor H' takes equivalences into equivalences.

Complexes K such that $\mathrm{id}_K \simeq 0$, or equivalently $K \simeq 0$, are called *contractible*. Clearly $K \simeq 0$ implies $HK = 0$ (by 3.5). As to the converse one has

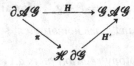

3.6 Proposition. *Let K be an acyclic complex, i.e., $HK=0$. Then $K\simeq 0$ if and only if for all n, $Z_n K$ is a direct summand of K_n.*

Proof. Assume $s\colon \mathrm{id}_K \simeq 0$, i.e. $\partial s + s\partial = \mathrm{id}_K$. Since $\partial|BK=0$ this implies $\partial s|BK = \mathrm{id}_{BK}$, hence the exact sequence $0\to ZK \xrightarrow{\subset} K \xrightarrow{\partial} BK \to 0$ splits, i.e. ZK is a direct summand. Conversely, assume there is $t\colon BK\to K$ with $\partial t = \mathrm{id}$, i.e. $K = ZK \oplus tBK = BK \oplus tBK$. Define s by $s|BK = t$, $s|tBK = 0$. Then $\partial s + s\partial|BK = \partial t = \mathrm{id}$, $\partial s + s\partial|tBK = s\partial|tBK = t\partial|tBK = \mathrm{id}$. ∎

An example K for which $HK=0$ but $K\not\simeq 0$ is as follows: $K_n = \mathbb{Z}_4$, $\partial_n =$ multiplication by 2 for all n.

Proposition 3.6 is particularly useful in connection with the following

3.7 Proposition. *If the mapping cone of $f\colon K\to L$ is contractible, $Cf\simeq 0$, then f is a homotopy equivalence.* (The converse is also true; cf. Exerc. 5.)

Proof. We show

I. If the inclusion $\iota\colon L\to Cf$, $\iota y = (y,0)$, is nulhomotopic, then f has a right homotopy inverse $g\colon L\to K$, $fg\simeq \mathrm{id}$.

II. If the projection $\kappa\colon Cf\to K^+$, $\kappa(y,x)=x$ is nulhomotopic then f has a left homotopy inverse $h\colon L\to K$, $hf\simeq\mathrm{id}$. This suffices since $Cf\simeq 0$ implies $\iota\simeq 0$, $\kappa\simeq 0$, and $h\simeq h(fg)=(hf)g\simeq g$.

I. Let $S\colon \iota\simeq 0$. Define $g\colon L\to K$, $\gamma\colon L\to L$ by $Sy = (\gamma(y), g(y))$; recall that $Cf = L\oplus K^+$ as a group (not as a complex! And γ is not a chain map!). Then $\partial S y + S\partial y = \iota y$ reads

$$(\partial\gamma y + fgy + \gamma\partial y,\ -\partial g y + g\partial y) = (y,0),$$

i.e., $\partial g = g\partial$ and $\partial\gamma + \gamma\partial = \mathrm{id} - fg$, as asserted.

II. Let $T\colon \kappa\simeq 0$. Define $h\colon L\to K$, $\eta\colon K\to K$ by $T(y,x) = h(y) + \eta(x)$. Then $\partial T + T\partial = \kappa$ reads $-\partial h y + h\partial y - \partial\eta x - \eta\partial x + hfx = x$ (recall that $\partial^{K^+} = -\partial^K$) i.e., $\partial h = h\partial$ and $\partial\eta + \eta\partial = hf - \mathrm{id}$. ∎

3.8 Exercises. *1.* The cone CK of every complex K is contractible, $CK\simeq 0$.

2*. If $(E)\colon 0\to K' \xrightarrow{i} K \xrightarrow{p} K'' \to 0$ is an exact sequence of chain maps, define $\rho\colon Ci\to K''$ by $\rho(x,x') = p(x)$. Prove that ρ is a chain map, $\rho_*\colon H(Ci)\cong HK''$, and the composite $HK'' \xrightarrow{\rho_*^{-1}} H(Ci) \xrightarrow{\kappa_*} (HK')^+$ coincides with $-\partial_*$. Formulate and prove dual results about $\sigma\colon (K')^+\to Cp$, $\sigma(x') = (0, ix')$. If the sequence (E) is direct then ρ and σ are homotopy equivalences.

3. Let $0 \to K' \xrightarrow{i} K \xrightarrow{p} K'' \to 0$ be an exact sequence of chain maps.

(a) If $i \simeq 0$, say s: $i \simeq 0$, then ps is a chain map $K'^{+} \to K''$, and $\partial_*(ps)_* = \mathrm{id}_{HK'}$.

(b) If t: $p \simeq 0$ then ti is a chain map $K'^{+} \to K''$, and $(ti)_* \partial_* = \mathrm{id}_{HK''}$.

4*. If $0 \to K' \xrightarrow{i} K \xrightarrow{p} K'' \to 0$ is a direct sequence of chain maps then

(a) $K' \simeq 0$ *or* $K'' \simeq 0 \Rightarrow K = K' \oplus K''$, i.e. the sequence splits.

(b) $K \simeq 0 \Leftrightarrow i \simeq 0$ *and* $p \simeq 0$.

5. Prove the converse of 3.7. There are at least two possibilities:

(i) Read the proof of 3.7 backwards and use exerc. 4b. (ii) Remark that $\mathrm{Hom}(X, f)$ is a homotopy equivalence hence (using 1.9) $\mathrm{Hom}(X, Cf)$ is acyclic hence $\mathrm{id}_{Cf} \in Z_0 \mathrm{Hom}(Cf, Cf)$ is homologous to zero.

6. If (E): $0 \to K' \to K \to K'' \to 0$ is exact and direct then

$$0 \to \mathrm{Hom}(L, K') \to \mathrm{Hom}(L, K) \to \mathrm{Hom}(L, K'') \to 0$$

is exact and direct for every complex L. If $L = K''$ then

$$\mathrm{id}_{K''} \in Z_0 \mathrm{Hom}(K'', K''),$$

and $\partial_*[\mathrm{id}_{K''}]$ is a homotopy class of chain maps $K'' \to (K')^+$. Show that the induced homomorphism $HK'' \to H(K')^+$ coincides with the connecting homomorphism of (E).

4. Free Complexes

These complexes have useful special properties, and they frequently come up in applications.

4.1 Definition. A complex K is called *free* if K_n is free for every $n \in \mathbb{Z}$.

4.2 Proposition. *In a free complex K the group of cycles $Z_n K$ is a direct summand of K_n.*

Proof. Subgroups of free groups are free (I, 2.23). Therefore $BK \subset K$ is free, therefore the exact sequence $0 \to ZK \to K \to BK \to 0$ splits (I, 2.24). ∎

4.3 Proposition. *If f: $K \to L$ is a chain map between free complexes such that f_*: $HK \cong HL$ then f is a homotopy equivalence.*

I.e., for free complexes the converse of 3.5 holds.

Proof. By Proposition 3.7 is suffices to prove that $Cf \simeq 0$. According to 3.6 we have to show that $HCf = 0$ and that the cycles ZCf are direct summands. The former holds by 2.14, the latter by 4.2. ∎

4.4 Definition

A complex K is called *short* if an integer n exists such that $K_i = 0$ for $i \neq n, n+1$, and $\partial_{n+1}: K_{n+1} \to K_n$ is monomorphic. (I.e. a complex is short if it is *essentially* concentrated in one dimension namely n.) If, moreover, $K_n \cong \mathbb{Z}$ then K is called *elementary*.

4.5 Proposition. *Every free complex K is a direct sum of short (free) complexes. If moreover every K_m is finitely generated, then K is a direct sum of elementary complexes.*

Proof. By 4.2 we can write K_m as a direct sum $K_m = Z_m K \oplus Z_m^\perp$. Put $K_i^{(m)} = 0$ for $i \neq m, m+1$, $K_m^{(m)} = Z_m K$, $K_{m+1}^{(m)} = Z_{m+1}^\perp$. Clearly, $K^{(m)}$ is a subcomplex, is short, and $K = \oplus_m K^{(m)}$.

If K_m is finitely generated then so are $Z_m K$ and $Z_m^\perp$. Moreover, there are bases $\{a_1^m, \ldots, a_r^m\}$ of $Z_m K$ and $\{b_1^{m+1}, \ldots, b_s^{m+1}\}$ of $Z_{m+1}^\perp$, $s \leq r$, such that $\partial_{m+1} b_i^{m+1} = \tau_i^m a_i^m$ with $\tau_i^m \subset \mathbb{Z}$, $i \leq s$ (view $Z_{m+1}^\perp$ as subgroup of Z_m via ∂_{m+1} and apply I, 2.25). Let $K^{(m,i)} \subset K$ the subcomplex generated by the pair (a_i^m, b_i^{m+1}) if $i \leq s$, and by the element a_i^m if $i > s$. Then $K^{(m,i)}$ is elementary and $K = \oplus_{i,m} K^{(m,i)}$. ∎

Remark. By I, 2.25, the base $\{a_i^m, b_j^n\}$ can even be so chosen that τ_i^m always divides τ_{i+1}^m (and all $\tau_i^m > 0$). It is then called a *canonical base* of K. The numbers $\tau_i^m > 1$ (or their primary parts) are called the *torsion coefficients* of K (or of HK); they are uniquely determined by HK, i.e., independent of the choice of the base $\{a_i^m, b_i^n\}$. For proofs and more details cf. Eilenberg-Steenrod V. 8, or Kurosh § 20.

4.6 Proposition. If *K is a free complex, L an arbitrary complex, and $\varphi_n: H_n K \to H_n L$, $n \in \mathbb{Z}$, a sequence of homomorphisms then there exists a chain map $f: K \to L$ such that $f_* = \varphi$.* I.e., every homomorphism $\varphi: HK \to HL$ of the homology of a free complex K can be realized by a chain map.

The proof is based on the following

4.7 Lemma. *Every commutative diagram*

$$\begin{array}{ccccc} F & \xrightarrow{\gamma_1} & G_0 & \xrightarrow{\gamma_0} & G_{-1} \\ \vdots g & & \downarrow g_0 & & \downarrow g_{-1} \\ G_1' & \xrightarrow[\gamma_1']{} & G_0' & \xrightarrow[\gamma_0']{} & G_{-1}' \end{array}$$

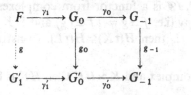

of abelian group homomorphism (without g as yet) whose second row is exact, whose first row is a complex (i.e., $\gamma_0 \gamma_1 = 0$), and where F is free, can be completed by a homomorphism g.

Proof of 4.7. If $a \in F$ then $\gamma'_0 g_0 \gamma_1 a = g_{-1} \gamma_0 \gamma_1 a = 0$, i.e. $g_0 \gamma_1 a \in \ker(\gamma'_0) = \operatorname{im}(\gamma'_1)$. Therefore, if $\{a_\mu\}$ is a base of F we can find elements $b_\mu \in G'_1$ with $\gamma'_1 b_\mu = g_0 \gamma_1 a_\mu$, and define g by $g a_\mu = b_\mu$. ∎

Proof of 4.6. Let $K = ZK \oplus Z^\perp$ as in the Proof 4.5. By Lemma 4.7 we can find first f_n^Z, then $f_{n+1}^\perp$ which make

$$\begin{array}{ccccc} Z_{n+1}^\perp & \xrightarrow{\partial} & Z_n K & \xrightarrow{\text{proj}} & H_n K \to 0 \\ \downarrow{\scriptstyle f_{n+1}^\perp} & & \downarrow{\scriptstyle f_n^Z} & & \downarrow{\scriptstyle \varphi_n} \\ L_{n+1} & \xrightarrow[\partial]{} & Z_n L & \xrightarrow[\text{proj}]{} & H_n L \to 0 \end{array}$$

commutative. Then $f: K \to L$, $f|ZK = f^Z$, $f|Z^\perp = f^\perp$ is a chain map as required. ∎

4.8 Corollary. *Let K, L be free complexes. Then $K \simeq L \Leftrightarrow HK \cong HL$.*

Proof. If $\varphi: HK \to HL$ is an isomorphism, it can be realized by a chain map $f: K \to L$ and f is then a homotopy equivalence by Proposition 4.3. The converse is contained in 3.5. ∎

4.9 Corollary. *If K is a free complex and HK is also free then $K \simeq HK$.* ∎

4.10 Exercises. *1.* a) For every abelian group A and integer n construct a free short complex K such that $H_n K \cong A$.
b) For every graded abelian group $G = \{G_n\}_{n \in \mathbb{Z}}$ construct a free complex K such that $HK \cong G$.

2. Construct a free complex K which is not a direct sum of elementary complexes. Hint: If K is a direct sum of elementary complexes then $H_n K$ is a direct sum of cyclic groups (is the converse true?).

3. If K is a free complex such that $H_i K = 0$ for $i < n$ then there exists a subcomplex $K' \subset K$ with $K'_i = 0$ for $i < n$ and $K' \simeq K$.

4. If $t: \partial\mathscr{A}\mathscr{G} \to \partial\mathscr{A}\mathscr{G}$ is a functor from complexes to complexes which *preserves homotopy* (i.e. $f \simeq g \Rightarrow t f \simeq t g$) and if K, L are free complexes such that $HK \cong HL$ then $H(tK) \cong H(tL)$. Construct examples of such functors.

5. If K is a free complex and $K \simeq HK$ then HK is free.

Chapter III

Singular Homology

1. Standard Simplices and Their Linear Maps

1.1 Definition. The *standard q-simplex* Δ_q consists of all points $x \in \mathbb{R}^{q+1}$ such that

(a) $0 \leq x_i \leq 1$, $i = 0, 1, \ldots, q$,

(b) $\sum_{i=0}^{q} x_i = 1$,

where $\mathbb{R}^{q+1}$ denotes euclidean space and $\{x_i\}$ are the coordinates of $x \in \mathbb{R}^{q+1}$. Clearly Δ_q is closed and bounded, hence *compact*. Because of (b) we can replace (a) by

(a′) $0 \leq x_i$, $i = 0, 1, \ldots, q$.

Therefore Δ_q is the intersection of the hyperplane $\sum_{i=0}^{q} x_i = 1$ with the positive "quadrant" $\{x_i \geq 0\}$. In particular, Δ_q is *convex* (i.e. any segment whose endpoints lie in Δ_q lies in Δ_q).

For instance, Δ_0 is a single point, Δ_1 is a segment, Δ_2 an equilateral triangle, Δ_3 a regular tetrahedron.

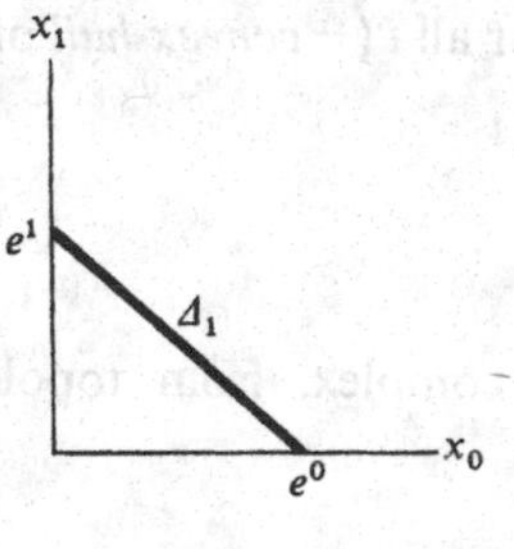

Fig. 1

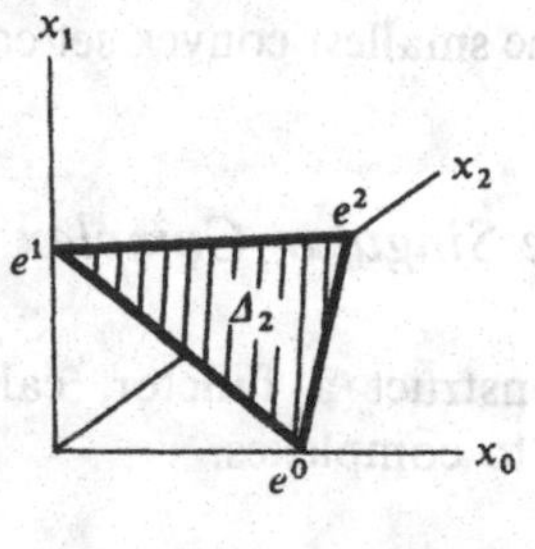

Fig. 2

The unit points $e^j = (0, \ldots, 0, 1, 0, \ldots, 0)$ of $\mathbb{R}^{q+1}$ lie in Δ_q; they are called the *vertices* of Δ_q.

1.2 Definition. A mapping f of Δ_q into $\mathbb{R}^n$ (or into a subset of $\mathbb{R}^n$) is called *linear* if a linear (in the usual sense) map $F\colon \mathbb{R}^{q+1} \to \mathbb{R}^n$ exists such that $F|\Delta_q = f$. If $P^0, P^1, \ldots, P^q \in \mathbb{R}^n$ are arbitrary points then there exists a unique linear map $f: \Delta_q \to \mathbb{R}^n$ such that $f(e^i) = P^i$, namely $f(x) = \sum_{i=0}^q x_i P^i$. The image $f(\Delta_q)$ consists of all points $P = \sum_{i=0}^n x_i P^i$ of $\mathbb{R}^n$ with $0 \le x_i \le 1$, $\sum x_i = 1$. Thus *linear maps of Δ_q are completely determined by their values on the vertices and these values can be prescribed.* In particular, we consider the linear maps

(1.3)
$$\varepsilon^j = \varepsilon^j_q: \Delta_{q-1} \to \Delta_q$$
$$\varepsilon^j(e^i) = e^i \quad \text{for} \quad i<j, \qquad \varepsilon^j(e^i) = e^{i+1} \quad \text{for} \quad i \ge j,$$

where $j = 0, 1, \ldots, q$. The image of ε^j_q consists of all points $x \in \Delta_q$ with $x_j = 0$; it is called the *j-th face* of Δ_q. The union of all faces of Δ_q is called the *boundary* of Δ_q and is denoted by $\dot{\Delta}_q$. It consists of all points $x \in \Delta_q$ with at least one vanishing coordinate.

For later use we note the

1.4 Lemma. $\varepsilon^j_{q+1} \varepsilon^k_q = \varepsilon^k_{q+1} \varepsilon^{j-1}_q$ *if* $k<j$.

Indeed, on both sides we have

$$e_i \mapsto e_i \quad \text{for} \quad i<k, \qquad e_i \mapsto e_{i+1} \quad \text{for} \quad k \le i < j-1,$$
$$e_i \mapsto e_{i+2} \quad \text{for} \quad i \ge j-1. \quad \blacksquare$$

1.5 Exercise. If $F: \mathbb{R}^{q+1} \to \mathbb{R}^n$ is a linear map and $K \subset \mathbb{R}^n$ is a convex set such that $F(e^i) \in K$, $i = 0, 1, \ldots, q$, then $F(\Delta_q) \subset K$. In particular, Δ_q is the smallest convex set containing e^i for all i (= *convex hull* of $\{e^i\}$).

2. The Singular Complex

We construct a functor, called singular complex, from topological spaces to complexes.

2.1 Definition. Let X be a topological space. A *singular q-simplex* of X is a continuous map $\sigma = \sigma_q: \Delta_q \to X$, $q \ge 0$. We consider the free abelian group $S_q X$ which is generated by the set of all singular q-simplices. The elements $c_q \in S_q X$ are called *singular q-chains* of X. By definition, every $c \in S_q X$ has a unique representation as finite linear combination of singular q-simplices σ, $c = \sum c_\sigma \cdot \sigma$, with integral coefficients c_σ. We shall not distinguish between a singular simplex σ and the chain c whose only non-zero coefficient is $c_\sigma = 1$. For $q<0$ we put $S_q X = 0$.

We define a homomorphism $\partial_q\colon S_qX \to S_{q-1}X$, $\partial_q(\sigma)=\sum_{j=0}^{q}(-1)^j(\sigma_q\varepsilon_q^j)$, where $\varepsilon_q^j\colon \Delta_{q-1}\to\Delta_q$ denotes the j-th face as in 1.3. Then

2.2 Proposition. *The sequence* $\cdots\leftarrow S_{q-1}X \xleftarrow{\partial_q} S_qX \xleftarrow{\partial_{q+1}} S_{q+1}X \leftarrow\cdots$ *is a complex, i.e.* $\partial_q\partial_{q+1}=0$. It is called the *singular complex of* X, and is denoted by SX.

Proof. For singular simplices σ we have

$$\begin{aligned}\partial\partial\sigma &= \partial\bigl(\textstyle\sum_j(-1)^j\sigma\varepsilon^j\bigr)=\sum_{j,k}(-1)^{j+k}\sigma\varepsilon^j\varepsilon^k\\ &= \textstyle\sum_{j\le k}(-1)^{j+k}\sigma\varepsilon^j\varepsilon^k+\sum_{j>k}(-1)^{j+k}\sigma\varepsilon^k\varepsilon^{j-1},\end{aligned}$$

the latter by 1.4. In the second sum we replace k by j and j by $k+1$; then corresponding terms of the two sums cancel. Thus $\partial\partial$ vanishes on a base $\{\sigma\}$, hence $\partial\partial=0$. ∎

If $f\colon X\to Y$ is a continuous map and $\sigma\colon \Delta_q\to X$ a singular simplex of X then the composite $f\sigma\colon \Delta_q\to Y$ is a singular simplex in Y, and we get a homomorphism

$$S_qf\colon S_qX\to S_qY,\qquad (S_qf)(\sigma)=f\sigma.$$

2.3 Proposition. *The sequence* $S_qf\colon S_qX\to S_qY$, $q\in\mathbb{Z}$, *is a chain map,* $Sf\colon SX\to SY$. Instead of Sf we usually write $f\colon SX\to SY$.

Proof. Multiplying $(f\sigma)\varepsilon^j=f(\sigma\varepsilon^j)$ with $(-1)^j$ and summing over j gives $\partial(f\sigma)=f(\partial\sigma)$. ∎

2.4 Proposition. $S(gf)=(Sg)(Sf)$, $S(\mathrm{id}_X)=\mathrm{id}_{SX}$ *(where* $g\colon Y\to Z$*), i.e.* S *is a functor from spaces to complexes,* $S\colon \mathcal{T}op\to\partial\mathcal{AG}$. ∎

2.5 We now generalize the preceding to pairs of spaces (X,A). If $i\colon A\to X$ is the inclusion map then $i\colon SA\to SX$ is monomorphic, hence SA can be thought of as a subcomplex of SX. The quotient $S(X,A)=SX/SA$ is called the *(relative) singular complex of* (X,A). If j denotes passage to quotients then

$$(2.6)\qquad 0\to SA\xrightarrow{\;i\;}SX\xrightarrow{\;j\;}S(X,A)\to 0$$

is an exact sequence of chain maps. *It splits in every dimension,* $S_qX=S_qA\oplus S_q(X,A)$. Indeed the base $\{\sigma\colon\Delta_q\to X\}$ of S_qX divides into two parts: the simplices in A, and those which are not in A. The former provide a base for S_qA, the latter for $S_q(X,A)$. Note that $S(X,\emptyset)=SX$.

A map $f\colon (X,A)\to(Y,B)$ *of pairs* (cf. I, 3.8) *induces a commutative diagram*

$$
(2.7)\qquad
\begin{array}{ccccccccc}
0 & \to & SA & \longrightarrow & SX & \longrightarrow & S(X,A) & \to & 0\\
 & & \Big\downarrow{\scriptstyle S(f|A)} & & \Big\downarrow{\scriptstyle Sf} & & \Big\downarrow{\scriptstyle \overline{Sf}} & & \\
0 & \to & SB & \longrightarrow & SY & \longrightarrow & S(Y,B) & \to & 0
\end{array}
$$

of chain maps with exact rows; the map $\overline{Sf}$ is obtained from Sf by passing to quotients. ∎

The functor properties 2.4 carry over to pairs. In fact we can view S as a *functor from pairs of spaces to short exact sequences of complexes.* We leave it to the reader to make this statement precise.

2.8 Exercise. Does the sequence $0\to S_qA\to S_qX\to S_q(X,A)\to 0$ split *naturally*?

3. Singular Homology

3.1 Definition. The (singular) *homology groups* of a space X resp. a pair of spaces (X,A) are, by definition, the homology groups of the singular complex SX resp. $S(X,A)$. We write $HX=HSX$, $H(X,A)=HS(X,A)$. The groups $H(X,A)$ are also called *relative* homology groups of X *mod* A, in contrast to the *absolute* groups HX. We say $z\in SX$ is a *cycle mod* A if $\partial z\in SA$, and z is a *boundary mod* A if $z=\partial x+y$ for some $x\in SX$, $y\in SA$. The relative homology group $H_q(X,A)$ is then isomorphic with the group of q-cycles mod A divided by the group of q-boundaries mod A, $H(X,A)\cong\dfrac{Z(X,A)}{B(X,A)}$.

If $f\colon (X,A)\to(Y,B)$ is a map of pairs then $Sf\colon S(X,A)\to S(Y,B)$ induces homomorphisms $Hf=f_*\colon\ H(X,A)\to H(Y,B)$. This turns singular homology into a functor from pairs of spaces to graded groups. By definition, it is composed of $\mathcal{T}\!op^{(2)}\xrightarrow{S}\partial\mathcal{AG}\xrightarrow{H}\mathcal{GAG}$.

The connecting homomorphism $\partial_*\colon H_{q+1}(X,A)\to H_qA$ of the sequence

$$0\to SA\xrightarrow{\ i\ }SX\xrightarrow{\ j\ }S(X,A)\to 0$$

is called the *connecting homomorphism* of (X,A), and the exact sequence (cf. II, 2.9)

$$(3.2)\qquad \cdots\xrightarrow{\partial_*}H_{q+1}A\xrightarrow{i_*}H_{q+1}X\xrightarrow{j_*}H_{q+1}(X,A)\xrightarrow{\partial_*}H_qA\xrightarrow{i_*}H_qX\xrightarrow{j_*}\cdots$$

is called the *homology sequence of* (X,A).

If $f\colon (X,A)\to(Y,B)$ is a map of pairs then

$$
\begin{array}{ccccccccc}
H_{q+1}A & \longrightarrow & H_{q+1}X & \longrightarrow & H_{q+1}(X,A) & \longrightarrow & H_qA & \longrightarrow & H_qX \\
\downarrow{\scriptstyle (f|A)_*} & & \downarrow{\scriptstyle f_*} & & \downarrow{\scriptstyle f_*} & & \downarrow & & \downarrow \\
H_{q+1}B & \longrightarrow & H_{q+1}Y & \longrightarrow & H_{q+1}(Y,B) & \longrightarrow & H_qB & \longrightarrow & H_qY
\end{array}
\tag{3.3}
$$

is a commutative diagram (II, 2.9 (a)) with exact rows.

Consider now a triple $B\subset A\subset X$ of spaces; one also writes (X,A,B). Inclusion i and projection j define an exact sequence

$$0\to S(A,B)\xrightarrow{i} S(X,B)\xrightarrow{j} S(X,A)\to 0$$

of chain maps. The resulting exact sequence

$$
\begin{aligned}
&\cdots\to H_{q+1}(A,B)\xrightarrow{i_*} H_{q+1}(X,B)\\
&\xrightarrow{j_*} H_{q+1}(X,A)\xrightarrow{\partial_*} H_q(A,B)\xrightarrow{i_*} H_q(X,B)\xrightarrow{j_*}\cdots
\end{aligned}
\tag{3.4}
$$

is called the *homology sequence of the triple* (X,A,B). For $B=\emptyset$ it reduces to 3.2.

3.5 Exercise. *1.* If (X,A,B) is a triple then the connecting homomorphism $\partial_*\colon H_{q+1}(X,A)\to H_q(A,B)$ coincides with the composite

$$H_{q+1}(X,A)\xrightarrow{\partial'_*} H_qA\xrightarrow{i'_*} H_q(A,B)$$

where ∂'_* is the connecting homomorphism of the pair (X,A).
2. If $B\subset A\subset X$ is a triple such that $i_*\colon HB\cong HA$ then $j_*\colon H(X,B)\cong H(X,A)$.

4. Special Cases

4.1 If P is *a single point* then there is just one singular simplex $\tau_q\colon \Delta_q\to P$ for every $q\geq 0$. We have $\tau_q\varepsilon^j=\tau_{q-1}$ for all $q>0$ and $0\leq j\leq q$, hence $\partial\tau_{2q}=\tau_{2q-1}$ for $q>0$ and $\partial\tau_{2q-1}=0$. Thus SP is the complex

$$\cdots 0\leftarrow \mathbb{Z}\xleftarrow{0}\mathbb{Z}\xleftarrow{\text{id}}\mathbb{Z}\xleftarrow{0}\mathbb{Z}\xleftarrow{\text{id}}\mathbb{Z}\xleftarrow{0}\cdots$$

and

$$H_0P=\mathbb{Z},\quad H_iP=0\quad\text{for}\quad i\neq 0. \tag{4.2}$$

4.3 Definition. For every space X the constant map $\gamma\colon X\to P$ (P = point) induces a homomorphism $\gamma_*=\gamma_*^X\colon HX\to HP$, called the *augmentation*. If $f\colon X\to Y$ is a map then $\gamma_*^Y f_*=\gamma_*^X$ (*naturality of* γ_*); in particular, f_* maps $\ker(\gamma_*^X)$ into $\ker(\gamma_*^Y)$. These groups are therefore functors of

$X \in \mathcal{T}op$; they are called the *reduced homology* and are denoted by $\tilde{H}_q X = \ker(\gamma_*: H_q X \to H_q P)$. If $q \neq 0$ then $\tilde{H}_q X = H_q X$ by 4.2.

If X is *not empty* then any map $\iota: P \to X$ is right inverse to γ, hence $\gamma_* \iota_* = \text{id}$. It follows that $H_0 X = \text{im}(\iota_*)_0 \oplus \ker(\gamma_*)_0 = \mathbb{Z} \oplus \tilde{H}_0 X$, i.e., in dimension zero, reduced and unreduced homology differ by a direct summand $\mathbb{Z}$. Moreover, the exact sequence $H_0 P \xrightarrow{\iota_*} H_0 X \xrightarrow{\kappa_*} H_0(X, P) \to 0$ of the pair (X, P) shows that $\kappa_*: \tilde{H}_0 X \cong H_0(X, P)$.

If (X, A) is a pair of spaces with $A \neq \emptyset$ then we have mappings $(X, A) \xrightarrow{\gamma} (P, P) \xrightarrow{\iota} (X, A)$, and $\gamma \iota = \text{id}$. It follows that ι_* maps the homology sequence of (P, P)—which is rather trivial—onto a direct summand of the homology sequence of (X, A); the other direct summand is $\ker(\gamma_*)$. Since $\ker(\gamma_*)$ is reduced homology this shows

4.4 Proposition. *If (X, A) is a pair of spaces with $A \neq \emptyset$ then we have an exact sequence.*

$$\cdots \xrightarrow{\partial_*} \tilde{H}_{q+1} A \xrightarrow{i_*} \tilde{H}_{q+1} X \xrightarrow{j_*} H_{q+1}(X, A) \xrightarrow{\partial_*} \tilde{H}_q A \xrightarrow{i_*} \tilde{H}_q X \xrightarrow{j_*} \cdots;$$

it is called the reduced homology sequence of (X, A).

4.5 The name *augmentation* is often used for the chain map $\eta = \eta^X$: $SX \to (\mathbb{Z}, 0)$, which takes every zero simplex σ_0 into $1 \in \mathbb{Z}$. This map is closely related to γ; in fact, $\eta^X = \eta^P \circ \gamma^X$. Moreover, the map η^P: $SP \to (\mathbb{Z}, 0)$ is a homotopy equivalence: $(\mathbb{Z}, 0)$ is a direct summand of SP, and the other direct summand is clearly nulhomotopic (cf. also 4.6). In particular, $\ker(\eta_*) = \ker(\gamma_*) = \tilde{H} X$. Therefore, the danger of confusing the two augmentations γ, η is not grave.—In the literature, the name "index" is also used for η.

After the one-point space we consider convex sets in $\mathbb{R}^n$. Their homology turns out be equally trivial.

4.6 Proposition. *If X is a non-empty convex subspace of euclidean space $\mathbb{R}^n$ then the augmentation η: $SX \to (\mathbb{Z}, 0)$ is a homotopy equivalence; in particular, $\tilde{H} X = 0$.*

Proof. The method of proof is known as "cone construction". Pick $P \in X$. For every σ_q: $\Delta_q \to X$, $q \geq 0$, define $(P \cdot \sigma_q)$: $\Delta_{q+1} \to X$ by

$$(P \cdot \sigma_q)(x_0, x_1, \ldots, x_{q+1})$$

$$(4.7) \qquad = \begin{cases} P & \text{if } x_0 = 1, \\ x_0 P + (1 - x_0)\, \sigma_q\left(\dfrac{x_1}{1 - x_0}, \ldots, \dfrac{x_{q+1}}{1 - x_0}\right) & \text{if } x_0 \neq 1. \end{cases}$$

This defines homomorphisms

$$P = P_q \colon S_q X \to S_{q+1} X, \quad P_q(\sigma) = P \cdot \sigma.$$

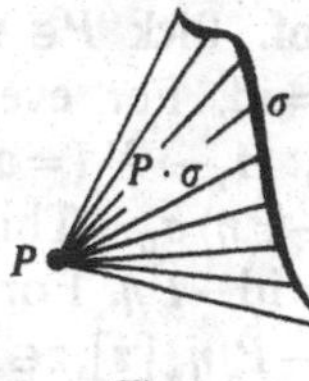

Fig. 3

Intuitively speaking, $P \cdot \sigma$ is obtained by projecting σ from the new vertex P, or by erecting the cone with vertex P over σ.

We compute the faces of $P \cdot \sigma$,

$$(P \cdot \sigma_q)\,\varepsilon^i(x_0, x_1, \ldots, x_q) = (P \cdot \sigma_q)(x_0, \ldots, x_{i-1}, 0, x_i, \ldots, x_q).$$

If $i=0$, this is $\sigma_q(x_0, \ldots, x_1)$; if $q=0$ and $i=1$ it is P, and if $q>0$ and $i>0$ it is

$$\begin{aligned} x_0 P + (1-x_0)\,\sigma_q&\left(\frac{x_1}{1-x_0}, \ldots, \frac{x_{i-1}}{1-x_0}, 0, \frac{x_i}{1-x_0}, \ldots, \frac{x_q}{1-x_0}\right) \\ &= x_0 P + (1-x_0)(\sigma_q \varepsilon^{i-1})\left(\frac{x_1}{1-x_0}, \ldots, \frac{x_q}{1-x_0}\right) \\ &= [P \cdot (\sigma_q \varepsilon^{i-1})](x_0, \ldots, x_q). \end{aligned}$$

If we define a chain map $\hat{P} \colon (\mathbb{Z}, 0) \to SX$ by $\hat{P}(m) = mP$, then we can express the result of the computation as follows

$$(4.8) \qquad \begin{aligned} &(P \cdot \sigma_q)\,\varepsilon^0 = \sigma_q, \quad (P \cdot \sigma_q)\,\varepsilon^{i+1} = P \cdot (\sigma_q \varepsilon^i) \quad \text{for } q>0, \\ &(P \cdot \sigma_0) \cdot \varepsilon^1 = (\hat{P}\eta)(\sigma_0). \end{aligned}$$

Taking alternating sums in 4.8 we get

$$(4.9) \qquad \partial_{q+1} P_q = \mathrm{id} - P_{q-1}\partial_q \quad \text{for} \quad q>0, \quad \text{and} \quad \partial_1 P_0 = \mathrm{id} - (\hat{P}\eta)_0,$$

i.e. $\{P_q\}$ is a homotopy $\mathrm{id} \simeq \hat{P}\eta$. Clearly $\eta\hat{P} = \mathrm{id}$. ∎

4.10 Corollary. *If $Y \subset \mathbb{R}^n$ is any non-empty subspace then $\partial_* \colon H_q(\mathbb{R}^n, Y) = \tilde{H}_{q-1} Y$.*

This follows from the reduced homology sequence 4.4 of $(\mathbb{R}^n, Y)$ because $\tilde{H}\,\mathbb{R}^n = 0$ by 4.6. ∎

We now show that $H_0 X = \mathbb{Z}$ for all pathwise connected spaces X. The reader might begin to suspect that H is a rather trivial functor altogether. He will have to wait until Chapter IV to see that this is not so.

4.11 Proposition. *If X is a non-empty pathwise connected space then the augmentation $\eta \colon SX \to (\mathbb{Z}, 0)$ induces an isomorphism $\eta_* \colon H_0 X = H_0(\mathbb{Z}, 0) = \mathbb{Z}$.*

Proof. Pick $P \in X$, and define $\hat{P}$: $(\mathbb{Z}, 0) \to SX$ by $\hat{P}(m) = mP$; clearly $\eta \hat{P} = \mathrm{id}$. For every 0-simplex σ_0: $\Delta_0 \to X$ we can find a 1-simplex $\pi \sigma_0$: $\Delta_1 \to X$ (= a path) with $(\pi \sigma_0) \varepsilon^0 = \sigma_0$, $(\pi \sigma_0) \varepsilon^1 = P$, hence $\partial(\pi \sigma_0) = (\mathrm{id} - \hat{P}\eta) \sigma_0$. This defines a homomorphism π: $S_0 X \to S_1 X$ with $\partial \pi = \mathrm{id} - \hat{P}\eta$. For homology classes it gives $0 = [\partial \pi z] = [z] - [\hat{P}\eta z] = [z] - \hat{P}_* \eta_* [z]$, $z \in Z_0 X$, i.e., $H_0 \hat{P}$ and $H_0 \eta$ are reciprocal isomorphisms. ∎

What about H_0 of non-connected spaces? This reduces to 4.11 via

4.12 Proposition. *Let X be an arbitrary space with path-components X_λ, $\lambda \in \Lambda$; let $A \subset X$ be a subspace and $A_\lambda = A \cap X_\lambda$. Then the inclusion maps $i_\lambda: (X_\lambda, A_\lambda) \to (X, A)$ induce a direct sum representation $\{i_\lambda\}: \oplus_{\lambda \in \Lambda} S(X_\lambda, A_\lambda) \cong S(X, A)$, hence* (II, 1.5) *$\{H i_\lambda\}: \oplus_{\lambda \in \Lambda} H(X_\lambda, A_\lambda) \cong H(X, A)$.*

In particular, $H_0 X$ is a free abelian group whose rank (cf. I, 2.29) *equals the number of path-components of X.*

Proof. Let s resp. s_λ denote the set of singular simplices of X resp. X_λ. Since every simplex $\sigma \in s$ has a pathwise connected image, this image must lie in some X_λ, hence $s = \bigcup_\lambda s_\lambda$. Every singular chain c has a unique representation

$$c = \sum_{\sigma \in s} c_\sigma \cdot \sigma = \sum_\lambda \sum_{\sigma \in s_\lambda} c_\sigma \cdot \sigma = \sum_\lambda c_\lambda, \quad c_\lambda \in S(X_\lambda),$$

hence $SX = \oplus_\lambda S(X_\lambda)$. Similarly, $SA = \oplus_\lambda S(A_\lambda)$, hence $SX/SA \cong \oplus_\lambda S(X_\lambda)/S(A_\lambda)$. ∎

4.13 Corollary. *If X is a discrete space then $H_i X = 0$ for $i \neq 0$, $H_0 X = \oplus_{x \in X} \mathbb{Z}$.* ∎

We conclude the discussion of special cases with some remarks on retracts.

4.14 Definition. If i: $A \subset X$ is a pair of spaces then A is called a *retract* of X if there is a map r: $X \to A$ such that $ri = \mathrm{id}$; any such r is called a *retraction*. For instance, every $P \in X$ is a retract of X; if B is any space and $Q \in B$, then $A \approx A \times Q \subset A \times B$ and r: $A \times B \to A \times Q$, $r(a, b) = (a, Q)$ is a retraction ("the factors of a product are retracts").

If (X, A) is as above then A is called a *neighborhood retract* (in X) if A has a neighborhood in X of which it is a retract. Every retract is a neighborhood retract but not conversely: If $X = [0, 1]$ is the unit interval and $A = \{0\} \cup \{1\}$ consists of the two end points then A is a neighborhood retract but not a retract (proof?).

For the moment we only discuss retracts; neighborhood retracts will become important later on (IV, VIII). If $r: X \to A$ is a retraction then $r: SX \to SA$ splits the exact sequence $0 \to SA \xrightarrow{i} SX \xrightarrow{j} S(X, A) \to 0$, hence (r, j): $SX \cong SA \oplus S(X, A)$, hence

$$(4.15) \qquad (r_*, j_*)\colon\ HX \cong HA \oplus H(X, A).$$

In other terms:

4.16 Proposition. *If A is a retract of X then the homology sequence of (X, A) decomposes into short exact sequences*

$$0 \xrightarrow{\partial_*=0} H_q A \xrightarrow{i_*} H_q X \xrightarrow{j_*} H_q(X, A) \to 0$$

which are split by r_*. ∎

4.17 Exercises. *1.* The homology sequence of the triple $P \in A \subset X$ is isomorphic with the reduced homology sequence of (X, A).

2. If X is a contractible space, $X \simeq P$, then η. $SX \to (\mathbb{Z}, 0)$ is a homotopy equivalence. Hint: Use a cone-construction as for 4.6.

3. Determine $H(\mathbb{R}, \mathbb{Q})$ where $\mathbb{Q} \subset \mathbb{R}$ is the subspace of the real line consisting of all rational numbers.

4. If $B \subset A \subset X$ is a triple such that A is a retract of X then $H(X, B) \cong H(X, A) \oplus H(A, B)$.

5. *Invariance under Homotopy*

We recall (I, 3.1) that two continuous maps $f, g\colon X \to Y$ are homotopic if there is a deformation $\Theta\colon [0, 1] \times X \to Y$ with $\Theta_0 = f$, $\Theta_1 = g$. Similarly for maps of pairs.

5.1 Proposition. *If $f, g\colon (X, A) \to (Y, B)$ are homotopic maps then $Sf, Sg\colon S(X, A) \to S(Y, B)(Y, B)$ are (chain-) homotopic.*

5.2 Corollary. *If $f, g\colon (X, A) \to (Y, B)$ are homotopic then $f_* = g_*\colon H(X, A) \to H(Y, B)$*—because homotopic chain maps induce the same homomorphism of homology (II, 3.4). ∎

5.3 Corollary. *If $(X, A) \simeq (Y, B)$ then $H(X, A) \cong H(Y, B)$.*

Proof. If $(X, A) \xrightarrow{f} (Y, B) \xrightarrow{f^-} (X, A)$ are reciprocal homotopy equivalences then $H(X, A) \xrightarrow{f_*} H(Y, B) \xrightarrow{f^-_*} H(X, A)$ are reciprocal isomorphisms by 5.2. ∎

5.4 Corollary. *If X is contractible, $X \simeq P$, then $\tilde{H}X = 0$. In fact, the augmentation η: $SX \to (\mathbb{Z}, 0)$ is a homotopy equivalence* (cf. II, 4.3). ∎

The situation is best illustrated by the following commutative functor diagram

$$\text{(5.5)}\qquad \begin{array}{ccccc} \mathcal{Top} & \xrightarrow{S} & \partial\mathcal{AG} & & \\ & & & \searrow^{H} & \\ \downarrow \pi & & \downarrow \pi & & \mathcal{GAG} \\ & & & \nearrow_{\bar{H}} & \\ \mathcal{Htp} & \overset{}{\underset{\bar{S}}{\cdots\cdots\to}} & \mathcal{H}\partial\mathcal{G} & & \end{array}$$

where π denotes passage to homotopy classes. Proposition 5.1 asserts that the dotted arrow $\bar{S}$ exists. In II, 3.4 the arrow $\bar{H}$ was shown to exist. Corollary 5.2 only says that $(\bar{H}\bar{S})$ exists and 5.3 remarks that $\bar{H}\bar{S}$ takes equivalences into equivalences (as any functor does).

5.6 Remark. If f: $(X, A) \to (Y, B)$ is a homotopy equivalence then so are f: $X \to Y$ and $f|A$: $A \to B$. The converse is not true; a counterexample is given by $X = Y = [0, 1]$, $A = \{0\} \cup \{1\}$, $B = [0, 1] - \{\frac{1}{2}\}$, $f =$ inclusion (proof?). On the chain level, however, the converse is true (II, 4.3).

The proof of 5.1 will be an easy consequence of the following

5.7 Proposition *If F^0, F^1: $SX \to S([0, 1] \times X)$ are natural chain maps such that the two composites $S\Delta_0 \xrightarrow{F^0, F^1} S([0, 1] \times \Delta_0) \xrightarrow{\eta} (\mathbb{Z}, 0)$* ($\Delta_0 =$ zero simplex, $\eta =$ augmentation 4.5) *coincide then there exists a natural homotopy s: $F^0 \simeq F^1$.* Naturality of $\varphi = F^0$, F^1, or s means, of course, that φ is defined for all spaces X and that

$$\text{(5.8)}\qquad \begin{array}{ccc} SX' & \xrightarrow{\varphi} & S([0, 1] \times X') \\ \downarrow h & & \downarrow \text{id} \times h \\ SX & \xrightarrow{\varphi} & S([0, 1] \times X) \end{array}$$

commutes for all continuous maps h: $X' \to X$.

Proof. We assume inductively that s_k: $S_k X \to S_{k+1}([0, 1] \times X)$ has already been found for $k < q$ and

$$\text{(5.9)}\qquad \partial s_k + s_{k-1}\partial = F_k^0 - F_k^1.$$

Let $\iota_q \in S_q(\Delta_q)$ denote the identity map of Δ_q. We compute

$$\begin{aligned} \partial\{F^0\iota_q - F^1\iota_q - s_{q-1}\partial\iota_q\} &= F^0\partial\iota_q - F^1\partial\iota_q - (\partial s_{q-1})(\partial\iota_q) \\ &= F^0\partial\iota_q - F^1\partial\iota_q - (F^0 - F^1 - s_{q-2}\partial)(\partial\iota_q) = 0. \end{aligned}$$

Thus $F^0\iota_q - F^1\iota_q - s_{q-1}\partial\iota_q$ is a q-cycle; if $q=0$ then its augmentation vanishes because $\eta F^0 = \eta F^1$. Therefore it is a boundary because $[0,1]\times\Delta_q$ is convex (4.6), i.e. we can find $b\in S_{q+1}([0,1]\times\Delta_q)$ with $\partial b = F^0\iota_q - F^1\iota_q - s_{q-1}\partial\iota_q$. Now define

(5.10) $\quad s_q\colon S_qX\to S_q([0,1]\times X),\quad s_q(\sigma)=(\mathrm{id}\times\sigma)\,b,$

where $\sigma\colon\Delta_q\to X$ ranges over all singular q-simplexes of X. We have to verify naturality 5.8, and formula 5.9 with $k=q$. Let $\sigma'\colon\Delta_q\to X'$. Then

$$(\mathrm{id}\times h)\,s_q\sigma' = (\mathrm{id}\times h)(\mathrm{id}\times\sigma')\,b = (\mathrm{id}\times h\sigma')\,b = s_q(h\sigma') = (s_q h)\,\sigma',$$

which proves naturality. Further

$$\begin{aligned}(\partial s_q)\,\sigma &= \partial(\mathrm{id}\times\sigma)\,b = (\mathrm{id}\times\sigma)\,\partial b\\ &= (\mathrm{id}\times\sigma)\{F^0\iota_q - F^1\iota_q - s_{q-1}\partial\iota_q\}\\ &= F^0\sigma\iota_q - F^1\sigma\iota_q - s_{q-1}\sigma\partial\iota_q\\ &= F^0\sigma - F^1\sigma - s_{q-1}\partial\sigma\iota_q = (F^0 - F^1 - s_{q-1}\partial)\,\sigma,\end{aligned}$$

which proves 5.9; naturality of F^0, F^1, s_{q-1} was used for the fourth equality. ∎

The preceding proof is typical for the method of "acyclic models" which is due to Eilenberg-MacLane (1953). We shall explain the general principle in VI, 11.

Proof of 5.1. For every space X the inclusions

$$F^t\colon X\to[0,1]\times X,\quad F^t(x)=(t,x),\quad 0\le t\le 1,$$

define natural chain maps $F^t\colon SX\to S([0,1]\times X)$, and by 5.7 there is a natural homotopy $s\colon F^0\simeq F^1$. If $A\subset X$ is a subspace then $F^t(SA)\subset S([0,1]\times A)$, and $s(SA)\subset S([0,1]\times A)$, the latter by naturality of s. Passing to quotients we get

$$\bar F^t\colon S(X,A)\to S([0,1]\times X,[0,1]\times A),\quad\text{and}\quad \bar s\colon \bar F^0\simeq\bar F^1.$$

Consider now a homotopy $\Theta\colon f\simeq g$, as assumed in 5.1. Clearly $\Theta_t=\Theta F^t$, hence $\bar\Theta_t=\bar\Theta\bar F^t\colon S(X,A)\to S(Y,B)$ by passage to quotients. Therefore $Sf=\bar\Theta_0=\bar\Theta\bar F^0\simeq\bar\Theta\bar F^1=\bar\Theta_1=Sg$. ∎

5.11 Examples. If $i\colon A\subset X$ is a pair of spaces then A is called a *deformation retract* (of X) if a homotopy $\Theta_t\colon X\to X$ exists with $\Theta_0=\mathrm{id}$, $\Theta_1(X)\subset A$ and $\Theta_1|A=\mathrm{id}$. Thus Θ_1 defines a retraction $r\colon X\to A$ with $ir=\Theta_1$; we have $ri=\mathrm{id}_A$, and $\Theta\colon\mathrm{id}_X\simeq ir$. In particular, i, r are reciprocal homotopy equivalences, hence $i_*\colon HA\cong HX$. If Θ can be so chosen that $\Theta_t|A=i$ for all t then A is called a *strong deformation retract*.

For instance, if $A \in X$ is a single point then A is a deformation retract if and only if X is contractible; we get $\tilde{H}X = 0$ as in 5.4. If $\mathbb{S}^{n-1} = \{x \in \mathbb{R}^n \mid \|x\| = 1\}$ denotes the *unit sphere* then $\mathbb{S}^{n-1}$ is a strong deformation retract of the *deleted euclidean space* $\mathbb{R}^n - \{0\}$: take $\Theta_t(x) = (1 - t + t/\|x\|)x$. The same deformation Θ shows that $\mathbb{S}^{n-1}$ is a strong deformation retract of the deleted unit ball $\mathbb{B}^n - \{0\}$ where $\mathbb{B}^n = \{x \in \mathbb{R}^n \mid \|x\| \leq 1\}$. In particular,

$$(5.12) \qquad H\mathbb{S}^{n-1} \cong H(\mathbb{B}^n - \{0\}) \cong H(\mathbb{R}^n - \{0\}).$$

5.13 Exercises. *1.* If $f\colon (X, A) \to (Y, B)$ is a map such that $f\colon X \simeq Y$ and $(f|A)\colon A \simeq B$ then $\bar{f}\colon S(X, A) \simeq S(Y, B)$; compare 5.6.

2. If A is a (strong) deformation retract of X then $A \times Y$ is a (strong) deformation retract of $X \times Y$. Draw pictures with $X = \mathbb{B}^2$, $A = \{0\}$, $Y = \mathbb{S}^1$.

3. The *cone* CX over X is obtained from $[0, 1] \times X$ by identifying the subspace $\{0\} \times X$ to one point v, the *vertex* of CX. Show: (i) CX is contractible, (ii) $H_q(CX, CX - \{v\}) \cong \tilde{H}_{q-1}X$.

4. Consider the solid torus, solid double-torus, solid triple-torus etc., as illustrated by

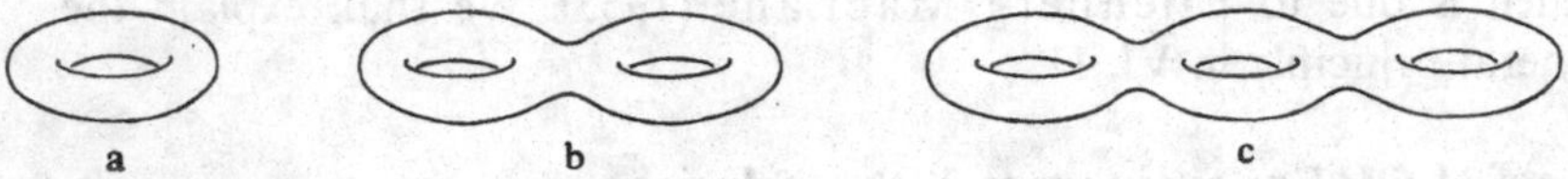

Fig. 4a–c

Show that they contain deformation retracts of the form

Fig. 5a–c

6. Barycentric Subdivision

This is a tool which will be used in §7.

6.1 Definition. For every space X we define homomorphisms $\beta_q\colon S_qX \to S_qX$, $q \geq 0$, called the *barycentric subdivision*, as follows:

$$(6.2) \quad \beta_0 = \mathrm{id}, \quad \beta_q \iota_q = B_q \cdot \beta_{q-1}(\partial \iota_q), \quad \beta_q(\sigma_q) = \sigma_q(\beta_q \iota_q), \quad q > 0,$$

where $\iota_q \in S_q \Delta_q$ denotes the identity map of Δ_q,

$$B_q = \left(\frac{1}{q+1}, \frac{1}{q+1}, \ldots, \frac{1}{q+1}\right) = \sum_{i=0}^{q} \frac{e_i}{q+1}$$

is the *barycenter* of Δ_q, $B_q \cdot$ is the cone construction as in 4.7 (recall that Δ_q is convex), and $\sigma_q \colon \Delta_q \to X$ is an arbitrary singular simplex.

Loosely speaking, the barycentric subdivision of σ_q is obtained by projecting the barycentric subdivision of $\partial\sigma_q$ from the center of σ_q. The reader is advised to draw some pictures. The crucial property of β is that it cuts simplices into smaller pieces, more precisely

6.3 Proposition. *The sequence $\beta_q \colon S_q X \to S_q X$, $q \geq 0$, is a natural chain map and has the following property: For every $q \geq 0$ and every real number $\varepsilon > 0$ there exists a number $N = N(\varepsilon, q)$ such that the chain $c = \beta^n(\iota_q) = \beta\beta \ldots \beta(\iota_q)$ for $n \geq N$ contains only simplices τ of diameter $\|\tau\| < \varepsilon$ (i.e., $\|\tau\| \geq \varepsilon \Rightarrow c_\tau = 0$).* The diameter of $\tau \colon \Delta_q \to \mathbb{R}^k$ is defined as $\|\tau\| = \operatorname{Max}\{\|\tau x - \tau y\| \mid x, y \in \Delta_q\}$.

Proof. If $f \colon X \to Y$ is a map then $(f\beta)\sigma_q = f(\beta\sigma_q) = (f\sigma_q)(\beta\iota_q) = \beta(f\sigma_q)$, which proves naturality. Next we verify $\partial\beta_q = \beta_{q-1}\partial$ by induction on q.

$$\begin{aligned}\partial(\beta_q\sigma_q) &= (\partial\sigma_q)(\beta_q\iota_q) = (\sigma_q\partial)(B_q \cdot \beta_{q-1}\partial\iota_q) = \sigma_q\partial(B_q \cdot \beta_{q-1}\partial\iota_q)\\ &= \sigma_q(\beta_{q-1}\partial\iota_q) = \beta_{q-1}\sigma_q\partial\iota_q = \beta_{q-1}\partial\sigma_q,\end{aligned}$$

where the fourth equality uses the boundary formula 4.9 and $\partial\beta_{q-1} = \beta_{q-2}\partial$.

It remains to find $N(\varepsilon, q)$. This is contained in the following more general

6.4 Lemma. *If $\sigma \colon \Delta_q \to \mathbb{R}^k$ is a linear simplex* (cf. 1.2) *then $\beta(\sigma)$ contains only linear simplices of diameter $\leq \frac{q}{q+1}\|\sigma\|$. In particular, $\beta^n(\iota_q)$ contains only simplices of diameter $\leq \left(\frac{q}{q+1}\right)^n \|\iota_q\|$.*

The proof of 6.4 uses

6.5 Lemma. *If $\sigma \colon \Delta_q \to \mathbb{R}^k$ is a linear simplex with vertices $P_0, P_1, \ldots, P_q$ then*
(a) $\|P - P'\| \leq \operatorname{Max}_{i=0}^{q} \|P - P_i\|$, *for all $P, P' \in \sigma(\Delta_q)$;*
(b) $\|\sigma\| \leq \operatorname{Max}_{i,j} \|P_i - P_j\|$.

Proof of 6.5. We have $P'=\sum_{i=0}^{q} x_i' P_i$ with $x_i' \geq 0$, $\sum x_i'=1$, hence

$$\begin{aligned}\|P-P'\| &= \|\sum(x_i'P-x_i'P_i)\| \leq \sum x_i' \|P-P_i\| \\ &\leq (\sum x_i')(\operatorname{Max}\|P-P_i\|) = \operatorname{Max}\|P-P_i\|.\end{aligned}$$

This proves (a); part (b) follows by applying (a) twice. ∎

Proof of 6.4. The following properties of the cone-construction are immediate from the Definition 4.7.

(i) *Given* $\tau\colon \Delta_r \to \mathbb{R}^l$, $P \in \mathbb{R}^l$, *and a linear map* $f\colon \mathbb{R}^l \to \mathbb{R}^k$ *then* $f(P\cdot\tau)=(fP)\cdot(f\tau)$.

(ii) *If* $\tau\colon \Delta_r \to \mathbb{R}^l$ *is linear with vertices* $Q_0, \ldots, Q_r$ *then* $P\cdot\tau\colon \Delta_{r+1} \to \mathbb{R}^l$ *is linear with vertices* $P, Q_0, \ldots, Q_r$.

Now

$$\begin{aligned}\beta\sigma=\sigma(B_q\cdot\beta\partial\iota_q) &= (\sigma B_q)\cdot(\sigma\beta\partial\iota_q), && \text{by (i)},\\ &= (\sigma B_q)\cdot(\beta\sigma\partial\iota_q), && \text{by naturality of } \beta,\\ &= \textstyle\sum_{j=0}^{q}(-1)^j(\sigma B_q)\cdot\beta(\sigma\varepsilon^j).\end{aligned}$$

Thus $\beta\sigma$ contains only simplices of the form $\sigma'=(\sigma B_q)\cdot\tau$ where τ is contained in some $\beta(\sigma\varepsilon^j)$. The diameter of σ' equals $\|P-Q\|$ where P, Q are vertices of σ' (by 6.5 (b)). These vertices are either vertices of τ or one of them equals σB_q (by (ii)). In the first case

$$\|\sigma'\|=\|P-Q\|\leq\|\tau\|\leq\frac{q-1}{q}\|\sigma\varepsilon^j\|\leq\frac{q-1}{q}\|\sigma\|\leq\frac{q}{q+1}\|\sigma\|,$$

the 3rd inequality by induction on q. In the second case, say $P=\sigma B_q$,

$$\|\sigma'\|=\|P-Q\|\leq\|P-P_i\|$$

for some i, by 6.5 (a), hence

$$\begin{aligned}\|\sigma'\|\leq\|\sigma B_q-P_i\| &= \left\|\textstyle\sum_{\mu=0}^{q}\frac{1}{q+1}P_\mu-P_i\right\|=\left\|\textstyle\sum_{\mu=0}^{q}\frac{1}{q+1}(P_\mu-P_i)\right\|\\ &\leq\frac{1}{q+1}\textstyle\sum_{\mu=0}^{q}\|P_\mu-P_i\|\leq\frac{q}{q+1}\|\sigma\|.\end{aligned}$$

This proves 6.5 and 6.3. ∎

We also need that $\beta\simeq\mathrm{id}$. This is contained in

6.6 Proposition. *If $\gamma^0, \gamma^1: SX \to SX$ are natural chain maps which agree in dimension zero, $\gamma_0^0 = \gamma_0^1$, then there exists a natural homotopy $s: \gamma^0 \simeq \gamma^1$.*

There is a direct proof by the method of acyclic models as for 5.7: the reader will find this an easy exercise. We reduce the problem to 5.7 by considering the composites $F^i: SX \xrightarrow{\gamma^i} SX \xrightarrow{J} S([0,1] \times X)$, where $J(x) = (0, x)$. Then a natural homotopy $u: F^0 \simeq F^1$ exists by 5.7. Composing with the projection $\pi: [0,1] \times X \to X$ gives a natural homotopy $s = \pi u: \pi F^0 \simeq \pi F^1$, and $\pi F^i = (\pi J) \gamma^i = \gamma^i$. ∎

6.7 Exercises. *1.* Let V_n be the set of vertices of Δ_n, and let $\mathscr{V}_n$ be the set of non-empty subsets of V_n. If $V \in \mathscr{V}_n$, define its barycentre $BV \in \Delta_n$ by $BV = \frac{1}{|V|} \sum_{v \in V} v$, where $|V|$ = cardinality of V. Show that every $x \in \Delta_n$ has a unique representation $x = \sum x_V \cdot BV$ such that (i) $0 \le x_V \le 1$, (ii) $\sum x_V = 1$, and (iii) $x_V \neq 0$, $x_W \neq 0 \Rightarrow V \subset W$ or $W \subset V$. The numbers $\{x_V\}$, $V \in \mathscr{V}_n$, are called the *derived (barycentric) coordinates* of x. We can think of $\{x_V\}$ as ordinary coordinates on Δ_N where $N = |\mathscr{V}_n| - 1 = 2^{n+1} - 2$; then $x \mapsto \{x_V\}$ maps Δ_n homeomorphically onto a union of certain lower-dimensional faces of Δ_N.

If $\varphi: V_n \to V_m$ is a map define the *derived map* $\varphi': \Delta_n \to \Delta_m$ by $(\varphi' x)_U = \sum_{\varphi V = U} x_V$, $U \in \mathscr{V}_m$. Show that this is well-defined, that φ' takes vertices into vertices, and that φ' is linear if and only if φ is injective.

2. Show that a sequence $\gamma_q: S_q X \to S_q X$ of natural homomorphisms ($X \in \mathscr{T}\!op$) is a chain map if and only if $\partial \gamma_q \iota_q = \gamma_{q-1} \partial \iota_q$ for all q, where $\iota_q = \mathrm{id}(\Delta_q)$.

7. *Small Simplices. Excision*

We show that in order to compute singular homology it suffices to consider small simplices (7.3). This implies that $H(X, A)$ is unchanged if one excises any part B of A which doesn't touch the boundary of A (7.4).

7.1 Definition. If X is a space and $\mathscr{U}$ is a set of subsets of X then $S\mathscr{U}$ denotes the smallest subcomplex of SX which contains all SU, $U \in \mathscr{U}$; i.e. $S\mathscr{U}$ is the subcomplex generated by $\{SU\}_{U \in \mathscr{U}}$. The chains of $S\mathscr{U}$ are linear combinations of simplexes $\sigma: \Delta_q \to X$ each of which maps Δ_q into some $U \in \mathscr{U}$, i.e. of simplexes which are "small of order $\mathscr{U}$".

If $A \subset X$ is a subspace we put $\mathscr{U} \cap A = \{U \cap A\}_{U \in \mathscr{U}}$, and define $S(\mathscr{U} \cap A)$, $S(\mathscr{U}, \mathscr{U} \cap A) = S\mathscr{U}/S(\mathscr{U} \cap A)$ accordingly. We have a commutative diagram of chain maps

$$
(7.2)\qquad
\begin{array}{ccccccccc}
0\to & S(\mathscr{U}\cap A) & \longrightarrow & S\mathscr{U} & \longrightarrow & S(\mathscr{U},\mathscr{U}\cap A) & \to 0\\
 & \downarrow{\scriptstyle\iota} & & \downarrow{\scriptstyle\iota} & & \downarrow{\scriptstyle\iota} & \\
0\to & SA & \longrightarrow & SX & \longrightarrow & S(X,A) & \to 0
\end{array}
$$

with exact rows whose vertical arrows ι are inclusions.

7.3 Proposition. *If every point of X is contained in the interior $\mathring{A}$ of A or in the interior $\mathring{U}$ of some $U\in\mathscr{U}$ then ι: $S(\mathscr{U},\mathscr{U}\cap A)\to S(X,A)$ is a homotopy equivalence, hence ι_*: $HS(\mathscr{U},\mathscr{U}\cap A)\cong H(X,A)$.*

If $\mathscr{U}$ consists of only one set Y this is known as the *excision-theorem.* The assumption then means $\mathring{Y}\cup\mathring{A}=X$, the conclusion is $S(Y,Y\cap A)\simeq S(X,A)$. In terms of complements $B=X-Y$, $\bar{B}=X-\mathring{Y}$, the assumption is $\bar{B}\subset\mathring{A}$, the conclusion $S(X-B,A-B)\simeq S(X,A)$. Thus

7.4 Corollary (Excision). *If (X,A) is a pair of spaces and $Y\subset X$ is such that $\mathring{Y}\cup\mathring{A}=X$ then j: $S(Y,Y\cap A)\simeq S(X,A)$ where $j=$ inclusion. If $B\subset A$ is such that $\bar{B}\subset\mathring{A}$ then j: $S(X-B,A-B)\simeq S(X,A)$. In particular j_*: $H(Y,Y\cap A)\cong H(X,A)$ resp. j_*: $H(X-B,A-B)\cong H(X,A)$.*

Proof of 7.3. Since $S(\mathscr{U},\mathscr{U}\cap A)$, $S(X,A)$ are free complexes it suffices to show (by II, 4.3) that ι_*: $HS(\mathscr{U},\mathscr{U}\cap A)=H(X,A)$. This, in turn, will follow from the homology sequence II, 2.9 provided we show

$$H_q\{S(X,A)/S(\mathscr{U},\mathscr{U}\cap A)\}=0 \quad \text{for all } q.$$

The elements $[z]$ of this group are represented by cycles z of X mod $S\mathscr{V}$ (where $\mathscr{V}=\mathscr{U}\cup\{A\}$), i.e. by chains $z\in S_qX$ such that $\partial z\in S\mathscr{V}$. We have to show that z is a boundary mod $S\mathscr{V}$, i.e. $z=\partial x+y$ with $x\in SX$, $y\in S\mathscr{V}$. We shall see below that:

(7.5) *If $n\in\mathbb{Z}$ is sufficiently large then $\beta^n(z)\in S\mathscr{V}$.*

Also, from 6.6, 6.2, we have a natural homotopy s: $\mathrm{id}\simeq\beta^n$, hence $z=\partial(sz)+s(\partial z)+\beta^n(z)$. This proves the assertion $[z]=0$ provided we can show $s(\partial z)\in S\mathscr{V}$. But $\partial z\in S\mathscr{V}$, and $s(S\mathscr{V})\subset S\mathscr{V}$ by naturality of s; in fact, if V is any element of $\mathscr{V}$ then $s(SV)\subset SV$ by naturality applied to $V\xrightarrow{\subset}X$.

It remains to prove 7.5. Since z is a finite linear combination of simplices σ: $\Delta_q\to X$ it suffices to show that for these σ we have $\beta^n(\sigma)\in S\mathscr{V}$ for large n. Now, the sets $\{\sigma^{-1}\mathring{V}\}_{V\in\mathscr{V}}$ form an open covering $\mathscr{W}$ of Δ_q. Choose $\varepsilon>0$ such that every subset of Δ_q whose diameter is less than ε

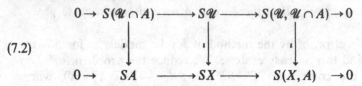

lies in some $\sigma^{-1}\mathring{V}$; this is possible because Δ_q is compact ("Lebesgue-number" of $\mathscr{W}$; cf. Schubert, I, 7.4). By Proposition 6.3, the chain $\beta^n \iota_q \in S_q(\Delta_q)$ consists of simplices of diameter $<\varepsilon$ only, provided n is large enough. But then $\beta^n \iota_q$ consists of simplices each of which lies in some $\sigma^{-1}\mathring{V}$, hence $\beta^n \sigma = \sigma(\beta^n \iota_q)$ consists of simplices each of which lies in some $\mathring{V} \subset V$, hence $\beta^n \sigma \in S\mathscr{V}$. ∎

7.6 Example. A pair of spaces (X, P) where P consists of a single point is called a *pointed space*, or *space with base point*. If (X, P) and (Y, Q) are pointed spaces then we define their *wedge* (or *one-point-union*) as

(7.7) $$X \vee Y = X \oplus Y / P \sim Q,$$

i.e. the topological sum with base points identified (this is then the natural base point for $X \vee Y$). We can think of X, Y as subspaces of $X \vee Y$ via $X, Y \subset X \oplus Y \to X \vee Y$; then $X \cup Y = X \vee Y$, $X \cap Y = P = Q$. Let $i: X \to X \vee Y \leftarrow Y: j$ denote the inclusion maps.

7.8 Proposition. *If the closure of P in X has a neighborhood U in X whose inclusion map $U \to X$ is homotopic rel. P to the constant map $U \to P$* (i.e. there is a deformation $d_t: U \to X$ with $d_0 =$ inclusion, $d_1(U) = P$, $d_t(P) = P$ for all $t \in [0, 1]$) *then*

(7.9) $$(i_*, j_*): H(X, P) \oplus H(Y, Q) \cong H(X \vee Y, P = Q).$$

By 4.3 we can also write $\tilde{H}X \oplus \tilde{H}Y \cong \tilde{H}(X \vee Y)$.

Proof. We can extend the deformation d to a deformation

$$D_t: U \vee Y \to X \vee Y \quad \text{by} \quad D_t | Y = j$$

(continuity of D is obvious if P and Q are closed points; the general case follows from V, 2.13). This deforms $U \vee Y$ into Y showing that

$$H(U \vee Y, Y) \to H(X \vee Y, Y)$$

is the zero-map. Consider then the commutative diagram

(7.10) $$\begin{array}{ccccccc} \xrightarrow{0} & H(X \vee Y, Y) & \longrightarrow & H(X \vee Y, U \vee Y) & \longrightarrow & H(U \vee Y, Y) \\ & \uparrow i_* & & \uparrow i_*' & & \uparrow i_*^U \\ & H(X, P) & \longrightarrow & H(X, U) & \longrightarrow & H(U, P) \end{array}$$

whose rows are parts of the exact homology sequences of the triples $(X \vee Y, U \vee Y, Y)$ resp. (X, U, P) and whose vertical arrows are induced by inclusions. All vertical arrows are monomorphic; in fact, they have left inverses because $i\colon X \to X \vee Y$ has a left inverse r, namely $r|X = \mathrm{id}$, $r(Y) = P$. The middle arrow i''_* is even isomorphic because $H(X \vee Y, U \vee Y) \cong H(X, U)$, by Excision 7.4. But then i'_* must also be isomorphic as can be seen from the five lemma or (simpler) by direct diagram chasing in 7.10.

Now, just as i has a left inverse so does $j\colon Y \to X \vee Y$, hence (4.16) we have a split exact sequence

$$0 \to \tilde{H}Y \xrightarrow{j_*} \tilde{H}(X \vee Y) \to H(X \vee Y, Y) \to 0. \tag{7.11}$$

This sequence is split by $H(X \vee Y, Y) \overset{i'_*}{\cong} \tilde{H}X \xrightarrow{i_*} \tilde{H}(X \vee Y)$, which proves the assertion. ▮

7.12 Exercises. *1.* If X is a metric space then the diameter $\|\sigma\|$ of a singular simplex $\sigma\colon \Delta_q \to X$ is defined by $\|\sigma\| = \mathrm{Max}\{\mathrm{dist}(\sigma x, \sigma y) | x, y \in \Delta_q\}$. Show that, for any $\varepsilon > 0$, the subgroups $S_q^\varepsilon X$ of $S_q X$, $q = 0, 1, 2, \ldots$, which are generated by all simplices of diameter less than ε form a subcomplex $S^\varepsilon X$, and $S^\varepsilon X \simeq SX$.

2. The wedge $X = \vee_\lambda X_\lambda$ of an arbitrary family of *pointed spaces* (X_λ, P_λ), $\lambda \in \Lambda$, is defined by taking the topological sum $\oplus_\lambda X_\lambda$ and identifying the set of base points $\oplus_\lambda P_\lambda$ to a single point, say P. Show: If the closure of P in X has a neighborhood U in X such that $\tilde{H}U \to \tilde{H}X$ is the zero-map then $\{i_{\lambda *}\}\colon \oplus_\lambda \tilde{H}X_\lambda \cong \tilde{H}(\vee_\lambda X_\lambda)$. Hint: Use the diagram

$$\begin{array}{ccccc}
0 \to H(X, P) & \longrightarrow & H(X, U) & \longrightarrow & H(U, P) \\
\uparrow & & \uparrow & & \uparrow \\
H(\oplus X_\lambda, \oplus P_\lambda) & \longrightarrow & H(\oplus X_\lambda, \oplus U_\lambda) & \longrightarrow & H(\oplus U_\lambda, \oplus P_\lambda)
\end{array}$$

where $U_\lambda = U \cap X_\lambda$, and prove, as in 7.10, that the left vertical arrow is isomorphic.

3. For any pair of pointed spaces (X, P), (Y, Q) there is a natural injection $J\colon X \vee Y \to X \times Y$, defined by $Jx = (x, Q)$, $Jy = (P, y)$ for $x \in X$, $y \in Y$. Show that if (7.9) holds then $J_*\colon H(X \vee Y) \to H(X \times Y)$ has a left inverse (hint: take the sum of the projections), hence the homology sequence of $(X \times Y, X \vee Y)$ splits into $H(X \times Y) \cong H(X \vee Y) \oplus H(X \times Y, X \vee Y)$—just as if $X \vee Y$ were a retract of $X \times Y$.

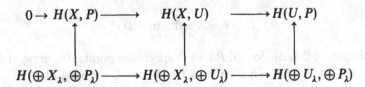

8. Mayer-Vietoris Sequences

A reader who at this point would rather study some interesting geometric *applications* instead of pursueing further the *theory* of singular homology can continue with Chapter IV, 1-5 now; the present section will not be needed before IV, 6.

Let X be a space and X_1, X_2 two subspaces. We denote this situation by $(X; X_1, X_2)$ and call it a *triad* (not to be confused with the more special triple of § 3); let $i_\nu: X_\nu \to X$ be the inclusions. We want to relate the groups $H(X_1)$, $H(X_2)$, $H(X_1 \cap X_2)$, $H(X_1 \cup X_2)$.

8.1 Proposition and Definition. *A triad* $(X; X_1, X_2)$ *is called* excisive *if one of the following equivalent conditions holds:*

(a) i_{1*}: $H(X_1, X_1 \cap X_2) \cong H(X_1 \cup X_2, X_2)$,

(b) i_{2*}: $H(X_2, X_1 \cap X_2) \cong H(X_1 \cup X_2, X_1)$,

(c) (i_{1*}, i_{2*}): $H(X_1, X_1 \cap X_2) \oplus H(X_2, X_1 \cap X_2) \cong H(X_1 \cup X_2, X_1 \cap X_2)$,

(d) i_*: $HS\{X_1, X_2\} \cong HS(X_1 \cup X_2) = H(X_1 \cup X_2)$,

(e) $\bar{i}_*$: $H[S\{X_1, X_2\}/S(X_1 \cap X_2)] \cong H[S(X_1 \cup X_2)/S(X_1 \cap X_2)]$
$= H(X_1 \cup X_2, X_1 \cap X_2)$,

(f) p_*: $H(X, X_1 \cup X_2) = H[SX/S(X_1 \cup X_2)] \cong H[SX/S\{X_1, X_2\}]$

where $S\{X_1, X_2\}$ *is the subcomplex of* $S(X_1 \cup X_2)$ *which is generated by* SX_1 *and* SX_2 (see 7.1), *and* i = *inclusion*, p = *projection.*

For instance, if X_1, X_2 are *open* in $X_1 \cup X_2$ then (d) holds by 7.3; these triads are excisive. Other important examples are CW-spaces and -subspaces (V, 4.5). A non-excisive triad is given by $X = \mathbb{R}$, $X_1 = (-\infty, 0]$, $X_2 = (0, +\infty)$.

Proof. We have the following exact sequences of chain maps

$$(8.2)\qquad 0 \to \frac{SX_1}{S(X_1 \cap X_2)} \xrightarrow{i_1} \frac{S(X_1 \cup X_2)}{SX_2} \longrightarrow \frac{S(X_1 \cup X_2)}{S\{X_1, X_2\}} \to 0,$$

$$(8.3)\qquad 0 \to S\{X_1, X_2\} \xrightarrow{i} S(X_1 \cup X_2) \longrightarrow \frac{S(X_1 \cup X_2)}{S\{X_1, X_2\}} \to 0,$$

$$(8.4)\qquad 0 \to \frac{S\{X_1, X_2\}}{S(X_1 \cap X_2)} \xrightarrow{\bar{i}} \frac{S(X_1 \cup X_2)}{S(X_1 \cap X_2)} \longrightarrow \frac{S(X_1 \cup X_2)}{S\{X_1, X_2\}} \to 0,$$

$$(8.5)\qquad 0 \to \frac{S(X_1 \cup X_2)}{S\{X_1, X_2\}} \longrightarrow \frac{SX}{S\{X_1, X_2\}} \xrightarrow{p} \frac{SX}{S(X_1 \cup X_2)} \to 0.$$

The homology sequence (II, 2.9) of 8.2 resp. 8.3 resp. 8.4 resp. 8.5 shows that i_{1*} resp. i_* resp. $\bar{i}_*$ resp. p_* is isomorphic if and only if

$$H[S(X_1 \cup X_2)/S\{X_1, X_2\}] = 0.$$

Thus (a), (d), (e), (f) are equivalent; by symmetry, (b), (d), (e), (f) are also equivalent. Equivalence of (c) and (e) follows from the commutative diagram

$$\begin{array}{ccc} SX_1/S(X_1 \cap X_2) \oplus SX_2/S(X_1 \cap X_2) & \xrightarrow[\cong]{(i_1, i_2)} & \dfrac{S\{X_1, X_2\}}{S(X_1 \cap X_2)} \\ & {\scriptstyle (i_1, i_2)} \searrow \quad \swarrow {\scriptstyle \bar{i}} & \\ & S(X_1 \cup X_2)/S(X_1 \cap X_2) & \end{array}$$

by passing to homology. ▮

8.6 Proposition and Definition. *For every triad* $(X; X_1, X_2)$ *the sequence*

$$(8.7)\quad 0 \to S(X_1 \cap X_2) \xrightarrow{(j_1, -j_2)} SX_1 \oplus SX_2 \xrightarrow{(i_1, i_2)} S\{X_1, X_2\} \to 0$$

is exact where i_ν, j_ν *are inclusions. If the triad is excisive then the homology sequence of* 8.7 *has the form*

$$(8.8)\quad \begin{aligned} \cdots \to H_{n+1}(X_1 \cup X_2) \xrightarrow{d_*} H_n(X_1 \cap X_2) &\xrightarrow{(j_{1*}, -j_{2*})} H_n X_1 \oplus H_n X_2 \\ &\xrightarrow{(i_{1*}, i_{2*})} H_n(X_1 \cup X_2) \xrightarrow{d_*} \cdots. \end{aligned}$$

This exact sequence is called the (absolute) *Mayer-Vietoris-sequence* of $(X; X_1, X_2)$.

We also have, for every triad $(X; X_1, X_2)$, *an exact sequence*

$$(8.9)\quad \begin{aligned} 0 \to SX/S(X_1 \cap X_2) &\xrightarrow{(j_1, -j_2)} SX/SX_1 \oplus SX/SX_2 \\ &\xrightarrow{(i_1, i_2)} SX/S\{X_1, X_2\} \to 0 \end{aligned}$$

where i_ν, j_ν *are projections. If the triad is excisive then the homology sequence of* 8.9 *has the form*

$$(8.10)\quad \begin{aligned} &\cdots \to H_{n+1}(X, X_1 \cup X_2) \xrightarrow{d_*} H_n(X, X_1 \cap X_2) \\ &\xrightarrow{(j_{1*}, -j_{2*})} H_n(X, X_1) \oplus H_n(X, X_2) \xrightarrow{(i_{1*}, i_{2*})} H_n(X, X_1 \cup X_2) \xrightarrow{d_*} \cdots. \end{aligned}$$

This exact sequence is called the *relative Mayer-Vietoris sequence* of $(X; X_1, X_2)$.

Proof. Clearly (i_1, i_2): $SX_1 \oplus SX_2 \to S\{X_1, X_2\}$ is epimorphic (by definition of $S\{X_1, X_2\}$), $(j_1, -j_2)$ is monomorphic, and $(i_1, i_2)(j_1, -j_2)=0$. If $(c_1, c_2) \in \ker(i_1, i_2)$ then, in SX, we have $i_1 c_1 + i_2 c_2 = 0$ which means that c_1 and $-c_2$ are the same chains of SX. But $c_1 \in SX_1$ and $c_2 \in SX_2$ hence $c = c_1 = -c_2 \in S(X_1 \cap X_2)$ and $(c_1, c_2) = (j_1, -j_2)(c)$. This proves exactness of 8.7. As to 8.8, one has only to use $HS\{X_1, X_2\} \cong H(X_1 \cup X_2)$ which is 8.1 (d).

In the second part, (i_1, i_2): $SX/SX_1 \oplus SX/SX_2 \to SX/S\{X_1, X_2\}$ is obviously epimorphic, $(j_1, -j_2)$ is monomorphic, and $(i_1, i_2)(j_1, -j_2)=0$. Let $(\bar{y}_1, \bar{y}_2) \in \ker(i_1, i_2)$ where $y_1, y_2 \in SX$ are representatives. Then $y_1 + y_2 \in S\{X_1, X_2\}$, i.e., $y_1 + y_2 = x_1 + x_2$ with $x_\nu \in SX_\nu$, hence $(y_1 - x_1) = -(y_2 - x_2)$ and $(\bar{y}_1, \bar{y}_2) = (j_1, -j_2)(\overline{y_1 - x_1})$. This proves exactness of 8.9. Finally, 8.10 follows from 8.1 (f). ∎

It is sometimes useful to know more about the boundary operator d_* of 8.8, 8.10:

8.11 Proposition. *The boundary operator d_* of the Mayer-Vietoris sequence 8.8 resp. 8.10 coincides with the following composition*

$$(8.12)\quad \begin{aligned} &H_{n+1}(X_1 \cup X_2) \to H_{n+1}(X_1 \cup X_2, X_2) \\ &\cong H_{n+1}(X_1, X_1 \cap X_2) \xrightarrow{\partial_*} H_n(X_1 \cap X_2) \end{aligned}$$

resp.

$$(8.13)\quad \begin{aligned} &H_{n+1}(X, X_1 \cup X_2) \xrightarrow{\partial_*} H_n(X_1 \cup X_2, X_1) \\ &\cong H_n(X_2, X_1 \cap X_2) \to H_n(X, X_1 \cap X_2) \end{aligned}$$

where all maps other than ∂_* are induced by inclusion.

Proof. Let $u \in H(X_1 \cup X_2) \cong HS\{X_1, X_2\}$ be represented by $x_1 + x_2 \in S\{X_1, X_2\}$ with $x_\nu \in SX_\nu$ and $0 = \partial(x_1 + x_2) = \partial x_1 + \partial x_2$. Then $d_* u$ is represented by $(j_1, -j_2)^{-1}(\partial x_1, \partial x_2) = (j_1, -j_2)^{-1}(\partial x_1, -\partial x_1) = \partial x_1$. But ∂x_1 is also representative for the image of u under 8.12.

For the second part we can choose a representative $z \in SX$ of $u \in H(X, X_1 \cup X_2) \cong H[SX/S\{X_1, X_2\}]$ with $\partial z \in S\{X_1, X_2\}$, hence $\partial z = x_1 + x_2$ with $x_\nu \in SX_\nu$. Then $d_* u$ is represented by $(j_1, -j_2)^{-1}(\partial z, 0) = (j_1, -j_2)^{-1}(x_2, 0) = x_2$. But this is also a representative for the image of u under 8.13. ∎

Mayer-Vietoris sequences are functorial, i.e.,

8.14 Proposition. *A map f: $(X; X_1, X_2) \to (Y; Y_1, Y_2)$ of excisive triads, i.e., a map f: $X \to Y$ with $f(X_\nu) \subset Y_\nu$, induces a homomorphism of the corresponding (absolute or relative) Mayer-Vietories sequences.*

We leave it to the reader to formulate this more rigorously and to prove it. ▮

For instance, any triad maps into $(P; P, P)$ where P is a single point. If $X_1 \cap X_2 \neq \emptyset$ then a map $(P; P, P) \xrightarrow{\iota} (X; X_1, X_2)$ exists and the composite $(P; P, P) \xrightarrow{\iota} (X; X_1, X_2) \xrightarrow{\gamma} (P; P, P)$ is the identity map. It follows that ι_* maps the (absolute) Mayer-Vietoris sequence of $(P; P, P)$ onto a direct summand of the Mayer-Vietoris-sequence of $(X; X_1, X_2)$; the other direct summand is $\ker(\gamma_*)$. Since $\ker(\gamma_*)$ is reduced homology this shows

8.15 Proposition (compare 4.4). *If $(X; X_1, X_2)$ is an excisive triad with $X_1 \cap X_2 \neq \emptyset$ then we have an exact sequence*

$$\cdots \to \tilde{H}_{n+1}(X_1 \cup X_2) \xrightarrow{d_*} \tilde{H}_n(X_1 \cap X_2) \xrightarrow{(j_{1*}, -j_{2*})} \tilde{H}_n X_1 \oplus \tilde{H}_n X_2 \xrightarrow{(i_{1*}, i_{2*})} \tilde{H}_n(X_1 \cup X_2) \xrightarrow{d_*} \cdots.$$

It is called the reduced Mayer-Vietoris *sequence of* $(X; X_1, X_2)$. ▮

8.16 Examples. Let $(X; X_1, X_2)$ be an excisive triad such that $X_1 \cap X_2 \neq \emptyset$.
1. If $\tilde{H}(X_1 \cap X_2) = 0$ then 8.15 shows

$$(i_{1*}, i_{2*}): \tilde{H}X_1 \oplus \tilde{H}X_2 \cong \tilde{H}(X_1 \cup X_2).$$

For instance, if $X = X_1 \vee X_2$ this result coincides with 7.9. The crucial part in the proof of 7.9 was to show that the triad satisfied the excisiveness-condition 8.1 (a).

2. If $\tilde{H}(X_1 \cup X_2) = 0$ then 8.15 shows

$$(j_{1*}, j_{2*}): \tilde{H}(X_1 \cap X_2) \cong \tilde{H}X_1 \oplus \tilde{H}X_2.$$

Consider, for instance, an open subset U of $\mathbb{R}^n$ which is homeomorphic with $\mathbb{S}^{n-1} \times (-1, 1)$, say $\Phi: \mathbb{S}^{n-1} \times (-1, 1) \approx U$. Such a subset is called a *thick sphere*, and $\Sigma^{n-1} = \Phi(\mathbb{S}^{n-1} \times \{0\}) \approx \mathbb{S}^{n-1}$ is called a spine of U. Let $V = \mathbb{R}^n - \Sigma$. Then $(\mathbb{R}^n; U, V)$ is an excisive (because open) triad with $\tilde{H}(U \cup V) = \tilde{H}\mathbb{R}^n = 0$, hence $\tilde{H}(U \cap V) \cong \tilde{H}U \oplus \tilde{H}V$. In particular, $\tilde{H}_0(U \cap V) \cong \tilde{H}_0 U \oplus \tilde{H}_0 V \cong \tilde{H}_0 V$ since U is connected. Obviously $U \cap V \approx \mathbb{S}^{n-1} \times (-1, 0) \cup \mathbb{S}^{n-1} \times (0, 1)$ has two components, hence $\tilde{H}_0 V = \tilde{H}_0(U \cap V) = \mathbb{Z}$; therefore $V = \mathbb{R}^n - \Sigma$ has two components. Thus

(8.17) *the spine Σ^{n-1} of any thick sphere in $\mathbb{R}^n$ divides $\mathbb{R}^n$ into two components, called* interior and exterior of Σ; $n > 1$.

This is a useful special case of the generalized Jordan-curve-theorem (IV, 7.2).

3. If $\tilde{H}X_1 = 0 = \tilde{H}X_2$ then 8.15 shows

$$d_*\colon \tilde{H}_{n+1}(X_1 \cup X_2) \cong \tilde{H}_n(X_1 \cap X_2).$$

An interesting example arises from the *suspension* ΣY of any space $Y \neq \emptyset$. The suspension is obtained from $[0,1] \times Y$ by identifying each of the subsets $\{0\} \times Y$ and $\{1\} \times Y$ to a point. More intuitively, it is the *double cone* over Y. The projection $[0,1] \times Y \to [0,1]$ defines a function $h\colon \Sigma Y \to [0,1]$ such that $h^{-1}\lambda \approx Y$ for $\lambda \neq 0, 1$ and $h^{-1}(0)$, $h^{-1}(1)$ are single points. Let $C_0 Y = h^{-1}(0,1]$, $C_1 Y = h^{-1}[0,1)$; these "open cones" are contractible (move vertically towards $h^{-1}(1)$ resp. $h^{-1}(0)$), hence $\tilde{H}C_0 Y = 0 = \tilde{H}C_1 Y$. Applying our isomorphism to the excisive triad $(\Sigma Y; C_0 Y, C_1 Y)$ shows

$$\text{(8.18)} \qquad \tilde{H}_{n+1}\Sigma Y \cong \tilde{H}_n(C_0 Y \cap C_1 Y) = \tilde{H}_n[(0,1) \times Y] \cong \tilde{H}_n Y,$$

the latter because $(0,1) \times Y \simeq Y$.

As an exercise, show that $\Sigma \mathbb{S}^i \approx \mathbb{S}^{i+1}$ and use this to compute $H\mathbb{S}^i$ inductively.

8.19 A Generalization. Consider a pair of triads $(A; A_1, A_2) \subset (X; X_1, X_2)$. The inclusion maps yield a commutative diagram of chain maps

$$\text{(8.20)} \qquad \begin{array}{ccccccccc} 0 \to & S\{A_1, A_2\} & \longrightarrow & S\{X_1, X_2\} & \longrightarrow & \dfrac{S\{X_1, X_2\}}{S\{A_1, A_2\}} & \to 0 \\ & \downarrow & & \downarrow & & \downarrow & \\ 0 \to & S(A_1 \cup A_2) & \longrightarrow & S(X_1 \cup X_2) & \longrightarrow & \dfrac{S(X_1 \cup X_2)}{S(A_1 \cup A_2)} & \to 0 \end{array}$$

with exact rows. If the triads are excisive then the first two vertical arrows induce isomorphisms on homology, and therefore also the third, by 2.10. Since the complexes are free we even get a *homotopy equivalence*

$$S\{X_1, X_2\}/S\{A_1, A_2\} \simeq S(X_1 \cup X_2)/S(A_1 \cup A_2).$$

Consider then the following sequence of chain maps

$$\text{(8.21)} \quad 0 \to \frac{S(X_1 \cap X_2)}{S(A_1 \cap A_2)} \xrightarrow{(j_1, -j_2)} \frac{SX_1}{SA_1} \oplus \frac{SX_2}{SA_2} \xrightarrow{(i_1, i_2)} \frac{S\{X_1, X_2\}}{S\{A_1, A_2\}} \to 0.$$

It is exact, just as 8.7. Its homology sequence has the form

$$(8.22)\quad \begin{aligned}\cdots &\to H_{n+1}(X_1\cup X_2, A_1\cup A_2)\xrightarrow{d_*} H_n(X_1\cap X_2, A_1\cap A_2)\xrightarrow{(j_{1*}-j_{2*})}\\ &\to H_n(X_1, A_1)\oplus H_n(X_2, A_2)\xrightarrow{(i_{1*}, i_{2*})}\\ &\to H_n(X_1\cup X_2, A_1\cup A_2)\xrightarrow{d_*} H_{n-1}(X_1\cap X_2, A_1\cap A_2)\to\cdots.\end{aligned}$$

This exact sequence is called the Mayer-Vietoris sequence of the pair of excisive triads $(X; X_1, X_2)\supset(A\colon A_1, A_2)$. It reduces to 8.8 if $A=\emptyset$, to 8.10 if $X_1=X_2=X$, to 8.16 if $A_1=A_2$ is a single point, to 3.4 if $X_1=X$ and $A_1\subset A_2=X_2$.

8.23 Exercises. *1**. Let $(X; X_1, X_2)$ be a triad, A a closed subset of X containing $X_1\cap X_2$ and such that X_1-A, X_2-A are open in $X_1\cup X_2$. Let W be a neighborhood of A, and put $W_1=W\cap X_1$, $W_2=W\cap X_2$ (for simplicity, assume first X_1, X_2 are closed and $A=X_1\cap X_2$).

a) Show that *if* $(W; W_1, W_2)$ *is excisive then so is* $(X; X_1, X_2)$, *and conversely.* Loosely speaking, this means that the property of being excisive depends only on the situation around the area of contact of X_1 and X_2. Hint: Compare the homology sequences of the triples $(X_1\cup X_2, W_1\cup X_2, X_2)$ and $(X_1, W_1, X_1\cap X_2)$.

b) Suppose there exists a retraction $r\colon W_2\to X_1\cap X_2=W_1\cap W_2$, and a deformation $D_t\colon W_1\cup X_2\to X_1\cup X_2$ such that $D_0=$inclusion, $D_1(W_1\cup X_2)\subset X_2$, $D_t(W_1)\subset X_1$, $D_t(X_2)\subset X_2$ for $0\le t\le 1$. Then $(X; X_1, X_2)$ is excisive. Hint: The deformation D shows that $H(W_1\cup X_2, X_2)\to H(X_1\cup X_2, X_2)$ and $H(W_1, X_1\cap X_2)\to H(X_1, X_1\cap X_2)$ are zero-maps. There results a commutative diagram

$$(8.24)\quad \begin{array}{ccccccccc} 0\to & H(X_1\cup X_2, X_2) & \longrightarrow & H(X_1\cup X_2, W_1\cup X_2) & \longrightarrow & H(W_1\cup X_2, X_2) & \to 0\\ & \uparrow{\scriptstyle\alpha} & & \uparrow{\scriptstyle\beta} & & \uparrow{\scriptstyle\gamma} & \\ 0\to & H(X_1, X_1\cap X_2) & \longrightarrow & H(X_1, W_1) & \longrightarrow & H(W_1, X_1\cap X_2) & \to 0\end{array}$$

with exact rows. Clearly α is monomorphic. Excise X_2 resp. $X_1\cap X_2$ to show that β is isomorphic. Excise X_2-W_2 and use r to show that γ is monomorphic. Diagram-chasing then shows that α is isomorphic.

Part (b) generalizes 7.8. Formulate and prove a corresponding generalization of 7.12 Exercise 2.

2. The absolute and the relative Mayer-Vietoris sequence of an excisive triad are closely related. Show that every term of the relative sequence 8.10 maps into the corresponding term of the absolute sequence 8.8 by a connecting homomorphism and that the resulting diagram commutes up to sign.

More generally, if

$$
(8.25)\qquad
\begin{array}{ccccccccc}
 & & 0 & & 0 & & 0 & & \\
 & & \downarrow & & \downarrow & & \downarrow & & \\
0 & \longrightarrow & K' & \longrightarrow & K & \longrightarrow & K'' & \longrightarrow & 0 \\
 & & \downarrow & & \downarrow & & \downarrow & & \\
0 & \longrightarrow & L' & \longrightarrow & L & \longrightarrow & L'' & \longrightarrow & 0 \\
 & & \downarrow & & \downarrow & & \downarrow & & \\
0 & \longrightarrow & M' & \longrightarrow & M & \longrightarrow & M'' & \longrightarrow & 0 \\
 & & \downarrow & & \downarrow & & \downarrow & & \\
 & & 0 & & 0 & & 0 & &
\end{array}
$$

is a commutative diagram of chain maps with exact rows and columns then the homology sequences of these rows and columns constitute a 2-dimensional lattice of group homomorphisms. It is commutative except for the $(\partial_* - \partial_*)$-squares which anticommute. Apply this to

$$
\begin{array}{ccccccccc}
 & & 0 & & 0 & & 0 & & \\
 & & \downarrow & & \downarrow & & \downarrow & & \\
0 & \longrightarrow & S(A_1 \cap A_2) & \longrightarrow & SA_1 \oplus SA_2 & \longrightarrow & S\{A_1, A_2\} & \longrightarrow & 0 \\
 & & \downarrow & & \downarrow & & \downarrow & & \\
0 & \longrightarrow & S(X_1 \cap X_2) & \longrightarrow & SX_1 \oplus SX_2 & \longrightarrow & S\{X_1, X_2\} & \longrightarrow & 0 \\
 & & \downarrow & & \downarrow & & \downarrow & & \\
0 & \longrightarrow & \dfrac{S(X_1 \cap X_2)}{S(A_1 \cap A_2)} & \longrightarrow & \dfrac{SX_1}{SA_1} \oplus \dfrac{SX_2}{SA_2} & \longrightarrow & \dfrac{S\{X_1, X_2\}}{S\{A_1, A_2\}} & \longrightarrow & 0 \\
 & & \downarrow & & \downarrow & & \downarrow & & \\
 & & 0 & & 0 & & 0 & &
\end{array}
$$

where $(X; X_1, X_2) \supset (A; A_1, A_2)$ is a pair of excisive triads. The above relations are then obtained by specialising $X_1 = X_2 = X$.

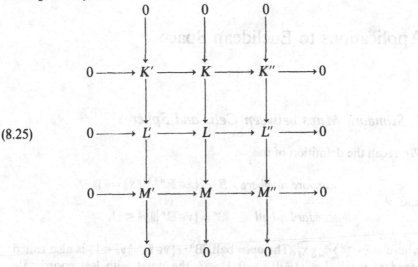

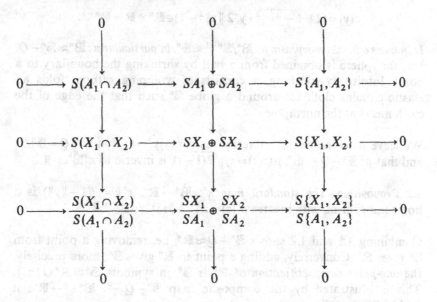

Chapter IV

Applications to Euclidean Space

1. Standard Maps between Cells and Spheres

We recall the definition of the

standard n-sphere $\mathbb{S}^n = \{x \in \mathbb{R}^{n+1} \mid \|x\| = 1\}$

and

standard n-ball $\mathbb{B}^n = \{y \in \mathbb{R}^n \mid \|y\| \leq 1\}$,

where $\|x\| = \sqrt{\sum_{i=0}^{n} x_i^2}$. The *open* ball, $\mathring{\mathbb{B}}^n = \{y \in \mathbb{R}^n \mid \|y\| < 1\}$ is also called *standard n-cell*. Let $Q = (0, \ldots, 0, 1) \in \mathbb{S}^n$, the point with last coordinate $Q_n = 1$.

1.1 Definition and Proposition. *The standard map* $\pi\colon (\mathbb{B}^n, \mathbb{S}^{n-1}) \to (\mathbb{S}^n, Q)$ *is defined by*

$$\pi(y) = (2\sqrt{1 - \|y\|^2} \cdot y, 2\|y\|^2 - 1) \in \mathbb{R}^n \times \mathbb{R} = \mathbb{R}^{n+1}.$$

It induces a homeomorphism $\bar{\pi}\colon \mathbb{B}^n/\mathbb{S}^{n-1} \approx \mathbb{S}^n$, *in particular* $\pi\colon \mathring{\mathbb{B}}^n \approx \mathbb{S}^n - Q$. I.e., the sphere is obtained from a ball by shrinking the boundary to a point. Intuitively speaking, π consists of wrapping without folds an elastic circular cloth $\mathbb{B}^n$ around a globe $\mathbb{S}^n$ such that the edge of the cloth meets at the northpole.

We leave it to the reader to verify that $\|\pi(y)\|^2 = 1$, that $\pi^{-1}Q = \mathbb{S}^{n-1}$, and that $\rho\colon \mathbb{S}^n - Q \to \mathring{\mathbb{B}}^n$, $\rho(z, t) = z/\sqrt{2(1-t)}$, is inverse to $\pi|\mathring{\mathbb{B}}^n$. ∎

1.2 Proposition. *The standard map* $\pi'\colon \mathring{\mathbb{B}}^n \to \mathbb{R}^n$, $\pi' y = y/(1 - \|y\|)$ is a homeomorphism, with inverse $\rho'(z) = z/(1 + \|z\|)$, $z \in \mathbb{R}^n$. ∎

Combining 1.1 and 1.2 shows $\mathbb{S}^n - Q \approx \mathbb{R}^n$, i.e. removing a point from $\mathbb{S}^n$ gives $\mathbb{R}^n$. Conversely, adding a point to $\mathbb{R}^n$ gives $\mathbb{S}^n$; more precisely, the *one-point compactification of* $\mathbb{R}^n$ *is* $\mathbb{S}^n$, in symbols, $\mathbb{S}^n \approx \mathbb{R}^n \cup \{\infty\}$. This is illustrated by the composite map $\mathbb{S}^n - Q \xrightarrow{\rho} \mathring{\mathbb{B}}^n \xrightarrow{\pi'} \mathbb{R}^n$; it "takes Q into ∞".

We want to show that simplices, cubes and their products are homeomorphic with balls. This is contained in

1.3 Proposition. *If $K \subset \mathbb{R}^n$ is a compact convex set which contains an n-ball B then there is a standard homeomorphism $(B, \dot{B}) \approx (K, \dot{K})$ where denotes the boundary.*

Proof. After parallel translation and multiplication with some $\vartheta > 0$ we can assume that B is the standard ball $\mathbb{B}^n$. Now, if $x \in K$ and $0 \le \lambda < 1$ then λx lies in the interior $\mathring{K}$ of K; in fact, λx lies in the open cone which is obtained by projecting $\mathring{\mathbb{B}}^n$ from x, and this cone lies in K because K is convex (Fig. 6). In particular, every ray from 0 contains exactly one point in the boundary $\dot{K}$ of K. Therefore the map

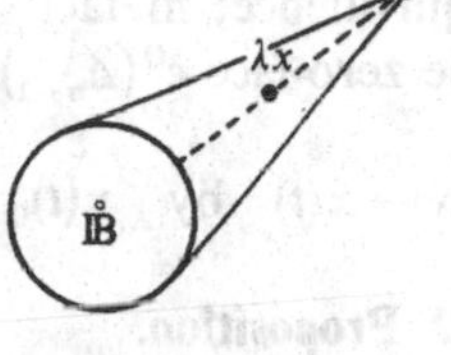

Fig. 6

$$v: \dot{K} \to \mathbb{S}^{n-1}, \quad v(y) = \frac{y}{\|y\|}$$

is bijective and hence homeomorphic (because $\dot{K}$ is compact). By radial extension we get the required homeomorphism

$$v: K \approx \mathbb{B}^n, \quad v(\lambda y) = \lambda \frac{y}{\|y\|}, \quad y \in \dot{K},\ 0 \le \lambda \le 1. \quad \blacksquare$$

1.4 In particular, 1.3 provides us with a standard homeomorphism between cube and ball $[-1,1]^n = [-1,1] \times \cdots \times [-1,1] \approx \mathbb{B}^n$. As to Δ_n we first define a linear embedding $\iota: \Delta_n \to \mathbb{R}^n$, $\iota(e^i) = e^i$ for $i < n$, $\iota(e^n) = -\sum_{i=0}^{n-1} e^i$. Then $\iota(\Delta_n) \approx \Delta_n$ is a convex set containing a ball around $0 = \iota\left(\frac{1}{n+1} \sum_{i=0}^{n} e^i\right)$, hence a standard homeomorphism $\Phi: \Delta_n \approx \mathbb{B}^n$. Under this homeomorphism Φ the *boundary* $\dot{\Delta}_n = \{x \in \Delta_n \mid x_i = 0$ for some $i \ge 0\}$ of Δ_n maps homeomorphically onto the boundary $\dot{\mathbb{B}}^n = \mathbb{S}^{n-1}$.

2. Homology of Cells and Spheres

Using the tools of Chapter III these groups are easily computed now. They lead to some of the best known theorems of topology such as 2.3–2.6.

2.1 Lemma. *Let Δ_n be the standard simplex, $\dot{\Delta}_n = \{x \in \Delta_n \mid x_j = 0$ for some $j \ge 0\}$ its boundary, and $\wedge_n = \{x \in \Delta_n \mid x_j = 0$ for some $j > 0\} =$ union of all* *faces but one. Then we have the following isomorphisms*

$$H_k(\Delta_n,\dot{\Delta}_n)\xrightarrow[\cong]{\partial_*}H_{k-1}(\dot{\Delta}_n,\wedge_n)\xleftarrow[\cong]{i_*}H_{k-1}(\dot{\Delta}_n-e^0,\wedge_n-e^0)$$
$$\xleftarrow[\cong]{\varepsilon^0_*}H_{k-1}(\Delta_{n-1},\dot{\Delta}_{n-1}),$$

where e^0 is the vertex $x_0=1$, $i=$ inclusion, and $n>0$.

Proof. The homotopy $x\mapsto(1-t)\,x+t\,e^0$ shows that $(\Delta_n,\wedge_n)\simeq(e^0,e^0)$. Hence $H(\Delta_n,\wedge_n)=0$, and the homology sequence III, 3.4 of the triple $(\Delta_n,\dot{\Delta}_n,\wedge_n)$ shows that ∂_* is isomorphic. The map i_* is isomorphic by excision 7.4. Finally, $\varepsilon^0\colon(\Delta_{n-1},\dot{\Delta}_{n-1})\to(\dot{\Delta}_n-e^0,\wedge_n-e^0)$ is a homotopy equivalence; in fact, we get a deformation retraction of $\dot{\Delta}_n-e^0$ onto the zero-face $\varepsilon^0(\Delta_{n-1})$, which takes $\wedge_n-e^0$ into $\varepsilon^0(\dot{\Delta}_{n-1})$, if we define

$$x\mapsto x(t)\quad\text{by}\quad x(t)_0=(1-t)\,x_0,\qquad x(t)_j=\frac{(1-t)\,x_0}{1-x_0}\,x_j\quad\text{for } j>0.\ ∎$$

2.2 Proposition.

(a) $\tilde{H}_k\mathbb{S}^n=\begin{cases}0 & \text{if } k\neq n\\ \mathbb{Z} & \text{if } k=n,\end{cases}$

(b) $H_k(\mathbb{B}^n,\mathbb{S}^{n-1})=\begin{cases}0 & \text{if } k\neq n\\ \mathbb{Z} & \text{if } k=n,\end{cases}$

(c) $H_k(\mathbb{R}^n,\mathbb{R}^n-P)=\begin{cases}0 & \text{if } k\neq n\\ \mathbb{Z} & \text{if } k=n,\end{cases}$ for any $P\in\mathbb{R}^n$.

Proof. By 1.4 and 2.1 we have

$$H_k(\mathbb{B}^n,\mathbb{S}^{n-1})\cong H_k(\Delta_n,\dot{\Delta}_n)\cong H_{k-1}(\Delta_{n-1},\dot{\Delta}_{n-1})\cong\cdots$$
$$\cong H_{k-n}(\Delta_0,\dot{\Delta}_0)=H_{k-n}\Delta_0.$$

Since Δ_0 is a point this proves part (b). Since $\mathbb{B}^{n+1}$ is contractible we have $\partial_*\colon H_{k+1}(\mathbb{B}^{n+1},\mathbb{S}^n)=\tilde{H}_k\mathbb{S}^n$; this reduces (a) to (b). As to (c) we can assume (after translation) that $P=0$. Then the inclusion $(\mathbb{B}^n,\mathbb{S}^{n-1})\to(\mathbb{R}^n,\mathbb{R}^n-P)$ induces homology isomorphisms for $\mathbb{B}^n$ and $\mathbb{S}^{n-1}$ (III, 5.12; both are deformation retracts); hence $H(\mathbb{B}^n,\mathbb{S}^{n-1})\cong H(\mathbb{R}^n,\mathbb{R}^n-P)$ by the five lemma. ∎

2.3 Corollary. *Spheres of different dimension are not homeomorphic. Euclidean spaces of different dimension are not homeomorphic.*

For spheres this is clear from 2.2(a). If $h\colon\mathbb{R}^m\approx\mathbb{R}^n$ then $h\colon(\mathbb{R}^m,\mathbb{R}^m-0)\approx(\mathbb{R}^n,\mathbb{R}^n-h(0))$, hence $m=n$ by 2.2(c). ∎

2.4 Corollary. *$\mathbb{S}^{n-1}$ is not a retract of $\mathbb{B}^n$.*

If $r: \mathbb{B}^n \to \mathbb{S}^{n-1}$ were a retraction, $r\,i = \mathrm{id}$, then the composite

$$\tilde{H}\mathbb{S}^{n-1} \xrightarrow{i_*} \tilde{H}\mathbb{B}^n \xrightarrow{r_*} H\mathbb{S}^{n-1}$$

would be the identity map. This is impossible because $\tilde{H}\mathbb{S}^{n-1} \neq 0$, $\tilde{H}\mathbb{B}^n = 0$. ∎

2.5 Corollary. *If $f: \mathbb{B}^n \to \mathbb{R}^n$ is continuous then either $f y = 0$ for some $y \in \mathbb{B}^n$ or $f z = \lambda z$ for some $z \in \mathbb{S}^{n-1}$, $\lambda > 0$.*

Proof. Define $\rho: \mathbb{B}^n \to \mathbb{R}^n$ as follows. $\rho x = (2\|x\| - 1)x - (2 - 2\|x\|) f(x/\|x\|)$ for $2\|x\| \geq 1$, $\rho x = -f(4\|x\|\,x)$ for $2\|x\| \leq 1$; in particular, $\rho z = z$ for $z \in \mathbb{S}^{n-1}$. Then $\rho|\mathring{\mathbb{B}}^n$ must assume the value 0 because otherwise $x \mapsto \dfrac{\rho x}{\|\rho x\|}$ would be a retraction of $\mathbb{B}^n$ onto $\mathbb{S}^{n-1}$. If $\rho x = 0$ and $2\|x\| \leq 1$ then $f y = 0$ where $y = 4\|x\|\,x$. If $\rho x = 0$ and $1 < 2\|x\| < 2$ then

$$f\left(\frac{x}{\|x\|}\right) = \frac{2\|x\| - 1}{2 - 2\|x\|}\,x,$$

hence $f z = \lambda z$ where $z = \dfrac{x}{\|x\|}$, $\lambda = \dfrac{2\|x\| - 1}{2 - 2\|x\|}\|x\|$. ∎

If in 2.5 we replace $f x$ by $g x - x$ we get the

2.6 Corollary (Brouwer fixed point theorem). *If $g: \mathbb{B}^n \to \mathbb{R}^n$ is continuous then either $g y = y$ for some $y \in \mathbb{B}^n$ or $g z = \mu z$ for some $z \in \mathbb{S}^{n-1}$, $\mu > 1$.* ∎

The reader might want to see a concrete (relative) cycle whose homology class generates $H_n(\mathbb{B}^n, \mathbb{S}^{n-1})$ resp. $\tilde{H}_n \mathbb{S}^n$; such cycles are usually called *fundamental cycles*. There is a very simple one, namely

2.7 Proposition. *The identity map $\iota_n: \Delta_n \to \Delta_n$ is a cycle mod $\dot{\Delta}_n$ whose homology class $[\iota_n]$ generates $H_n(\Delta_n, \dot{\Delta}_n) = \mathbb{Z}$. Its boundary $\partial \iota_n$ is a cycle on $\dot{\Delta}_n$ whose homology class generates $\tilde{H}_{n-1} \dot{\Delta}_n = \mathbb{Z}$.*

Proof. Since $\partial_*: H_n(\Delta_n, \dot{\Delta}_n) \cong \tilde{H}_{n-1} \dot{\Delta}_n$ the two assertions are equivalent. Clearly $[\iota_0]$ generates $H_0 \Delta_0 = H_0(\Delta_0, \dot{\Delta}_0)$, so we can proceed by induction on n. In the notation of 2.1, it is clear that $\varepsilon^0: \Delta_{n-1} \to \dot{\Delta}_n$ is a representative for both $\partial_*[\iota_n] \in H_{n-1}(\dot{\Delta}_n, \wedge_n)$ and $i_* \varepsilon^0_*[\iota_{n-1}] \in H_{n-1}(\dot{\Delta}_n, \wedge_n)$; hence $[\iota_{n-1}] = (\varepsilon^0_*)^{-1} i_*^{-1} \partial_*[\iota_n]$, which proves the assertion. ∎

Lemma 2.1 and its consequence 2.2 can be generalized by multiplying all pairs and maps with an extra space Y; the proofs remain the same, with Y playing a dummy role:

2.8 Proposition. *Let $P \in \mathbb{S}^n$. There are natural isomorphisms*

(a) $H_k(\mathbb{S}^n \times Y, P \times Y) \cong H_{k-n} Y$,

(b) $H_k(\mathbb{B}^n \times Y, \mathbb{S}^{n-1} \times Y) \cong H_{k-n} Y$.

Proof. In the notation of 2.1 and with the same reasoning as in 2.1 (multiplied by Y) we get isomorphisms

$$(2.9)\quad \begin{aligned} &H_k[(\Delta_n, \dot{\Delta}_n) \times Y] \overset{\partial_*}{\cong} H_{k-1}[(\dot{\Delta}_n, \wedge_n) \times Y] \\ &\overset{(i\times \mathrm{id})_*^{-1}}{\cong} H_{k-1}[(\dot{\Delta}_n - e^0, \wedge_n - e^0) \times Y] \overset{(\varepsilon^0 \times \mathrm{id})_*^{-1}}{\cong} H_{k-1}[(\Delta_{n-1}, \dot{\Delta}_{n-1}) \times Y], \end{aligned}$$

hence, by iteration $H_k[(\Delta_n, \dot{\Delta}_n) \times Y] \cong H_{k-n}[(\Delta_0, \dot{\Delta}_0) \times Y] \cong H_{k-n} Y$. This proves part (b) of 2.8 because $(\Delta_n, \dot{\Delta}_n) \approx (\mathbb{B}^n, \mathbb{S}^{n-1})$.

Consider now the triple $(\mathbb{B}^{n+1} \times Y, \mathbb{S}^n \times Y, P \times Y)$. Since $\mathbb{B}^{n+1} \times Y \simeq P \times Y$ we get $H(\mathbb{B}^{n+1} \times Y, P \times Y) = 0$, and the homology sequence of triples shows $\partial_*\colon H_{k+1}(\mathbb{B}^{n+1} \times Y, \mathbb{S}^n \times Y) \cong H_k(\mathbb{S}^n \times Y, P \times Y)$. This reduces (a) to (b). ∎

2.10 Corollary. *There are natural isomorphisms*

$$(\rho, q_*)\colon H_k(\mathbb{S}^n \times Y) \cong H_{k-n} Y \oplus H_k Y$$

where $q\colon \mathbb{S}^n \times Y \to Y$ is the projection, $q(x, y) = y$.

Indeed, $q\colon \mathbb{S}^n \times Y \to Y = P \times Y$ is a retraction, and therefore we have an exact sequence

$$0 \to H(P \times Y) \to H(\mathbb{S}^n \times Y) \to H(\mathbb{S}^n \times Y, P \times Y) \to 0$$

which is split by q_* (see III, 4.16). The assertion now follows from 2.8 (a). ∎

We can now compute the homology of any finite products $\mathbb{S}^{n_1} \times \cdots \times \mathbb{S}^{n_r}$ of spheres (using 2.10 and 2.2). In particular, we find $H_n(\mathbb{S}^n \times \mathbb{S}^n) \cong \mathbb{Z} \oplus \mathbb{Z}$ for $n > 0$. In order to describe generators we consider the maps

$$(2.11)\quad H_n \mathbb{S}^n \oplus H_n \mathbb{S}^n \xrightarrow{(i_{1*},\, i_{2*})} H_n(\mathbb{S}^n \times \mathbb{S}^n) \xrightarrow{(p_{1*},\, p_{2*})} H_n \mathbb{S}^n \oplus H_n \mathbb{S}^n,$$

where $i_1, i_2\colon \mathbb{S}^n \to \mathbb{S}^n \times \mathbb{S}^n$ are injections, $i_1(x) = (x, P)$, $i_2(x) = (P, x)$, and $p_1, p_2\colon \mathbb{S}^n \times \mathbb{S}^n \to \mathbb{S}^n$ are the two projections. The composite 2.11 is the identity map, hence (i_{1*}, i_{2*}) maps $H_n \mathbb{S}^n \oplus H_n \mathbb{S}^n = \mathbb{Z} \oplus \mathbb{Z}$ isomorphically onto a direct summand of $H_n(\mathbb{S}^n \times \mathbb{S}^n) = \mathbb{Z} \oplus \mathbb{Z}$. The only such summand is the whole group, hence

2.12 Proposition. *Both maps* 2.11 *are isomorphisms.* ∎

2.13 Exercises. *1.* Compute $H(\mathbb{R}^n - F)$ where F is a finite set. Hint: Compute $H(\mathbb{R}^n, \mathbb{R}^n - F)$ first, using a suitable excision.

2. If $g: \mathbb{B}^n \to \mathbb{R}^n$, $n>1$, is a map without fixed point ($g\,y \neq y$) then the angle $\sphericalangle(0, z, g\,z)$ assumes all values from 0 to π as z varies in $\mathbb{S}^{n-1}$.

3. Prove: The k-th homology group of the n-dimensional torus $\mathbb{S}^1 \times \cdots \times \mathbb{S}^1$ (n factors) is free of rank $\binom{n}{k}$.

4. Let $D = D_g$ be the space which is obtained from $\mathbb{B}^2$ by removing the interiors of g disjoint (closed) disks inside $\mathbb{B}^2$; thus D_g is a disk with g holes. Take two copies D_g^+, D_g^- of D_g and identify their boundaries. The resulting space S_g is called orientable *surface of genus g*; we have $S_g = D_g^+ \cup D_g^-$, and $D_g^+ \cap D_g^-$ is the disjoint union of $g+1$ circles. For instance, $S_0 \approx \mathbb{S}^2$, $S_1 \approx \mathbb{S}^1 \times \mathbb{S}^1$.

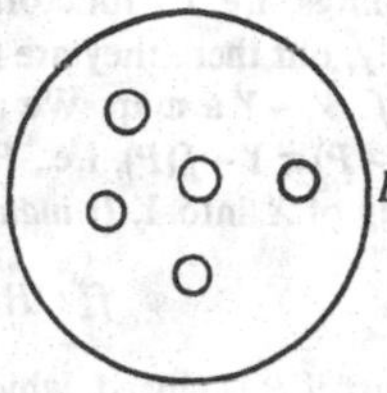

Fig. 7

Prove: $H_0 S_g \cong \mathbb{Z} \cong H_2 S_g$, $H_1 S_g \cong \mathbb{Z} \oplus \mathbb{Z} \oplus \cdots \oplus \mathbb{Z}$ ($2g$ summands) and $H_i S_g = 0$ if $i>2$. Describe generators of $H_1 S_g$. *Hint.* $D_g^{\cdot}$ is a retract of S_g, hence $HS_g = H(D_g^{\cdot}) \oplus H(S_g, D_g^+)$. Compute $H(D^+)$ as in Exercise 1. Use excision and a homotopy to prove $H(S_g, D_g^+) = H(D_g^-, \dot{D}_g^-)$. The homology of $D_g^- = D_g$ and its boundary $\dot{D}_g$ is known; determine the inclusion map $H\dot{D}_g \to HD_g$, and get $H(D_g, \dot{D}_g)$ from the homology sequence. – The Mayer-Vietoris sequence of $(S_g; D_g^+, D_g^-)$ can also be used to compute HS_g.

5. Generalize 4. to higher dimensions replacing $\mathbb{B}^2$ by $\mathbb{B}^n$, $n>2$. You will find $H_0 S_g \cong \mathbb{Z} \cong H_n S_g$, $H_1 S_g \cong g \cdot \mathbb{Z} \cong H_{n-1} S_g$, $H_i S_g = 0$ otherwise.

3. *Local Homology*

Homology groups are *global* invariants; spaces with different homology can still be locally homomorphic, e.g. $\mathbb{S}^n$ and $\mathbb{R}^n$. However, some relative homology groups turn out to be *local invariants*, as we shall see now.

3.1 Definition. Let X be a space and $P \in X$. The groups $H(X, X-P)$ are called *local homology groups of X at P*.

The adjective "local" is justified by

3.2 Proposition. *If P is closed in X, $P = \bar{P}$, (e.g. if X is a T_1-space) and if V is any neighborhood of P then $H(V, V-P) \cong H(X, X-P)$ under inclusion.* I.e., local homology $H(X, X-P)$ can be computed in any neighborhood V of P.

Proof. The pair $(V, V-P)$ is obtained from $(X, X-P)$ by excising $B = X - V$. Because V is a neighborhood of P we have $P \in \mathring{V}$ (= interior of V), hence $\bar{B} = X - \mathring{V} \subset X - P = X - \bar{P} = (X-P)^\circ$. Therefore the excision theorem III, 7.4 applies. ∎

3.3 For a better understanding of local homology one has to study its behavior under mappings, i.e. its functorial properties. Because of 3.2, the maps need only be defined locally, but there they are not quite arbitrary. More precisely, let X, Y be spaces, $P \in V \subset X$, and $f\colon V \to Y$ a map. We assume that a neighborhood U of P exists such that $U \subset V$ and $f(U-P) \subset Y - f(P)$, i.e., P is an isolated counterimage of $Q = f(P)$. Such an f is called a *P-map* of X into Y. *It induces homomorphisms of local homology at P*

$$f_*^P\colon H(X, X-P) \cong H(U, U-P) \xrightarrow{(f|U)_*} H(Y, Y-Q), \tag{3.4}$$

at least if P is closed, which we always assume. This homomorphism does not depend on the choice of U; more generally

3.5 Proposition. *Any two P-maps $f\colon V \to Y$, $f'\colon V' \to Y$ which agree in a neighborhood of P induce the same homomorphism of the local homology groups.*

Proof. If U, U' are the neighborhoods of P which were used to define $f_*^P, f_*'^P$ then we can find a neighborhood W of P such $f|W = f'|W$ and $W \subset U \cap U'$; in particular, $f(W-P) \subset (Y-Q)$. Consider the commutative diagram

$$\begin{array}{ccc} & H(U, U-P) & \\ {\scriptstyle\cong}\swarrow & \uparrow & \searrow{\scriptstyle (f|U)_*} \\ H(X, X-P) & & H(Y, Y-Q) \\ {\scriptstyle\cong}\nwarrow & & \nearrow{\scriptstyle (f|W)_*} \\ & H(W, W-P) & \end{array}$$

The upper row is f_*^P, the lower row depends only on $f|W$. ∎

3.6 Proposition 3.5 suggests the following definitions: Two P-maps of X into Y are *P-equivalent* if they agree in a neighborhood of P. The equivalence class of f is denoted by f^P; it is called the *germ of f at P*. If $f\colon V \to Y$ is a P-map of X into Y, and $g\colon W \to Z$ is a $Q = f(P)$-map of Y into Z then $gf\colon f^{-1} W \to Z$ is a P-map whose germ at P depends only on the germs of f and g. Therefore $g^Q \circ f^P = (gf)^P$ defines a composition law for germs; we get a category whose objects are pairs (X, P) (*pointed spaces*; $P = \bar{P} \in X$), and the morphisms $(X, P) \to (Y, Q)$ are germs of P-maps. Local homology groups then are functors of this category (by 3.5). In particular, equivalent objects have isomorphic local homology, i.e.,

3.7 Corollary to 3.2. *If $P \in X$, $Q \in Y$ are closed points such that $(V, P) \approx (W, Q)$ for suitable neighborhoods V, W then $H(X, X-P) \cong H(Y, Y-Q)$.*

Indeed, $H(X, X-P) \cong H(V, V-P) \cong H(W, W-Q) \cong H(Y, Y-Q)$. ∎

This is illustrated by the following Theorems 3.8, 3.9 of Brouwer.

3.8 Proposition (Invariance of Dimension). *If $P \in \mathbb{R}^m$, $Q \in \mathbb{R}^n$ have neighborhoods V, W such that $(V, P) \approx (W, Q)$*[1] *then $m = n$.*

[1] In fact, already $V \approx W$ implies $m = n$ (see 7.4).

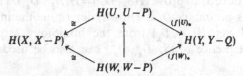

This holds because $H(\mathbb{R}^m, \mathbb{R}^m - P) \cong H(\mathbb{R}^n, \mathbb{R}^n - Q)$ implies $m = n$ (2.2(c)). ∎

3.9 Proposition (Invariance of the Boundary). *Let $\mathbb{R}^n_+ = \{x \in \mathbb{R}^n \mid x_0 \geq 0\}$ denote the "upper half" of $\mathbb{R}^n$. If $P, Q \in \mathbb{R}^n_+$ have neighborhoods V, W such that $(V, P) \approx (W, Q)$ then either both of P, Q lie on the boundary $\dot{\mathbb{R}}^n_+ = \{x \in \mathbb{R}^n \mid x_0 = 0\}$ or both lie in the interior $\mathring{\mathbb{R}}^n_+ = \{x \in \mathbb{R}^n \mid x_0 > 0\}$ of $\mathbb{R}^n_+$.* I.e., a local homeomorphism never takes a boundary point into an interior point, or vice versa.

Proof. If $P \in \mathbb{R}^n_+$ is a boundary point then $(\mathbb{R}^n_+, \mathbb{R}^n_+ - P)$ is contractible to (S, S) where $S \in \mathring{\mathbb{R}}^n_+$ (just deform radially towards S); hence

$$H(\mathbb{R}^n_+, \mathbb{R}^n_+ - P) \cong H(S, S) = 0.$$

On the other hand, $Q \in \mathring{\mathbb{R}}^n_+$ implies $H_n(\mathbb{R}^n_+, \mathbb{R}^n_+ - Q) \cong H_n(\mathbb{R}^n, \mathbb{R}^n - Q)$ by excision III, 7.4, and this is not zero by 2.2(c). ∎

3.10 We now show that in many spaces local homology groups can be expressed in terms of suitable absolute groups; this gives the connection to older definitions of local homology.

Let X be a space and $P \in X$ a closed point. We assume that P has a neighborhood V such that the inclusion map $i: V \to X$ is homotopic to a constant map (if this is true for all $P \in X$ then X is said to be *semi-locally contractible*). Then $\tilde{H} V \to \tilde{H} X$ is zero, and the diagram

$$\begin{array}{ccccccc}
\tilde{H}_k V & \xrightarrow{j_*} & H_k(V, V-P) & \xrightarrow{\partial_*} & \tilde{H}_{k-1}(V-P) & \to & \tilde{H}_{k-1} V \\
\downarrow{\scriptstyle 0} & & \downarrow{\scriptstyle \cong} & & & & \\
\tilde{H}_k X & \longrightarrow & H_k(X, X-P) & & & &
\end{array}$$

shows $j_* = 0$. The first row being exact this implies

(3.11) $$H_k(X, X-P) \cong \ker[\tilde{H}_{k-1}(V-P) \to \tilde{H}_{k-1} V].$$

If V is contractible or at least acyclic, $\tilde{H} V = 0$, then we get

(3.12) $$H_k(X, X-P) = \tilde{H}_{k-1}(V-P).$$

The right side is, essentially, the local homology group as used in Seifert-Threlfall § 32.

3.13 Exercises. *1.* Show that $H_0(X, X-P) = \mathbb{Z}$ if P is a full pathcomponent, and $H_0(X, X-P) = 0$ otherwise.

2. Construct a space X, a point $P \in X$, and a neighborhood V of P such that $H(X, X-P) \not\cong H(V, V-P)$.

3. If $Q \in Y$ is a closed point then

$$H_k[\mathbb{R}^n \times Y, \mathbb{R}^n \times Y-(P,Q)] \cong H_{k-n}(Y, Y-Q), \quad P \in \mathbb{R}^n. \tag{3.14}$$

Hint: Write $[\mathbb{R}^n \times Y, \mathbb{R}^n \times Y-(P,Q)] = (\mathbb{R}^n, \mathbb{R}^n - P) \times (Y, Y-Q)$ and proceed in analogy with the proof of 2.8(b).

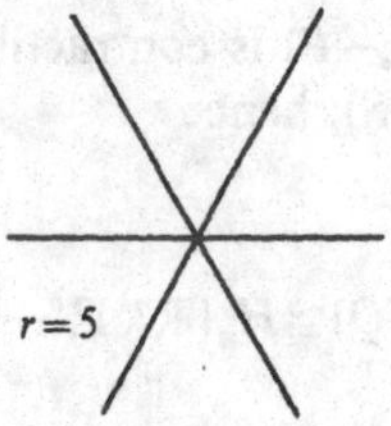

Y, for $r=5$

In particular, let $Y = \{z \in \mathbb{C} \mid z^{r+1} \geq 0\}$; i.e., a star with $(r+1)$ rays. Let $Q=0$. The product $\mathbb{R}^{n-1} \times Y$ is sometimes called *branched* euclidean space.

It is the union of $r+1$ half-spaces $\mathbb{R}^n_+$ which intersect at the *branch point locus* $\mathbb{R}^{n-1} \times Q$. The number r is called the *branch point order*. The n-th local homology group at a branch point turns out (by 3.14) to be free of rank r. It follows that *any local homeomorphism preserves the branch point order (invariance of branch point order)*.

4. Compute the local homology of a suspension space ΣY at a vertex $\{0\} \times Y$ (cf. III, 8.16, example 3).

4. The Degree of a Map

Every endomorphism φ of a free cyclic group is given by an integer, i.e., $\varphi(x) = dx$ for some uniquely determined $d \in \mathbb{Z}$. Applying this remark to homology groups defines the notion of degree in algebraic topology, which has many applications (e.g. 4.4, 4.8).

4.1 Definition. If $f\colon \mathbb{S}^n \to \mathbb{S}^n$ resp. $f\colon (\mathbb{B}^{n+1}, \mathbb{S}^n) \to (\mathbb{B}^{n+1}, \mathbb{S}^n)$ is a map then the induced endomorphism f_* of $\tilde{H}_n \mathbb{S}^n \cong \mathbb{Z}$ resp. $H_{n+1}(\mathbb{B}^{n+1}, \mathbb{S}^n) \cong \mathbb{Z}$ is given by $f_*(x) = \deg(f) \cdot x$, where $\deg(f) \in \mathbb{Z}$ is a uniquely determined integer. This integer is called the *degree of f*.

Some elementary properties of the degree are as follows.

4.2 Proposition.

(i) $\deg(\mathrm{id}) = +1$.

(ii) $\deg(f \circ f') = \deg(f) \cdot \deg(f')$.

(iii) $f \simeq f' \Rightarrow \deg(f) = \deg(f')$.[2]

(iv) The degree of a homotopy equivalence is ± 1.[2]

(v) If $f: (\mathbb{B}^{n+1}, \mathbb{S}^n) \to (\mathbb{B}^{n+1}, \mathbb{S}^n)$ then $\deg(f) = \deg(f|\mathbb{S}^n)$.

Indeed, (i), (ii) just express the functor properties of f_*, and (iii), (iv) the homotopy invariance. Property (v) follows from the commutative diagram

$$
\begin{array}{ccc}
H_{n+1}(\mathbb{B}^{n+1}, \mathbb{S}^n) & \xrightarrow{\ f_*\ } & H_{n+1}(\mathbb{B}^{n+1}, \mathbb{S}^n) \\
\partial_* \downarrow \cong & & \partial_* \downarrow \cong \\
\tilde{H}_n \mathbb{S}^n & \xrightarrow{\ (f|\mathbb{S}^n)_*\ } & \tilde{H}_n \mathbb{S}_n .
\end{array} \quad \blacksquare
$$

4.3 Example. *The degree of a linear map* $\beta: (\Delta_n, \dot{\Delta}_n) \to (\Delta_n, \dot{\Delta}_n)$ *which permutes the vertices equals the signature of the permutation,* $\deg(\beta) = \mathrm{sign}(\beta | \{e^0, e^1, \ldots, e^n\})$. *The degree of an orthogonal map* $\alpha: \mathbb{S}^n \to \mathbb{S}^n$ *equals the determinant,* $\deg(\alpha) = \det(\alpha)$. *The antipodal map* $x \mapsto -x$, *for instance, has degree* $(-1)^{n+1}$.

Proof. Let $\nu = (\nu_0, \nu_1, \ldots, \nu_r)$ denote the linear map $\Delta_r \to \Delta_n$ which takes e^i into e^{ν_i} $(0 \le \nu_i \le n)$. We want to show that $[\nu_0, \ldots, \nu_n] = \mathrm{sign}(\nu)[0, 1, \ldots, n]$ if ν is a permutation $(r = n)$ and $[\ ]$ denotes homology classes in $H_n(\Delta_n, \dot{\Delta}_n)$; recall (2.7) that $[0, 1, \ldots, n]$ generates this group. Suppose first ν is a transposition of consecutive vertices $i, i+1$. Let $\mu: \Delta_{n+1} \to \Delta_n$ be the map $(0, 1, \ldots, i-1, i+1, i, i+1, \ldots, n)$; it is obtained from id by inserting $i+1$ in front of i. Then

$$\partial \mu = (-1)^i (0, 1, \ldots, n) + (-1)^{i+2}(0, \ldots, i-1, i+1, i, i+2, \ldots, n) + R$$

where the remainder R consists of terms which omit one vertex; thus $R \in S(\dot{\Delta}_n)$. Passing to homology mod $\dot{\Delta}_n$ therefore gives

$$0 = [0, \ldots, n] + [0, \ldots, i-1, i+1, i, i+2, \ldots, n],$$

as asserted. Since every permutation ν is a product of such transpositions, say $\nu = \tau_1 \ldots \tau_q$, and $\mathrm{sign}(\nu) = (-1)^q$, the first part of 4.3 follows from 4.2(ii).

Every orthogonal map $\alpha: \mathbb{S}^n \to \mathbb{S}^n$ with determinant $+1$ is homotopic to the identity, hence $\deg(\alpha) = +1$. If $\det(\alpha) = -1$ then α is homotopic to the reflection ρ at any hyperplane containing 0. As in 1.4, consider the linear $(n+1)$ simplex s in $\mathbb{R}^{n+1}$ with vertices $(e^0, \ldots, e^n, -\sum_{i=0}^n e^i)$.

---

[2] The converse is also true; cf. Spanier 7.5.7.

Its homology class $[s]$ generates $H_{n+1}(\mathbb{R}^{n+1}, \mathbb{R}^{n+1}-0)$, and $[\partial s]$ generates $\tilde{H}_n(\mathbb{R}^{n+1}-0)=\tilde{H}_n\mathbb{S}^n$; cf. 2.7. There is a reflection ρ which interchanges e^0, e^1 and leaves e^i fixed for $i>1$. The simplex ρs has vertices $(e^1, e^0, e^2, e^3, \ldots, e^n, -\sum e^i)$, hence $[\rho s]=-[s]$ by part one, hence $\rho_*[\partial s]=[\rho\,\partial s]=-[\partial s]$, hence $\deg(\rho)=-1$. ∎

As an application we prove

4.4 Proposition. *If $f\colon \mathbb{S}^n \to \mathbb{S}^n$ has no fixed point then* $\deg(f)=(-1)^{n+1}$. *If $f\colon \mathbb{S}^n \to \mathbb{S}^n$ has no antipodal point ($fx \neq -x$) then* $\deg(f)=+1$. *In particular, every map $f\colon \mathbb{S}^{2k} \to \mathbb{S}^{2k}$ has a fixed point or an antipodal point.*

Proof. If f has no fixed point then $d_t x=(1-t)(fx)-tx \neq 0$ for $0 \leq t \leq 1$, hence $D_t x=(d_t x)/\|d_t x\|$ is a deformation of f into the antipodal map $x \mapsto -x$, hence $\deg(f)=(-1)^{n+1}$ by 4.3. If $fx \neq -x$ then $gx=-fx$ has no fixed point, hence $(-1)^{n+1}\deg(f)=\deg(g)=(-1)^{n+1}$ ∎

The following is a slight variation of the notion of degree.

4.5 Definition. If $\mu\colon \mathbb{S}^n \times \mathbb{S}^n \to \mathbb{S}^n$, $n>0$, is a map then the induced homomorphism

$$H_n\mathbb{S}^n \oplus H_n\mathbb{S}^n \overset{2.12}{\cong} H_n(\mathbb{S}^n \times \mathbb{S}^n) \xrightarrow{\mu_*} H_n\mathbb{S}^n$$

has the form $\mu_*(x_1, x_2)=d_1 x_1+d_2 x_2$ where d_1, d_2 are uniquely determined integers. The pair (d_1, d_2) is called *the bidegree of* μ. Its properties are analoguous to 4.2. In particular (analogue of 4.2(ii)).

4.6 Proposition. *If $f_1, f_2\colon \mathbb{S}^n \to \mathbb{S}^n$ are maps then the degree of the composition $\mathbb{S}^n \xrightarrow{(f_1, f_2)} \mathbb{S}^n \times \mathbb{S}^n \xrightarrow{\mu} \mathbb{S}^n$ is given by*

$$\deg[\mu(f_1, f_2)]=d_1 \cdot \deg(f_1)+d_2 \cdot \deg(f_2).$$

Proof. Let $p_1, p_2\colon \mathbb{S}^n \times \mathbb{S}^n \to \mathbb{S}^n$ denote the two projections. Then $p_\nu(f_1, f_2)=f_\nu$, and the direct sum representation 2.11 shows

$$(p_{1*}, p_{2*})(f_1, f_2)_*(x)=(f_{1*}x, f_{2*}x)=(\deg(f_1)\,x, \deg(f_2)\,x) \quad \text{for } x \in H_n\mathbb{S}^n.$$

Hence $\mu_*(f_1, f_2)_*(x)=[d_1 \cdot \deg(f_1)+d_2 \cdot \deg(f_2)](x)$. ∎

Intuitively, we think of μ as a multiplicative structure on $\mathbb{S}^n$, such as the multiplication of complex numbers ($n=1$) or quaternions ($n=3$); in these cases $\mu(z_1, z_2)=z_1 \cdot z_2$ has bidegree $(1, 1)$, and we find

4.7 Corollary. *The mapping $p_k\colon z \mapsto z^k$, $k \in \mathbb{Z}$, of the group $\mathbb{S}^1$ resp. $\mathbb{S}^3$ of unit complex numbers resp. quaternions has degree k.*

Indeed, $\deg(p_k) = \deg(p_{k-1}) + \deg(\mathrm{id}) = \deg(p_{k-1}) + 1$ by 4.6, and $\deg(p_1) = \deg(\mathrm{id}) = 1$. ∎

As an application we prove

4.8 Proposition (Fundamental Theorem of Algebra). *Every complex polynomial $p(z) = z^k + c_1 z^{k-1} + \cdots + c_k$, $k > 0$, has a zero.*

Proof. For every p which has no zero on $\mathbb{S}^1$ we define

$$\hat{p}\colon \mathbb{S}^1 \to \mathbb{S}^1, \qquad \hat{p}(z) = \frac{p(z)}{\|p(z)\|},$$

and we prove 4.8 in two steps:

(i) If p has no zero z with $\|z\| \le 1$ then $\deg(\hat{p}) = 0$.

(ii) If p has no zero z with $\|z\| \ge 1$ then $\deg(\hat{p}) = k$.

For case (i) we consider the deformation

$$\hat{p}_t\colon \mathbb{S}^1 \to \mathbb{S}^1, \qquad \hat{p}_t(z) = \frac{p(t\,z)}{\|p(t\,z)\|}.$$

Clearly $\hat{p}_1 = \hat{p}$, $\hat{p}_0 =$ constant, hence $\deg(\hat{p}) = 0$. For case (ii) we consider the deformation $\hat{p}_t(z) = \dfrac{q(z,t)}{\|q(z,t)\|}$ where

$$(4.9) \qquad q(z,t) = t^k\, p\left(\frac{z}{t}\right) = z^k + t(c_1 z^{k-1} + t\, c_2 z^{k-2} + \cdots + t^{k-1} c_k).$$

The right side of 4.9 shows that $q(z,t)$ is continuous (even where $t = 0$). Clearly $\hat{p}_1 = \hat{p}$ and $\hat{p}_0(z) = z^k$, hence $\deg(\hat{p}) = \deg(\hat{p}_0) = k$ by 4.7. ∎

This result 4.8 and its proof generalize to other multiplications $\mu\colon \mathbb{S}^n \times \mathbb{S}^n \to \mathbb{S}^n$ on spheres with bidegree (α, β) such that $\alpha > 0$, $\beta > 0$ (exercise!). We shall see in VII, 10.1 that $\alpha\beta \neq 0$ implies that n is odd.

4.10 Exercises. *1.* Every map $\mathbb{S}^0 \to \mathbb{S}^0$ resp. $(\mathbb{B}^1, \mathbb{S}^0) \to (\mathbb{B}^1, \mathbb{S}^0)$ has degree 0, or ± 1.

2. If $f\colon X \to Y$ is a map then $\mathrm{id} \times f\colon [0,1] \times X \to [0,1] \times Y$ takes $\{t\} \times X$ into $\{t\} \times Y$ and therefore induces a map $\Sigma f\colon \Sigma X \to \Sigma Y$ of suspensions (III, 8.16, example 3). In particular, if $f\colon \mathbb{S}^n \to \mathbb{S}^n$ then $\Sigma f\colon \mathbb{S}^{n+1} \to \mathbb{S}^{n+1}$.

Prove that $\deg(\Sigma f)=\deg(f)$ (hint: use naturality of III, 8.18). *Corollary: For $n>0$ there exist maps $\mathbb{S}^n\to\mathbb{S}^n$ of arbitrary degree.*

3. If $m, n\geq 0$ then every point z of $\mathbb{S}^{m+n+1}\subset\mathbb{R}^{m+n+2}=\mathbb{R}^{m+1}\times\mathbb{R}^{n+1}$ can be represented in the form $z=\cos(t)\cdot x+\sin(t)\cdot y$ with $x\in\mathbb{S}^m$, $y\in\mathbb{S}^n$, $0\leq t\leq\pi/2$, and this representation is unique except that x resp. y is undetermined when $t=\pi/2$ resp. 0. Given $f\colon \mathbb{S}^m\to\mathbb{S}^m$, $g\colon \mathbb{S}^n\to\mathbb{S}^n$ define their *join* $f*g\colon \mathbb{S}^{m+n+1}\to\mathbb{S}^{m+n+1}$ by $(f*g)(z)=\cos(t)\cdot f(x)+\sin(t)\cdot g(y)$, and prove $\deg(f*g)=\deg(f)\cdot\deg(g)$. Hint: Use $f*g=(f*\mathrm{id})(\mathrm{id}*g)$, and prove $\deg(f*\mathrm{id})=\deg(f)$ by induction on n; start the induction with exercise 2.

4. A *tangent vector field* on $\mathbb{S}^n$ is a continuous function v which assigns to every $x\in\mathbb{S}^n$ a vector $v(x)\in\mathbb{R}^{n+1}$ which is tangent to $\mathbb{S}^n$ at x. For example, if $n=2k+1$ then $v(x)=(x_1, -x_0, x_3, -x_2, \ldots, x_{2k+1}, -x_{2k})$ is a tangent vector field on $\mathbb{S}^{2k+1}$ which is nowhere zero. Prove: If n is even then every vector field on $\mathbb{S}^n$ vanishes somewhere. Hint: Move x slightly in direction $v(x)$. This gives a map of degree $+1$ which must have a fixed point.

5. If a complex polynomial p has no zero on $\mathbb{S}^1=\{z\in\mathbb{C}\mid\|z\|=1\}$ and has m zeros (counted with multiplicity) inside $\mathbb{S}^1$ then the map $\hat{p}\colon\mathbb{S}^1\to\mathbb{S}^1$, $\hat{p}(z)=\dfrac{p(z)}{\|p(z)\|}$ has degree m.

5. *Local Degrees*

This notion will show that the degree of §4 can be determined locally (with respect to the range) namely as "number of counterimages of a point", each counterimage counted with its multiplicity.

5.1 Definition. Let $V\subset\mathbb{S}^n$, $n>0$, be an open set, $f\colon V\to\mathbb{S}^n$ a map and $Q\in\mathbb{S}^n$ a point such that $f^{-1}(Q)$ is compact. Consider the composite

$$(5.2)\qquad \begin{aligned} H_n\mathbb{S}^n &\xrightarrow{j_*} H_n(\mathbb{S}^n, \mathbb{S}^n-f^{-1}Q)\overset{\text{exc}}{\cong} H_n(V, V-f^{-1}Q)\\ &\xrightarrow{f_*} H_n(\mathbb{S}^n, \mathbb{S}^n-Q)\cong H_n\mathbb{S}^n \end{aligned}$$

where *exc* is an excision isomorphism (III, 7.4) and the last isomorphism is given by

$$(5.3)\qquad H_n(\mathbb{S}^n, \mathbb{S}^n-Q)\cong H_n(\mathbb{S}^n, P)\cong \tilde{H}_n\mathbb{S}^n=H_n\mathbb{S}^n, \qquad P\in\mathbb{S}^n-Q,$$

because $\mathbb{S}^n-Q\simeq P$.

The composition 5.2 has the form $x \mapsto (\deg_Q f)\,x$ where $(\deg_Q f) \in \mathbb{Z}$ is a uniquely determined integer. This integer is called the (local) *degree of f over Q*. Note that the degree over Q is only defined if $f^{-1}Q$ is compact.

5.4 Examples. *If $Q \notin \operatorname{im}(f)$ then $\deg_Q(f)=0$. If $f\colon V \to \mathbb{S}^n$ is the inclusion map then $\deg_Q(f)=1$ for all $Q \in V$. If f is a homeomorphism onto an open set $fV \subset \mathbb{S}^n$ then $\deg_Q(f)=\pm 1$ for all $Q \in fV$.*

Proof. The first and second assertions are clear from the definition. In the third case, 5.2 becomes a sequence of isomorphisms

$$\begin{aligned} H_n\mathbb{S}^n &\cong H_n(\mathbb{S}^n, \mathbb{S}^n - f^{-1}Q) \cong H_n(V, V - f^{-1}Q) \cong H_n(fV, fV - Q) \\ &\cong H_n(\mathbb{S}^n, \mathbb{S}^n - Q) \cong H_n\mathbb{S}^n. \quad \blacksquare \end{aligned}$$

5.5 Proposition. *If $f^{-1}Q \subset K \subset U \subset V$ where K is compact and U is a neighborhood of K then the degree of f over Q is also given by the composite*

$$H_n\mathbb{S}^n \to H_n(\mathbb{S}^n, \mathbb{S}^n - K) \overset{\text{exc}}{\cong} H_n(U, U - K) \xrightarrow{f_*} H_n(\mathbb{S}^n, \mathbb{S}^n - Q) \cong H_n\mathbb{S}^n.$$

I.e., we can replace $f^{-1}Q$ by any larger compact set inside V and/or we can cut down V to any neighborhood of $f^{-1}Q$. For instance, we can cut down to $f^{-1}V$ where V is any neighborhood of Q; this justifies the adjective "local".

The proof follows immediately from the commutative diagram

$$\begin{array}{ccccc} & \nearrow & H_n(\mathbb{S}^n, \mathbb{S}^n - f^{-1}Q) \cong H_n(V, V - f^{-1}Q) & \overset{f_*}{\searrow} & \\ H_n\mathbb{S}^n & & \uparrow \qquad\qquad\qquad \uparrow & & H_n(\mathbb{S}^n, \mathbb{S}^n - Q) \\ & \searrow & H_n(\mathbb{S}^n, \mathbb{S}^n - K) \;\cong\; H_n(U, U - K) & \underset{f_*}{\nearrow} & \end{array} \quad \blacksquare$$

5.6 Corollary. *If $f\colon \mathbb{S}^n \to \mathbb{S}^n$ then $\deg(f)=\deg_Q(f)$ for any $Q \in \mathbb{S}^n$. If $f\colon (\mathbb{B}^n, \mathbb{S}^{n-1}) \to (\mathbb{B}^n, \mathbb{S}^{n-1})$ then $\deg(f)=\deg_Q(f \mid \mathring{\mathbb{B}}^n)$ for any $Q \in \mathbb{B}^n$ such that $f^{-1}Q \subset \mathring{\mathbb{B}}^n$.*

In particular, *deg*(f) is of a local nature with respect to the range.

Proof. The first part follows from 5.5 with $K = \mathbb{S}^n = U$. For the second part we think of $\mathbb{S}^n$ as $\mathbb{R}^n \cup \{\infty\}$, and we extend $f\colon (\mathbb{B}^n, \mathbb{S}^{n-1}) \to (\mathbb{B}^n, \mathbb{S}^{n-1})$ radially to a map $F\colon (\mathbb{S}^n, \mathbb{S}^n - \mathring{\mathbb{B}}^n) \to (\mathbb{S}^n, \mathbb{S}^n - \mathring{\mathbb{B}}^n)$. Then

$$\deg(F) = \deg_Q(F) = \deg_Q(F \mid \mathring{\mathbb{B}}^n) = \deg_Q(f \mid \mathring{\mathbb{B}}^n)$$

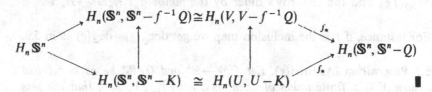

by part one and 5.5. On the other hand we have $\deg(F) = \deg(f)$ from the following diagram

$$\begin{array}{ccccccc}
(\mathbb{B}^n, \mathbb{S}^{n-1}) & \longrightarrow & (\mathbb{R}^n, \mathbb{R}^n - \mathring{\mathbb{B}}^n) & \xrightarrow{\text{exc}} & (\mathbb{S}^n, \mathbb{S}^n - \mathring{\mathbb{B}}^n) & \longleftarrow & (\mathbb{S}^n, \infty) \\
\downarrow{\scriptstyle f} & & \downarrow{\scriptstyle F} & & \downarrow{\scriptstyle F} & & \downarrow{\scriptstyle F} \\
(\mathbb{B}^n, \mathbb{S}^{n-1}) & \longrightarrow & (\mathbb{R}^n, \mathbb{R}^n - \mathring{\mathbb{B}}^n) & \xrightarrow{\text{exc}} & (\mathbb{S}^n, \mathbb{S}^n - \mathring{\mathbb{B}}^n) & \longleftarrow & (\mathbb{S}^n, \infty).
\end{array} \quad \blacksquare$$

As to functorial properties of the local degree we only mention

5.7 Proposition. *If $f\colon V \to \mathbb{S}^n$ and $Q \in \mathbb{S}^n$ are as in 5.1, and $g\colon \mathbb{S}^n \to \mathbb{S}^n$ is any map then the local degree of $fg\colon g^{-1}V \to \mathbb{S}^n$ over Q is defined and* $\deg_Q(fg) = \deg_Q(f) \cdot \deg(g)$.

Proof. Consider the commutative diagram

$$\begin{array}{ccccccc}
H_n\mathbb{S}^n & \to & H_n(\mathbb{S}^n, \mathbb{S}^n - K) & \cong & H_n(V, V-K) & \xrightarrow{f_*} & \\
\uparrow{\scriptstyle g_*} & & \uparrow{\scriptstyle g_*} & & \uparrow{\scriptstyle g_*} & & H_n(\mathbb{S}^n, \mathbb{S}^n - Q) \cong H_n\mathbb{S}^n \\
H_n\mathbb{S}^n & \to & H_n(\mathbb{S}^n, \mathbb{S}^n - g^{-1}K) & \cong & H_n(g^{-1}V, g^{-1}V - g^{-1}K) & \xrightarrow{(fg)_*} &
\end{array}$$

where $K = f^{-1}Q$. The upper row defines $\deg_Q(f)$, the lower row defines $\deg_Q(fg)$, and the two rows differ by the factor $g_*\colon H_n\mathbb{S}^n \to H_n\mathbb{S}^n$. ∎

For instance, if f is the inclusion map we get $\deg_Q(g) = \deg(g)$ as in 5.6.

5.8 Proposition (Additivity). *Let $f\colon V \to \mathbb{S}^n$ and $Q \in \mathbb{S}^n$ be as in 5.1 and assume V is a finite union of open sets, $V = \bigcup_{\lambda=1}^r V_\lambda$ such that the sets $f_\lambda^{-1}Q$, $f_\lambda = f|V_\lambda$, are mutually disjoint, $(f_\lambda^{-1}Q) \cap (f_\mu^{-1}Q) = \emptyset$ if $\lambda \neq \mu$. Then* $\deg_Q(f) = \sum_{\lambda=1}^r \deg_Q(f_\lambda)$.

This often allows us to compute $\deg_Q(f)$. Suppose, for instance, that $f^{-1}Q$ is a finite set, $f^{-1}Q = \{P_1, \ldots, P_r\}$. Then we can choose open sets V_λ such that $P_\lambda \in V_\lambda$, $P_\mu \notin V_\lambda$ for $\mu \neq \lambda$, and we are left with computing $\deg_Q(f_\lambda)$. This number is sometimes called the *multiplicity* of the counterimage point P_λ; thus, $\deg_Q(f)$ *equals the number of points in $f^{-1}Q$, counted with their multiplicities*. The multiplicity of P_λ can be determined in any neighborhood of P_λ (5.5); if f is locally homeomorphic then all multiplicities are ± 1, by 5.4.

Proof of 5.8. Choose open neighborhoods U_λ of $f_\lambda^{-1}Q$ in V_λ such that $U_\lambda \cap U_\mu = \emptyset$ if $\lambda \neq \mu$, and put $U = \bigcup_{\lambda=1}^{r} U_\lambda$. Consider the diagram

$$
\begin{array}{ccccccc}
H_n\mathbb{S}^n & \xrightarrow{j_*} & H_n(\mathbb{S}^n, \mathbb{S}^n - f^{-1}Q) & \overset{\text{exc}}{\cong} & H_n(U, U - f^{-1}Q) & \xrightarrow{f_*} & \\
\downarrow \{\text{id}\} & & \downarrow \{i_{\lambda *}\} & & \cong \uparrow \{i'_{\lambda *}\} & & \\
\oplus H_n\mathbb{S}^n & \xrightarrow{\oplus j_{\lambda *}} & \oplus H_n(\mathbb{S}^n, \mathbb{S}^n - f_\lambda^{-1}Q) & \overset{\text{exc}}{\cong} & \oplus H_n(U_\lambda, U_\lambda - f_\lambda^{-1}Q) & \xrightarrow{\oplus f_{\lambda *}} &
\end{array}
$$

(5.9)

$$
\begin{array}{ccc}
\rightarrow H_n(\mathbb{S}^n, \mathbb{S}^n - Q) & \cong & H_n\mathbb{S}^n \\
\uparrow \{\text{id}\} & & \uparrow \{\text{id}\} \\
\rightarrow \oplus H_n(\mathbb{S}^n, \mathbb{S}^n - Q) & \cong & \oplus H_n\mathbb{S}^n
\end{array}
$$

where all sums $\oplus$ extend over $\lambda = 1, 2, \ldots, r$, where $\{\text{id}\}$ is a map all of whose components are identity maps, and i_λ, i'_λ denote inclusions. The map $\{i'_{\lambda *}\}$ is isomorphic because U is the disjoint union of $\{U_\lambda\}$; cf. III, 4.12. Commutativity of 5.9 is clear except perhaps for the second square; there it asserts that the composite

$$
\begin{aligned}
H_n(U_\lambda, U_\lambda - f_\lambda^{-1}Q) &\xrightarrow{i'_{\lambda *}} H_n(U, U - f^{-1}Q) \\
&\rightarrow H_n(\mathbb{S}^n, \mathbb{S}^n - f^{-1}Q) \xrightarrow{i_{\mu *}} H_n(\mathbb{S}^n, \mathbb{S}^n - f_\mu^{-1}Q)
\end{aligned}
$$

agrees with the inclusion for $\lambda = \mu$ (this is obvious), and is zero if $\lambda \neq \mu$ (this follows because $U_\lambda \subset \mathbb{S}^n - f_\mu^{-1}Q$).

By 5.5, the upper row of 5.9 defines $\deg_Q(f)$, the lower row defines $\{\deg_Q(f_\lambda)\}$. Therefore composition of the lower row with the two outside vertical arrows gives $\deg_Q(f) = \sum \deg_Q(f_\lambda)$. ∎

5.10 Example. *For every* $k \in \mathbb{Z}$ *and every* $n > 0$ *we construct a map* $f: \mathbb{S}^n \to \mathbb{S}^n$ *with* $\deg(f) = k$. Think of $\mathbb{S}^n$ as $\mathbb{R}^n \cup \{\infty\}$.

Define $g: \mathbb{S}^n \to \mathbb{S}^n$ by

$$
g(x) = \begin{cases} x & \text{if } \|x\| \leq 1 \\ x(2 - \|x\|)^{-1} & \text{if } 1 \leq \|x\| \leq 2 \\ \infty & \text{if } \|x\| \geq 2. \end{cases}
$$

Clearly $\deg(g) = \deg_0(g) = +1$. For every $P \in \mathbb{R}^n$ consider also the parallel translation $\tau_P: \mathbb{S}^n \to \mathbb{S}^n$ where $\tau_P(x) = x - P$ if $x \in \mathbb{R}^n$ and $\tau_P(\infty) = \infty$. Since $\tau_P \simeq \text{id}$ we have $\deg_0(g\tau_P) = \deg_0(g)\deg(\tau_P) = \deg_0(g) = +1$ by 5.7.

Suppose now that $k \geq 0$, and choose k points $P_1, \ldots, P_k \in \mathbb{R}^n$ whose mutual distance is >4; if B_λ denotes the ball with center P_λ and radius 2 then $B_\lambda \cap B_\mu = \emptyset$ for $\lambda \neq \mu$. Define

$$f\colon \mathbb{S}^n \to \mathbb{S}^n, \quad f|B_\lambda = (g\,\tau_{P_\lambda})|B_\lambda, \quad f(x) = \infty \quad \text{if } x \notin \bigcup_{\lambda=1}^k B_\lambda.$$

Clearly $f^{-1}(0) = \{P_1, P_2, \ldots, P_k\}$, and by 5.8 we have

$$\deg(f) = \sum \deg_0(f|B_\lambda) = \sum \deg_0(g\,\tau_{P_\lambda}) = k.$$

If $k<0$ we construct f' of degree $-k$ first and put $f = r f'$ where r is the reflection at a hyperplane; then

$$\deg(f) = \deg(r)\deg(f') = (-1)(-k) = k. \quad \blacksquare$$

Note, incidentally, that for $k>0$ the map f is such that every point $Q \in \mathbb{R}^n$ has exactly k counterimages, whereas ∞ has infinitely many. Still, $\deg_\infty(f) = k$ because $\deg_\infty(f) = \deg(f)$.

How does $\deg_Q(f)$ depend on Q? We give the answer in a simple case; some deeper results are contained in the next §6.

5.11 Definition. If $V \subset \mathbb{S}^n$ is an open set, and $f\colon V \to \mathbb{S}^n$ a map we say that f is *proper over* $W \subset \mathbb{S}^n$ if $f^{-1}(L)$ is compact for every compact set $L \subset W$. If W is a single point this is just the condition that $\deg_W(f)$ be defined. For instance, if $f\colon V \approx f(V)$ then f is proper over $f(V)$. Every $f\colon \mathbb{S}^n \to \mathbb{S}^n$ is proper over $\mathbb{S}^n$. The inclusion $(0,1) \to \mathbb{R}$ is *not* proper over $(0,2)$.

5.12 Proposition and Definition. *If W is a connected open subset of $\mathbb{S}^n$, $n>0$, and if $f\colon V \to \mathbb{S}^n$ is proper over W then* $\deg_Q(f)$ *is defined and is the same for all $Q \in W$.* If $f\colon V \approx W = f(V)$ then this number $\deg_Q(f)$, $Q \in W$, equals $+1$ or -1; in the first case f is called *orientation preserving*, in the other case *orientation reversing*.

Proof. Consider first any great (=geodesic) arc A in W; then $f^{-1}A$ is compact. Since $\mathbb{S}^n - A$ is contractible (deform radially from a point in A) we have $H_n(\mathbb{S}^n, \mathbb{S}^n - A) \cong H_n\mathbb{S}^n$. Let $Q \in A$ and consider the commutative diagram

$$\begin{array}{ccccc}
 & & & H_n(\mathbb{S}^n, \mathbb{S}^n - Q) & \\
 & & \overset{f_*}{\nearrow} & & \searrow\!\!= \\
H_n\mathbb{S}^n \to H_n(\mathbb{S}^n, \mathbb{S}^n - f^{-1}A) \cong H_n(V, V - f^{-1}A) & & & & H_n\mathbb{S}^n \\
 & & \underset{f_*}{\searrow} & & \nearrow\!\!= \\
 & & & H_n(\mathbb{S}^n, \mathbb{S}^n - A) &
\end{array}$$

By 5.5 the upper row defines $\deg_Q(f)$. Therefore the lower row also defines $\deg_Q(f)$. Since the lower row does not depend on Q the number $\deg_Q(f)$ is the same for all $Q \in A$. Because W is connected, any two points $P, Q \in W$ can be connected by a polygon consisting of great arcs, hence $\deg_P(f) = \deg_Q(f)$. ∎

5.13 Exercises. *1.* Define p_k: $\mathbb{C} \to \mathbb{C}$ by $p_k(z) = z^k$ and prove that $\deg_0(p_k) = k$. Note that $p_k^{-1}(0)$ consists of just one point.

2. If f: $V \to \mathbb{R}^m$, g: $V' \to \mathbb{R}^n$ are maps such that $\deg_0(f)$, $\deg_0(g)$ are defined then the local degree of $f \times g$: $V \times V' \to \mathbb{R}^m \times \mathbb{R}^n = \mathbb{R}^{m+n}$ over 0 is defined and equals $\deg_0(f) \cdot \deg_0(g)$. Hint: Use 4.10 Exercise 3.

3. If f_t: $V \to \mathbb{S}^n$ is a deformation and $Q \in \mathbb{S}^n$ is a point such that $\bigcup_{0 \le t \le 1} f_t^{-1}(Q)$ is compact then $\deg_Q(f_0) = \deg_Q(f_1)$. Hint: Use 5.5 with $K = \bigcup_t f_t^{-1}(Q)$.

4. Let $V \subset \mathbb{R}^k$ be an open set, f: $V \to \mathbb{R}^k$ a map which is continuously differentiable ("of class C^1"), and $Q \in \mathbb{R}^k$ a point such that the jacobian of f is non-zero at all points of $f^{-1}(Q)$ (Q is a "*regular value*"). Prove: If $f^{-1}(Q)$ is finite then $\deg_Q(f) = p - n$ where p (resp. n) is the number of points in $f^{-1}(Q)$ at which the jacobian is positive (resp. negative). Hint: By additivity 5.8 one can assume that $f^{-1}Q$ is a single point, say $Q = 0$, $f^{-1}Q = 0$. Apply Exercise 3 to the deformation $f_t = (1-t)f + tf'(0)$ where $f'(0)$: $\mathbb{R}^k \to \mathbb{R}^k$ is the derivative of f at 0.

5. If V, W are open subsets of $\mathbb{S}^n$, and f: $V \to W$, g: $W \to \mathbb{S}^n$ are maps such that $\deg_P(f)$, $\deg_Q(g)$, $\deg_Q(gf)$ are defined for some $P \in W$, $Q = g(P)$, is it always true that $\deg_Q(gf) = \deg_P(f) \cdot \deg_Q(g)$?

6. *Homology Properties of Neighborhood Retracts in* $\mathbb{R}^n$

In Section 5 the local degree $\deg_Q(f)$ of f: $V \to \mathbb{S}^n$ over Q was defined as the image under f_*: $H_n(V, V-K) \to H_n(\mathbb{S}^n, \mathbb{S}^n - Q) \cong \mathbb{Z}$ of a certain element $y \in H_n(V, V-K)$. In this section we consider inclusion maps f only but we replace y by an arbitrary element of $H_n(V, V-K)$ and we study the resulting function of $Q \in K$. It turns out that $H_n(V, V-K)$ can be expressed in terms of such functions, and $H_i(V, V-K) = 0$ if $i > n$. Similar results for more general spaces (manifolds) will be proved in VIII, 3.

We begin with arbitrary subsets $B \subset A \subset \mathbb{S}^n$. For every $P \in A$ the inclusions induce maps

$$H_n(\mathbb{S}^n - B, \mathbb{S}^n - A) \xrightarrow{j_P} H_n(\mathbb{S}^n, \mathbb{S}^n - P) \xleftarrow[\cong]{i_P} \tilde{H}_n \mathbb{S}^n,$$

where i_P is isomorphic because $\mathbb{S}^n - P$ is contractible.

6.1 Lemma. *For every* $y \in H_n(\mathbb{S}^n - B, \mathbb{S}^n - A)$ *the mapping*

$$Jy\colon A \to \tilde{H}_n \mathbb{S}^n, \qquad (Jy)(P) = i_P^{-1} j_P\, y$$

is continuous, i.e. locally constant. Further, $(Jy)|B = 0$.

Proof. Let $(Jy)(P)=x$, i.e., $j_P\, y=i_P\, x$. We have to construct a neighborhood U of P such that $Q\in U\cap A$ implies $j_Q\, y=i_Q\, x$ for all $Q\in U\cap A$. If $\zeta\in S(\mathbb{S}^n-B)\subset S(\mathbb{S}^n)$ resp. $z\in S(\mathbb{S}^n)$ are representative (relative) cycles, then the assumption $j_P[\zeta]=i_P[z]$ means that there are chains $c\in S(\mathbb{S}^n)$, $c'\in S(\mathbb{S}^n-P)$ such that $\zeta-z=\partial c+c'$. Now c' is a linear combination of finitely many singular simplices σ each of which avoids P, and hence avoids a neighborhood U_σ of P (because $\operatorname{im}(\sigma)$ is closed). Therefore $c'\in S(\mathbb{S}^n-U)\subset S(\mathbb{S}^n-Q)$ where $U=\bigcap_\sigma U_\sigma$ and $Q\in U\cap A$, hence $j_Q[\zeta]=i_Q[z]$.

The second assertion, $Jy|B=0$, follows because for $P\in B$ the map j_P factors through a zero-group:

$$H(\mathbb{S}^n-B, \mathbb{S}^n-A)\to H(\mathbb{S}^n-B, \mathbb{S}^n-B)\to H(\mathbb{S}^n, \mathbb{S}^n-P). \quad \blacksquare$$

This leads to the following

6.2 Definition. Let $B\subset A\subset\mathbb{S}^n$. Let $\Gamma(A,B)$ denote the (additive) group of continuous (=locally constant) functions $A\to\tilde{H}_n\mathbb{S}^n$ which are zero on B, and put $\Gamma(A,\emptyset)=\Gamma A$. Lemma 6.1 defines a homomorphism

$$J=J(A,B)\colon\ H_n(\mathbb{S}^n-B,\mathbb{S}^n-A)\to\Gamma(A,B).$$

Clearly

6.3 Proposition. *The homomorphism J is natural with respect to inclusions, i.e., if $(A_1,B_1)\subset(A_2,B_2)$ then the diagram*

$$\begin{array}{ccc} H_n(\mathbb{S}^n-B_2,\mathbb{S}^n-A_2) & \xrightarrow{i_*} & H_n(\mathbb{S}^n-B_1,\mathbb{S}^n-A_1) \\ \downarrow J & & \downarrow J \\ \Gamma(A_2,B_2) & \xrightarrow{i'} & \Gamma(A_1,B_1) \end{array}$$

is commutative where i_*, i' are induced by inclusion. $\blacksquare$

The importance of J stems from the following

6.4 Proposition. *If $X\subset Y$ are subsets of $\mathbb{S}^n$ which are neighborhood retracts* (e.g. if X, Y are open) *then*

(a) $H_i(Y,X)=0$ *for* $i>n$

(b) $J\colon\ H_n(Y,X)\cong\Gamma(\mathbb{S}^n-X,\mathbb{S}^n-Y)$.

(Recall that X is a neighborhood retract if there exists an open set in $\mathbb{S}^n$ of which X is a retract; cf. III, 4.14.)

6.5 Corollary. *If $X \subset \mathbb{S}^n$ is a neighborhood retract, $X \neq \mathbb{S}^n$, $n > 0$, then $H_i X = 0$ for $i \geq n$, and $\tilde{H}_{n-1} X \cong \tilde{\Gamma}(\mathbb{S}^n - X)$ where $\tilde{\Gamma} = \Gamma / C$ is Γ modulo the subgroup C of constant functions, and $\tilde{H}$ = reduced homology.*

Proof. Because $X \neq \mathbb{S}^n$ the inclusion map $X \to \mathbb{S}^n$ is nulhomotopic, hence $\tilde{H} X \to \tilde{H} \mathbb{S}^n$ is the zero-map and the homology sequence of $(\mathbb{S}^n, X)$ decomposes into short exact sequences

$$0 \to \tilde{H}_{i+1} \mathbb{S}^n \to H_{i+1}(\mathbb{S}^n, X) \to \tilde{H}_i X \to 0.$$

If $i \geq n$ the first two terms vanish (6.4(a)), hence $H_i X = \tilde{H}_i X = 0$. If $i = n-1$ we apply J and get a diagram

(6.6)
$$\begin{array}{ccccccccc}
0 \to & H_n \mathbb{S}^n & \longrightarrow & H_n(\mathbb{S}^n, X) & \longrightarrow & \tilde{H}_{n-1} X & \to 0 \\
& J \downarrow \cong & & J \downarrow \cong & & \tilde{J} \vdots \cong & \\
0 \to & \Gamma \mathbb{S}^n & \xrightarrow{\ r\ } & \Gamma(\mathbb{S}^n - X) & \longrightarrow & \operatorname{coker}(r) = \tilde{\Gamma}(\mathbb{S}^n - X) & \to 0
\end{array}$$

whose first two vertical arrows are isomorphic by 6.4(b) (r = restriction). Because the first square commutes (6.3) we can fill in the dotted arrow $\tilde{J}$. ∎

The following lemma is a crucial tool in proving 6.4.

6.7 Lemma. *Let (Y, X_1, X_2) be an excisive triad in $\mathbb{S}^n$. If 6.4(a), (b) hold for (Y, X_1), (Y, X_2) and $(Y, X_1 \cup X_2)$ then also for $(Y, X_1 \cap X_2)$.*

Proof. Consider the following portion of the relative $M-V$ sequence (III, 8.10)

$$H_{i+1}(Y, X_1 \cup X_2) \to H_i(Y, X_1 \cap X_2) \to H_i(Y, X_1) \oplus H_i(Y, X_2).$$

If $i > n$ then the outside terms vanish by assumption, therefore also the middle term. This proves (a). A similar argument works for (b). One considers the diagram

(6.8)
$$\begin{array}{ccccc}
0 = H_{n+1}(Y, X_1 \cup X_2) & \longrightarrow & H_n(Y, X_1 \cap X_2) & \xrightarrow{(j_{1*},\, -j_{2*})} & \\
\| & & \downarrow J & & \\
0 & \longrightarrow & \Gamma(X_1' \cup X_2', Y') & \xrightarrow[(j_1',\, -j_2')]{} & \\
& & & & \\
& \to & H_n(Y, X_1) \oplus H_n(Y, X_2) & \xrightarrow{(i_{1*},\, i_{2*})} & H_n(Y, X_1 \cup X_2) \\
& & \cong \downarrow J \oplus J & & \cong \downarrow J \\
& \to & \Gamma(X_1', Y') \oplus \Gamma(X_2', Y') & \xrightarrow[(i_1',\, i_2')]{} & \Gamma(X_1' \cap X_2', Y')
\end{array}$$

in which the first row is part of the relative Mayer-Vietoris sequence; j', i' in the second row denote restriction maps, and $X' = \mathbb{S}^n - X$. The diagram is commutative, by 6.3.

Clearly, $(j'_1, -j'_2)$ maps $\Gamma(X'_1 \cup X'_2, Y')$ monomorphically into the kernel of (i'_1, i'_2). Therefore we have a sequence of monomorphisms

(6.9) $\ker(i_{1*}, i_{2*}) \cong H_n(Y, X_1 \cap X_2) \xrightarrow{J} \Gamma(X'_1 \cup X'_2, Y') \xrightarrow{(j'_1, -j'_2)} \ker(i'_1, i'_2)$

whose composite is isomorphic (because the two vertical arrows of 6.8 on the right are isomorphic). Hence, 6.9 consists of isomorphisms. ∎ (In particular, the second row of 6.8 is exact.)

Proof of 6.4. We proceed in several steps. As before we abbreviate $X' = \mathbb{S}^n - X$, and we think of $\mathbb{S}^n$ as $\mathbb{R}^n \cup \{\infty\}$. We assume $n > 0$.

Step 1. $Y = \mathbb{S}^n$, $X = \mathbb{S}^n$ *or* $X = \mathbb{S}^n - P$ *or* $X = \emptyset$ (P=a point). If $X = \mathbb{S}^n$ then $H(Y, X) = 0 = \Gamma(\emptyset, \emptyset)$. The cases $X = \mathbb{S}^n - P$ and $X = \emptyset$ have been settled before (IV, 2); we have $H_n(\mathbb{S}^n, \mathbb{S}^n - P) \cong H_n \mathbb{S}^n \cong \Gamma(\mathbb{S}^n) \cong \Gamma(P)$ for $n > 0$.

Step 2. $Y = \mathbb{S}^n$, $X = \mathbb{S}^n - \square$ *where* $\square$ *is a closed rectilinear cube in* $\mathbb{R}^n$, $0 \leq \dim \square \leq n$.

If B is an open ball containing $\square$ and with center $P \in \square$ then $\mathbb{S}^n - P \simeq \mathbb{S}^n - B \simeq \mathbb{S}^n - \square$ (by radial deformation) hence $H(\mathbb{S}^n, \mathbb{S}^n - \square) \cong H(\mathbb{S}^n, \mathbb{S}^n - P)$. Also $\Gamma \square \cong \Gamma P$, so that step 2 reduces to step 1.

Step 3. $Y = \mathbb{S}^n$, $X = \mathbb{S}^n - F$ *where* F *is a finite union of cubes of a fixed lattice.* A *lattice* in $\mathbb{R}^n$ is given by n positive numbers $(\mu_1, \ldots, \mu_n)$; its cubes have the form

$$\square = \{x \in \mathbb{R}^n \mid m_i \mu_i \leq x_i \leq n_i \mu_i \text{ for all } i\}$$

with fixed $m_i \in \mathbb{Z}$, and $n_i = m_i$ or $n_i = m_i + 1$.

We proceed by induction on the number of cubes in F. Let $F_1 \subset F$ be a cube of maximal dimension and let F_2 denote the closure of $F - F_1$. Then we can apply Step 2 or the inductive hypothesis to F_1, F_2, and $F_1 \cap F_2$, i.e., 6.4(a), (b) hold for $(\mathbb{S}^n, \mathbb{S}^n - F_1)$, $(\mathbb{S}^n, \mathbb{S}^n - F_2)$ and

$$(\mathbb{S}^n, \mathbb{S}^n - F_1 \cap F_2) = (\mathbb{S}^n, (\mathbb{S}^n - F_1) \cup (\mathbb{S}^n - F_2)).$$

Therefore, by Lemma 6.7 they hold for

$$(\mathbb{S}^n, (\mathbb{S}^n - F_1) \cap (\mathbb{S}^n - F_2)) = (\mathbb{S}^n, \mathbb{S}^n - F_1 \cup F_2) = (\mathbb{S}^n, \mathbb{S}^n - F).$$

Step 4. $Y=\mathbb{S}^n$, *X open.* We can assume that $X\neq\emptyset$ (by Step 1), and then $\infty\in X$, i.e., $X'=\mathbb{S}^n-X\subset\mathbb{R}^n$. We first show that $J\colon H_n(\mathbb{S}^n,X)\to\Gamma X'$ is surjective. If $s\in\Gamma X'$ then the compact set X' decomposes into a finite number of disjoint compact pieces X'_k such that $s|X'_k$ is constant (n.b. s is locally constant). Let $\varepsilon>0$ be smaller than the minimal distance between any two pieces and choose a lattice L (see Step 3) whose cubes have diameter less than $\varepsilon/2$. Let F be the union of all cubes of L which meet X'. Then $X'\subset F$, and we can extend s to a function $t\in\Gamma F$ such that $t|\square$, for $\square\subset F$, is the constant $s(\square\cap X')$. By Step 3, there exists $y\in H_n(\mathbb{S}^n,\mathbb{S}^n-F)$ such that $Jy=t$, therefore, by naturality 6.3 of J, we have $J(i_*y)=t|X'=s$ where $i_*\colon H(\mathbb{S}^n,\mathbb{S}^n-F)\to H(\mathbb{S}^n,X)$. This proves surjectivity.

Let now $[z]\in H_i(\mathbb{S}^n,X)$, $i\geq n$; also assume $J[z]=0$ if $i=n$. We have to show $[z]=0$. The simplices of ∂z lie in $X=\mathbb{S}^n-X'$, and by compactness they avoid a whole neighborhood V of X'; thus we can consider the homology class $[z]_V$ of z in $H_i(\mathbb{S}^n,\mathbb{S}^n-V)$. Also, if $i=n$ we have $J([z]_V)|X'=0$ hence $J([z]_V)$ is zero in a whole neighborhood $W\subset V$ of X' (because it is locally constant); if $i>n$, put $W=V$. Choose a lattice (see Step 3) whose cubes have diameter less than *distance*$(X',\mathbb{S}^n-W)$, and let F be the union of all cubes which meet X'. Then $X'\subset F\subset W$, and $j_*([z]_F)=[z]$, where $j_*\colon H(\mathbb{S}^n,\mathbb{S}^n-F)\to H(\mathbb{S}^n,X)$; also $J([z]_F)=0$ if $i=n$. By Step 3 we know $[z]_F=0$, hence $[z]=0$ as asserted.

Step 5. $Y=\mathbb{S}^n$, $n>0$, $X\neq\mathbb{S}^n$ *an arbitrary neighborhood retract.* Let $U\neq\mathbb{S}^n$ be an open set of which X is a retract, $i\colon X\to U$, $r\colon U\to X$, $ri=\mathrm{id}$, hence $r_*i_*=\mathrm{id}$. For $p>n$ we have a commutative diagram

$$\begin{array}{ccc} H_p(\mathbb{S}^n,X) & \longrightarrow & H_p(\mathbb{S}^n,U)=\text{zero}, \quad \text{by Step 4} \\ \downarrow{\scriptstyle \partial_*\,\cong} & & \downarrow{\scriptstyle \partial_*\,\cong} \\ H_{p-1}X & \xrightarrow[\subset]{i_*} & H_{p-1}U, \end{array}$$

in which the ∂_* are isomorphic as follows from the homology sequence (see proof of 6.5). The diagram proves $H_p(\mathbb{S}^n,X)=0$ as asserted in 6.4 (a).

For part (b) we consider the diagram (6.6),

$$\begin{array}{ccccccc} 0\to H_n\mathbb{S}^n & \longrightarrow & H_n(\mathbb{S}^n,X) & \longrightarrow & \tilde{H}_{n-1}X & \to 0 \\ \| & & \downarrow{\scriptstyle J} & & \downarrow{\scriptstyle \tilde{J}} & \\ 0\to \Gamma\mathbb{S}^n & \longrightarrow & \Gamma X' & \longrightarrow & \tilde{\Gamma}X' & \to 0, \end{array}$$

where $\tilde{\Gamma}$ is Γ modulo the group of constant functions. The five lemma shows that J is isomorphic if and only if $\tilde{J}$ is isomorphic; we shall prove the latter. First we see from

$$
(6.10)\qquad
\begin{array}{ccc}
\tilde H_{n-1}X & \xrightarrow[\subset]{i_*} & \tilde H_{n-1}U \\
\Big\downarrow{\scriptstyle \tilde J} & & \Big\downarrow{\scriptstyle \tilde J\,\cong} \\
\tilde\Gamma X' & \xrightarrow{i'} & \tilde\Gamma U'
\end{array}
$$

that $\tilde J$ is monomorphic. Surjectivity is more delicate: Since $U \neq \mathbb{S}^n$ we can assume $U \subset \mathbb{R}^n$, so that we can speak of "segments" in U. For every $Q \in X'$, let V_Q consist of all points $P \in U$ such that the whole segment $\overline{P, r(P)}$ lies in $U-Q$. Clearly $X \subset V_Q \subset U-Q$, V_Q is open, and the inclusion $k_Q\colon V_Q \to U-Q$ is homotopic to the composite $V_Q \xrightarrow{r|V_Q} X \xrightarrow{i_Q} U-Q$; a homotopy is given by $P \mapsto (1-t)P + tr(P)$. It follows that $k_{Q*} = i_{Q*}(r|V_Q)_*$. For easier reference we record the whole situation in the following commutative diagram, in which all horizontal maps are induced by inclusions.

$$
\begin{array}{ccccccc}
\tilde H_{n-1}X & \xrightarrow{j_{Q*}} & \tilde H_{n-1}V_Q & \xrightarrow{k_{Q*}} & \tilde H_{n-1}(U-Q) & & i_{Q*}=k_{Q*}j_{Q*} \\
\Big\downarrow{\scriptstyle \tilde J} & & \Big\downarrow{\scriptstyle \tilde J\,\cong} & & \Big\downarrow{\scriptstyle \tilde J\,\cong} & & \\
\tilde\Gamma X' & \xrightarrow{j'_Q} & \tilde\Gamma V'_Q & \xrightarrow{k'_Q} & \tilde\Gamma(U-Q)' & \xrightarrow{l'_Q} & \tilde\Gamma U', \quad i'_Q=k'_Q j'_Q, \quad i'=l'_Q i'_Q.
\end{array}
$$

Let $\rho_Q = (r|V_Q)_*\,\tilde J^{-1} j'_Q\colon \tilde\Gamma X' \to \tilde H_{n-1}X$. Then

$$
(6.11)\qquad i'_Q \tilde J \rho_Q = i'_Q,
$$

because

$$
\begin{aligned}
i'_Q \tilde J \rho_Q &= i'_Q \tilde J (r|V_Q)_*\, \tilde J^{-1} j'_Q = \tilde J\, i_{Q*}(r|V_Q)_*\, \tilde J^{-1} j'_Q \\
&= \tilde J k_{Q*} \tilde J^{-1} j'_Q = \tilde J \tilde J^{-1} k'_Q j'_Q = i'_Q.
\end{aligned}
$$

Composing (6.11) with l'_Q gives $(i'\tilde J)\rho_Q = i'$. The right side of this does not depend on Q, and $i'\tilde J$ is monomorphic (see 6.10), hence $\rho = \rho_Q$ is independent of Q. We claim $\tilde J \rho = \mathrm{id}$, in particular $\tilde J$ is epimorphic.

We can identify $\tilde\Gamma X', \tilde\Gamma U' \ldots$ with $\Gamma(X', \infty), \Gamma(U', \infty) \ldots$ (every coset has a unique representative which is zero at ∞). Given $f \in \Gamma(X', \infty)$, then $i'_Q \tilde J \rho(f) = i'_Q(f)$ by 6.11, i.e., the two functions f and $\tilde J \rho(f)$ agree in $(U-Q)'$, in particular on Q. Since $Q \in X'$ is arbitrary, f and $\tilde J \rho(f)$ agree on X', as asserted.

Step 6. $X \subset Y \neq \mathbb{S}^n$ arbitrary neighborhood retracts in $\mathbb{S}^n$.

The homology sequence of $(\mathbb{S}^n, Y, X)$ contains the following bit: $H_{p+1}(\mathbb{S}^n, Y) \to H_p(Y, X) \to H_p(\mathbb{S}^n, X)$. If $p > n$ the outside terms vanish by Step 5, hence $H_p(Y, X) = 0$ as asserted in 6.4(a). For part (b) we consider the commutative diagram

$$\begin{array}{ccccccc} 0=H_{n+1}(\mathbb{S}^n, Y) & \longrightarrow & H_n(Y,X) & \longrightarrow & H_n(\mathbb{S}^n,X) & \longrightarrow & H_n(\mathbb{S}^n,Y) \\ & & \downarrow J & & \cong\downarrow J & & \cong\downarrow J \\ 0 & \longrightarrow & \Gamma(X',Y') & \longrightarrow & \Gamma X' & \longrightarrow & \Gamma Y' \end{array}$$

where the first row is part of the homology sequence and the second row is induced by inclusions. The two vertical arrows on the right are isomorphic by Step 5, hence $J\colon H_n(Y,X)\cong\Gamma(X',Y')$. ∎

6.12 Exercises. *1.* If X is a neighborhood retract in $\mathbb{R}^n$ then $\tilde{H}_{n-1}X\cong\Gamma_b(\mathbb{R}^n-X)=$ group of locally constant functions $\mathbb{R}^n-X\to\mathbb{Z}$ which vanish outside some bounded set. Hint: By 6.5, $\tilde{H}_{n-1}X\cong\tilde{\Gamma}(\mathbb{S}^n-X)\cong\Gamma(\mathbb{S}^n-X,\infty)$.

2. If $\alpha\colon\mathbb{S}^{n-1}\to\mathbb{R}^n$ is a mapping let $W_\alpha\in\Gamma_b(\mathbb{R}^n-\alpha\mathbb{S}^{n-1})$ denote the image of the generator of $\tilde{H}_{n-1}\mathbb{S}^{n-1}$ under the composition

$$\tilde{H}_{n-1}\mathbb{S}^{n-1}\xrightarrow{\alpha_*}\tilde{H}_{n-1}(\alpha\mathbb{S}^{n-1})\xrightarrow{\tilde{J}}\tilde{\Gamma}(\mathbb{S}^n-\alpha\mathbb{S}^{n-1})\cong\Gamma_b(\mathbb{R}^n-\alpha\mathbb{S}^{n-1}).$$

The value $W_\alpha(P)$ at a point $P\in\mathbb{R}^n-\alpha\mathbb{S}^{n-1}$ is called *winding number of* α *at* P. Discuss the formal properties of W and study some examples.

*3**. Prove: If A is an arbitrary subset of $\mathbb{S}^n$ whose complement $\mathbb{S}^n-A$ is not connected then every neighborhood of A contains a neighborhood U such that $\tilde{H}_{n-1}U\neq0$. Construct an example such that $\tilde{H}_{n-1}A=0$ but $\mathbb{S}^n-A$ is not connected, and another example where A is open, $\tilde{H}_{n-1}A=0$ but $\mathbb{S}^n-A$ is not arcwise connected (hint: use the graph of $\sin(1/x)$).

4. Let $A_n\subset\mathbb{R}^2\subset\mathbb{S}^2$ be the circle with radius $1/n$ and center $(0,1/n)$, and let $A=\bigcup_{n=1}^{\infty}A_n$. According to H. B. Griffith there are non-zero elements $a\in H_1A$ such that, for every neighborhood V of $(0,0)$ in A, a lies in the image of $H_1V\to H_1A$ (they can be thought of as being infinite products of commutators in the fundamental group π_1A). Show that $a\in\ker(\tilde{J}\colon H_1A\to\tilde{\Gamma}(\mathbb{S}^2-A))$, hence $J\colon H_2(\mathbb{S}^2,A)\to\Gamma(\mathbb{S}^2-A)$ is not monomorphic.

5. Construct a triad $(Y;X_1,X_2)$ in $\mathbb{S}^n$ for which the second row of 6.8 is not exact.

*6**. Is the group $H_{n-1}X$ always *free* (abelian) if X is a neighborhood retract in $\mathbb{S}^n$? The reader who wants to tackle this rather difficult problem is advised to consult Specker.

7. *Jordan Theorem, Invariance of Domaine*

A locally constant function $A \to H_n\mathbb{S}^n$ is constant on every component of A. The size of ΓA (see 6.2) therefore gives information about the number of components of A. More precisely

7.1 Lemma. *Let* $A \subset \mathbb{S}^n$. *The rank* (I, 2.29) *of* ΓA *equals* $c(A)$ = *number of components of* A (an integer or ∞; if we were to distinguish between different infinite cardinals the lemma would have to be formulated differently).

Proof. Let $A = A_1 \cup \cdots \cup A_r$ a decomposition of A into pairwise disjoint non-empty relatively open sets. The subgroup $\Gamma_r \subset \Gamma A$ of functions which are constant on each A_i is isomorphic to $\mathbb{Z} \oplus \mathbb{Z} \cdots \oplus \mathbb{Z} = r \cdot \mathbb{Z}$ hence $\operatorname{rank}(\Gamma) \geq \operatorname{rank}(\Gamma_r) = r$. If $c(A) = \infty$ we can make r arbitrarily large, hence $\operatorname{rank}(\Gamma) = \infty$. If $c(A) < \infty$ then we can find a decomposition with $r = c(A)$ and each A_i connected, hence $\Gamma_r = \Gamma$, hence $\operatorname{rank}(\Gamma) = r$. ∎

7.2 Proposition (Jordan Theorem). (a) *If* $X \subset \mathbb{S}^n$ *is homeomorphic to* $\mathbb{B}^n$ *then* $\mathbb{S}^n - X$ *is connected,* i.e., $c(\mathbb{S}^n - X) = 1$.

(b) *If* $X \subset \mathbb{S}^n$ *is homeomorphic to* $\mathbb{S}^{n-1}$ *then* $\mathbb{S}^n - X$ *has two components,* $c(\mathbb{S}^n - X) = 2$.

Proof. We shall see in a moment (remark after 7.3) that X is a neighborhood retract. Therefore, by 7.1 and 6.5,

$$c(\mathbb{S}^n - X) = \operatorname{rank}\Gamma(\mathbb{S}^n - X) = 1 + \operatorname{rank}\tilde{\Gamma}(\mathbb{S}^n - X) = 1 + \operatorname{rank}\tilde{H}_{n-1}X,$$

as asserted. ∎

7.3 Lemma (compare 8.5). *Let* $A \subset N$ *be a closed subset of the normal space* N, *and let* $f\colon A \to X$ *be a continuous map.*

(a) *If* $X \approx \mathbb{B}^n$ *then* f *admits an extension* $h\colon N \to X$.

(b) *If* $X \approx \mathbb{S}^{n-1}$ *then* f *admits an extension* $g\colon V \to X$ *to some open neighborhood* V *of* A *in* N.

Assuming the lemma, we can take $N = \mathbb{S}^n$, $A = X$ (as in 7.2), and $f = \mathrm{id}$. The extension which the lemma guarantees is then a retraction, and so X is a neighborhood retract.

To prove the lemma remark that $\mathbb{B}^n \approx [0,1] \times \cdots \times [0,1] = [0,1]^n$. A map into $\mathbb{B}^n$ is then an n-tuple of functions with values in $[0,1]$. Such functions can always be extended from A to N by Tietze's extension lemma; this proves (a). To prove (b) view f as a map into $\mathbb{B}^n$ (since

$\mathbb{S}^{n-1}\subset\mathbb{B}^n$), choose an extension $h\colon N\to\mathbb{B}^n$, put $V=h^{-1}(\mathbb{B}^n-\{0\})$, and $g(z)=h(z)/\|h(z)\|$. ∎

7.4 Proposition (Invariance of Domain). *If $X\subset\mathbb{R}^n$ is open and $f\colon X\to\mathbb{R}^n$ is an injective continuous map then $f(X)\subset\mathbb{R}^n$ is also open. In other words, every injective continuous map $f\colon X\to\mathbb{R}^n$ is open.*

7.5 Corollary. *If $X\subset\mathbb{R}^n$ is open, $X\neq\emptyset$, and $g\colon X\to\mathbb{R}^m$ is an injective continuous map then $m\geq n$.*—Because otherwise the map $f\colon X\to\mathbb{R}^m\times\mathbb{R}^{n-m}\approx\mathbb{R}^n$, $f(x)=(g(x),0)$, although being injective, would have a non-open image $f(X)\subset\mathbb{R}^m\times\mathbb{R}^{n-m}\approx\mathbb{R}^n$. This refines our earlier result that euclidean spaces of different dimension cannot be homeomorphic.

Proof of 7.4. Let $P\in X$, choose $\varepsilon>0$ such that $B^n=\{x\in\mathbb{R}^n\,|\,\|P-x\|\leq\varepsilon\}$ is contained in X, and put $S^{n-1}=\{x\in B^n\,|\,\|P-x\|=\varepsilon\}$. Since f is injective and B^n is compact, $fB^n\approx B^n\approx\mathbb{B}^n$ and $fS^{n-1}\approx\mathbb{S}^{n-1}$. By 7.2, $\mathbb{R}^n-fB^n$ is connected, and $\mathbb{R}^n-fS^{n-1}=f(B^n-S^{n-1})\cup(\mathbb{R}^n-fB^n)$ has two components. Since $f(B^n-S^{n-1})$ is connected it must be a component of $\mathbb{R}^n-fS^{n-1}$, and since this set is open, $f(B^n-S^{n-1})$ must also be an open subset of $\mathbb{R}^n$. Now $fP\in f(B^n-S^{n-1})\subset fX$ shows that every point in fX has a neighborhood in fX, i.e. fX is open. ∎

7.6 Exercises. *1.* There is no injective continuous map $\mathbb{S}^n\to\mathbb{R}^n$.

2. Let $C_k\subset\mathbb{R}^2$ denote the circle with radius $1/k$ and center $(0,1/k)$, and let $A_r=\bigcup_{k=1}^{r-1}C_k$, $r=2,3,\ldots$. Prove: If $A\subset\mathbb{S}^2$ is homeomorphic with A_r, $r\leq\infty$, then $\mathbb{S}^2-A$ has r components.

3. If $\alpha\colon\mathbb{S}^{n-1}\to\mathbb{R}^n$ is an injective continuous map then the component of $\mathbb{S}^n-\alpha(\mathbb{S}^{n-1})$ which contains ∞ is called the *exterior* of $\alpha\mathbb{S}^{n-1}$, the other component is called the *interior*. Show that the winding number (6.12, Exerc. 2) of α is ±1 at every interior point and 0 at every exterior point.

8. Euclidean Neighborhood Retracts (ENRs)

The results of §6 suggest a more careful study of subsets of euclidean space which are neighborhood retracts. We deduce a few simple results about these sets here. We show that the property of being a neighborhood retract (of some $\mathbb{R}^n$) is topologically invariant, and we provide some criteria for a space to be (homeomorphic with) a euclidean neighborhood retract.

Clearly open sets are neighborhood retracts, and neighborhood retracts of neighborhood retracts are neighborhood retracts. Not every subset of $\mathbb{R}^n$ is a neighborhood retract: it has to be locally closed (8.1) and locally contractible (8.7), and these properties are also sufficient (8.12).

8.1 Proposition. *If $X \subset \mathbb{R}^n$ is a neighborhood retract then X is of the form $X = C \cap O$ where C is closed and O is open.*

Proof. Let O be an open set of which X is a retract. The retraction can be viewed as a map $r\colon O \to O$, and $X = \{P \in O \mid rP = P\}$ is clearly closed in O, hence $X = \overline{X} \cap O$. ∎

Sets of the form $C \cap O$ are called *locally closed.* They can always be realized as *closed* subsets of euclidean space; more precisely,

8.2 Lemma. *Every locally closed subset X of $\mathbb{R}^n$ is homeomorphic with a closed subset of $\mathbb{R}^{n+1}$.*

Proof. If $O \subset \mathbb{R}^n$ is an open set then

$$j\colon O \to \mathbb{R}^n \times \mathbb{R}, \quad j(P) = (P, 1/d(P, \mathbb{R}^n - O)), \quad d = \text{distance},$$

is an embedding of O into $\mathbb{R}^{n+1}$ (the projection $(P, t) \mapsto P$ is an inverse of j) whose image is closed; indeed $jO = \{(Q, t) \in \mathbb{R}^n \times \mathbb{R} \mid t \cdot d(Q, \mathbb{R}^n - O) = 1\}$. If $X \subset O$ is closed in O then $jX \approx X$ is closed in jO, and hence in $\mathbb{R}^{n+1}$. ∎

8.3 Lemma. *The following properties of $X \subset \mathbb{R}^n$ are equivalent.*

(i) *X is locally closed, i.e. $X = C \cap O$ where C is closed and O is open.*

(ii) *Every point $P \in X$ has an open neighborhood U in $\mathbb{R}^n$ such that $X \cap U$ is closed in U.*

(iii) *Every point $P \in X$ has a compact neighborhood in X, i.e., X is locally compact.*

Since (iii) is an intrinsic property of X this implies

8.4 Corollary. *If $X \subset \mathbb{R}^m$ is locally closed and $Y \subset \mathbb{R}^n$ is homeomorphic with X then Y is locally closed.* ∎

Proof of 8.3. (iii) ⇒ (ii): Given $P \in X$, let $K \subset X$ be a compact neighborhood in X, hence $K = X \cap V$ for some neighborhood V in $\mathbb{R}^n$. Let $U = \mathring{V}$, then $X \cap U = K \cap U$ is closed in U.

(ii) ⇒ (i): For every $P \in X$ and neighborhood $U = U_P$ as in (ii) we have $X \cap U = \overline{X} \cap U$, hence $X = X \cap (\bigcup_P U_P) = \bigcup_P (\overline{X} \cap U_P) = \overline{X} \cap (\bigcup_P U_P)$, which proves (i).

(i) $\Rightarrow$ (iii): Given $P \in X = C \cap O$, let $K \subset O$ be a compact neighborhood of P in $\mathbb{R}^n$ then $K \cap X = K \cap C$ is a compact neighborhood of P in X. ∎

Remark. In 8.3, $\mathbb{R}^n$ can replaced by any locally compact space (cf. Bourbaki I, 9.7).

8.5 Proposition and Definition. *If $X \subset \mathbb{R}^m$ is a neighborhood retract and $Y \subset \mathbb{R}^n$ is homeomorphic with X then Y is a neighborhood retract.* I.e., the property of being a neighborhood retract of a euclidean space is intrinsic, it does not depend on the embedding. We define therefore: A topological space Y is called a *euclidean neighborhood retract* (*ENR*) if a neighborhood retract $X \subset \mathbb{R}^n$ exists which is homeomorphic with Y. Any other $X' \subset \mathbb{R}^k$ which is homeomorphic with Y will then also be a neighborhood retract. For instance, $\mathbb{S}^{n-1}$ is a retract of $\mathbb{R}^n - \{0\}$, and $\mathbb{B}^n$ is a retract of $\mathbb{R}^n$. *Therefore any subset of $\mathbb{R}^k$ which is homeomorphic with $\mathbb{S}^{n-1}$ or $\mathbb{B}^n$ is a neighborhood retract* (cf. 7.3).

Proof. By assumption on X we have $X \xrightarrow{i} U \xrightarrow{r} X$, $ri = \mathrm{id}$, where $j: U \xrightarrow{\subset} \mathbb{R}^m$ is open; in particular, X is locally closed (8.1). Further $h: Y \approx X$, hence (8.4) Y is locally closed, i.e., $Y = C \cap O$ is a closed subset of some open set O. By Tietze's extension lemma there exists a map $g: O \to \mathbb{R}^m$ such that $g|Y = jih$. Then the set $g^{-1}U$ is open (in O hence in $\mathbb{R}^n$) and $h^{-1}rg: g^{-1}U \to Y$ is a retraction. ∎

8.6 Proposition. *Let X be an ENR. If $f_0, f_1: Y \to X$ are mappings and $B \subset Y$ is a subset such that $f_0|B = f_1|B$ then there exists an open neighborhood W of B in Y and a homotopy $\Theta: f_0|W \simeq f_1|W$ with $\Theta_t|B = f_0|B$ for all t.*

Proof. We have $X \xrightarrow{i} O \xrightarrow{r} X$ where O is open in $\mathbb{R}^n$ and $ri = \mathrm{id}$. Let $W \subset Y$ consist of all points $y \in Y$ such that the whole segment from $if_0(y)$ to $if_1(y)$ lies in O. Clearly, W is open and $B \subset W$. Define $\Theta: W \times [0,1] \to X$ by $\Theta(y,t) = r[(1-t)\, if_0(y) + t\, if_1(y)]$. ∎

For instance, the two projections $f_0, f_1: X \times X \to X$ agree on the diagonal $B = \{(x_1, x_2) \in X \times X \mid x_1 = x_2\}$; the conclusion of 8.6 in this special case is called *uniform local contractibility*. It easily implies the general result 8.6 (exercise!).

8.7 Corollary. *If $B \subset X$ are ENRs then B is a neighborhood retract in X (obviously). If $r: V \to B$ is such a retraction then B has an open neighborhood W in V such that $i(r|W) \simeq j$ where $i: B \to V$, $j: W \to V$ are inclusions.*

For instance, if B is a point this asserts that X is locally contractible. In general, it asserts that B is "almost a neighborhood-deformation-retract". For the proof one can assume that V is open, hence an ENR, and one can apply 8.6 to $f_0 = ir, f_1 = \mathrm{id}_V$. ∎

If we want to know whether a given space X is an ENR we can first ask whether X can be embedded into some $\mathbb{R}^k$, and then whether a given subspace Y of $\mathbb{R}^k$ is a neighborhood retract. We give useful, although crude, answers to these questions (8.8, 8.10), and we point out some finer results (8.9, 8.11, 8.12).

8.8 Proposition. *If a Hausdorff space X can be covered by finitely many locally compact open sets X_i, $i=1,2,\dots,r$, such that each X_i is homeomorphic with a subset of a euclidean space then X itself is homeomorphic with a closed subset of some euclidean space.*

Proof. Choose embeddings $h_i\colon X_i \to \mathbb{R}^{m_i}$ with closed image $h_i X_i$; this is possible by 8.3 and 8.2. Define

$$H_i\colon X \to \mathbb{S}^{m_i} = \mathbb{R}^{m_i} \cup \{\infty\}, \qquad H_i | X_i = h_i, \qquad H_i(X - X_i) = \infty.$$

This is continuous: If $A \subset \mathbb{S}^{m_i}$ is a closed set and if $\infty \notin A$ then $H_i^{-1}(A) \approx A \cap h_i X_i$ is compact, hence closed. If $\infty \in A$ then $H_i^{-1}(\mathbb{S}^{m_i} - A) = h_i^{-1}(\mathbb{S}^{m_i} - A)$ is open in X_i, hence open in X, and therefore its complement, namely $H_i^{-1}(A)$ is closed. It follows that

$$H = \{H_i\}\colon X \to \prod_{i=1}^{r} \mathbb{S}^{m_i} \subset \mathbb{R}^N, \qquad HP = (H_1 P, H_2 P, \dots, H_r P),$$

is continuous $(N = r + \Sigma m_i)$; moreover, since H_i is an embedding of X_i, H is an embedding of $\bigcup_i X_i = X$. Because $HX \approx X$ is locally compact it is locally closed in $\mathbb{R}^N$ (8.3), hence closed in $\mathbb{R}^{N+1}$ (8.2). ∎

8.9 Remark. If a Hausdorff space X is covered by a sequence X_i, $i=1,2,\dots$, of locally compact open sets such that each X_i is homeomorphic with a subset of a *fixed* euclidean space then the conclusion of 8.8 still holds. Indeed, X is then a countable union of compact sets and has finite covering dimension (cf. Hurewicz-Wallman, Chapter V), hence the proof of 8.8 can be adapted as is done, for instance, by Bos. Finer embedding theorems can be found in Hurewicz-Wallman, Chapter V.

8.10 Proposition (Compare Hanner, Theorem 3.3). *If a Hausdorff space X is a finite union of ENR's, $X = \bigcup_{i=0}^{r} X_i$, and if each X_i is open in X then X itself is an ENR.*

Proof. By 8.8 we can assume that X is a closed subset of $\mathbb{R}^n$, and by induction we can assume $r=1$, i.e., $X = X_0 \cup X_1$. Let then $r_i\colon O_i \to X_i$, $i=0,1$, be neighborhood retractions (O_i open in $\mathbb{R}^n$). Put

$O_{01}=r_0^{-1}(X_0\cap X_1)\cap r_1^{-1}(X_0\cap X_1)$, then $r_0, r_1\colon O_{01}\to X_0\cap X_1$ are neighborhood retractions. Since $X_0\cap X_1$ is an ENR (it is open in X_0) O_{01} contains an open neighborhood U_{01} of $X_0\cap X_1$ in which r_0, r_1 are homotopic retractions (by 8.6), say $r_t\colon U_{01}\to X_0\cap X_1$, $r_t|X_0\cap X_1=\mathrm{id}$, $0\le t\le 1$.

Let $U_0\subset O_0$, $U_1\subset O_1$ be open neighborhoods of $X-X_1$, $X-X_0$, such that $\bar{U}_0\cap\bar{U}_1=\emptyset$ (this is possible because $X-X_i$ is closed), and let $\tau\colon \mathbb{R}^n\to[0,1]$ be a continuous function such that $\tau|U_0=0$, $\tau|U_1=1$ (e.g., $\tau P=d(P,U_0)/(d(P,U_0)+d(P,U_1))$, where $d=$distance). Put $U=U_0\cup U_1\cup U_{01}$. This is an open neighborhood of X and the following $\rho\colon U\to X$ is a retraction:

$$\rho|U_0=r_0|U_0,\quad \rho|U_1=r_1|U_1,\quad \rho(P)=r_{\tau P}(P)\quad \text{if } P\in U_{01}. \blacksquare$$

8.11 Remark. Just as 8.8 extends to countable unions $X=\bigcup_{i=1}^{\infty}X_i$, so does 8.10. One assumes that each X_i is homeomorphic with a neighborhood retract of some fixed $\mathbb{R}^n$, hence $X\subset\mathbb{R}^k$, by 8.9. One can arrange the $\{X_i\}$ to be locally finite and then prove the result by infinite but locally finite iteration of the argument for 8.10. A finer result, essentially due to Borsuk (cf. also Hanner, Theorems 5.1 and 4.2; or Kuratowski, Chapter VII, §48, IV) is as follows.

8.12 Proposition.

If $X\subset\mathbb{R}^n$ is locally compact and locally contractible then X is a neighborhood retract, hence an ENR.

We shall only sketch the proof. Since X is locally compact we can assume it is *closed* in $\mathbb{R}^n$ (8.3, 8.2). Decompose $\mathbb{R}^n-X$ into convex cells; for more precision, take a cubical lattice L in $\mathbb{R}^n$ (cf. V, 3.4), and successive refinements $L', L''\ldots$ of it, say by halving the generating vectors of $L, L'\ldots$. Among the open n-cubes of $L, L',\ldots$ consider those whose closure lies in $\mathbb{R}^n-X$ and which are maximal in this respect; call them *admissible*. Their closures cover $\mathbb{R}^n-X$ and intersect only on lower dimensional faces. An (open) k-cube of $L, L',\ldots$ with $0\le k<n$ is called admissible if it is the face of an admissible n-cube of the same lattice and is maximal in this respect. Every point of $\mathbb{R}^n-X$ then lies in an admissible cube, and has a neighborhood which meets only finitely many admissible cubes.

For every k from 0 to n we shall now define a subset A_k of $\mathbb{R}^n-X$ and a map $\rho_k\colon A_k\to X$ such that A_k is the union of A_{k-1} with certain admissible k-cubes, and $\rho_k|A_{k-1}=\rho_{k-1}$. For A_0 we take the set of all admissible 0-cubes (vertices), and let $\rho_0(a)$ be any point in X whose distance from $a\in A_0$ is minimal. Let A_k be the union of A_{k-1} with all admissible k-cubes e whose boundary $\bar{e}-e$ lies in A_{k-1} and such that ρ_{k-1} can be extended to a map $A_{k-1}\cup e\to X$. Choose an extension $\rho_e\colon A_{k-1}\cup e\to X$ such that the diameter of $\rho_e(\bar{e})$ is essentially minimal ($<$twice the inf of all diameters of extensions), and define ρ_k by $\rho_k|A_{k-1}\cup e=\rho_e$. Finally, put $V=A_n\cup X$, and define $\rho\colon V\to X$, by $\rho|A_n=\rho_n$, $\rho|X=\mathrm{id}$. *Claim:* ρ is a neighborhood retraction. It is clear that $\rho|A_n=\rho_n$ is continuous (because the set of admissible cubes is locally finite). If $P\in X$, and W is a spherical neighborhood of P, choose spherical neighborhoods $W=W_{2n}\supset W_{2n-1}\supset\cdots\supset W_1\supset W_0$ of P in X such that W_j is contractible in W_{j+1} and has radius at most one tenth as big as W_{j+1}. Let U be a spherical neighborhood of P in $\mathbb{R}^n$ whose radius is at most one tenth that of W_0. Define $U_0 = A_0 \cap U$ and, inductively, let U_k be the union of U_{k-1} with all admissible k-cubes in U whose boundary lies in U_{k-1}. By minimality, $\rho_0(U_0) \subset W_0$. Assume by induction that ρ_k is defined on U_k, and $\rho(U_k) \subset W_{2k}$. If e is a $(k+1)$-cube of U_{k+1} then ρ_k extends to a map $U_k \cup e \to W_{2k+1} \subset X$ because W_{2k} is contractible in W_{2k+1}, hence ρ_{k+1} is defined on e; further, $\rho_{k+1}(\bar{e}) \subset W_{2k+2}$ by minimality. With $k=n$ we see that ρ is defined on U_n, and $\rho(U_n) \subset W$. Now $U_n \cup X$ is a neighborhood of P in $\mathbb{R}^n$ (admissible cubes which come close to P lie in U_n), hence ρ is defined in a neighborhood of P, and is continuous at P. ∎

8.13 Exercises. *1.* Let X be an ENR. Show that for every normal space Y, closed subset A of Y, and every map $f: A \to X$ there exists an extension of f to a neighborhood of A (in Y). Any space X with this property is called *absolute neighborhood retract* (=ANR), so ENR ⇒ ANR. Conversely, locally compact separable metric ANR's of finite dimension are ENR's (compare Hurewicz-Wallman, Chapter V).

2. Let X be an ENR (ANR). Show that for every binormal space Y (i.e., $Y \times [0,1]$ is normal), closed subspace A of Y, and every pair of maps $F_0, F_1: Y \to X$ such that $F_0|A \simeq F_1|A$ there exists a neighborhood V of A such that $F_0|V \simeq F_1|V$; in fact, any homotopy $F_0|A \simeq F_1|A$ can be extended to a neighborhood of A.

3.* A pair of spaces $Y \subset X$ is said to have the *homotopy extension property* (HEP) if the following holds: Given any map $F: X \to Z$ and any homotopy $d_t: Y \to Z$, $0 \leq t \leq 1$, such that $d_0 = F|Y$ there exists a homotopy $D_t: X \to Z$ such that $D_0 = F$ and $D_t|Y = d_t$. If X is an ENR show that $Y \subset X$ has the HEP if and only if Y is a closed neighborhood retract in X.

4. A space X is called *locally n-connected* if every neighborhood V of every point $P \in X$ contains a neighborhood W such that any map $\varphi: \mathbb{S}^j \to W$, $j \leq n$, admits an extension $\Theta: \mathbb{B}^{j+1} \to V$, $\Theta|\mathbb{S}^j = \varphi$. Clearly, locally contractible spaces are locally n-connected for all n. Check that the proof of 8.12 only assumes X to be locally $(n-1)$-connected (and locally compact); these properties of $X \subset \mathbb{R}^n$ therefore imply local contractibility.

5.* If $A \subset X$ are ENRs and A is compact then X/A (obtained from X by identifying all of A to one point) is also an ENR. Hint: Choose a closed embedding $h: X - A \to \mathbb{R}^n$ and extend it to a map $H: X \to \mathbb{S}^n$ where $HA = \infty$. If X is compact this induces an embedding $X/A \subset \mathbb{S}^n$, and 8.12 implies that X/A is a neighborhood retract. If X is not compact, one first embeds V/A where V is a compact neighborhood of A.

6.* If $A \subset X$ are ENRs and A is a closed subset of X then the projection map induces an isomorphism $H(X,A) \cong \tilde{H}(X/A)$. This can be shown using 8.7 and excision arguments. A more adequate proof uses limits and excision (compare VIII, 6.12 and 6.20).

Chapter V

Cellular Decomposition and Cellular Homology

1. Cellular Spaces

It is often possible to decompose a space X whose homology one wants to compute into simple pieces whose homology properties are known, and thereby deduce information about HX. An instructive example is the decomposition of a suspension into two cones (III, 8.16, example 3). In this section we discuss a general class of decompositions, "cellular" ones, and show how they can be used to simplify the computation of HX. The most important examples of cellular decompositions are CW-decompositions which will be studied in the succeeding sections.

1.1 Definition. A *filtration* of a topological space X is a sequence of subspaces $X^n \subset X$, $n \in \mathbb{Z}$, such that $X^n \subset X^{n+1}$ for all n. A filtration is called *cellular* if

(i) $H_i(X^n, X^{n-1}) = 0$ for $i \neq n$;
(ii) $SX = \bigcup_{n \in \mathbb{Z}} SX^n$,

i.e. every singular simplex of X lies in some X^n. In particular $X = \bigcup X^n$ (because $S_0 X = \bigcup S_0 X^n$). A space together with a cellular filtration is called a *cellular space.*

If X, Y are cellular spaces a *cellular map* $f: X \to Y$ is a continuous map such that $f(X^n) \subset Y^n$ for all n. Cellular spaces and maps then form a category.

For instance, if X is a space and $\pi: X \to \mathbb{R}$ a continuous function then $X^n = \{x \in X \mid \pi(x) \leq n\}$ defines a filtration. Condition (ii) is satisfied, but (i) requires additional assumptions on π. This type of example is important in differential topology, in particular in Morse theory (cf. Milnor 1963). Other examples will be given in §3.

1.2 Definition. For every cellular (or even filtered) space X put $W_n X = H_n(X^n, X^{n-1})$, and let $\partial_n: W_n X \to W_{n-1} X$ denote the composite

$$H_n(X^n, X^{n-1}) \xrightarrow{\partial_*} H_{n-1} X^{n-1} \to H_{n-1}(X^{n-1}, X^{n-2}).$$

Then $\partial_{n-1}\partial_n=0$ because already the composite

$$H_{n-1}X^{n-1}\to H_{n-1}(X^{n-1},X^{n-2})\xrightarrow{\partial_*}H_{n-2}X^{n-2}$$

vanishes. Therefore $WX=\{W_nX,\partial_n\}_{n\in\mathbb{Z}}$ is a complex, called the *cellular complex* of X. A cellular map $f\colon X\to Y$ clearly induces a chain map $Wf\colon WX\to WY$, and W thus becomes a covariant functor from cellular spaces and-maps to the category $\partial\mathscr{A}\mathscr{G}$ of complexes.

1.3 Proposition. *For cellular spaces X there is a natural isomorphism*

$$\Theta\colon HWX\cong H(X,X^{-1}).$$

1.4 Remarks. It looks as if X^{-1} played a very special role in 1.3. But $H(X,X^{-1})\cong H(X,X^{-2})\cong H(X,X^{-3})\cong\cdots$, as follows from the homology sequence of the appropriate triples because $H(X^n,X^{n-1})=0$ for $n<0$. In many examples X^{-1} will be empty.—

The definition of WX applies to arbitrary filtered spaces (not only cellular ones) but, in general, $HWX\not\cong H(X,X^{-1})$. Certain relations between the groups $\{H_p(X^q,X^{q-1})\}$ and HX, however, always exist and are usually expressed by the *spectral sequence* associated with the *exact couple* $\{H_p(X^q,X^{q-1}),H_pX^q\}$, or the *filtered complex* $\{SX^p\}$; cf. Hu 1959. The assumption of cellularity implies that the spectral sequence both *converges* and *degenerates*, and the following proof of 1.3 is an extract of a standard spectral sequence argument (compare, for instance, Godement, I.4.4).

1.5 Lemma. $H_n(X^p,X^q)=0$ for $p\geq q\geq n$ or $n>p\geq q$.

Proof by induction on $p-q$. For $p-q=0$ the assertion is trivial. For $p-q>0$ the homology sequence of the triple (X^p,X^{q+1},X^q) contains the following portion

$$H_n(X^{q+1},X^q)\to H_n(X^p,X^q)\to H_n(X^p,X^{q+1}).$$

The left term is zero by 1.1 (i), the right term by induction, hence also the middle term, as asserted. ∎

1.6 Lemma. $H_n(X,X^q)=0$ for $q\geq n$.

1.7 Corollary. $H_n(X^q,X^r)\cong H_n(X,X^r)$ provided $q>n$ and $q\geq r$.

The corollary follows from 1.6 and the homology sequence of the triple (X,X^q,X^r).

Proof of 1.6. Let $[z]\in H_n(X, X^q)$ where $z\in S_n X$ is a representative (relative) cycle. Because $SX=\bigcup_p SX^p$ there exists $p\geq q$ such that $z\in S_n X^p$, hence $[z]\in \mathrm{im}[H_n(X^p, X^q)\to H_n(X, X^q)]$, and this group is zero by 1.5. ∎

Proof of 1.3. Let $k\leq n-2$, and consider the diagram

$$\begin{array}{ccccccc}
 & & H_{n+1}(X^{n+1}, X^n) & & & & 0 \\
 & & \downarrow{\scriptstyle \partial_*} & \searrow{\scriptstyle \partial_{n+1}} & & & \downarrow \\
(1.8)\quad 0 & \longrightarrow & H_n(X^n, X^k) & \xrightarrow[i_*]{} & H_n(X^n, X^{n-1}) & \xrightarrow{\partial_*} & H_{n-1}(X^{n-1}, X^k) \\
 & & \downarrow & & & \searrow{\scriptstyle \partial_n} & \downarrow{\scriptstyle j_*} \\
 & & H_n(X^{n+1}, X^k) & & & & H_{n-1}(X^{n-1}, X^{n-2}) \\
 & & \downarrow & & & & \\
 & & 0 & & & &
\end{array}$$

where the two columns and the middle row are portions of the exact homology sequence of the appropriate triples; the zeros which appear are justified by Lemma 1.5. The two triangles are commutative by naturality of ∂_*. Now

$$\begin{aligned}
H_n(X, X^k) &\cong H_n(X^{n+1}, X^k) && \text{by 1.7}\\
&\cong H_n(X^n, X^k)/\mathrm{im}(\partial_*) && \text{because the left column is exact}\\
&\cong \mathrm{im}(i_*)/\mathrm{im}(i_*\partial_*) && \text{because } i_* \text{ is monomorphic}\\
&= \ker(\partial_*)/\mathrm{im}(\partial_{n+1}) && \text{because the row is exact}\\
&= \ker(j_*\partial_*)/\mathrm{im}(\partial_{n+1}) && \text{because } j_* \text{ is monomorphic}\\
&= \ker(\partial_n)/\mathrm{im}(\partial_{n+1}) = H_n WX.
\end{aligned}$$

Thus $H_n WX\cong H_n(X, X^{-1})$ if $n>0$, $H_0 WX\cong H_0(X, X^{-2})\cong H_0(X, X^{-1})$, and $H_n WX=0=H_n(X, X^{-1})$ for $n<0$ because then $W_n X=0$. ∎

It is sometimes useful to have a description of the isomorphism Θ: $H_n WX\cong H(X, X^{-1})$ in terms of representative chains. This is easily extracted from the proof of 1.3; reading the sequence of isomorphisms there from bottom to top one finds

1.9 Proposition. *If* $y\in H_n WX$, $n\geq 0$, *is represented by*

$$z\in Z_n WX\subset H_n(X^n, X^{n-1})$$

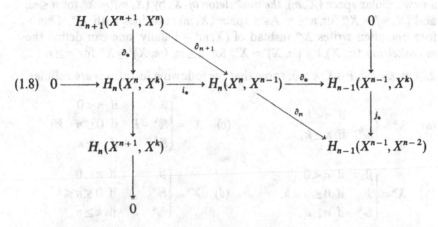

then $z \in H_n(X^n, X^{n-1})$ *has a representative* $\zeta \in S_n X^n$ *with* $\partial\zeta \in S_n(X^{-1})$ *(this uses exactness of the row* 1.8) *hence* ζ *is an n-cycle of* X mod X^{-1}. *Its homology class* $[\zeta] \in H_n(X, X^{-1})$ *agrees with* $\Theta(y)$. ∎

1.10 Exercises. *1.* For every cellular space X and integer m one defines a new cellular space (X, m), *the m-skeleton of* X, by $(X,m)^n = X^n$ for $n \leq m$, and $(X,m)^n = X^m$ for $n \geq m$. As a space (X, m) coincides with X^m. Therefore one often writes X^m instead of (X, m).—Dually, one can define the *m-coskeleton* (m, X) by $(m,X)^n = X^m$ for $n \leq m$, $(m,X)^n = X^n$ for $n \geq m$.

2. Let $X = \mathbb{S}^k$, $P \in X$, $k > 0$; show that the following filtrations are cellular.

$$(a)\quad X^n = \begin{cases} P & \text{if } n<k \\ \mathbb{S}^k & \text{if } n \geq k, \end{cases} \qquad (c)\quad X^n = \begin{cases} \emptyset & \text{if } n<0 \\ \mathbb{S}^k - P & \text{if } 0 \leq n < k \\ \mathbb{S}^k & \text{if } n \geq k, \end{cases}$$

$$(b)\quad X^n = \begin{cases} \emptyset & \text{if } n<0 \\ P & \text{if } 0 \leq n < k \\ \mathbb{S}^k & \text{if } n \geq k, \end{cases} \qquad (d)\quad X^n = \begin{cases} \emptyset & \text{if } n<0 \\ \mathbb{S}^n & \text{if } 0 \leq n \leq k \\ \mathbb{S}^k & \text{if } k \geq n, \end{cases}$$

where $\mathbb{S}^n = \{x \in \mathbb{S}^k \mid x_i = 0 \text{ for } n < i \leq k\}$.

3. If X is a cellular space, define $V_n X = \{\xi \in S_n X^n \mid \partial\xi \in SX^{n-1}\}$. Show that $VX = \{V_n X\}$ is a subcomplex of SX containing SX^{-1} and that

$$i_*\colon H(VX/SX^{-1}) \cong H(SX/SX^{-1}) = H(X, X^{-1}).$$

Define $p\colon V_n X/S_n X^{-1} \to W_n X = H(X^n, X^{n-1})$ by passage to homology classes. Show that p is a chain map and $p_*\colon H(VX/SX^{-1}) \cong HWX$. Prove that the isomorphism Θ of 1.3 coincides with $i_* \, p_*^{-1}$ (compare Schubert; IV, 3.4).

Corollary. *If* WX *is a free complex then there exists a homotopy equivalence* $\vartheta\colon WX \simeq S(X, X^{-1})$ *which is natural up to homotopy.*

2. *CW-Spaces*

In homology theory the most useful cellular decompositions are *CW*-decompositions, as introduced by J. H. C. Whitehead 1949. Their role here is essentially that of a tool for computation; they are much more basic for other parts of topology, in particular for homotopy theory. In this § we discuss *CW*-spaces from the point of view of general topology; their homological properties will be studied in § 3.

2.1 Definition. Let X be a Hausdorff space. A *CW-decomposition* of X is a set $\mathscr{E}$ of subspaces of X with the following properties (i)–(iv).

(i) $X=\bigcup_{e\in\mathscr{E}} e,\quad e\neq e' \Rightarrow e\cap e'=\emptyset$,

i.e., $\mathscr{E}$ is a covering of X by pairwise disjoint sets.

(ii) *Every $e\in\mathscr{E}$ is homeomorphic to some euclidean space $\mathbb{R}^{|e|}$*

The number $|e|$ is well determined (by invariance of dimension IV, 2.3); it is called the *dimension* of e. The sets $e\in\mathscr{E}$ which are homeomorphic with $\mathbb{R}^n$ are the *n-cells*, and the union $X^n=\bigcup_{|e|\leq n} e$ is the *n-skeleton* of the CW-decomposition.

(iii) *For every n-cell $e\in\mathscr{E}$ there exists a continuous map Φ_e: $(\mathbb{B}^n, \mathbb{S}^{n-1})\to(X^{n-1}\cup e, X^{n-1})$ such that Φ_e: $\mathbb{B}^n-\mathbb{S}^{n-1}\approx e$.* As usual $\mathbb{B}^n=\{x\in\mathbb{R}^n \mid \|x\|\leq 1\}$ $=n$-ball and $\mathbb{S}^{n-1}=\{x\in\mathbb{B}^n \mid \|x\|=1\}=(n-1)$-sphere.

This condition refines (ii): not only is e homeomorphic with $\mathbb{R}^n\approx\mathbb{B}^n-\mathbb{S}^{n-1}$ but a homeomorphism can be chosen which extends to the boundary $\mathbb{S}^{n-1}$. On $\mathbb{S}^{n-1}$, Φ_e need not be homeomorphic but

$$\Phi_e(\mathbb{S}^{n-1})\subset X^{n-1}.$$

Φ_e is called a *characteristic map* for e, and $\varphi_e=\Phi_e|\mathbb{S}^{n-1}$: $\mathbb{S}^{n-1}\to X^{n-1}$ an *attaching map* for e (this name is explained by 2.9).

In many important examples the set $\mathscr{E}$ is finite (*finite CW-decomposition*) and then (i)–(iii) is all we require. In general, there are two more conditions.

(iv) *The closure $\bar{e}$ of every cell is contained in a finite union of cells* (Closure finiteness).

(v) *A subset $A\subset X$ is closed (in X) if and only if $A\cap\bar{e}$ is closed in $\bar{e}$ for every cell $e\in\mathscr{E}$* (Weak topology). Equivalently: A map f: $X\to Y$ is continuous if every $f|\bar{e}$ is continuous.

It is conditions (iv)–(v) to which the notation CW refers.—A Hausdorff space X together with a CW-decomposition $\mathscr{E}$ is called a *CW-space* (originally "CW-complex"). The *dimension* of a CW-space, dim X, is the least integer n such that $X^n=X$; if no such n exists then dim $X=\infty$.

Given a subset $\mathscr{E}'\subset\mathscr{E}$, put $X'=\bigcup_{e\in\mathscr{E}'} e$. If $\mathscr{E}'$ is a CW-decomposition of X' then $(X', \mathscr{E}')$ is called a *CW-subspace* of $(X, \mathscr{E})$; often, we shall simply say "X' is a CW-subspace of X".

We now deduce a few basic properties of CW-spaces.

2.2 *Let Φ_e: $\mathbb{B}^n\to X$ be a characteristic map for $e\in\mathscr{E}$. Then $\bar{e}=\Phi(\mathbb{B})$.* (Note: the proof will not use (iv) or (v).)

Proof. By continuity $\Phi(\mathbb{B})=\Phi(\overline{\mathbb{B}-\mathbb{S}})\subset\overline{\Phi(\mathbb{B}-\mathbb{S})}=\bar{e}$. Conversely, $\mathbb{B}$ is compact, hence $\Phi(\mathbb{B})$ is compact and therefore closed (X being hausdorff), hence $\bar{e}\subset\Phi(\mathbb{B})$. ∎

2.3 *Let $\mathscr{E}'\subset\mathscr{E}$ be a finite set of cells. Then $X'=\bigcup_{e\in\mathscr{E}'} e$ is a CW-subspace if and only if X' is closed.* Consequence: *Finite unions and arbitrary intersections of finite CW-subspaces* (*i.e. having finitely many cells*) *are again CW-subspaces.* (We shall see in 2.7 that this generalizes to arbitrary *CW*-subspaces.)

Proof. If X' is a *CW*-subspace then every cell e of X' has a characteristic map $\Phi\colon \mathbb{B}\to X'$ and $\bar{e}=\Phi(\mathbb{B})\subset X'$, hence $X'=\bigcup_{e\in\mathscr{E}'}\bar{e}$ is closed. Conversely, if X' is closed and $\Phi\colon \mathbb{B}\to X$ is a characteristic map for $e\subset X'$ then $\Phi(\mathbb{B})=\bar{e}\subset X'$, so $\Phi\colon \mathbb{B}\to X'$. This proves condition (iii), whereas (i) and (ii) are obvious. ∎

2.4 *The closure $\bar{e}$ of every cell is contained in a finite CW-subspace.*

Proof by induction on $n=|e|$. From (iii) and 2.2 we see that $\bar{e}-e\subset X^{n-1}$, i.e. $\bar{e}-e$ meets only cells of dimension $<n$, say $e_1,\ldots,e_r$; their number is finite by (iv). By induction, every $\bar{e}_i$ lies in a finite *CW*-subspace X_i; hence $\bar{e}\subset e\cup X_1\cup X_2\ldots\cup X_r$ which is a *CW*-subspace by 2.3. ∎

2.5 *A subset $A\subset X$ is closed if and only if A intersects every finite CW-subspace in a closed set, i.e., X has the weak topology with respect to finite CW-subspaces.*

This follows from (v) because every $\bar{e}$ lies in a finite *CW*-subspace. ∎

2.6 *Every compact set $K\subset X$ is contained in a finite CW-subspace. In particular, X is compact if and only if it consists of finitely many cells.*

Proof. In every cell e which meets K pick a point $k_e\in e\cap K$. The set k, consisting of all k_e, is closed because its intersection with every finite *CW*-subspace is finite (hence closed). Similarly, every subset of k is closed, hence k is discrete. But it is also compact, being a closed subset of a compact set K. Therefore k is finite, i.e., K meets only finitely many cells, hence the result by 2.4 and 2.3. ∎

2.7 Proposition. *Let $\mathscr{E}'\subset\mathscr{E}$ be a set of cells and put $X'=\bigcup_{e\in\mathscr{E}'} e$. The following are equivalent.*

(a) *X' is a CW-subspace,*
(b) *X' is closed,*
(c) *$e\subset X'\Rightarrow\bar{e}\subset X'$.*

Consequence: Arbitrary unions or intersections of CW-subspaces are again CW-subspaces.

(For unions one uses (a)$\Leftrightarrow$(c), for intersections (a)$\Leftrightarrow$(b).)

Proof of 2.7. The implication (b)$\Rightarrow$(c) is obvious, and (a)$\Rightarrow$(c) follows from 2.2. Assuming (c), we now prove

(d) *If $A \subset X'$ is such that $\bar{e} \cap A$ is closed in $\bar{e}$, for every cell e of X' then A is closed in X.*

Letting $A = X'$ this shows (c)$\Rightarrow$(b). Letting $A \subset X'$ be arbitrary again it shows that $(X', \mathscr{E}')$ satisfies condition (2.1(v)). Because (2.1(i), (ii), (iv)) are obvious and (iii) follows from 2.2 we get (c)$\Rightarrow$(a).

In order to prove (d) let X_α be any finite CW-subspace of X. Then $X_\alpha \cap X'$ consists of finitely many cells, say $e_1, \ldots, e_r$, and $\bar{e}_i \subset X_\alpha \cap X'$ because X_α is closed and (c) holds. Hence

$$X_\alpha \cap A = (X_\alpha \cap X') \cap A = \left(\bigcup_{i=1}^r \bar{e}_i\right) \cap A = \bigcup_{i=1}^r (\bar{e}_i \cap A)$$

is closed, and therefore A is closed in X by 2.5. ∎

2.8 *Let $\mathscr{E}' \subset \mathscr{E}$ be a set of n-cells. Then $X' = X^{n-1} \cup \bigcup_{e \in \mathscr{E}'} e$ is closed, i.e. is a CW-subspace. In particular, the skeletons X^{n-1} are CW-subspaces (put $\mathscr{E}' = \emptyset$) and the n-cells e are open in X^n (put $\mathscr{E}'$ = set of n-cells $\neq e$).* This follows from 2.7 because X' satisfies condition (c) (use 2.2 and 2.1(iii)). ∎

2.9 Proposition. *Let $\mathscr{E}^i \subset \mathscr{E}$ be the set of i-cells, and consider $\bigoplus_{i=0}^\infty \mathscr{E}^i \times \mathbb{B}^i$ where $\oplus$ denotes the topological sum and $\mathscr{E}^i$ has the discrete topology* (this space is then the topological sum of as many standard balls as there are cells in X). *For every cell $e \in \mathscr{E}$ choose a characteristic map $\Phi_e: \mathbb{B}^{|e|} \to X$, and define*

(a) $\Phi: \bigoplus_{i=0}^\infty \mathscr{E}^i \times \mathbb{B}^i \to X$, $\Phi(e, y) = \Phi_e(y)$ for $(e, y) \in \{e\} \times \mathbb{B}^{|e|}$;

(b) $\Phi^n: \bigoplus_{i=0}^n \mathscr{E}^i \times \mathbb{B}^i \to X^n$, $\Phi^n = \Phi | \bigoplus_{i=0}^n \mathscr{E}^i \times \mathbb{B}^i$;

(c) $\Phi^{(n)}: X^{n-1} \oplus (\mathscr{E}^n \times \mathbb{B}^n) \to X^n$, $\Phi^{(n)} | X^{n-1} =$ inclusion, $\Phi^{(n)} | \mathscr{E}^n \times \mathbb{B}^n = \Phi | \mathscr{E}^n \times \mathbb{B}^n$.

Claim: These maps are identification maps. Thus, X can be obtained by suitably pasting standard balls, and X^n can be obtained from X^{n-1} by attaching standard n-balls $\{e\} \times \mathbb{B}^n \approx \mathbb{B}^n$ via the attaching maps $\varphi_e = \Phi_e | \mathbb{S}^{n-1}$.

Proof. By condition (2.1(v)), a map $f: X \to Y$ is continuous if $f | \bar{e}$ is continuous for all cells e. Since $\Phi_e: B^{|e|} \to \bar{e}$ is an identification map (2.2), f is continuous if $f \Phi_e$ is continuous for all e, i.e., if $f \Phi$ is continuous.

This just says that Φ is an identification map. Similarly for Φ^n, replacing X by X^n. Furthermore, we can factor Φ^n as follows,

$$\Phi^n\colon \oplus_{i=0}^n \mathscr{E}^i \times \mathbb{B}^i \xrightarrow{\Phi^{n-1}\oplus \mathrm{id}} X^{n-1} \oplus (\mathscr{E}^n \times \mathbb{B}^n) \xrightarrow{\Phi^{(n)}} X^n,$$

hence $\Phi^{(n)}$ is an identification map (being the second factor of an identification map). ∎

Conversely,

2.10 Proposition. *Let $(X', \mathscr{E}')$ a CW-space with $\dim X' < n$, let $\{\varphi_\lambda\colon \mathbb{S}^{n-1} \to X'\}_{\lambda \in \Lambda}$ a family of continuous maps, form $\mathscr{X} = X' \oplus (\Lambda \times \mathbb{B}^n)$, where Λ has the discrete topology, and identify each $(\lambda, y) \in \Lambda \times \mathbb{S}^{n-1} \subset \mathscr{X}$ with $\varphi_\lambda(y) \in X' \subset \mathscr{X}$. Let X denote the resulting space and $\Phi\colon \mathscr{X} \to X$ the identification map. The sets $\Phi(e)$ with $e \in \mathscr{E}'$, and $e_\lambda = \Phi(\{\lambda\} \times \mathring{\mathbb{B}}^n)$ with $\lambda \in \Lambda$, then form a CW-decomposition of X, $\dim X \leq n$, $X^{n-1} = \Phi(X') \approx X'$, and the map $\mathbb{B}^n \approx \{\lambda\} \times \mathbb{B}^n \to X$ is characteristic for the n-cell e_λ.*

This, of course, provides a convenient recursive method for constructing CW-spaces, starting with a discrete space X^0.

Proof. If $A \subset X'$ is closed then

$$\begin{aligned}\Phi^{-1}\Phi(A) &= (X' \cap \Phi^{-1}\Phi(A)) \cup (\Lambda \times \mathbb{B}^n \cap \Phi^{-1}\Phi(A)) \\ &= A \cup (\Phi|\Lambda \times \mathbb{S}^{n-1})^{-1}\Phi A = A \cup (\textstyle\bigcup_\lambda \{\lambda\} \times \varphi_\lambda^{-1} A)\end{aligned}$$

is also closed, hence ΦA is closed (by definition of the identification topology), hence $\Phi|X'\colon X' \to \Phi X'$ is a closed map. Since it is also continuous and bijective we have $\Phi\colon X' \approx \Phi(X')$. Similarly, if $O \subset \Lambda \times \mathring{\mathbb{B}}^n$ is open then $\Phi^{-1}\Phi(O) = O$, hence $\Phi(O)$ is open, hence $\Phi\colon \Lambda \times \mathring{\mathbb{B}}^n \approx \Phi(\Lambda \times \mathring{\mathbb{B}}^n)$. This proves condition (2.1 (ii)). Moreover, it shows that characteristic maps of X' followed by Φ give characteristic maps for cells of dimension less than n, and $\mathbb{B}^n \approx \{\lambda\} \times \mathbb{B}^n \to X$ is characteristic for the n-cell e_λ, hence condition (2.1 (iii)) holds.

Next we show that X is hausdorff. For every pair $P, Q \in \mathscr{X}$ such that $\Phi(P) \neq \Phi(Q)$ we have to find disjoint open neighborhoods $U, V \subset \mathscr{X}$ such that $\Phi^{-1}\Phi U = U$, $\Phi^{-1}\Phi V = V$; then ΦU, ΦV will be disjoint neighborhoods of $\Phi(P)$, $\Phi(Q)$. If $P, Q \in X'$ then they have disjoint open neighborhoods U', V' in X', and we put

$$U = U' \cup \left\{(\lambda, x) \in \Lambda \times \mathbb{B}^n \,\middle|\, \|x\| > 0,\ \varphi_\lambda\left(\frac{x}{\|x\|}\right) \in U'\right\},$$

$$V = V' \cup \left\{(\lambda, x) \in \Lambda \times \mathbb{B}^n \,\middle|\, \|x\| > 0,\ \varphi_\lambda\left(\frac{x}{\|x\|}\right) \in V'\right\}.$$

If $P \in X'$, $Q=(\lambda_0, x_0) \in \Lambda \times \mathring{\mathbb{B}}^n$ we put

$$U = X' \cup \{(\lambda, x) \in \Lambda \times \mathbb{B}^n \mid \|x\| > \tfrac{1}{2}(1+\|x_0\|)\},$$

$$V = \{(\lambda, x) \in \Lambda \times \mathbb{B}^n \mid \|x\| < \tfrac{1}{2}(1+\|x_0\|)\}.$$

If both $P, Q \in \Lambda \times \mathring{\mathbb{B}}^n$ let U, V be arbitrary disjoint open neighborhoods contained in $\Lambda \times \mathring{\mathbb{B}}^n$.

Now to condition (iv). It is clear for cells of dimension $<n$. For n-cells e_λ we have (2.2)

$$\bar{e}_\lambda = \Phi(\{\lambda\} \times \mathbb{B}^n) = \Phi(\{\lambda\} \times \mathring{\mathbb{B}}^n) \cup \varphi_\lambda(\mathbb{S}^{n-1}) = e_\lambda \cup \varphi_\lambda(\mathbb{S}^{n-1}).$$

Since $\varphi_\lambda(\mathbb{S}^{n-1}) \subset X'$ is compact it meets only finitely many cells (2.6), and $\bar{e}_\lambda$ meets only one more. In order to prove (v), assume $A \subset X$ intersects the closure of every cell in a closed set. Then $\Phi^{-1}(A) \cap X' \approx A \cap \Phi(X')$ is closed because $\Phi(X')$ is a CW-space. Further,

$$(\Phi^{-1}A) \cap (\{\lambda\} \times \mathbb{B}^n) = \Phi^{-1}[A \cap \Phi(\{\lambda\} \times \mathbb{B}^n)] \cap (\{\lambda\} \times \mathbb{B}^n)$$

is closed, for every $\lambda \in \Lambda$, because $A \cap \Phi(\{\lambda\} \times \mathbb{B}^n) = A \cap \bar{e}_\lambda$ is closed by assumption, and Φ is continuous. Since $\mathfrak{X}$ is the topological sum of X' and the $\{\lambda\} \times \mathbb{B}^n$ it follows that $\Phi^{-1}A$ is closed, hence A is closed in the identification topology. ∎

2.11 Proposition. *CW-spaces are normal spaces.* (In fact, they are even paracompact; cf. Miyazaki, or Mather.)

Proof. Let A, B be disjoint closed sets in a CW-space X; we must find a function $\rho\colon X \to [0,1]$ such that $\rho|A=0$, $\rho|B=1$. By induction, we shall construct functions $\rho_n\colon X^n \to [0,1]$, $n=0,1,\ldots$, such that $\rho_n|A \cap X^n = 0$, $\rho_n|B \cap X^n = 1$, $\rho_n|X^{n-1} = \rho_{n-1}$; and we define ρ by $\rho|X^n = \rho_n$.

Suppose then we already have ρ_{n-1}, $n>0$; the start, ρ_0, being obvious. For every n-cell e take a characteristic map $\Phi_e\colon (\mathbb{B}^n, \mathbb{S}^{n-1}) \to (X^n, X^{n-1})$ and choose a function $\rho_e\colon \mathbb{B}^n \to [0,1]$ with

$$\rho_e|\mathbb{S}^{n-1} = \rho_{n-1}\Phi_e|\mathbb{S}^{n-1}, \quad \rho_e|\Phi_e^{-1}A = 0, \quad \rho_e|\Phi_e^{-1}B = 1;$$

such a function ρ_e exists by Tietze's extension lemma. Now define ρ_n by $\rho_n|X^{n-1} = \rho_{n-1}$, $\rho_n\Phi_e = \rho_e$. ∎

We conclude this § with a technical result which is needed in V, 4.

2.12 Proposition. *Let X be a CW-space, $Y \subset X$ a CW-subspace, and $M \subset X^n - (X^{n-1} \cup Y)$ a set which meets every n-cell of $X - Y$ in exactly one point. Then $X^{n-1} \cup Y$ is a strong deformation retract of $X^n \cup Y - M$.*

Proof. For each n-cell $e \subset X - Y$ choose a characteristic map $\Phi_e\colon \mathbb{B}^n \to X^n$ such that $\Phi_e(0) = M \cap e$. A deformation $D\colon I \times (X^n \cup Y - M) \to X^n \cup Y - M$ as required is then given by

$$D(t,x) = \begin{cases} x & \text{if } x \in X^{n-1} \cup Y \\ \Phi_e\left[(1-t)\zeta + t\dfrac{\zeta}{\|\zeta\|}\right] & \text{if } x = \Phi_e(\zeta),\ \zeta \in \mathbb{B}^n - \{0\},\ e \not\subset Y, \end{cases}$$

i.e. $X^{n-1} \cup Y$ remains fixed and the deleted n-cells $e - M$ are deformed radially onto the boundary. The only question is whether D is continuous. Since Y remains fixed (and $Y \times I$ is closed) it suffices to prove continuity of $D|I \times (X^n - M)$. Consider then the identification map $\Phi^{(n)}\colon X^{n-1} \oplus (\mathscr{E}^n \times \mathbb{B}^n) \to X^n$ of Proposition 2.9 ($\mathscr{E}^n$ = set of n-cells).

The map $\mathrm{id} \times \Phi^{(n)}\colon I \times [X^{n-1} \oplus (\mathscr{E}^n \times \mathbb{B}^n)] \to I \times X^n$ is also an identification map because I is compact (cf. 2.13), hence its restriction $I \times (\Phi^{(n)})^{-1}(X^n - M) \to I \times (X^n - M)$ is an identification map (from the definition of the identification topology because M is closed). Now clearly $\mathscr{D}\colon I \times (\Phi^{(n)})^{-1}(X^n - M) \to (\Phi^{(n)})^{-1}(X^n - M)$,

$$\mathscr{D}(t,\xi) = \begin{cases} \xi & \text{if } \xi \in (\Phi^{(n)})^{-1}(X^{n-1} \cup Y) \\ \left[e, (1-t)\zeta + t\dfrac{\zeta}{\|\zeta\|}\right] & \text{if } \xi = (e,\zeta) \in \mathscr{E}^n \times (\mathbb{B}^n - \{0\}),\ e \not\subset Y, \end{cases}$$

is continuous, and $D|I \times (X^n - M)$ is obtained from $\mathscr{D}$ by passing to quotients. ∎

2.13 Lemma. *If $\Phi\colon A \to B$ is an identification map and C is a locally compact (Hausdorff-)space then $\mathrm{id} \times \Phi\colon C \times A \to C \times B$ is an identification map.*

Proof (D. Epstein). If $U \subset C \times B$ is such that $V = (\mathrm{id} \times \Phi)^{-1} U$ is open then we must show that U is open. Let $(c,b) \in U$, and pick $(c,a) \in V$ such that $\Phi a = b$. Every neighborhood of c contains a compact neighborhood (cf. Schubert; I, 7.5); in particular, c has a compact neighborhood K such that $K \times \{a\} \subset V$. Let $W = \{x \in A | K \times \{x\} \subset V\} = \{x | K \times \{\Phi x\} \subset U\}$. Then W is an open set, and $\Phi^{-1}(\Phi W) = W$, hence ΦW is open (Φ being an identification). It follows that U contains a neighborhood of (c,b), namely $K \times \Phi W$; hence U is open. ∎

2.14 Exercises. *1.* If $X \supset Y$ is a pair of Hausdorff-spaces then a *CW-decomposition of* X mod Y is a set $\mathscr{E}$ of disjoint cells in X whose union is $X - Y$ and such that: (iii) Every n-cell $e \in \mathscr{E}$ admits a characteristic map $(\mathbb{B}^n, \mathbb{S}^{n-1}) \to (X^{n-1} \cup e, X^{n-1})$ where $X^{n-1} = Y \cup \bigcup_{|e|<n} e$. (iv) The

closure $\bar{e}$ of every cell lies in a finite union of cells with Y. (v) A subset $A \subset X$ is closed if and only if each of $\{A \cap \bar{e}\}_{e \in \mathscr{E}}$ and $A \cap Y$ is closed. The triple $(X, Y; \mathscr{E})$ is called a *relative CW-space*. Generalize the preceding results to relative *CW*-spaces.

2. Let $(X, \mathscr{E})$ a *CW*-space, and consider the identification map

$$\Phi: \oplus_{i=0}^{\infty} \mathscr{E}^i \times \mathbb{B}^i \to X$$

of 2.9(a). Show that the following are equivalent. (i) Φ is a closed map, (ii) X is locally compact, (iii) every point of X has a neighborhood which is a finite *CW*-space.

3. Show that *every CW-subspace Y of a CW-space X has an open neighborhood in X of which it is a strong deformation retract.* This follows by iterating 2.12: We know that $X^{n-1} \cup Y$ is a strong deformation retract of $X^n \cup Y - M^n$ where $M = M^n$, as in 2.12, is the set of centers of n-cells in $X - Y$. Let $r_n: X^n \cup Y - M^n \to X^{n-1} \cup Y$ denote the retraction. Define $V_0 = Y$, and $V_n = r_n^{-1}(V_{n-1})$ for $n > 0$. Then $V = \bigcup_{n=0}^{\infty} V_n$ is open in X, and $r: V \to X$, $r|V_n = r_1 r_2 \ldots r_n$, is a strong deformation retraction (in order to prove the last assertion it is convenient to use $[0, \infty]$ as parameter-intervall for the deformation, and to place the given deformation $V_n \simeq V_{n-1}$ in $[n-1, n]$; continuity has only to be checked on finite skeletons, i.e., on $V_n \times [0, \infty]$).

In fact, this construction proves more: If $W \subset Y$ then $r|r^{-1}W: r^{-1}W \to W$ is a strong deformation retract (and if W is relatively open in Y then $r^{-1}W$ is open in X). Also: *If W is relatively open in Y then every neighborhood U of W in X contains an open neighborhood V_U of W, of which W is a strong deformation retract.* Indeed, one can take $V_U = \{v \in r^{-1}W \mid$ the deformation path of v lies in $\mathring{U}\}$.

Corollary 1. *In a CW-space X every cell has an open neighborhood in X of which it is a strong deformation retract*—because every cell is relatively open in some *CW*-subspace Y. Since every point lies in a cell this implies

Corollary 2. *In a CW-space every point has an open neighborhood of which it is a strong deformation retract.*

3. Examples

3.1 A *zerodimensional CW*-space is the same as a discrete space. A *one-dimensional CW*-space X is often called a *graph*.

3.2 The n-sphere $\mathbb{S}^n = \{x \in \mathbb{R}^{n+1} \mid \|x\| = 1\}$ admits a CW-decomposition into one zero-cell $e^0 = (0, 0, \ldots, 0, 1)$ and one n-cell $e^n = \mathbb{S}^n - e^0$. The standard map $\pi\colon (\mathbb{B}^n, \mathbb{S}^{n-1}) \to (\mathbb{S}^n, e^0)$ of IV, 1.1 is characteristic for e^n.

Another CW-decomposition of $\mathbb{S}^n$ has two i-cells e^i_+, e^i_- for every i with $0 \leq i \leq n$, namely

$$e^i_+ = \{x \in \mathbb{S}^n \mid x_n = x_{n-1} = \cdots = x_{i+1} = 0,\ x_i > 0\},$$

$$e^i_- = \{x \in \mathbb{S}^n \mid x_n = x_{n-1} = \cdots = x_{i+1} = 0,\ x_i < 0\}.$$

This decomposition has the advantage of being invariant under the antipodal map $A\colon x \mapsto -x$. In fact, $A(e^i_+) = e^i_-$, $A(e^i_-) = e^i_+$. A characteristic map $\Phi^i_+\colon \mathbb{B}^i \to \mathbb{S}^n$ for e^i_+ is given by

$$\Phi^i_+(y_0, y_1, \ldots, y_{i-1}) = (y_0, \ldots, y_{i-1}, +\sqrt{1 - \sum y_j^2}, 0, 0, \ldots, 0),$$

for e^i_- by $\Phi^i_- = A \circ \Phi^i_+$.

3.3 A CW-decomposition of the n-ball $\mathbb{B}^n$ is obtained by decomposing first the boundary sphere $\mathbb{S}^{n-1}$ and then adding one n-cell $e^n = \mathring{\mathbb{B}}^n = \mathbb{B}^n - \mathbb{S}^{n-1}$. The identity map of $\mathbb{B}^n$ is characteristic for e^n. In particular, we can decompose $\mathbb{B}^n$ into three cells e^0, e^{n-1}, e^n.

The n-simplex Δ_n, which is homeomorphic with $\mathbb{B}^n$, decomposes into cells $e^{i_0 i_1 \ldots i_k}$, $0 \leq i_0 < i_1 < \cdots i_k \leq n$, as follows:

$$e^{i_0 \ldots i_k} = \{x \in \Delta_n \mid x_{i_0} x_{i_1} \ldots x_{i_k} > 0,\ x_{i_0} + x_{i_1} + \cdots + x_{i_k} = 1\}.$$

If we identify $\Delta_k \approx \mathbb{B}^k$ then the linear map $\Phi^{i_0 \ldots i_k}\colon \Delta_k \to \Delta_n$ which takes the ν-th vertex $e_\nu \in \Delta_k$ into $e_{i_\nu} \in \Delta_n$ is characteristic for $e^{i_0 \ldots i_k}$. *This CW-decomposition of Δ_n is invariant under linear maps $\Delta_n \to \Delta_n$ which permute the vertices.*—The closed sets $\bar{e}^{i_0 \ldots i_k} = \operatorname{im}(\Phi^{i_0 \ldots i_k})$ are called the *k-faces* of Δ_n; there are $\binom{n+1}{k+1}$ of them.

3.4 Every base $b_1, b_2, \ldots, b_n$ of the vector space $\mathbb{R}^n$ defines a CW-subdivision (*lattice*) of $\mathbb{R}^n$ as follows: The k-cells are the sets

$$e^{i_1 \ldots i_k}_{j_1 \ldots j_n} = \{\textstyle\sum_{\mu=1}^n j_\mu b_\mu + \sum_{\nu=1}^k t_\nu b_{i_\nu} \mid 0 < t_\nu < 1\}$$

where $j_\mu \in \mathbb{Z}$ are arbitrary integers, and i_ν are integers such that $1 \leq i_1 < i_2 \cdots < i_k \leq n$. The map $[0,1]^k \to \mathbb{R}^n$, $(t_1 \ldots t_k) \mapsto \sum j_\mu b_\mu + \sum t_\nu b_{i_\nu}$ is characteristic for $e^{i_1 \ldots i_k}_{j_1 \ldots j_n}$ (using $[0,1]^k \approx \mathbb{B}^k$). If the basic vectors b_μ are mutually orthogonal and of equal length then all cells are cubes (*cubical lattice*).

3.5 If F is a (not necessarily commutative) field then the *projective n-space* $P_n F$, $n \geq 0$, is the set of all 1-dimensional linear subspaces of the (left) vector space F^{n+1}. Any non-zero vector $(\xi_0, \ldots, \xi_n) \in F^{n+1}$ generates a 1-dimensional subspace which we denote by $[\xi_0, \ldots, \xi_n]$. The scalars ξ_i are called the *homogeneous coordinates* of $[\xi_0, \ldots, \xi_n] \in P_n F$; they are only determined up to a common (left) factor $\lambda \in F^* = F - \{0\}$ = multiplicative group of F. We can therefore think of $P_n F$ as being obtained from $F^{n+1} - \{0\}$ by identifying vectors which are proportional: $P_n F = (F^{n+1} - \{0\})/F^*$ = *set of orbits of* F^* *in* $F^{n+1} - \{0\}$.

If F is the field of real numbers $\mathbb{R}$, complex numbers $\mathbb{C}$, or quaternions $\mathbb{H}$ then $F^{n+1} - \{0\}$ is a topological space and we can equip $P_n F = (F^{n+1} - \{0\})/F^*$ with the *identification topology*; let $\pi: F^{n+1} - \{0\} \to P_n F$ be the identification map. We show that $P_n F$ is a *Hausdorff space*: If $\xi = [\xi_0, \ldots, \xi_n]$, $\eta = [\eta_0, \ldots, \eta_n]$ are two different points then there are indices i, j such that (ξ_i, ξ_j) and (η_i, η_j) are not proportional. We can assume that ξ_i, η_i are *real* numbers; then $\xi_i \eta_j - \xi_j \eta_i \neq 0$ (this is a determinant if $F = \mathbb{R}$ or $\mathbb{C}$; if $F = \mathbb{H}$ it still has the same properties because ξ_i, η_i are in the center of $\mathbb{H}$). Let V (resp. W) consist of all points $\zeta = [\zeta_0, \ldots, \zeta_n] \in P_n F$ such that $\|\zeta_j \xi_i - \zeta_i \xi_j\|$ is smaller (resp. greater) than $\|\zeta_j \eta_i - \zeta_i \eta_j\|$; then $\pi^{-1} V$, $\pi^{-1} W$ are disjoint open sets, hence V, W are disjoint neighborhood of ξ, η.

Every 1-dimensional subspace of F^{n+1} meets the sphere $\mathbb{S}^{(n+1)d-1} = \{x \in F^{n+1} \mid \|x\| = 1\}$ where $d = \dim(F) = 1, 2$, or 4. Therefore $\rho = \pi \mid \mathbb{S}^{(n+1)d-1}$ is a surjective map $\mathbb{S}^{(n+1)d-1} \to P_n F$, and because spheres are compact ρ is even an identification map (called *Hopf-map*). Thus $P_n F$ can be obtained from $\mathbb{S}^{(n+1)d-1}$ by identifying points which differ only by a (left) factor $\lambda \in F^*$; this factor must have absolute value $\|\lambda\| = 1$, i.e., $\lambda \in \mathbb{S}^{d-1}$, hence $P_n F \approx \mathbb{S}^{(n+1)d-1}/\mathbb{S}^{d-1}$ = space of orbits of $\mathbb{S}^{d-1}$ in $\mathbb{S}^{(n+1)d-1}$. In particular, $P_n F$ *is compact*.—In the real case, $F = \mathbb{R}$, the sphere $\mathbb{S}^{d-1} = \mathbb{S}^0$ consists of the numbers $+1$, -1 only; thus $P_n \mathbb{R}$ is obtained from $\mathbb{S}^n$ by identifying antipodal points.

A CW-decomposition of $P_n F$ is as follows: Put

$$e^k = \{[\xi_0, \ldots, \xi_n] \in P_n F \mid \xi_k \neq 0,\ \xi_j = 0 \text{ for } j > k\}, \quad k = 0, 1, \ldots, n;$$

i.e., e^k is obtained from $P_k F = \{\xi \mid \xi_j = 0 \text{ for } j > k\}$ by removing the hyperplane at infinity ($\xi_k = 0$). Thus, e^k is homeomorphic with affine space $F^k \approx \mathbb{R}^{dk}$; a homeomorphism is given by

$$[\xi_0, \ldots, \xi_n] \mapsto (\xi_k^{-1} \xi_0, \ldots, \xi_k^{-1} \xi_{k-1}).$$

Clearly $P_n F = e^0 \cup e^1 \cup \cdots \cup e^n$ is a decomposition of $P_n F$ into disjoint cells of dimension $0, d, 2d \ldots nd$; it remains to find characteristic a map

Φ^k: $\mathbb{B}^{dk} \to P_n F$ for e^k. Let $\mathbb{B}^{dk} = \{z \in F^k \mid \|z\| \le 1\}$, and define

(3.6) Φ^k: $\mathbb{B}^{dk} \to P_n F$, $\Phi^k(z_0, \ldots, z_{k-1}) = [z_0, z_1, \ldots, z_{k-1}, 1 - \|z\|, 0, \ldots, 0]$.

The composition $(\mathbb{B}^{dk} - \mathbb{S}^{dk-1}) \xrightarrow{\Phi^k} e^k \approx F^k$ takes $(z_0, \ldots, z_{k-1})$ into $(z_0/(1-\|z\|), \ldots, z_{k-1}/(1-\|z\|))$, and is clearly a homeomorphism (IV, 1.2); therefore Φ^k is a characteristic map.

It is interesting to note that the characteristic map Φ^n: $\mathbb{B}^{dn} \to P_n F$ is surjective and hence an identification map. The attaching map $\Phi^n | \mathbb{S}^{dn-1}$ agrees with the *Hopf-map* $(z_0, \ldots, z_{n-1}) \mapsto [z_0, \ldots, z_{n-1}]$. Thus $P_n F$ can be obtained from $\mathbb{B}^{dn}$ by identifying points on the boundary $\mathbb{S}^{n-1}$ which differ only by a factor $\lambda \in \mathbb{S}^{d-1} \subset F^*$. In particular, $P_n \mathbb{R}$ is obtained from $\mathbb{B}^n$ by identifying antipodal points on the boundary $\mathbb{S}^{n-1}$.

3.7 *If X is a CW-space and $X' \subset X$ a CW-subspace then the quotient* $X'' = X/X'$ (which is obtained by identifying all of X' to a single point) *inherits a CW-structure from X, and the identification map ρ: $X \to X''$ maps cells onto cells.* In fact, *if $\mathscr{E}' \subset \mathscr{E}$ denotes the set of all cells in X' then $\mathscr{E}'' = \{X'\} \cup \{\rho(e) \mid e \in \mathscr{E} - \mathscr{E}'\}$ is a CW-decomposition of X''.*

Proof. X'' is hausdorff: If $P \in X'' - \{X'\}$ then, by normality 2.11, there is a function τ: $X \to [0, 1]$ such that $\tau(\rho^{-1}P) = 0$, $\tau | X' = 1$; passing to quotients it induces a function τ'': $X'' \to [0, 1]$ which separates P and $\{X'\}$. If P, Q are different points in $X'' - \{X'\}$ then $\rho^{-1}P$, $\rho^{-1}Q$ have disjoint neighborhoods V, W in $X - X'$, hence ρV, ρW are disjoint neighborhoods of P, Q.—If Φ: $\mathbb{B}^n \to X$ is a characteristic map for $e \in \mathscr{E} - \mathscr{E}'$ then $\rho \Phi$ is characteristic for $\rho(e)$, proving 2.1 (iii). Clearly, closure-finiteness passes from X to X''. Finally, if $A \subset X''$ meets every $\overline{\rho(e)} = \rho \Phi_e(\mathbb{B}^n)$, $e \in \mathscr{E} - \mathscr{E}'$, in a closed set then $\rho^{-1}A$ meets every $X' \cup \Phi_e(\mathbb{B}^n)$ in a closed set, hence $\rho^{-1}A$ is closed, hence A is closed, proving 2.1 (v). ∎

For instance, $P_k F = \{[\xi_0, \ldots, \xi_n] \in P_n F \mid \xi_j = 0 \text{ for } j > k\}$, $k \le n$, is a CW-subspace of $P_n F$ (see 3.5); the quotient $P_n F / P_k F$ is known as *stunted projective space*; it decomposes into cells $e^0, e^{k+1}, \ldots, e^n$ of dimension 0, $d(k+1), \ldots, dn$ where $d = 1, 2, 4$ as $F = \mathbb{R}, \mathbb{C}, \mathbb{H}$. If $k = n-1$ then $P_n F / P_{n-1} F \approx \mathbb{S}^{dn}$ (compare 3.2).

3.8 If $\{X_\lambda\}_{\lambda \in \Lambda}$ is a family of CW-spaces then the topological sum $X = \bigoplus_{\lambda \in \Lambda} X_\lambda$ is also a CW-space; this is quite obvious from the definitions. If e_λ^0 is a zero-cell of X_λ then $X' = \bigcup_\lambda e_\lambda^0$ is a (discrete) CW-subspace of X. The quotient X/X' is the *wedge* (compare III, 7; Exerc. 2) of the spaces X_λ (with base points e_λ^0), $X/X' = \vee_\lambda X_\lambda$. In particular, *any wedge of CW-spaces (with zero-cells as base points) is again a CW-space.*

3.9 If X, Y are spaces with CW-decomposition $\mathscr{A}=\{a\}$, $\mathscr{B}=\{b\}$ then $\mathscr{A}\times\mathscr{B}=\{a\times b\}$ is a decomposition of $X\times Y$ into disjoint cells. *Is it a CW-decomposition?* Axioms 2.1 (i)–(iv) are easily verified; one finds that $(X\times Y)^n=\bigcup_{i+j=n}X^i\times Y^j$, and that products of characteristic maps are characteristic (using $\mathbb{B}^i\times\mathbb{B}^j\approx\mathbb{B}^{i+j}$). In particular, the answer is affirmative if X, Y are compact. Axiom 2.1 (v), however, fails to hold, in general (Exerc. 5). *It is satisfied if one of X, Y is locally compact*, i.e., *$\mathscr{A}\times\mathscr{B}$ is a CW-decomposition if one of X, Y is locally compact.*

Proof. Let $\Phi_a\colon \mathbb{B}^{|a|}\to X$, $\Phi_b\colon \mathbb{B}^{|b|}\to Y$ be characteristic maps for $a\in\mathscr{A}$, $b\in\mathscr{B}$. We have to show that

$$\{\Phi_a\times\Phi_b\}\colon \oplus_{a,b}\mathbb{B}^{|a|}\times\mathbb{B}^{|b|}\to X\times Y$$

is an identification map (by 2.2 this is equivalent with 2.1 (v); cf. also proof of 2.9 (a)). This map factors as follows:

$$\oplus\mathbb{B}^{|a|}\times\mathbb{B}^{|b|}=(\oplus\mathbb{B}^{|a|})\times(\oplus\mathbb{B}^{|b|})\xrightarrow{\{\Phi_a\}\times\mathrm{id}}X\times(\oplus\mathbb{B}^{|b|})\xrightarrow{\mathrm{id}\times\{\Phi_b\}}X\times Y.$$

The first arrow is an identification map because $\oplus\mathbb{B}^{|b|}$ is locally compact (and $\{\Phi_a\}$ is an identification map), the second arrow is an identification map if X is locally compact (cf. 2.13). Hence the result, because identification maps compose. ∎

3.10 The unit interval $[0,1]$ is compact and has the CW-decomposition $[0,1]=\{0\}\cup\{1\}\cup(0,1)$. If X is any CW-space then, by 3.9, $[0,1]\times X$ is a CW-space with cells $\{0\}\times e$, $\{1\}\times e$, $(0,1)\times e$ where e ranges over all cells of X. The *suspension* ΣX of X is obtained from $[0,1]\times X$ by shrinking each of the CW-subspaces $\{0\}\times X$, $\{1\}\times X$ to a point. By 3.7 is has a CW-decomposition into cells $(0,1)\times e$, plus the two zero-cells $\{0\}\times X$, $\{1\}\times X$.

3.11 Exercises. *1.* As in IV, 2, Exerc. 4, let D_h be the space which is obtained from the 2-sphere by removing the interiors of $h+1$ disjoint discs, i.e., by puncturing $h+1$ holes. Take two copies of D_h and identify corresponding points on the boundary circles. The resulting space S_h is called *orientable surface of genus h.* Show (say by induction on h) that S_h admits a CW-decomposition consisting of one 0-cell, $2h$ 1-cells $a_1,\dots,a_h$, $b_1,\dots,b_h$, and one 2-cell e^2 whose attaching map $\varphi\colon\mathbb{S}^1\to S_h^1$ is as follows: Subdivide $\mathbb{S}^1$ into $4h$ equal consecutive segments

$\alpha_1, \beta_1, \alpha_1^-, \beta_1^-, \ldots, \alpha_h, \beta_h, \alpha_h^-, \beta_h^-$, map α_i, α_i^- linearly (but with opposite orientation) onto a_i; similarly for β_i, β_i^-, b_i.

Clearly φ is surjective, hence S_h is obtained from $\mathbb{B}^2$ by identifying points on the boundary $\mathbb{S}^1$ which have the same φ-image (see Fig. 8). Show that $S_0 \approx \mathbb{S}^2$, $S_1 \approx \mathbb{S}^1 \times \mathbb{S}^1$. Compare this and the next exercise with Seifert-Threlfall §§ 37–38.

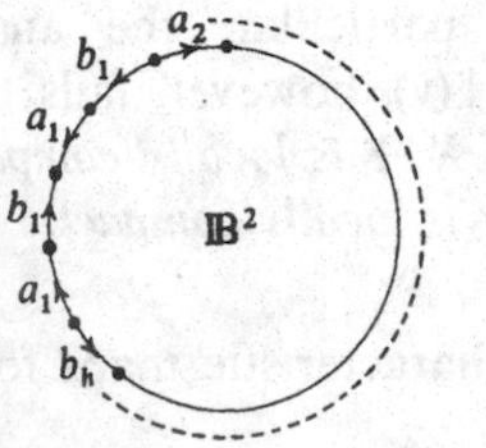

Fig. 8

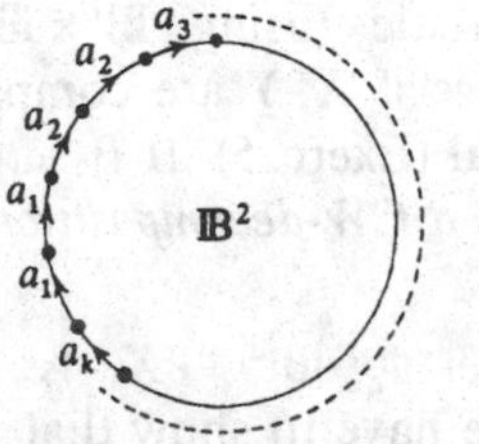

Fig. 9

2. As in *1.* let D_{k-1} be the 2-sphere with k holes, $k>0$. On the boundary circle of each hole identify antipodal points. The resulting space P_k is called *non-orientable surface of genus k*. Show that P_k admits a CW-decomposition consisting of one 0-cell, k 1-cells $a_1, \ldots, a_k$, and one 2-cell e^2 whose attaching map is described by the symbol $\alpha_1 \alpha_1 \alpha_2 \alpha_2 \ldots \alpha_k \alpha_k$ (analoguous to Exerc. 1), hence P_k is obtained from $\mathbb{B}^2$ by identifying points on the boundary as indicated in Fig. 9. Show that $P_1 \approx P_2 \mathbb{R}$.

3. Let π_q be the group of complex numbers $\zeta \in \mathbb{C}$ such that $\zeta^q = 1$ (it is cyclic of order q with generator $t = e^{2\pi i/q}$). This group π_q operates on $\mathbb{S}^{2n-1} = \{z \in \mathbb{C}^n \mid \|z\| = 1\}$ by ordinary scalar multiplication. The *orbit space* $L_q^{2n-1} = \mathbb{S}^{2n-1}/\pi_q$ (obtained by identifying z with tz) is called *lens space*. Show that the following cells form a CW-decomposition of $\mathbb{S}^{2n-1}$ which is compatible with the operation of π_q and which (by projection) induces a CW-decomposition of L_q^{2n-1}.

$$e_r^{2k} = \left\{(z_0, z_1, \ldots, z_{n-1}) \in \mathbb{S}^{2n-1} \mid z_j = 0 \text{ for } j > k, \arg(z_k) = r\frac{2\pi}{q}\right\},$$

$$e_r^{2k+1} = \left\{(z_0, \ldots, z_{n-1}) \in \mathbb{S}^{2n-1} \mid z_j = 0 \text{ for } j > k, r\frac{2\pi}{q} < \arg(z_k) < (r+1)\frac{2\pi}{q}\right\},$$

$$r = 0, 1, \ldots, q-1, \quad k = 0, 1, \ldots, n-1.$$

More generally, let $(l_1, \ldots, l_{n-1})$ be integers prime to q and let π_q operate on $\mathbb{S}^{2n-1}$ by $t(z_0, \ldots, z_{n-1}) = (t z_0, t^{l_1} z_1, \ldots, t^{l_{n-1}} z_{n-1})$. The orbit space

$L_q^{2n-1}(l_1, \dots, l_{n-1})$ is still called *lens space*. Construct CW-decompositions as above.

4*. The product $\mathscr{A} \subset \mathscr{B}$ of two countable CW-decompositions is again a CW-decomposition (Milnor 1956, Lemma 2.1). If $X = \bigvee [0,1]$ is a wedge of uncountably many unit segments (base point 0) then $X \times X$ is not a CW-space (Dowker § 5).

4. Homology Properties of CW-Spaces

We show that the filtration of a CW-space X by skeletons X^n is cellular, determine the cellular chain groups $W_n X$, and deduce consequences for HX.

4.1 Proposition. *Let X be a CW-space, $Y \subset X$ a CW-subspace* (e.g., $Y = \emptyset$), *and put $X_Y^n = X^n \cup Y$ (in particular, $X_Y^n = Y$ for $n < 0$). Let $M_Y^n \subset (X_Y^n - X_Y^{n-1})$ be a set which meets every n-cell of $X - Y$ in exactly one point. Then*

$$H_i(X_Y^n, X_Y^{n-1}) \cong H_i(X_Y^n, X_Y^n - M_Y^n) \cong H_i(X_Y^n - X_Y^{n-1}, X_Y^n - X_Y^{n-1} - M_Y^n)$$

$$\cong \oplus_e H_i(e, e - M_Y^n) = \begin{cases} 0 & \text{if } i \neq n \\ \mathbb{Z}\mathscr{E}^n(X-Y), \end{cases} \tag{4.2}$$

where the sum $\oplus$ ranges over all n-cells e in $\mathscr{E}^n(X-Y)$ = set of n-cells in $X - Y$, and $\mathbb{Z}\mathscr{E}^n(X-Y)$ is the free abelian group generated by $\mathscr{E}^n(X-Y)$.

In particular, $\{X_Y^n\}$ is a cellular filtration of X (cf. 2.6 *for condition* 1.1 (ii)), *and the homology $H(X, Y) = H(X, X_Y^{-1})$ is naturally isomorphic* (1.3) *to the homology of the cellular complex $W(X, Y)$, where $W_n(X, Y) = H_n(X_Y^n, X_Y^{n-1}) \cong \mathbb{Z}\mathscr{E}^n(X-Y)$.*

Proof. The first isomorphism follows because X_Y^{n-1} is a deformation retract of $X_Y^n - M_Y^n$ (see 2.12), the second by excision (III, 7.4), the third because $X_Y^n - X_Y^{n-1}$ is the *disjoint union* of the open n-cells $e \in \mathscr{E}^n(X-Y)$, and the fourth because $H(e, e - M_Y^n) \cong H(\mathbb{R}^n, \mathbb{R}^n - 0) \cong (\mathbb{Z}, n)$. ∎

4.3 Corollary. *If X is a compact CW-space then $H_i X$ is finitely generated for all i, and $H_i X = 0$ for $i > \dim X$. More generally, if $Y \subset X$ is a CW-subspace such that $X - Y$ contains only finitely many n-cells (resp. no n-cells) then $H_n(X, Y)$ is finitely generated (resp. $H_n(X, Y) = 0$).*

Indeed, already $W_n(X, Y)$ is finitely generated (resp. zero), and $H_n(X, Y) = H_n W(X, Y)$. ∎

4.4 Corollary. *If (X, Y) is a pair of CW-spaces then the identification map $\rho\colon (X, Y) \to (X/Y, \{Y\})$ induces isomorphisms*

$$\rho_*\colon H(X, Y) \cong H(X/Y, \{Y\}) = \tilde{H}(X/Y).$$

Indeed, by 3.7, X/Y is a CW-space and ρ is a cellular map which maps the cells of $X-Y$ homeomorphically onto the corresponding cells of $X/Y-\{Y\}$; since $W(X, Y)$ depends only on the cells in $X-Y$ (cf. 3rd terms in 4.2) we get $W\rho\colon W(X, Y) \cong W(X/Y, \{Y\})$. ∎

4.5 Corollary. *If (X, Y), (X', Y') are pairs of CW-spaces and $f\colon (X, Y) \to (X', Y')$ is a continuous (not necessarily cellular) map which, by passing to quotients, induces a homeomorphism $\bar{f}\colon X/Y \approx X'/Y'$ then $f_*\colon H(X, Y) \cong H(X', Y')$.*

This is a strong excision theorem. It follows immediately from 4.4. ∎

4.6 Corollary. *If X is a CW-space and $X_1, X_2 \subset X$ are CW-subspaces then $(X; X_1, X_2)$ is an excisive triad* (see III, 8.1).

Indeed, the inclusion $(X_1, X_1 \cap X_2) \to (X_1 \cup X_2, X_2)$ fulfils the hypothesis of 4.5 and is therefore a homology isomorphism. ∎

4.7 Proposition. *If $Z \subset Y \subset X$ are CW-spaces and subspaces then the inclusions $(Y, Z) \to (X, Z) \to (X, Y)$ induce an exact sequence*

$$0 \to W(Y, Z) \to W(X, Z) \to W(X, Y) \to 0$$

of chain maps. Under the isomorphisms $\Theta\colon HW(X, Y) \cong H(X, Y)$ of 1.3 *the connecting homomorphism d_* of this sequence transforms into the ordinary connecting homomorphism ∂_* of the triple (X, Y, Z)* (cf. III, 3.4).

Proof. Exactness follows from third terms in 4.2 because $(X_Z^n - X_Z^{n-1}) = (Y_Z^n - Y_Z^{n-1}) \oplus (X_Y^n - X_Y^{n-1})$, a disjoint union. In order to prove $\partial_* \Theta = \Theta\, d_*$ we use 1.9. This implies that every $y \in H_n W(X, Y)$ has a representative $\zeta \in S_n X^n$ with $\partial\zeta \in S_{n-1} Y$ and that $\Theta\, y = [\zeta]$ for any such ζ. It follows that $d_*\, y$ is represented by $\partial\zeta$, hence $\Theta(d_*\, y) = [\partial\zeta] \in H_{n-1}(Y, Z)$. But also, $[\partial\zeta] = \partial_*[\zeta]$ by the very definition of ∂_*; hence $\partial_*\, \Theta = \Theta\, d_*$. ∎

4.8 Example (cf. 3.5). The projective spaces $P_n F$, $n \geq 0$, over the fields $F = \mathbb{R}, \mathbb{C}, \mathbb{H}$ admit CW-decompositions $P_n F = e^0 \cup e^1 \cup \cdots \cup e^n$ into cells of dimension $0, d, 2d, \ldots, nd$ where $d = \dim(F) = 1, 2, 4$. In case $F = \mathbb{C}, \mathbb{H}$ there are no cells of odd dimensions, hence $W_{2i+1}(P_n F) = 0$, hence the boundary ∂ of $W(P_n F)$ vanishes, hence $HP_n F = WP_n F$, i.e., *if $F = \mathbb{C}$ or $\mathbb{H}$*

then

$$(4.9)\qquad H_j(P_n F) \cong \begin{cases} \mathbb{Z} & \text{if } j=0, d, 2d, \ldots, nd \\ 0 & \text{otherwise}. \end{cases}$$

In order to compute the homology of real projective spaces $P_n\mathbb{R}$ we have to determine the boundary operator $\partial\colon W_j \to W_{j-1}$; this will be done in 6.13. But even without knowing ∂ we can assert that every $H_j P_n \mathbb{R}$ is cyclic—because it is a quotient of a subgroup of a cyclic group.

The inclusion mappings $i\colon P_n F \to P_m F$, $n \leq m$, are cellular. In fact, $P_n F$ is the $[d(n+1)-1]$-skeleton of $P_m F$. The induced map of cellular complexes is therefore isomorphic up to dimension $(n+1)d-1$ (included), hence

$$(4.10)\qquad i_*\colon H_j P_n F \cong H_j P_m F \quad \text{for } j<(n+1)d-1,\ n \leq m.$$

Analogous results hold for stunted projective spaces (3.7); we leave it to the reader to formulate and prove them, as exercises.

Spaces which are retracts of CW-spaces inherit some of their homology properties. As an example we show

4.11 Proposition. *If Y is a compact ENR (euclidean neighborhood retract, cf. IV, 8) then $H_i Y$ is finitely generated for all i, and $H_i Y = 0$ for sufficiently large i.*

Proof. We can assume $Y \subset \mathbb{R}^n$. Let $Y \xrightarrow{i} O \xrightarrow{r} Y$ be a neighborhood retraction, $ri = \mathrm{id}$. Choose a lattice decomposition (3.4) of $\mathbb{R}^n$ which is so fine that every closed cell which meets Y lies in O. Let $X \subset O$ be the union of all closed cells which meet Y. Then X is a compact CW-space, $Y \subset X$, and $r|X\colon X \to Y$ is a retraction. Hence HY is a direct summand of HX (cf. III, 4.15), and the assertion follows from 4.3. ∎

4.12 Exercises. *1.* If $\{X_\lambda\}_{\lambda \in \Lambda}$ is a family of CW-spaces with base points $e_\lambda^0 \in X_\lambda^0$ then the wedge $X = \vee_{\lambda \in \Lambda} X_\lambda$ is also a CW-space by 3.8. Show that $W(X, e^0) \cong \oplus_\lambda W(X_\lambda, e_\lambda^0)$ where $e^0 \in X^0$ is the base point of the wedge. This implies $\tilde{H}X \cong \oplus_\lambda \tilde{H}X_\lambda$.

2. A connected graph $Y \neq \emptyset$ is called a *tree* if $Y - e$ is disconnected for every 1-cell $e \subset Y$. Show that every tree is contractible. Show that every graph X contains a tree Y with $Y^0 = X^0$ (construct Y starting at one zero-cell and letting Y branch out). Using $\tilde{H}Y = 0$, prove $HX \cong H(X/Y)$. Because X/Y is a wedge of circles this gives HX by Exerc. 1.

3. Let $C \subset \mathbb{R}^2$ be the union of all circles C_n, $n = 1, 2, \ldots$, with radius $1/n$ and center $(0, 1/n)$. Then C is compact but HC is not finitely generated (hint: C retracts onto arbitrarily large finite wedges of circles), hence C admits no CW-decomposition (and is not an ENR).

5. The Euler-Poincaré Characteristic

Euler's polyhedron formula is perhaps easier to explain to a non-mathematician than any other non-trivial result of algebraic topology. Roughly speaking it asserts that $\alpha_0 - \alpha_1 + \alpha_2 = 2$ for every decomposition of $\mathbb{S}^2$ into disjoint cells where α_i is the number of i-cells. More generally, we shall see that for any finite CW-space the number $\sum (-1)^i \alpha_i$ is independent of the decomposition.

5.1 Definition. If $G = \{G_i\}_{i \in \mathbb{Z}}$ is a graded abelian group such that $\operatorname{rank}(G_i)$ is finite for all i and equals zero for almost all i then

$$\chi(G) = \sum_{i \in \mathbb{Z}} (-1)^i \operatorname{rank}(G_i)$$

is defined and is called the *Euler-Poincaré-characteristic* of G (recall that $\operatorname{rank}(A)$ = maximal number of linearly independent elements in A; see I, 2.29). If K is a complex then we define $\chi(K)$ to be the Euler-Poincaré-characteristic of the underlying graded group (i.e. we ignore ∂^K).

5.2 Proposition. *If K is a complex such that $\chi(K)$ is defined then $\chi(HK)$ is also defined, and $\chi(HK) = \chi(K)$.*

Proof. For every abelian group G and subgroup $G' \subset G$ we have $\operatorname{rank}(G) = \operatorname{rank}(G') + \operatorname{rank}(G/G')$ (see I, 2.28–2.29); in particular, $\operatorname{rank}(H_i K) \leq \operatorname{rank}(Z_i K) \leq \operatorname{rank}(K_i)$, hence $\chi(HK)$ is defined. Moreover, $\operatorname{rank}(K_i) = \operatorname{rank}(Z_i K) + \operatorname{rank}(B_{i-1} K)$, $\operatorname{rank}(Z_i K) = \operatorname{rank}(B_i K) + \operatorname{rank}(H_i K)$. Multiply both equations with $(-1)^i$, sum over i, and get $\chi(K) = \chi(ZK) - \chi(BK)$, $\chi(ZK) = \chi(BK) + \chi(HK)$. Substitute and get $\chi(K) = \chi(HK)$. ∎

5.3 Corollary. *If $\cdots \leftarrow G_{i-1} \leftarrow G_i \leftarrow G_{i+1} \leftarrow \cdots$ is an exact sequence then $\chi\{G_i\} = 0$—provided it is defined.*

Proof. View the sequence as a complex G. Then $HG = 0$, hence $\chi(G) = \chi(HG) = 0$. ∎

5.4 Corollary. *Let G', G, G'' be graded abelian groups which admit an exact sequence*

$$(5.5) \qquad \cdots \leftarrow G_{i-1} \leftarrow G'_{i-1} \leftarrow G''_i \leftarrow G_i \leftarrow G'_i \leftarrow G''_{i+1} \leftarrow \cdots.$$

If any two of $\chi(G')$, $\chi(G)$, $\chi(G'')$ are defined then so is the third, and

$$\chi(G) = \chi(G') + \chi(G'').$$

(In most examples, 5.5 is the homology sequence of an exact sequence $0 \to K' \to K \to K'' \to 0$ of complexes.)

Proof. Suppose, for instance, $\chi(G')$, $\chi(G)$ are defined. Because 5.5 is exact, we get $\operatorname{rank}(G_i'') \le \operatorname{rank}(G_i) + \operatorname{rank}(G_{i-1}')$, hence $\chi(G'')$ is defined. We can then apply 5.3 to the exact sequence 5.5 and get

$$\sum(-1)^{3i}\operatorname{rank}(G_i) + \sum(-1)^{3i-1}\operatorname{rank}(G_i'') + \sum(-1)^{3i+1}\operatorname{rank}(G_i') = 0.$$

But this is just the assertion, $\chi G - \chi G'' - \chi G' = 0$. ∎

5.6 Definition. *The Euler-Poincaré characteristic of a space Y or a pair of spaces (Y, A)* is, by definition, the Euler-Poincaré characteristic of its homology, $\chi(Y, A) = \chi H(Y, A)$—provided the latter is defined, i.e. if $\operatorname{rank}(\oplus_i H_i(Y, A)) < \infty$.

Applying 5.4 to the homology sequence of (Y, A) shows

5.7 Proposition. *If (Y, A) is a pair of spaces such that two of the numbers $\chi(A)$, $\chi(Y)$, $\chi(Y, A)$ are defined then so is the third, and*

$$\chi(Y) = \chi(A) + \chi(Y, A).\ ∎$$

Similarly, 5.4 applies to Mayer-Vietoris sequences:

5.8 Proposition. *If $(Y; Y_1, Y_2)$ is an excisive triad and if two of the numbers $\chi(Y_1 \cup Y_2)$, $\chi(Y_1 \cap Y_2)$, $\chi(Y_1) + \chi(Y_2)$ are defined then so is the third, and*

$$\chi(Y_1) + \chi(Y_2) = \chi(Y_1 \cup Y_2) + \chi(Y_1 \cap Y_2).\ ∎$$

For instance, Y could be a CW-space and Y_1, Y_2 two CW-subspaces; such a triad is always excisive by 4.6. If $Y_1 \cup Y_2$ has only a finite number of cells then 5.8 is also clear from the following generalization of Euler's polyhedron formula.

5.9 Proposition. *If (Y, A) is a pair of CW-subspaces such that $Y - A$ contains only finitely many cells then $\chi(Y, A)$ is defined and*

$$\chi(Y, A) = \sum_{i=0}^{\infty}(-1)^i \alpha_i$$

where α_i is the number of i-cells in $Y - A$. In particular, this number $\sum(-1)^i \alpha_i$ depends only on $H(Y, A)$, not on the CW-decomposition.

Proof. The cellular chain group $W_i(Y, A)$ is free on α_i generators (4.2) hence $\operatorname{rank}(W_i(Y, A)) = \alpha_i$, hence $\sum(-1)^i \alpha_i = \chi W(Y, A) = \chi HW(Y, A) = \chi(Y, A)$, by 5.2 and 4.1. ∎

5.10 Exercises. *1.* Verify the formulas

$$\chi(\mathbb{B}^n)=1, \quad \chi(\mathbb{S}^n)=1+(-1)^n, \quad \chi(P_n\mathbb{R})=\tfrac{1}{2}(1+(-1)^n),$$
$$\chi(P_n\mathbb{C})=n+1=\chi(P_n\mathbb{H}), \quad \chi(S_h)=2-2h,$$
$$\chi(P_k)=2-k, \quad \chi(L_q^{2n-1})=0$$

where S_h, P_k are the surfaces of 3.11, Exerc. 1, 2, and L_q^{2n-1} the lens space of 3.11 Exerc. 3.

2. If Y is a finite CW-space and $\pi\colon \tilde{Y}\to Y$ is a q-sheeted covering then $\tilde{Y}$ is also a CW-space (cf. Schubert, III 6.9), and $\chi(\tilde{Y})=q\cdot\chi(Y)$, provided $q<\infty$.

3. Let $\mathscr{W}$ be the set of homeomorphism classes of compact CW-spaces. Define $\Phi\colon \mathscr{W}\to\mathbb{Z}$, $\Phi Y=\chi Y-1$, and verify that $\Phi Y=\Phi A+\Phi(Y/A)$ for every pair (Y,A) in $\mathscr{W}$. Conversely, if G is an abelian group and $\Psi\colon\mathscr{W}\to G$ is a map such that $\Psi Y=\Psi A+\Psi(Y/A)$ for every pair (Y,A) in $\mathscr{W}$ show that $\Psi Y=(\Phi Y)\cdot(\Psi S^0)$ for all $Y\in\mathscr{W}$. Hint: Take $Y=\mathbb{B}^n$ and $A=\mathbb{S}^{n-1}$ or $A=\{x\in\mathbb{B}^n \mid \|x\|\geq\frac{1}{2}\}$; comparing gives $\Psi\mathbb{S}^{n-1}=\Psi(\mathbb{S}^{n-1}\times[0,1])$. Next, take $Y=\mathbb{S}^{n-1}\times[0,1]$, $A=\mathbb{S}^{n-1}\times\{0\}$, and get $\Psi\mathbb{B}^n=0$ for $n>0$; prove $\Psi\mathbb{B}^0=0$ separately. Now proceed by induction on the number of cells in $Y\in\mathscr{W}$. Cf. Watts.

6. Description of Cellular Chain Maps and of the Cellular Boundary Homomorphism

We give simple geometric interpretations for the matrices of $Wf\colon WX\to WY$ and $\partial\colon W_n X\to W_{n-1}X$; this can be used for actual computation.

6.1 If X is a CW-space then X^n and X^n/X^{n-1} are also CW-spaces (3.7), and the natural maps $X\supset X^n\to X^n/X^{n-1}$ induce isomorphisms $W_nX\cong W_nX^n\cong W_n(X^n/X^{n-1})=\tilde{H}^n(X^n/X^{n-1})$ because all n-cells are mapped homeomorphically (cf. fourth term in 4.2). If Y is also a CW-space then every continuous map $f\colon (X^n,X^{n-1})\to(Y^n,Y^{n-1})$ (such a map will be called *n-cellular*) induces a cellular map $\bar{f}\colon X^n/X^{n-1}\to Y^n/Y^{n-1}$ and homomorphisms W_nf, $W_n\bar{f}$ such that the diagram

$$\begin{array}{ccc} W_nX & \xrightarrow{W_nf} & W_nY \\ \| \wr & & \| \wr \\ W_n(X^n/X^{n-1}) & \xrightarrow[W_n\bar{f}]{} & W_n(Y^n/Y^{n-1}) \end{array}$$

commutes, i.e., the *isomorphism* $W_n X = W_n(X^n/X^{n-1})$ *is natural with respect to n-cellular maps.* In particular, it commutes with cellular maps $f: X \to Y$ because they are n-cellular for all n.

We want to give a description of $W_n f \sim W_n \bar{f}$ which is practical for actual computation. Let us remark first that

(6.2) $$W_n(X^n/X^{n-1}) = \tilde{H}_n(X^n/X^{n-1}) \cong \oplus_e \tilde{H}_n(X^n/X^n - e),$$

where e ranges over the set $\mathscr{E}^n$ of all n-cells. This isomorphism is induced by the projections $p^e\colon X^n/X^{n-1} \to X^n/X^n - e$, or the inclusions $i^e = X^n/X^n - e \to X^n/X^{n-1}$ ($i^e = \mathrm{id}$ on e, constant outside). Indeed, it is clear that i^e_* maps $\tilde{H}_n(X^n/X^n - e)$ isomorphically onto the summand $H_n(e, e - M)$ of 4.2, and that $p^e i^e = \mathrm{id}$, $p^e i^{e'} = \text{constant}$ if $e' \neq e$. Thus

$$\tilde{H}_n(X^n/X^n - e) \xrightarrow{i^e_*} \tilde{H}_n(X^n/X^{n-1}) \xrightarrow{p^e_*} \tilde{H}_n(X^n/X^n - e)$$

are the inclusion and projection mappings of the direct sum decomposition 6.2. The map $W_n \bar{f}\colon \tilde{H}_n(X^n/X^{n-1}) \to \tilde{H}_n(Y^n/Y^{n-1})$ is therefore given by the matrix whose entries are the maps

(6.3) $$f^a_b = (p^b \bar{f} i^a)_*\colon \tilde{H}_n(X^n/X^n - a) \xrightarrow{i^a_*} \tilde{H}_n(X^n/X^{n-1}) \xrightarrow{W_n \bar{f}} \\ \to \tilde{H}_n(Y^n/Y^{n-1}) \xrightarrow{p^b_*} \tilde{H}_n(Y^n/Y^n - b),$$

where a resp. b range over the n-cells of X resp. Y.

These f^a_b are homomorphisms between free cyclic groups, hence are integers, defined up to sign. In order to remove the ambiguity of signs one has to specify isomorphisms $\tilde{H}_n(X^n/X^n - a) \cong \tilde{H}^n(Y^n/Y^n - b)$. This is usually done with the aid of characteristic maps $\Phi^a\colon (\mathbb{B}^n, \mathbb{S}^{n-1}) \to (X^n, X^n - a)$, etc. In fact, Φ^a is n-cellular and induces a bijective hence *homeomorphic* map $\bar{\Phi}^a\colon \mathbb{B}^n/\mathbb{S}^{n-1} \to X^n/X^n - a$; therefore

$$X^n/X^n - a \overset{\bar{\Phi}^a}{\approx} \mathbb{B}^n/\mathbb{S}^{n-1} \overset{\bar{\Phi}^b}{\approx} Y^n/Y^n - b.$$

Thus we get

6.4 Proposition. *Under the isomorphisms*

$$W_n X \cong \oplus_a \tilde{H}_n(X^n/X^n - a) \overset{\oplus \bar{\Phi}^a_*}{\cong} \oplus_a \tilde{H}_n(\mathbb{B}^n/\mathbb{S}^{n-1}), \quad a \in \mathscr{E}^n(X)$$

$$W_n Y \cong \oplus_b \tilde{H}_n(Y^n/Y^n - b) \overset{\oplus \bar{\Phi}^b_*}{\cong} \oplus_b \tilde{H}_n(\mathbb{B}^n/\mathbb{S}^{n-1}), \quad b \in \mathscr{E}^n(Y)$$

the map $W_n f\colon W_n X \to W_n Y$ *transforms into the homomorphism*

$$\oplus_a \tilde{H}_n(\mathbb{B}^n/\mathbb{S}^{n-1}) \to \oplus_b \tilde{H}_n(\mathbb{B}^n/\mathbb{S}^{n-1})$$

whose matrix-component $f_b^a \in \mathbb{Z}$ *is the degree of the composite map*

$$\begin{aligned}(6.5)\qquad & \mathbb{B}^n/\mathbb{S}^{n-1} \xrightarrow{\Phi^a} X^n/X^n - a \xrightarrow{i^a} X^n/X^{n-1} \xrightarrow{f} \\ & \to Y^n/Y^{n-1} \xrightarrow{p^b} Y^n/Y^n - b \xrightarrow{(\Phi^b)^{-1}} \mathbb{B}^n/\mathbb{S}^{n-1}. \quad \blacksquare\end{aligned}$$

We can compute this degree over any point Q of $\mathbb{B}^n/\mathbb{S}^{n-1}$ (IV, 5.6). In particular, we can choose $Q \in \mathring{\mathbb{B}}^n$ (= interior of $\mathbb{B}^n$). But over $\mathring{\mathbb{B}}^n$ the maps 6.5 have the following form (up to homeomorphism)

$$(\Phi^a)^{-1} f^{-1}(b) \overset{\Phi^a}{\approx} a \cap f^{-1}(b) \xrightarrow{\subset} f^{-1}(b) \xrightarrow{f} b \overset{p^b}{\approx} b \overset{\Phi^b}{\approx} \mathring{\mathbb{B}}^n,$$

hence

6.6 Corollary. *The matrix component* f_b^a *of the homomorphism* $W_n f$ *agrees with the degree (in the sense of* IV, 5) *of the composite map*

$$(\Phi^a)^{-1} f^{-1}(b) \overset{\Phi^a}{\approx} a \cap f^{-1}(b) \xrightarrow{f} b \overset{\Phi^b}{\approx} \mathring{\mathbb{B}}^n$$

(over any point $Q \in \mathring{\mathbb{B}}^n$*).* ∎

In particular, if $Q \in \mathring{\mathbb{B}}$ is such that $a \cap f^{-1}(b)$ is finite, then f_b^a is the number of points in $a \cap f^{-1}(b)$, each one counted with its multiplicity (see comment after IV, 5.8). For instance, $f_b^a = 0$ if $b \not\subset f(a)$, and $f_b^a = \pm 1$ if f maps $a \cap f^{-1}(b)$ homeomorphically onto b.

6.7 We now discuss the cellular boundary homomorphism $\partial\colon W_n X \to W_{n-1} X$. Since $W_i X = H_i(X^i, X^{i-1}) \cong \bigoplus_{e \in \mathscr{E}^i} \tilde{H}_i(X^i/X^i - e)$ is a direct sum of free cyclic groups the cellular boundary homomorphism ∂ can be described by integral matrix components ∂_b^a similar to $W_n f$ (cf. 6.3). Their geometric meaning can be seen from the diagram

$$(6.8)\qquad \begin{array}{ccccc}
\tilde{H}_n(X^n/X^n - a) & \xrightarrow{i_*^a} & \tilde{H}_n(X^n/X^{n-1}) \cong H_n(X^n, X^{n-1}) & \xrightarrow{\partial = \partial_*} & \\
\cong \uparrow \Phi_*^a & & \uparrow \Phi_*^a \qquad\qquad \uparrow \Phi_*^a & & \\
\tilde{H}_n(\mathbb{B}^n/\mathbb{S}^{n-1}) & = & \tilde{H}_n(\mathbb{B}^n/\mathbb{S}^{n-1}) \cong H_n(\mathbb{B}^n, \mathbb{S}^{n-1}) & \overset{\partial_*}{\cong} & \\
 & & \to H_{n-1}(X^{n-1}, X^{n-2}) & \xrightarrow{p_*^b} & \tilde{H}_{n-1}(X^{n-1}/X^{n-1} - b) \\
 & & \uparrow \varphi_*^a & & \cong \uparrow \Phi_*^b \\
 & & \cong \quad \tilde{H}_{n-1}(\mathbb{S}^{n-1}) & & \tilde{H}_{n-1}(\mathbb{B}^{n-1}/\mathbb{S}^{n-2}),
\end{array}$$

where a, b are n- resp. $(n-1)$-cells of X with characteristic maps Φ^a, Φ^b, and attaching map $\varphi^a = \Phi^a | \mathbb{S}^{n-1}$. The maps i_*^a, p_*^b are defined as after 6.2; they are inclusions resp. projections of direct sum representations.

The composite top row of 6.8 is the component ∂_b^a of ∂. The diagram shows

6.9 Proposition. *Under the isomorphisms*

$$W_n X \cong \oplus_a H_n(X^n, X^n - a) \overset{\oplus \Phi_*^a}{\cong} \oplus_a H_n(\mathbb{B}^n, \mathbb{S}^{n-1}) \overset{\oplus \partial_*}{\cong} \oplus_a \tilde{H}_{n-1} \mathbb{S}^{n-1}, \; a \in \mathscr{E}^n,$$

$$W_{n-1} X \cong \oplus_b \tilde{H}_{n-1}(X^{n-1}/X^{n-1} - b) \overset{\oplus \Phi_*^b}{\cong} \oplus_b \tilde{H}_{n-1}(\mathbb{B}^{n-1}/\mathbb{S}^{n-2})$$
$$\cong \oplus_b \tilde{H}_{n-1} \mathbb{S}^{n-1}, \quad b \in \mathscr{E}^{n-1},$$

the cellular boundary $\partial \colon W_n X \to W_{n-1} X$ *transforms into the homomorphism* $\oplus_a \tilde{H}_{n-1} \mathbb{S}^{n-1} \to \oplus_b \tilde{H}_{n-1} \mathbb{S}^{n-1}$ *whose (matrix-) component* $[a:b] \in \mathbb{Z}$ *is the degree of the composite map*

$$(6.10) \quad \mathbb{S}^{n-1} \xrightarrow{\varphi^a} X^{n-1} \to X^{n-1}/X^{n-1} - b \xrightarrow{(\Phi^b)^{-1}} \mathbb{B}^{n-1}/\mathbb{S}^{n-2} \approx \mathbb{S}^{n-1}. \quad \blacksquare$$

The integer $[a:b]$ is often called *incidence number of a and b*; up to sign it is determined by a and b alone; the sign depends on the choice of characteristic maps Φ^a, Φ^b, or rather on the choice of an isomorphism $\tilde{H}_n(X^n/X^n - a) \cong \tilde{H}_{n-1}(X^{n-1}/X^{n-1} - b)$.

Computing the degree of 6.10 over a point in $\mathring{\mathbb{B}}^{n-1}$ shows (compare proof of 6.6)

6.11 Corollary. *The incidence number* $[a:b]$ *agrees with the degree (in the sense of* IV, 5) *of the map*

$$(\Phi^b)^{-1} \circ \varphi^a \colon (\varphi^a)^{-1} b \to \mathring{\mathbb{B}}^{n-1}$$

(over any point $Q \in \mathring{\mathbb{B}}^{n-1}$). $\blacksquare$

In particular, if $Q \in \mathring{\mathbb{B}}^{n-1}$ is such that $(\varphi^a)^{-1} Q$ is finite then $[a:b]$ is the number of points in $(\varphi^a)^{-1} Q$, each one counted with its multiplicity. For instance, $[a:b] = 0$ if $b \not\subset \varphi^a(\mathbb{S}^{n-1})$, and $[a:b] = \pm 1$ if $\varphi^a \colon (\varphi^a)^{-1} b \approx b$. If $\varphi^a \colon (\varphi^a)^{-1} b \to b$ is locally homeomorphic then every counterimage point has multiplicity ± 1.

6.12 Orientation of Cells. If X is a CW-space and $e \subset X$ is an n-cell then $\tilde{H}_n(X^n/X^n - e) \cong \mathbb{Z}$. Any such isomorphism (equivalently: a generator of $\tilde{H}_n(X^n/X^n - e)$) is called an *orientation* of e. For oriented cells the components f_b^a (see 6.3) of $W_n f$ or ∂_b^a (see 6.7) of the cellular boundary ∂ can be viewed as integers (without having characteristic maps intervene). In practice, cells are always oriented by choosing a homeomorphism $X^n/X^n - e \approx \mathbb{S}^n$ (usually some $\overline{\Phi^e}$) and picking a generator in $\tilde{H}_n \mathbb{S}^n$. The *standard choice* for generators s^n in $\tilde{H}_n \mathbb{S}^n$ is as follows: $s^0 \in \tilde{H}_0(\mathbb{S}^0)$ is the homology class of $\{+1\}-\{-1\}$. For $n>0$ let $b^n \in H_n(\mathbb{B}^n, \mathbb{S}^{n-1})$ be the generator such that $\partial_* b^n = s^{n-1}$ (inductively) and let $s^n \in H_n \mathbb{S}^n$ correspond to b^n under the standard map π: $(\mathbb{B}^n, \mathbb{S}^{n-1}) \to (\mathbb{S}^n, \text{point})$ of IV, 1.1.

6.13 Example (cf. 3.5). We compute the cellular boundary in real projective n-space $P_n\mathbb{R}$, $n>0$. There is one cell e^i in every dimension i such that $0 \le i \le n$. The attaching map φ^i: $\mathbb{S}^{i-1} \to P_{i-1}\mathbb{R}$ for e^i, $i>0$, agrees with the Hopf map $(x_0, \ldots, x_{i-1}) \mapsto [x_0, \ldots, x_{i-1}]$, i.e. it is the twofold covering of $P_{i-1}\mathbb{R}$ by $\mathbb{S}^{i-1}$. The counterimage $(\varphi^i)^{-1}[x]$ of every point $[x] \in e^{i-1}$ (in fact of every $[x] \in P_{i-1}\mathbb{R}$) consists of two points, namely x and $-x$, and φ^i is locally homeomorphic. The multiplicities $\mu(\pm x)$ of these counterimage points are therefore ± 1 and the incidence number $[e^i : e^{i-1}]$ is 0 or ± 2 depending on whether $\mu(x) = -\mu(-x)$ or $\mu(x) = \mu(-x)$. If A is the antipodal map on $\mathbb{S}^{i-1}$ then $\varphi^i = \varphi^i A$; therefore, by IV, 5.7, we have $\mu(x) = \deg(A) \cdot \mu(-x) = (-1)^i \mu(-x)$, hence

$$\partial(e^i) = \pm[1 + (-1)^i]\, e^{i-1}, \quad \text{for } i>0. \tag{6.14}$$

Thus $WP_n\mathbb{R}$ is the complex

$$W_0 = \mathbb{Z} \xleftarrow{0} \mathbb{Z} \xleftarrow{2} \mathbb{Z} \xleftarrow{0} \mathbb{Z} \xleftarrow{2} \cdots \xleftarrow{1+(-1)^n} \mathbb{Z} \leftarrow 0 = W_{n+1}$$

and

$$\tilde{H}_i(P_n\mathbb{R}) \cong \begin{cases} 0 & \text{if } i \text{ is even or } i>n. \\ \mathbb{Z}_2 & \text{if } i \text{ is odd and } 0<i<n. \\ \mathbb{Z} & \text{if } i=n \text{ is odd.} \end{cases} \tag{6.15}$$

6.16 Exercises. *1.* Let $X_k = \mathbb{S}^n \vee \mathbb{S}^n \vee \cdots \vee \mathbb{S}^n$ be a wedge of k n-spheres, decomposed into one 0-cell and k n-cells, $n>0$. If f: $X_k \to X_l$ is a cellular map then $W_n f$ is given by an integral $k \times l$ matrix $\{f_j^i\}$. Show that every integral $k \times l$ matrix $\{\alpha_j^i\}$ belongs to some map f.

Hint: Reduce to $k=1$. Let Y denote the union of l disjoint open balls on $\mathbb{S}^n$ and π: $\mathbb{S}^n \to \mathbb{S}^n/\mathbb{S}^n - Y \approx X_l$ the projection. Given $\alpha_1, \alpha_2 \ldots \alpha_l \in \mathbb{Z}$ choose g_j: $\mathbb{S}^n \to \mathbb{S}^n$ of degree α_j and put $f = \{g_j\} \circ \pi$: $\mathbb{S}^n \to X_l$.

2. Let R, F be free abelian groups with bases A, B, and let β: $R \to F$ be a homomorphism, $\beta(a) = \sum_{b \in B} \beta_b^a \cdot b$, $a \in A$. Use Exercise 1 to construct a cellular map φ^a: $\mathbb{S}^n \to \bigvee_{b \in B} \mathbb{S}^n$, $n>0$, whose matrix is $\{\beta_b^a\}_{b \in B}$. For every $a \in A$, use φ^a to attach an $(n+1)$-cell e_a^n to $\bigvee_{b \in B} \mathbb{S}^n$. The resulting CW-space X_β has only n- and $(n+1)$-cells and ∂: $W_{n+1}X \to W_n X$ is isomorphic with β: $R \to F$; in particular $H_{n+1}(X_\beta) = \ker(\beta)$, $H_n(X_\beta) = \operatorname{coker}(\beta)$.

If G is an arbitrary abelian group, choose an exact sequence $0 \to R \xrightarrow{\beta} F \to G \to 0$; then $H_n(X_\beta) = G$ and $\tilde{H}_k(X_\beta) = 0$ for $k \neq n$. Using wedges of such spaces construct a space X whose homology groups $H_k X$, $k > 0$, agree with prescribed abelian groups G_k.

3. a) All incidence numbers are zero in the CW-decomposition 3.11, Exerc. 1, of the orientable surface S_h. Therefore $H_1(S_h)$ is free on $2h$ generators, $H_2(S_h) \cong \mathbb{Z}$, $H_i(S_h) = 0$ for $i > 2$.

b) In the CW-decomposition 3.11 Exerc. 2 of the non-orientable surface P_k the 2-cell e^2 has incidence number 2 with every 1-cell. Therefore $H_1(P_k) \cong \mathbb{Z}_2 \oplus$ free group on $(k-1)$ generators, $H_i(P_k) = 0$ for $i > 1$.

4. Consider the second CW-decomposition of $\mathbb{S}^n$ which is described in 3.2. Show that (with suitable orientations of cells) the following formulas hold:

$$\partial(e_+^{2j}) = e_+^{2j-1} + e_-^{2j-1} = \partial(e_-^{2j}),$$

$$\partial(e_+^{2j+1}) = e_+^{2j} - e_-^{2j} = -\partial(e_-^{2j+1}).$$

Remark that the covering map $\mathbb{S}^n \to P_n\mathbb{R}$ is cellular and use the above formulas to give another prove of 6.14.

5. Prove the following formulas for the CW-decomposition 3.11 Exerc. 3 of $\mathbb{S}^{2n-1}$

$$\partial(e_r^{2k}) = \textstyle\sum_{i=0}^{q-1} e_i^{2k-1}, \quad \partial(e_r^{2k+1}) = e_r^{2k} - e_{r+1}^{2k}$$

if the cells are suitably oriented, and $e_q^{2k} = e_0^{2k}$. The projection $\rho: \mathbb{S}^{2n-1} \to L_q^{2n-1}$ onto the lens space is cellular; applying ρ to the above formula yields the following boundaries in L_q^{2n-1}: $\partial(e^{2k}) = q \cdot e^{2k-1}$, $\partial(e^{2k+1}) = 0$. Compute $H(L_q^{2n-1})$.

7. *Simplicial Spaces*

A simplicial structure is a CW-structure with additional features: The characteristic maps form part of the structure, they are injective and they are inter-related by *linear changes of coordinates*. Historically, homology theory started with simplicial spaces and simplicial homology (see § 8).

7.1 Definition. Let X be a Hausdorff space. For every $n = 0, 1, \ldots$ let $\mathscr{S}_n$ be a set of continuous maps $s: \Delta_n \to X$. Then $\{\mathscr{S}_n\}$ or $\mathscr{S} = \bigcup_{n=0}^{\infty} \mathscr{S}_n$ is called a *simplicial atlas* if the following conditions (i)–(iv) hold.

(i) $X = \bigcup_{s \in \mathscr{S}} \operatorname{im}(s)$.

(ii) *Every $s \in \mathscr{S}$ is injective.* Since X is hausdorff and Δ_n compact, $s(\Delta_n)$ is closed and homeomorphic with Δ_n, $s: \Delta_n \approx s(\Delta_n) = \operatorname{im}(s)$.

(iii) *Any two* $s, t \in \mathscr{S}$ *are linearly related.* By this we mean the following: If $s\colon \Delta_m \to X$, $t\colon \Delta_n \to X$ then $\Delta_m^{st} = s^{-1} t(\Delta_n)$ resp. $\Delta_n^{ts} = t^{-1} s(\Delta_m)$ is a face (3.3) of Δ_m resp. Δ_n and $t^{-1} s\colon \Delta_m^{st} \to \Delta_n$ resp. $s^{-1} t$ is a linear map; since s, t are injective we have in fact a linear isomorphism $t^{-1} s\colon \Delta_m^{st} \approx \Delta_n^{ts}$.

If $t^{-1} s$, $s^{-1} t$ also preserve the *order of the vertices* then s, t are said to be *order-linearly related*, and if this holds for all s, t of an atlas $\mathscr{S}$ then $\mathscr{S}$ is called an *ordered simplicial atlas.*

If $s, t \in \mathscr{S}$ and $\operatorname{im}(s) \subset \operatorname{im}(t)$ (equivalently: $\Delta_m^{st} = \Delta_m$) then s is called a *face of* t.

(iv) *A set* $A \subset X$ *is closed if and only if* $A \cap \operatorname{im}(s)$ *is closed for all* $s \in \mathscr{S}$, i.e., X has the *weak topology* with respect to $\{\operatorname{im}(s)\}_{s \in \mathscr{S}}$. Equivalently, $f\colon X \to Z$ *is continuous if and only if* fs *is continuous for all* $s \in \mathscr{S}$. (Note: There is also a *strong topology* which is used in some connections; see 7.14.)

An example is provided in 3.3; there we described a CW-decomposition of Δ_n and characteristic maps which form an ordered simplicial atlas. But the identity map $\mathrm{id}\colon \Delta_n \to \Delta_n$ alone is also an ordered simplicial atlas on Δ_n.

7.2 Proposition. *Every (ordered) simplicial atlas* $\mathscr{S}$ *is contained in a unique maximal (ordered) simplicial atlas* $\mathscr{T}$. *In fact,* $\mathscr{T}_n$, $n = 0, 1, \ldots$, *is the set of all injective maps* $t\colon \Delta_n \to X$ *which are (order-) linearly related (as in (iii)) to all* $s \in \mathscr{S}$.

Proof. Take $\mathscr{T}_n$ as described. By (iii) we have $\mathscr{S} \subset \mathscr{T}$, and every atlas containing $\mathscr{S}$ is contained in $\mathscr{T}$. It suffices, therefore, to show that $\mathscr{T}$ is an atlas. Conditions (i), (ii) and (iv) are obvious; for (iii), let $t\colon \Delta_n \to X$ be in $\mathscr{T}$, pick $P \in \mathring{\Delta}_n = \Delta_n - \dot{\Delta}_n = \{x \in \Delta_n | x_i > 0 \text{ for all } i\}$, and choose $s\colon \Delta_k \to X$ in $\mathscr{S}$ such that $tP \in \operatorname{im}(s)$. Then $P \in t^{-1} s(\Delta_k)$, hence $t^{-1} s(\Delta_k) = \Delta_n$ (because it is a face of Δ_n), hence $t\Delta_n \subset s\Delta_k$. If also $t'\colon \Delta_m \to X$ is in $\mathscr{T}$ then $t^{-1} t'(\Delta_m) = (t^{-1} s)(s^{-1} t')\Delta_m$ is a face of Δ_n because (s, t), (s, t') are linearly related, and $t^{-1} t' = (t^{-1} s)(s^{-1} t')$ is a linear map. This proves (iii) for $\mathscr{T}$. ∎

7.3 Definition. An *(ordered) simplicial structure* on X (also: *a triangulation of* X) is a maximal (ordered) simplicial atlas $\mathscr{T}$. By 7.2, every (ordered) simplicial atlas $\mathscr{S}$ defines a unique (ordered) simplicial structure $\mathscr{T}$, and two atlasses $\mathscr{S}, \mathscr{S}'$ define the same structure if and only if $\mathscr{S} \cup \mathscr{S}'$ is again an atlas. A Hausdorff space X together with an (ordered) simplicial structure $\mathscr{T}$ on X is called an *(ordered) simplicial space.* If $X' \subset X$ and

$\mathscr{T}' \subset \mathscr{T}$ is a triangulation of X' then $(X', \mathscr{T}')$ is called a *simplicial subspace* of $(X, \mathscr{T})$.

The elements $v \in \mathscr{T}_0$, and also their images $v(\Delta_0) \in X$, are called *vertices* (of $\mathscr{T}$, or of X); in general, $t \in \mathscr{T}_n$, or $\operatorname{im}(t) \subset X$, is called a *simplex* of dimension n, of $\mathscr{T}$ or of X. If $t \in \mathscr{T}_n$, $v \in \mathscr{T}_0$ and $v(\Delta_0) \in t(\Delta_n)$ then v is called a *vertex of* t. Every $t \in \mathscr{T}_n$ has exactly $(n+1)$-vertices, namely $t(e^i)$, $i=0,\ldots,n$, where e^i is the i-th vertex of Δ_n. If $\mathscr{T}$ is ordered, and $t, t' \in \mathscr{T}_n$ have the same vertices then $t=t'$ (because $t^{-1}t'(\Delta_n)$ contains all vertices, hence $t^{-1}t'(\Delta_n)=\Delta_n$, and $t^{-1}t'\colon \Delta_n \to \Delta_n$ is order preserving, hence $t^{-1}t'=\mathrm{id}$). If $\mathscr{T}$ is not ordered and $t \in \mathscr{T}_n$ then there are exactly $n!$ simplices $\tau \in \mathscr{T}_n$ with the same set of vertices as t, namely all $\tau = t \circ \pi$, where π denotes a permutation of $(0,1,\ldots,n)$ and also the linear isomorphism $\Delta_n \to \Delta_n$ which permutes the vertices of Δ_n accordingly.

For example, the set $\mathscr{T}$ of maps $\Phi^{i_0 \ldots i_k}\colon \Delta_k \to \Delta_n$ of 3.3 is an ordered triangulation of Δ_n called *standard triangulation* of Δ_n. If we remove $\Phi^{01\ldots n} = \mathrm{id}_{\Delta_n}$ from $\mathscr{T}$ then the rest, $\dot{\mathscr{T}}$, is an ordered triangulation of $\dot{\Delta}_n$, hence $(\dot{\Delta}_n, \dot{\mathscr{T}})$ is a simplicial subspace of $(\Delta_n, \mathscr{T})$. Since $\dot{\Delta}_n \approx \mathbb{S}^{n-1}$ this provides triangulations of spheres.

7.4 Proposition. *Every maximal simplicial atlas $\mathscr{T}$ contains an ordered maximal atlas $\mathscr{S}$.*

Proof. Choose a complete order on $\mathscr{T}_0$, put

$$\mathscr{S}_n = \{t \in \mathscr{T}_n \mid t(e^0) < t(e^1) < \cdots < t(e^n)\},$$

and verify that $\{\mathscr{S}_n\}$ is a maximal ordered atlas. ▮ Conversely,

$$\mathscr{T}_n = \{s \circ \pi \mid s \in \mathscr{S}_n, \text{ and } \pi\colon \Delta_n \approx \Delta_n \text{ a linear isomorphism}\}.$$

7.5 Proposition. *Let $\mathscr{S}$ be an ordered simplicial atlas on X. Then the following are equivalent*

(a) *$\mathscr{S}$ is maximal.*

(b) $s \in \mathscr{S}_n \Rightarrow s_n \varepsilon_n^i \in \mathscr{S}_{n-1}$ *for* $i=0,1,\ldots,n$, *where* $\varepsilon_n^i\colon \Delta_{n-1} \to \Delta_n$ *is defined as in* III, 1.3.

(c) *Given $P \in X$ there is a unique $s \in \mathscr{S}$, say $s \in \mathscr{S}_n$, such that* $P \in s(\mathring{\Delta}_n)$. This s is called the *carrier of* P.

Proof. (a) ⇒ (b): Given $s \in \mathscr{S}_n$, let $\mathscr{S}' = \mathscr{S} \cup \{s\varepsilon_n^i\}$ = union of $\mathscr{S}$ with $s\varepsilon_n^i$. For every $t \in \mathscr{S}$ we know that $t^{-1}s$ is order preserving, hence also $t^{-1}s\varepsilon_n^i$, hence $\mathscr{S}'$ is an ordered atlas. But $\mathscr{S}$ is maximal, hence $\mathscr{S}' = \mathscr{S}$, i.e., $s\varepsilon_n^i \in \mathscr{S}$.

(b) ⇒ (c): Given $P \in X$, choose $t: \Delta_m \to X$ in $\mathcal{S}$ such that $P \in \operatorname{im}(t)$, say $P = t(x)$ where $x = \sum_{i=0}^m x_i e^i$, $x_i \geq 0$, $\sum x_i = 1$. Let $0 \leq i_1 < i_2 < \cdots i_{m-n} \leq m$ be all indices such that $x_{i_\nu} = 0$, i.e., $i \neq i_1, \ldots, i_{m-n} \Rightarrow x_i > 0$. Then $x = \varepsilon^{i_{m-n}} \ldots \varepsilon^{i_1}(y)$ for some $y \in \mathring{\Delta}_n$, hence $P \in (t\, \varepsilon^{i_{m-n}} \ldots \varepsilon^{i_1})(\mathring{\Delta}_n)$, and $s = t\, \varepsilon^{i_{m-n}} \ldots \varepsilon^{i_1} \in \mathcal{S}$ by assumption (b). If also $r: \Delta_l \to X$ is in $\mathcal{S}$ and $P \in r(\mathring{\Delta}_l)$ then $y \in s^{-1} r(\mathring{\Delta}_l)$ hence $s^{-1} r(\Delta_l) = \Delta_n$ (because it is a face of Δ_n); similarly $\Delta_l = r^{-1} s(\Delta_n)$, hence $l = n$. Because $r^{-1} s$ is order preserving, $r^{-1} s = \mathrm{id}$, hence $r = s$.

(Remark: We did not use (b) to show uniqueness, i.e. *the uniqueness part of c holds for arbitrary ordered atlasses.*)

(c) ⇒ (a): Let $\mathcal{T}$ an ordered atlas containing $\mathcal{S}$, pick $t: \Delta_m \to X$ in $\mathcal{T}$, $x \in \mathring{\Delta}_m$, and choose $s: \Delta_n \to X$ in $\mathcal{S}$ such that $t(x) \in s(\mathring{\Delta}_n)$; this is possible by assumption (c). By the remark above (in parenthesis) this implies $t = s$, hence $\mathcal{T} \subset \mathcal{S}'$. ∎

7.6 Proposition. *Let $\mathcal{T}$ be a triangulation of X, and let $\mathscr{E}^n$ be the set of all subsets of X of the form $t(\mathring{\Delta}_n)$ where $t \in \mathcal{T}_n$. Then $\mathscr{E} = \bigcup_n \mathscr{E}^n$ is a CW-decomposition of X, and $t \in \mathcal{T}_n$ is a characteristic map for $t(\mathring{\Delta}_n) \in \mathscr{E}^n$ (using $\Delta_n \approx \mathbb{B}^n$). If $\mathcal{S} \subset \mathcal{T}$ is an ordered triangulation then the correspondence $\mathcal{S}_n \to \mathscr{E}^n$, $s \mapsto s(\mathring{\Delta}_n)$ is bijective.*

Proof. By 7.5(c), the sets $t(\mathring{\Delta}_n)$ cover X, and the correspondence $s \mapsto s(\mathring{\Delta}_n)$ is bijective. If $t(\mathring{\Delta}_n) \cap t'(\mathring{\Delta}_{n'}) \neq \emptyset$ then $n = n'$ and t differs from t' only by a permutation of the vertices (compare proof of 7.5(b) ⇒ (c), 2nd part) hence $t(\mathring{\Delta}_n) = t'(\mathring{\Delta}_n)$. This proves condition 2.1(i). The remaining conditions 2.1(ii)–(v) are obvious. ∎

We now discuss maps between simplicial spaces.

7.7 Definition. If $(X, \mathcal{S}), (Y, \mathcal{T})$ are simplicial spaces, and $s \in \mathcal{S}_m$ then a map $f: \operatorname{im}(s) \to Y$ is called *linear* if some $t \in \mathcal{T}$ exists (say $t \in \mathcal{T}_n$) such that $\operatorname{im}(f) \subset \operatorname{im}(t)$ and $t^{-1} f s: \Delta_m \to \Delta_n$ is linear. This definition does not depend on the choice of s and t: If $\operatorname{im}(s') = \operatorname{im}(s)$, and $\operatorname{im}(f) \subset \operatorname{im}(t')$ then $t'^{-1} f s' = (t'^{-1} t)(t^{-1} f s)(s^{-1} s')$ is also linear because (s, s') and (t, t') are linearly related. By the same argument, if $f: \operatorname{im}(s) \to Y$ is linear and $\operatorname{im}(s') \subset \operatorname{im}(s)$ then $f|\operatorname{im}(s'): \operatorname{im}(s') \to Y$ is also linear.

7.8 Proposition. *If $(X, \mathcal{S}), (Y, \mathcal{T})$ are simplicial spaces, $s \in \mathcal{S}_m$, $t \in \mathcal{T}_n$, and $y^0, y^1, \ldots, y^m$ are arbitrary points in $\operatorname{im}(t)$ then there is a unique linear map $f: \operatorname{im}(s) \to Y$ such that $f s(e^i) = y^i$.* I.e., a linear map of $\operatorname{im}(s)$ is determined by its values on the vertices of s, and these values can be prescribed with the sole restriction that they must lie in some simplex $\operatorname{im}(t)$.

Proof. There is a unique linear map $g\colon \Delta_m \to \Delta_n$ such that $g(e^i) = t^{-1}(y^i)$, and $f = t g s^{-1}$. ∎

7.9 Proposition and Definition. *Let $(X, \mathscr{S}), (Y, \mathscr{T})$ be simplicial spaces. The following properties of a map $f\colon X \to Y$ are equivalent.*

(a) *f maps every simplex $\mathrm{im}(s) \subset X$, linearly onto some simplex $\mathrm{im}(t) \subset Y$.*

(b) *f maps vertices into vertices and is linear on each simplex $\mathrm{im}(s)$ of X.*

Such a map is called *simplicial*. If $\mathscr{S}, \mathscr{T}$ are ordered and all $t^{-1} f s$ preserve the order of the vertices then f is an *ordered simplicial map*. Composites of simplicial maps are again simplicial, and identy maps are simplicial. Simplicial spaces and maps then form a category which we denote by $\mathscr{S}\!\mathit{pl}$.

Proof. (a) ⇒ (b) is clear: f maps every vertex $v \in X$ onto a simplex, $f(v) = t(\Delta_n)$, hence $n = 0$ and $f(v)$ is a vertex.

Conversely, if (b) holds and $s \in \mathscr{S}_m$ then f maps $\mathrm{im}(s)$ linearly into some simplex $\tau(\Delta_n)$, $\tau \in \mathscr{T}_n$, and maps vertices of s into vertices of τ. Therefore $\tau^{-1} f s(\Delta_m)$ is a face of Δ_n, and $f s(\Delta_m) = \tau[\tau^{-1} f s(\Delta_m)]$ is a simplex of Y. ∎

7.10 Proposition. *Simplicial maps $f\colon X \to Y$ are CW-maps with respect to the CW-decomposition $\mathscr{E}^n = \{t(\mathring{\Delta}_n) \mid t \in \mathscr{T}_n\}$ of 7.6.*

Proof. Surjective linear maps never raise dimensions. Therefore f maps $X^n = \bigcup_{|s| \le n} \mathrm{im}(s)$ (where s denotes simplices of X, and $|\,| =$ dimension) into Y^n. ∎

7.11 Proposition. *Given simplicial spaces $(X, \mathscr{S}), (Y, \mathscr{T})$ and a map $\varphi\colon \mathscr{S}_0 \to \mathscr{T}_0$ such that $\{\varphi v^0, \varphi v^1, \ldots, \varphi v^m\}$ are vertices of a simplex in Y whenever $\{v^0, \ldots, v^m\}$ are vertices of a simplex in X. Then there exists a unique simplicial map $f\colon X \to Y$ such that $f(v) = \varphi(v)$ for $v \in \mathscr{S}_0$.* I.e., simplicial maps are determined by their values on vertices and these values can be prescribed with the only restriction that vertices of a simplex go into vertices of a simplex.

Proof. For every $s \in \mathscr{S}$ there is, by 7.8, a unique linear map $f^s\colon \mathrm{im}(s) \to Y$ such that $f^s(v) = \varphi(v)$ for all vertices v of s. Uniqueness insures that $f^s, f^{s'}$ agree on $\mathrm{im}(s) \cap \mathrm{im}(s')$ (note that $\mathrm{im}(s) \cap \mathrm{im}(s') = \mathrm{im}(s'')$ for some $s'' \in \mathscr{S}$), hence (by 7.1(iv)) there is a unique map $f\colon X \to Y$ such that $f \mid \mathrm{im}(s) = f^s$, and this f is simplicial by 7.9(b). ∎

7.12 Example and Definition. Let $Y = [0, 1]$ be the unit interval with the obvious simplicial structure (the linear map $\Delta_1 \to [0, 1]$, $e^0 \mapsto 0$, $e^1 \mapsto 1$

is an atlas). For every simplicial space $(X, \mathscr{S})$ and every vertex $v \in \mathscr{S}_0$ there is, by 7.11, a unique simplicial map $\hat{v}: X \to [0, 1]$ such that $\hat{v}(v) = 1$, and $\hat{v}(w) = 0$ if $w \neq v$, $w \in \mathscr{S}_0$. This map is called the *barycentric v-coordinate.* For every $x \in X$ the numbers $\{x_v = \hat{v}(x)\}$, $v \in \mathscr{S}_0$, are called the *barycentric coordinates of* x. They have the following properties:

(7.13) $\quad x_v \geq 0$; *for fixed* $x \in X$ *almost all* x_v *are zero*; $\sum_v x_v = 1$.

To see the second and third property choose $s \in \mathscr{S}_n$ such that $x \in \operatorname{im}(s)$. Then $x_v = 0$ if v is not a vertex of s, and $s^{-1}(x) = \sum_{i=0}^n x_v \cdot e^i$, where $v^i = s(e^i)$ are the vertices of s. This also justifies the notation "barycentric coordinate" and shows that x is determined by its barycentric coordinates.

7.14 Remark. In some situations it is advantageous to introduce the *strong topology* in a simplicial space X. By this is meant the coarsest topology under which all barycentric coordinates $\hat{v}: X \to [0, 1]$ are continuous. Then a map $g: Z \to X$ (Z any topological space) is continuous if and only if all composites $\hat{v} \circ g$ are continuous. If X is locally finite, i.e. if every vertex occurs in a finite number of simplices only, then the weak and strong topology coincide. Otherwise they don't, but in any case the two topologies define homotopy equivalent spaces (cf. A. 2.9).

Proposition 7.11 suggests the following

7.15 Definition. A *vertex schema* is a set V together with a set $\mathscr{D}$ of finite subsets of V, called the *distinguished subsets*, such that

(a) for every $v \in V$, $\{v\} \in \mathscr{D}$, i.e., *all singletons are distinguished,*

(b) $D \in \mathscr{D}$, $D' \subset D \Rightarrow D' \in \mathscr{D}$, i.e., *subsets of distinguished sets are distinguished.*

For example, if $\mathscr{S}$ is a triangulation of X, let $V = \mathscr{S}_0$ and call $D \subset \mathscr{S}_0$ distinguished, if the points in D form the vertices of a simplex $s \in \mathscr{S}$. We denote this vertex schema by $S(X, \mathscr{S})$.

A *map* $(V, \mathscr{D}) \to (V', \mathscr{D}')$ *of vertex schemata* is a set theoretic map φ: $V \to V'$ which takes distinguished sets into distinguished sets, i.e., $D \in \mathscr{D} \Rightarrow \varphi(D) \in \mathscr{D}'$.

Under ordinary composition these maps form a category, denoted by $\mathscr{V}\mathscr{S}$.

For example, if $f: (X, \mathscr{S}) \to (Y, \mathscr{T})$ is a simplicial map then the induced map $\mathscr{S}_0 \to \mathscr{T}_0$ is a map of vertex schemata which we denote by Sf: $S(X, \mathscr{S}) \to S(Y, \mathscr{T})$.

If we associate with every simplicial space $(X, \mathscr{S})$ its vertex schema $S(X, \mathscr{S})$, and with every simplicial map $f: (X, \mathscr{S}) \to (Y, \mathscr{T})$ the induced map Sf of vertex schemata then $S: \mathscr{Spl} \to \mathscr{VS}$ is a covariant functor, and in fact,

7.16 Proposition. S: *$\mathscr{Spl} \to \mathscr{VS}$ is an equivalence of categories*, i.e., there exists a functor $R: \mathscr{VS} \to \mathscr{Spl}$ such that $RS \sim \mathrm{Id}_{\mathscr{Spl}}$, $SR \sim \mathrm{Id}_{\mathscr{VS}}$.

Proof. The proof is suggested by 7.12–7.13. Given a vertex schema $(V, \mathscr{D})$, let X denote the set of all functions $x: V \to \mathbb{R}$ such that

(a) $\{v \in V \mid x(v) \neq 0\} \in \mathscr{D}$, i.e., the set of points where x does not vanish is distinguished, in particular finite;

(b) $x(v) \geq 0$, $\sum_{v \in V} x(v) = 1$.

If $D \in \mathscr{D}$ has $n+1$ elements, and $\alpha: D \to (0, 1, \ldots, n)$ is a bijection define $s_\alpha: \Delta_n \to X$ by

$$(s_\alpha y)(v) = \begin{cases} y_{\alpha(v)} = \alpha(v)\text{-th barycentric coordinate of } y \in \Delta_n \text{ if } v \in D, \\ 0 \text{ if } v \in V - D. \end{cases}$$

Obviously, s_α is injective. Further, every $x \in X$ is of the form $s_\alpha y$ for some α and y (take $D = \{v \mid x(v) \neq 0\}$). Introduce in X the weak topology with respect to the maps s_α, i.e., the finest topology for which all s_α are continuous. Then $f: X \to Z$ is continuous (Z any topological space) if and only if all $f s_\alpha$ are continuous. In particular, the maps $\hat{v}: X \to \mathbb{R}$, $\hat{v}(x) = x(v)$, $v \in V$, are continuous.

If $x \neq x'$ then $x(v) \neq x'(v)$ for some $v \in V$, hence $\hat{v}(x) \neq \hat{v}(x')$, hence X is hausdorff. We claim, $\mathscr{S} = \{s_\alpha\}$ is a triangulation of X. Conditions 7.1 (i), (ii) hold as remarked above, and (iv) holds by definition of the topology in X. If $s_\alpha, s_\beta \in \mathscr{S}$ then $s_\beta^{-1} s_\alpha$ is that linear map (defined on a face of Δ_n) which sends the vertex e^j into $e^{\beta\alpha^{-1}(j)}$, hence condition (iii). If $s: \Delta_l \to X$ is linearly related with all $s_\alpha \in \mathscr{S}$, then $s(\Delta_l) \subset s_\alpha(\Delta_n)$ for some $\alpha: D \approx (0, 1, \ldots, n)$ (pick $P \in \mathring{\Delta}_l$ and choose s_α with $sP \in \mathrm{im}(s_\alpha)$). Let

$$D' = \{v \in D \mid s_\alpha(e^{\alpha(v)}) \in s(\Delta_l)\},$$

and define $\beta: D' \to (0, 1, \ldots, l)$ by $s(e^{\beta(v)}) = s_\alpha(e^{\alpha(v)})$. Then $s = s_\beta \in \mathscr{S}$, hence $\mathscr{S}$ is maximal, i.e. a triangulation of X. We put $R(V, \mathscr{D}) = (X, \mathscr{S})$. The vertices of $R(V, \mathscr{D})$ correspond to bijections $D \approx \{0\}$, i.e. to singletons $D = \{v\} \in \mathscr{D}$; and $\{v^0\}, \ldots, \{v^n\}$ are vertices of a simplex if and only if $\{v^0, \ldots, v^n\}$ is distinguished.

If $\varphi: (V, \mathscr{D}) \to (V', \mathscr{D}')$ is a map of vertex schemata we can therefore (cf. 7.11) define a simplicial map $R\varphi: R(V, \mathscr{D}) \to R(V, \mathscr{D}')$ by $(R\varphi)\{v\} =$

$\{\varphi v\}$, and thus get a functor $R: \mathscr{V}\mathscr{S} \to \mathscr{S}pl$. Further

$$(V, \mathscr{D}) \to SR(V, \mathscr{D}), \quad v \mapsto \{v\}, \quad v \in V,$$

is a natural equivalence of vertex schemata; and on the other side, in $\mathscr{S}pl$, there is a simplicial equivalence (using 7.11)

$$(X, \mathscr{S}) \to RS(X, \mathscr{S}), \quad v \mapsto \hat{v}, \quad v \in \mathscr{S}_0. \quad \blacksquare$$

7.17 Remark. In order to establish the preceding homeomorphism $(X, \mathscr{S}) \approx RS(X, \mathscr{S})$ we never really used that X is hausdorff. But $RS(X, \mathscr{S})$ was shown to be hausdorff, hence X is, i.e., *if a space X admits a simplicial atlas with Properties* (i)–(iv) *of* 7.1 *then X is hausdorff.*

7.18 Exercises. *1**. Let $b_1, b_2, \ldots, b_n$ be a base of the vector space $\mathbb{R}^n$. For every permutation π of $(1, 2, \ldots, n)$ consider the linear simplex $s^\pi: \Delta_n \to \mathbb{R}^n$ with vertices 0, $b_{\pi(1)}$, $b_{\pi(1)} + b_{\pi(2)}$, $b_{\pi(1)} + b_{\pi(2)} + b_{\pi(3)}, \ldots, \sum_i b_{\pi(i)}$. Show that these simplices form a simplicial atlas of the basic parallelepiped $P = \{\sum_{i=1}^n t_i b_i \mid 0 \le t_i \le 1\}$. By parallel translation with $v \in \mathbb{R}^n$ one gets a simplicial atlas for $v + P$, and if v varies over all integral linear combinations of $(b_1, \ldots, b_n)$ one gets a simplicial atlas on $\mathbb{R}^n$ which is invariant under translation with b_i.

*2**. For each $n = 0, 1, \ldots$, let $\mathscr{U}^n$ be a simplicial atlas on Δ_n. We say $\mathscr{U} = \{\mathscr{U}^n\}$ is *compatible* if every $\varepsilon^j: \Delta_{n-1} \to \Delta_n$ (cf. III, 1.3) is simplicial with respect to the triangulations defined by $\mathscr{U}^{n-1}, \mathscr{U}^n$. For every $u: \Delta_k \to \Delta_n$ in $\mathscr{U}^n$, $u^{-1}\varepsilon^j(\Delta_{n-1})$ is then a simplicial subspace of Δ_k (in the standard triangulation 3.3). If this subspace $u^{-1}\varepsilon^j(\Delta_{n-1})$ is always a face (cf. 3.3) of Δ_k then we say $\mathscr{U}$ is *strongly compatible*. Show: If $\mathscr{U} = \{\mathscr{U}^n\}$ is strongly compatible and $\mathscr{S}$ is any simplicial atlas on X then the union of the sets

$$(\mathscr{S}\mathscr{U})_n = \{s \circ u \mid s \in \mathscr{S}_n, u \in \mathscr{U}^n\}, \quad n = 0, 1, \ldots,$$

is also a simplicial atlas on X. If $f: (X, \mathscr{S}) \to (X', \mathscr{S}')$ is an *injective* simplicial map then $f: (X, \mathscr{S}\mathscr{U}) \to (X', \mathscr{S}'\mathscr{U})$ is also simplicial.

For instance, the barycentric subdivision $B(\iota_n)$ of $\iota_n = \mathrm{id}(\Delta_n)$, as defined in III, 6.1, is a linear combination of linear simplices $\Delta_n \to \Delta_n$ which form a simplicial atlas $\mathscr{B}^n$ on Δ_n. The sequence $\{\mathscr{B}^n\}$ is strongly compatible, and $\mathscr{S}\mathscr{B}$ is called the *barycentric subdivision of* $\mathscr{S}$.

3. Let $(X, \mathscr{S})$ be a simplicial space such that the set $\mathscr{S}_0$ of vertices is finite, say $\mathscr{S}_0 = \{v_0, v_1, \ldots, v_N\}$.

(i) Define a simplicial map $I: X \to \Delta_N$ by $I(v_k) = e^k$ and show that I maps X isomorphically onto a simplicial subspace of Δ_N.

(ii) If $\mathscr{S}_j=\emptyset$ for $j>n$, i.e., if $\dim(X)\leq n$, choose points $w_0, w_1, \ldots, w_N$ in Δ_{2n+1} such that any r of them are linearly independent (in $\mathbb{R}^{2n+2}$), for $r\leq 2n+2$. There is a unique map $J: X\to\Delta_{2n+1}$ which is linear on each simplex of X and takes v_i into w_i. Show that J is injective. In particular, *any X of* $\dim\leq n$ *embeds "rectilinearly" into* Δ_{2n+1}.

4. Let $(X,\mathscr{S}),(Y,\mathscr{T})$ be simplicial spaces. A map $f\colon X\to Y$ is called *direct* if for every vertex $v\in X$ there exists a vertex $v'\in Y$ such that $\hat{v}(x)>0\Rightarrow\hat{v}'(f(x))>0$. (Note: The set $\{x\in X\,|\,\hat{v}(x)>0\}$ is called the *open star* of v; thus, f is direct if it maps open stars into open stars.) If f is direct and $v_0, v_1, \ldots, v_n$ are vertices of a simplex of X then, for some $x\in X$, all $\hat{v}_i(x)$ are positiv, hence all $\hat{v}_i'(f(x))>0$, hence $\{v_i'\}$ are vertices of a simplex of Y (e.g. the carrier of $f(x)$). One can then define a simplicial map $f'\colon X\to Y$ by $f'(v)=v'$. This is called a *direct simplicial approximation* of f. There is a unique deformation $\vartheta\colon f\simeq f'$ such that $\hat{w}\vartheta(x,t)=(1-t)\,\hat{w}(f(x))+t\,\hat{w}(f'(x))$ for all vertices $w\in Y$.

If $f\colon X\to Y$ is any map and $\mathscr{S}$ is finite (equivalently: X compact) then f is direct with respect to a triangulation $\mathscr{S}^{\mathscr{U}}$ where $\mathscr{U}=\mathscr{B}\mathscr{B}\ldots\mathscr{B}$ is an iterated barycentric subdivision in the sense of Exerc. 2 (hint: use III, 6.4 which implies that open stars become arbitrarily small under iterated barycentric subdivision). The resulting simplicial map $f'\colon(X,\mathscr{S}^{\mathscr{U}})\to(Y,\mathscr{T})$ is called a *simplicial approximation of* f. Compare Spanier, 3.4–3.5.

5. If Y is any set and $\{Y_v\}_{v\in V}$ any family of non-empty subsets of Y, define a vertex schema $(V,\mathscr{D})$ as follows: A finite subset $D\subset V$ is in $\mathscr{D}$ if and only if $\bigcap_{v\in D}Y_v\neq\emptyset$. The corresponding (7.16) simplicial space $R(V,\mathscr{D})$ is called the *nerve of* $\{Y_v\}$. Nerves of open coverings $\{Y_v\}$ of topological spaces Y are used in Čech (co-)homology theory (cf. 8.8 Exerc. 3 and A. 3.5).

6. An *ordered vertex schema* is a vertex schema $(V,\mathscr{D})$ together with a (partial) order on V such that $v,w\in V$ are comparable if and only if $\{v,w\}\in\mathscr{D}$. Show (as in 7.16) that ordered simplicial maps and maps of ordered vertex schemata form equivalent categories.

8. *Simplicial Homology*

Simplicial homology is closer to intuition than others: simplicial chains can be thought of as chunks of spaces (counted with multiplicities), and cycles are chunks without boundaries. However, for actual computation, CW-decompositions (or other means) are more adequate: a triangulation is too rich a structure (for this purpose), and it is often hard to find one. Computing homology with simplicial chains is like computing integrals ${}_a\!\int^b f(x)\,dx$ with approximating Riemann-sums.

8.1 Definition. Let $(X, \mathcal{S})$ be a simplicial space. If $\sigma: \Delta_n \to X$ is a simplicial map (with respect to the standard triangulation of Δ_n) then every composite $\Delta_{n-1} \xrightarrow{e^i} \Delta_n \xrightarrow{\sigma} X$ is also simplicial. Simplicial maps $\sigma: \Delta_n \to X$, $n=0, 1, \ldots$, therefore generate a subcomplex $\mathrm{Sp}(X)$ of the singular complex $S(X)$. Clearly Sp is functorial with respect to simplicial maps $X \to Y$, and the inclusion $\mathrm{Sp}\,X \subset SX$ is natural.

If $\sigma: \Delta_n \to X$ is simplicial then $\sigma(\Delta_n) \subset X^n = n$-skeleton of X (cf. 7.6), hence $\sigma \in S(X^n)$, and $(\partial\sigma) \in S(X^{n-1})$. We can therefore form the homology class $[\sigma] \in H_n(X^n, X^{n-1}) = W_n X$, and we can define a chain map

$$\gamma: \mathrm{Sp}\,X \to WX \quad \text{by} \quad \nu(\sigma) = [\sigma]. \tag{8.2}$$

From the Definition 1.2 of the boundary operator in WX it is clear, indeed, that $\gamma\partial = \partial\gamma$.

8.3 Proposition. *The chain map* $\gamma: \mathrm{Sp}\,X \to WX$ *is epimorphic, and the kernel of* $\gamma_n: \mathrm{Sp}_n X \to W_n X$ *is generated by all elements of the following two types:*

(a) *non-injective maps* $\tau: \Delta_n \to X$,

(b) $[\sigma\pi\text{-}\mathrm{sign}(\pi)\,\sigma]$,

where $\sigma: \Delta_n \to X$ *is simplicial, and* π *is a permutation of* $(0, 1, \ldots, n)$; *as before, we use the same letter* π *to denote the linear isomorphism* $\Delta_n \to \Delta_n$ *which takes* e^i *into* $e^{\pi(i)}$. In other words, degenerate simplicices are annihilated, and simplicices which differ only by a permutation π (of coordinates) are identified up to sign(π).

In particular, the elements (a), (b) generate a subcomplex of $\mathrm{Sp}\,X$ *(namely* $\{(a), (b)\} = \ker(\gamma)$), *and* $\mathrm{Sp}\,X/\{(a), (b)\} \cong WX$.

8.4 Definition. The complex $SP(X) = \mathrm{Sp}\,X/\{(a), (b)\} = \mathrm{Sp}\,X/\ker(\gamma)$ is called the *simplicial complex* of $(X, \mathcal{S})$. If $X' \subset X$ is a simplicial subspace then the inclusion $\mathrm{Sp}\,X' \subset \mathrm{Sp}\,X$ induces an inclusion $SP(X') \subset SP(X)$, and the quotient $SP(X, X') = SP(X)/SP(X')$ is the *simplicial complex of the pair* (X, X'). A simplicial map $f: X \to Y$ induces $\mathrm{Sp}(f): \mathrm{Sp}(X) \to \mathrm{Sp}(Y)$, and by passage to quotients, $SP(f): SP(X) \to SP(Y)$; similarly for maps of pairs. Thus $SP: \mathcal{S}pl \to \partial\mathcal{AG}$ is a functor from simplicial spaces to complexes. By 8.3, it depends only on the underlying CW-structure: γ induces a natural isomorphism $SP(X) \cong W(X)$.

Since non-injective simplicial maps $\Delta_n \to X$ can be neglected, $SP(X)$ can also be described as follows: $SP_n(X)$ is generated by $\mathcal{S}_n$ (these are precisely the injective simplicial maps $\Delta_n \to X$) with defining relations $\{s\pi = \mathrm{sign}(\pi)\, s\}$, $s \in \mathcal{S}_n$, π a permutation of $(0, 1, \ldots, n)$. The boundary

$\partial: SP_n X \to SP_{n-1} X$ is induced by the usual boundary of singular simplices. If $f: (X, \mathscr{S}) \to (Y, \mathscr{T})$ is simplicial and $s \in \mathscr{S}_n$ then

$$(SP_n f)(s) = \begin{cases} f s & \text{if } f s \text{ is injective, i.e., } (f s) \in \mathscr{T}_n, \\ 0 & \text{otherwise.} \end{cases}$$

8.5 Corollary to 8.3 (Invariance of simplicial homology). *In the category of simplicial pairs (X, X') and maps there is a natural isomorphism $HSP(X, X') \cong H(X, X')$. In particular, $HSP(X, X')$ is independent of the triangulation.*

Proof. By 8.3 we have $SPX \cong WX$, $SPX' \cong WX'$, hence (4.7) $SP(X, X') \cong W(X, X')$, hence $HSP(X, X') \cong HW(X, X') \cong H(X, X')$ by 4.1. ∎

Proof of 8.3. We know that the homology class $[\iota_n]$ of $\iota_n = \text{id}: \Delta_n \to \Delta_n$ generates $H_n(\Delta_n, \dot{\Delta}_n) \cong \mathbb{Z}$ (cf. IV, 2.7), and that $[\pi] = \text{sign}(\pi)[\iota_n]$ (cf. IV, 4.3). Further, if we choose one characteristic map $\Phi^e: (\Delta_n, \dot{\Delta}_n) \to (X^n, X^{n-1})$ for every n-cell e of X then $\{\Phi^e_*[\iota_n]\}$ is a base for $W_n X = H_n(X^n, X^{n-1})$ (cf. fourth terms in 4.2). But $s \in \mathscr{S}_n$ is characteristic for $e = s(\dot{\Delta}_n)$, and $s_*[\iota_n] = [s] = \gamma(s)$. Therefore γ maps every $s \in \mathscr{S}_n$ onto an element of this base (up to sign), and

$$\gamma(s\pi) = s_* \pi_* [\iota_n] = s_*(\text{sign}(\pi)[\iota_n]) = \text{sign}(\pi)\, \gamma(s).$$

All non-injective $\tau: \Delta_n \to X$ map into zero under γ because $\tau(\Delta_n) \subset X^{n-1}$. The result now follows because $\text{Sp}_n X$ has a base consisting of (i) all non-injective simplicial maps $\Delta_n \to X$, and (ii) all $s \in \mathscr{S}_n$. ∎

In the case of an *ordered* simplicial space $(X, \mathscr{S})$ the connection between simplicial and singular homology is even more direct: Let $SP'_n X$ be the free abelian group generated by $\mathscr{S}_n$; clearly, $SP'_n X \subset S_n X$. Since $s \in \mathscr{S}_n$ implies $s\varepsilon^i \in \mathscr{S}_{n-1}$, by 7.5, these groups form a subcomplex of the singular complex SX. Consider then the chain maps

$$SX \xleftarrow{\;j\;} SP'X \xrightarrow[\cong]{\;\nu\;} SPX \xrightarrow[\cong]{\;\bar{\gamma}\;} WX \tag{8.6}$$

where j = inclusion, $\bar{\gamma}$ is induced by γ, and ν is the composite $SP'X \subset \text{Sp}X \to \text{Sp}X/\{(a), (b)\} = SPX$. Clearly ν is isomorphic (just recall the definition of $\{(a), (b)\}$). The chain map $\mu = j\nu^{-1}\bar{\gamma}^{-1}: WX \to SX$ takes every $z \in W_n X = H_n(X^n, X^{n-1})$ into a representative $\zeta \in S(X^n)$. The induced homology homomorphism μ_* therefore takes $[z] \in HWX$ into $[\zeta] \in HX$, hence μ_* coincides with the isomorphism $\Theta: HWX \cong HX$ (cf. 1.9). In particular, j_* is isomorphic. We record these facts as

8.7 Proposition. (i) $SP'X \overset{\nu}{\cong} SPX \overset{\bar{\gamma}}{\cong} WX$. *In particular, $SP'X$ depends only on the CW-decomposition of X; not on the triangulation, or even on the ordering of the triangulation.*

(ii) j_*: $HSP'X \cong HSX = HX$.

Similar results hold for pairs (X, X'); they follow from the absolute case by the five lemma. ∎

It should be noted that this can be used to realize the isomorphism $HSPX \cong HX$ of 8.5 (for *non-ordered* triangulations $\mathscr{T}$ of X) by a chain map j': $SPX \to SX$—just choose an ordered triangulation $\mathscr{S}$ in $\mathscr{T}$ (see 7.4), and put $j' = j\nu^{-1}$.

8.8 Exercises. *1.* Triangulate the projective plane $X = P_2\mathbb{R}$ and compute $HSPX$.

2. Prove $H(\mathrm{Sp}X) \cong H(SPX)$. (Hint: Generalize to pairs. Treat $(\Delta_n, \dot{\Delta}_n)$ first, then (X^n, X^{n-1}), then proceed by induction on dimension and use the five lemma.)

*3**. If Y is a topological space and $\mathscr{U}, \mathscr{V}$ are open coverings of Y then $\mathscr{U}$ is said to *refine* $\mathscr{V}$, in symbols $\mathscr{U} < \mathscr{V}$, if every $U \in \mathscr{U}$ is contained in some $V \in \mathscr{V}$. Choose a function ψ: $\mathscr{U} \to \mathscr{V}$ such that $U \subset \psi(U)$ for all $U \in \mathscr{U}$, and define a simplicial map Ψ: $\mathrm{nerv}\,\mathscr{U} \to \mathrm{nerv}\,\mathscr{V}$ (cf. 7.18, Exerc. 5) which on vertices agrees with ψ (cf. 7.11). Show that, up to homotopy, Ψ is independent of the choice of ψ. Let Ω be the set of all open coverings of Y. A *Čech homology class y* of Y is a family $\{y_{\mathscr{U}} \in H(\mathrm{nerv}\,\mathscr{U})\}_{\mathscr{U} \in \Omega}$ such that $\mathscr{U} < \mathscr{V} \Rightarrow y_{\mathscr{V}} = \Psi_*(y_{\mathscr{U}})$. Under the addition $(y + y')_{\mathscr{U}} = y_{\mathscr{U}} + y'_{\mathscr{U}}$, Čech classes form a graded group, called the *Čech homology of Y*. Turn Čech homology into a functor and study its properties [cf. Eilenberg-Steenrod, Chap. IX].

Chapter VI

Functors of Complexes

If $T: \partial\mathcal{AG} \to \partial\mathcal{AG}$ is a functor from complexes to complexes then $X \mapsto TSX$ provides a generalization of the singular complex SX which may yield new useful topological invariants. We study this question (§§ 2–7), at least if T is the (dimension-wise) prolongation of an additive functor $t: \mathcal{AG} \to \mathcal{AG}$. We find that for every abelian group G there is, essentially, one covariant and one contravariant t such that $t\mathbb{Z} = G$. The resulting groups $HTSX$ are the *homology respectively cohomology groups of X with coefficients in G*. The functors t are also useful in studying product spaces; these questions are discussed in §§ 8–12.

In many applications the group G has some module structure which is inherited by TSX. We can then compose t resp. T with functors defined on modules. In order to avoid repetition we study functors from modules to groups, $\mathcal{Mod} \to \mathcal{AG}$, rightaway (we do not treat the—obvious—generalization $\mathcal{Mod} \to \mathcal{Mod}'$ because we want to keep the notation simple).

The reader who is familiar with basic facts about modules, additive functors, ⊗, Tor, Hom, Ext may skip §§ 1–6 and 8, although he might find the treatment of ⊗, Hom in §§ 5–6, 8 interesting.

1. Modules

The notion of module generalizes both "abelian groups" and "vector spaces"; abelian groups are $\mathbb{Z}$-modules, vector spaces over the field k are k-modules. In general, we consider an arbitrary ring R (which we always assume to have a unit element 1). An R-module is then an abelian group on which R operates in an additive fashion. More formally,

1.1 Let M be an abelian group and $\operatorname{End}(M)$ the ring of endomorphisms of M. A *left R-structure* in M is a ring homomorphism (preserving units) $\Theta: R \to \operatorname{End}(M)$. An abelian group together with a left R-structure is called a *left R-module.* The endomorphisms $\Theta(r)\in \operatorname{End}(M)$ are sometimes called *homotheties.*

If we put $rx=[\Theta(r)]\,x$, $r\in R$, $x\in M$, then $(r,x)\mapsto rx$ is a mapping $R\times M\to M$ with the following properties.

$$(1.2)\qquad \begin{aligned} &r(x_1+x_2)=r\,x_1+r\,x_2, &&(r_1+r_2)\,x=r_1\,x+r_2\,x,\\ &(r_1 r_2)\,x=r_1(r_2\,x), && 1x=x; \end{aligned}$$

the first equation asserts that $\Theta(r)\in\operatorname{End}(M)$, the others say that Θ is a ring homomorphism. Conversely, any map $R\times M\to M$ which satisfies 1.2 (a "structure map") defines a left R-structure Θ, by $[\Theta(r)]\,x=r\,x$.

A homomorphism $f\colon L\to M$ between R-modules is called *R-homomorphism* (or module homomorphism) if $f\circ\Theta(r)=\Theta(r)\circ f$ for all $r\in R$, i.e. if $f(r\,x)=r\,f(x)$ for $r\in R$, $x\in L$.

Clearly left R-modules and R-homomorphisms form a category; we denote it by R-$\mathcal{Mod}$.

If L is an abelian group again and $\Theta'\colon R\to\operatorname{End}(L)$ is an antihomomorphism, i.e. satisfies $\Theta'(r_1 r_2)=\Theta'(r_2)\,\Theta'(r_1)$, then Θ' is called a *right R-structure* and (L,Θ') a *right R-module.* If we put $x\,r=[\Theta'(r)]\,x$, $r\in R$, $x\in L$, then we get formulas analoguous to 1.2 which express the structure properties. There is really no essential difference between left- and right-modules: If we define the *opposite ring* R^{op} to coincide with R as an additive group but having the multiplication reversed, $r \underset{\mathrm{op}}{\cdot} s=s\cdot r$, then left R^{op}-modules are right R-modules, and vice versa. The category of right R-modules is denoted by $\mathcal{Mod}$-$R=R^{\mathrm{op}}$-$\mathcal{Mod}$.

Every abelian group M has a unique $\mathbb{Z}$-module structure $\Theta\colon\mathbb{Z}\to\operatorname{End}(M)$, $\Theta(n)=n\cdot(\mathrm{id}_M)$. Thus $\mathbb{Z}$-$\mathcal{Mod}=\mathcal{AG}$. Also, every abelian group M can be viewed as a left $\operatorname{End}(M)$-Module; indeed, the identity map $\Theta=\mathrm{id}\colon\operatorname{End}(M)\to\operatorname{End}(M)$ is a left $\operatorname{End}(M)$-structure.

1.3 If $f_1,f_2\colon L\to M$ are two R-homomorphisms then $f_1+f_2\colon L\to M$, $(f_1+f_2)\,x=f_1(x)+f_2(x)$, is also an R-homomorphism. Under this addition the set of R-homomorphisms $L\to M$ is an abelian group which we denote by $\operatorname{Hom}_R(L,M)$. It is a subgroup of $H_{\mathbb{Z}}(L,M)$, the group of all group-homomorphisms from L to M. If $g\colon L'\to L$, $h\colon M\to M'$ are R-homomorphisms then

$$\operatorname{Hom}_R(g,h)\colon \operatorname{Hom}_R(L,M)\to\operatorname{Hom}_R(L',M'),\quad [\operatorname{Hom}_R(g,h)](f)=h\circ f\circ g,$$

is a homomorphism of groups. In this way Hom_R is a functor from modules to abelian groups, contravariant in the first variable and covariant in the second.

1.4 If M is an R-module and $M' \subset M$ is a subgroup such that $rM' \subset M'$ for all $r \in R$ then M' with the induced R-structure is called a *submodule* of M. In this case the quotient group M/M' inherits a module structure, namely $r\bar{x} = \overline{rx}$, where $x \in M$, $\bar{x}$ its class in M/M'; we say M/M' is the *quotient module* of M by M'.

1.5 Many notions and results now carry over from abelian groups to modules. For instance, if $f: L \to M$ is a module homomorphism then $\ker(f)$, $\mathrm{im}(f)$, $\mathrm{coker}(f) = M/\mathrm{im}(f)$, $\mathrm{coim}(f) = L/\ker(f)$ are defined as groups but are sub- resp. quotient-modules. Similarly the notions *direct sum* or *product, exact sequence, complex, homology of a complex* generalize, and the *exact homology sequence* (of a short exact sequence of R-complexes) consists of R-homomorphisms. The category of (left) R-complexes and R-chain-maps is denoted by $\partial R\text{-}\mathcal{M}od$.

1.6 If $L' \subset L$ is a submodule and $f: L \to M$ is an R-homomorphism such that $f|L' = 0$ then there exists a unique R-homomorphism $\bar{f}: L/L' \to M$ such that $\bar{f}(\bar{x}) = f(x)$, where $x \in L$, $\bar{x}$ its class in L/L' (*passage to quotients*). This is clear for abelian groups, and one has only to check that $\bar{f}$ is an R-map. But $\bar{f}(r\bar{x}) = \bar{f}(\overline{rx}) = f(rx) = rf(x) = r\bar{f}(\bar{x})$.

We can state this result as follows: *If $0 \to L' \to L \to L'' \to 0$ is an exact sequence of R-modules then*

$$0 \to \mathrm{Hom}_R(L'', M) \to \mathrm{Hom}_R(L, M) \to \mathrm{Hom}_R(L', M) \tag{1.7}$$

is also exact, for every R-module M.

1.8 The ring R is itself an R-module with respect to the structure map $R \times R \to R$, $(r, s) \mapsto r \cdot s$. In fact, this defines a left R-structure and a right one, simultaneously. The homotheties are the left respectively right translations of R. The right translations are left R-homomorphisms, and vice versa. An R-homomorphism $f: R \to M$ is entirely determined by the image of 1. Indeed, $f(r) = f(r1) = rf(1)$; similarly for right modules. Conversely, for every $x \in M$ the map $\hat{x}: R \to M$, $\hat{x}(r) = rx$ is an R-map. Thus

$$\mathrm{Hom}_R(R, M) \cong M, \quad f \mapsto f(1). \tag{1.9}$$

1.10 A (left) R-module L is called *free* if it is isomorphic with a direct sum of the form $\oplus_{\gamma \in \Gamma} R$. If $\{i_\gamma: R \to L\}_{\gamma \in \Gamma}$ is a direct sum representation then the set of elements $\{x_\gamma = i_\gamma(1)\}_{\gamma \in \Gamma}$ is called a *base of L*. For instance, if R is a field then every module (= vector space) is free (= has a base). *If $B \subset L$ is a base of L, and if $y = \{y_b \in M\}_{b \in B}$ is any family of elements in any R-module M then there is a unique R-homomorphism $\hat{y}: L \to M$*

such that $\hat{y}(b)=y_b$ for $b\in B$. Indeed, by 1.9 there are unique R-homomorphisms $\hat{y}_b\colon R\to M$ such that $\hat{y}_b(1)=y_b$; hence $\{\hat{y}_b\}_{b\in B}\colon \oplus_{b\in B}R\to Y$, by the direct sum definition, and $\hat{y}$ is the composite $L\cong\oplus_{b\in B}R\to Y$. ∎

We shall often use this principle to define R-homomorphisms of free modules.

1.11 Some of our results on free complexes (II, 4) used the fact that every subgroup of a free abelian group is free. In order to generalize these we have to assume that *every submodule of a free R-module is free.* A ring R for which this is true will be called *hereditary*[3]. All fields are, of course, hereditary. A commutative ring is hereditary if and only if it is a principal ideal domain.

An example of a non-hereditary ring is $R=\mathbb{Z}/4\mathbb{Z}$: the submodule of R which is generated by the class (2) is not free.

For hereditary rings the results and proofs of II, 4 on free complexes carry over almost verbatim. In particular, *every free complex C is a direct sum of short complexes* (=everywhere zero except in two consecutive dimensions n, $n-1$; ∂_n monomorphic), *every homology homomorphism $HC\to HD$ is realized by a chain map* (C free, D arbitrary), *and $C\simeq C'\Leftrightarrow HC\cong HC'$ if both C and C' are free.*

1.12 Exercises. *1.* An *action* of a (not necessarily abelian) group π on a (left) R-module M is a function ϑ which to every $\omega\in\pi$ assigns an R-automorphism $\vartheta(\omega)$ of M such that $\vartheta(\omega_1\omega_2)=(\vartheta\omega_1)\circ(\vartheta\omega_2)$. Let $\Omega=R\pi$ be the *group ring of π over R*: as an additive group Ω agrees with the free R-module generated by the elements of π, the multiplication in Ω is $(\sum r_\omega\cdot\omega)(\sum r_{\omega'}\cdot\omega')=\sum(r_\omega r_{\omega'})\cdot(\omega\omega')$. Show that the notions *π-action* and *Ω-structure* are equivalent. If $\pi=\mathbb{Z}$ is free cyclic then giving a π-action is equivalent to giving an R-automorphism α $(=\vartheta(1))$ of M.

2. Let $\Omega=R[u]$ denote the ring of polynomials in one indeterminate u, and coefficients in R. Show that an Ω-structure Φ on M is the same as an R-module structure Θ together with an R-endomorphism β $(=\Phi(u))$ of M.

*3**. If $R=\mathbb{Z}/n\mathbb{Z}$, $n>0$, then an R-module is the same as an abelian group M such that $nx=0$ for all $x\in M$. Show that every R-module

[3] These rings are more special than hereditary rings in Cartan-Eilenberg (compare also Cohn). However, there is no serious danger of confusion because the results which we prove for hereditary rings are also valid with the more general definition; the reader who is familiar with the technique of projective modules will be able to generalize the proofs.

is a direct sum of modules of the form $\mathbb{Z}/m\mathbb{Z}$ where m divides n (cf. Kaplansky, Thm. 6).

4. In II, 3.6 an example of a free complex K over $R=\mathbb{Z}/4\mathbb{Z}$ was given such that $HK=0$ but $K\not\simeq 0$. It shows that not all results of II, 4 generalize to arbitrary rings R. However, if R is any ring, C a free R-complex such that $HC=0$ *and* $C_i=0$ for $i<0$ then $C\simeq 0$. Prove this (construct the nulhomotopy $s_n\colon C_n\to C_{n+1}$ by induction) and deduce from it (cf. proof of II, 4.3) that any chain map $f\colon C\to C'$ between free R-complexes such that $Hf\colon HC\cong HC'$ and $C_i=0=C'_i$ for $i<0$, is a homotopy equivalence (R arbitrary). *Corollary:* If C and HC is free, and $C_i=0$ for $i<0$, then $C\simeq HC$.

2. Additive Functors

We consider functors t from the category R-$\mathcal{M}od$ of left R-modules to the category $\mathcal{AG}$ of abelian groups. Both, covariant and contravariant functors play a role, but there is no essential difference between them (they are dual). In fact, if we were to replace $\mathcal{AG}$ by an arbitrary abelian category $\mathcal{A}$ then covariant and contravariant functors R-$\mathcal{M}od\to\mathcal{A}$ would be equivalent notions (cf. I, 1.5). We can not use this formal equivalence here but still we shall often treat covariant functors only and shall rely on the reader's ability to dualize the treatment. As a help we mark these numbers by ⁊; in order to dualize, the reader has to replace covariant by contravariant, to reverse every arrow of the form $t\varphi$ (where φ is an R-homomorphism) and every composition of the form $(t\varphi)(t\psi)$, and to interchange the following pairs in $\mathcal{AG}$: sum-product, left-right, epi-mono, ker-coker, im-coim.

2.1⁊ Definition. A functor $t\colon R$-$\mathcal{M}od\to\mathcal{AG}$ is called *additive* if $t(\alpha+\beta)=t\alpha+t\beta$ holds for all R-modules M,N and all $\alpha,\beta\in\mathrm{Hom}_R(M,N)$. In other words, $t\colon \mathrm{Hom}_R(M,N)\to\mathrm{Hom}_{\mathbb{Z}}(tM,tN)$ is a homomorphism; in particular $t0=0$.

2.2⁊ Remark. If R is a *commutative* ring then for every $a\in R$ and every R-module M multiplication with a,

$$\Theta_a\colon M\to M,\qquad \Theta_a(x)=a\,x,$$

is a module homomorphism. Indeed, $\Theta_a(r\,x)=a(r\,x)=r(a\,x)=r\,\Theta_a(x)$ for all $r\in R$. Applying an additive functor t gives a homomorphism $t\Theta_a\colon tM\to tM$. We can then define an R-structure on tM by $a\,y=(t\Theta_a)\,y$, $a\in R$, $y\in tM$. The identities 1.2 follow from the (obvious) equations

$\Theta_1 = \mathrm{id}$, $\Theta_{ab} = \Theta_a \Theta_b$, $\Theta_{a+b} = \Theta_a + \Theta_b$. If $f\colon M \to M'$ is a module homomorphism then $f\Theta_a = \Theta_a f$, hence $(tf)(t\Theta_a) = (t\Theta_a)(tf)$, i.e., $tf\colon tM \to tM'$ is a module homomorphism. Altogether this shows that any additive functor $t\colon R\text{-}\mathcal{M}od \to \mathcal{AG}$ can, automatically, be viewed as a functor from R-modules to R-modules, $t\colon R\text{-}\mathcal{M}od \to R\text{-}\mathcal{M}od$.

If R is not commutative then the multiplications Θ_a are still R-homomorphisms provided a lies in the center cR of R. For every additive t we get $t\colon R\text{-}\mathcal{M}od \to cR\text{-}\mathcal{M}od$.

2.3⁊ Proposition. *If $t\colon R\text{-}\mathcal{M}od \to \mathcal{AG}$ is additive and $\{i_\mu\colon M_\mu \to M\}$, $\mu = 1, 2, \ldots, r$, is a direct sum representation in $R\text{-}\mathcal{M}od$ then $\{t\,i_\mu\colon tM_\mu \to tM\}$ is a direct sum representation in $\mathcal{AG}$. I.e., t takes finite direct sums into direct sums.*

Proof. If $p_\nu\colon M \to M_\nu$, $\nu = 1, 2, \ldots, r$, are the projections, defined by $p_\nu i_\mu = 0$ for $\nu \neq \mu$ and $p_\nu i_\nu = \mathrm{id}$, then $\sum_\nu i_\nu p_\nu = \mathrm{id}$. Applying t gives the direct sum relations $(t\,p_\nu)(t\,i_\mu) = 0$ for $\nu \neq \mu$, $(t\,p_\nu)(t\,i_\nu) = \mathrm{id}$, $\sum_\nu (t\,i_\nu)(t\,p_\nu) = \mathrm{id}$. ∎

2.4⁊ Definition. An additive functor $t\colon R\text{-}\mathcal{M}od \to \mathcal{AG}$ will, in general, not commute with infinite sums (Exerc. 3). If it does it is called strongly additive. More precisely, t is *strongly additive* if the mapping

$$\{t\,i_\gamma\}\colon \oplus_{\gamma\in\Gamma} tM_\gamma \to t(\oplus_{\gamma\in\Gamma} M_\gamma)$$

is isomorphic for every family $\{M_\gamma\}_{\gamma\in\Gamma}$ of R-modules ($i_\nu\colon M_\nu \to \oplus_\gamma M_\gamma$ the inclusion).

In the covariant case, strong additivity follows from surjectivity of $\{t\,i_\gamma\}$, i.e.,

2.5 Proposition. *For every covariant additive $t\colon R\text{-}\mathcal{M}od \to \mathcal{AG}$ and every family $\{M_\gamma\}_{\gamma\in\Gamma}$ of R-modules the map*

$$\{t\,i_\gamma\}\colon \oplus_\gamma tM_\gamma \to t(\oplus_\gamma M_\gamma)$$

is monomorphic ($i_\nu\colon M_\nu \to \oplus_\gamma M_\gamma$ the inclusion).

Proof. For every finite subset K of Γ consider the commutative diagram

$$\begin{array}{ccc}
\oplus_{\kappa\in K} tM_\kappa & \xrightarrow{\ \subset\ } & \oplus_{\gamma\in\Gamma} tM_\gamma \\
{\scriptstyle\{t\,i_\kappa\}}\big\downarrow \cong & & \big\downarrow{\scriptstyle\{t\,i_\gamma\}} \\
t(\oplus_{\kappa\in K} M_\kappa) & \xrightarrow[t\,i_K]{} & t(\oplus_{\gamma\in\Gamma} M_\gamma),
\end{array}$$

where $i_K\colon \oplus_{k\in K} M_k \to \oplus_{\gamma\in\Gamma} M_\gamma$ denotes the inclusion of the partial sum. The left vertical map is isomorphic by 2.3. The map i_K has a left inverse π, hence $t\,i_K$ has the left inverse $t\,\pi$; in particular, $t\,i_K$ is monomorphic. The diagram shows then that $\{t\,i_\gamma\}$ restricted to the partial sum $\oplus_k t\,M_k$ is monomorphic. Since every element of $\oplus_\gamma t\,M_\gamma$ lies in some finite partial sum the whole map $\{t\,i_\gamma\}$ is monomorphic. ∎

2.6† Definition and Proposition. *If $t\colon R\text{-}\mathcal{M}od \to \mathcal{AG}$ is additive and*

$$C\colon \cdots \leftarrow C_i \xleftarrow{\partial} C_{i+1} \xleftarrow{\partial} C_{i+2} \leftarrow \cdots$$

is a complex of R-homomorphisms then

$$t\,C\colon \cdots \leftarrow t\,C_i \xleftarrow{t\partial} t\,C_{i+1} \xleftarrow{t\partial} t\,C_{i+2} \leftarrow \cdots$$

is a complex (because $(t\,\partial)(t\,\partial)=t(\partial\,\partial)=0$). *If $f=\{f_i\colon C_i\to C_i'\}_{i\in\mathbb{Z}}$ is a chain map then $t f=\{t f_i\colon t\,C_i\to t\,C_i'\}$ is a chain map. If $s\colon f\simeq g$ is a chain homotopy, $\partial s+s\partial=f-g$, then $(t\,\partial)(t\,s)+(t\,s)(t\,\partial)=t f-t\,g$, hence $t\,s\colon t f\simeq t\,g$.* In this fashion, every additive $t\colon R\text{-}\mathcal{M}od\to\mathcal{AG}$ extends to a homotopy preserving functor $t\colon \partial R\text{-}\mathcal{M}od\to\partial\mathcal{AG}$, which we denote by the same letter. Since t preserves homotopies it takes homotopy equivalent complexes into homotopy equivalent complexes. ∎

2.7 Convention. If t is *contravariant* we assign to $t(C_i)$ the dimension $-i$, so $(t\,C)_i=t(C_{-i})$. We also write $(t\,C)^i=(t\,C)_{-i}=t(C_i)$, similarly for cycles, boundaries, etc.; e.g., $H^i\,t\,C=H_{-i}\,t\,C$.

In general, a homology isomorphism $HC\cong HC'$ does not imply $H\,t\,C\cong H\,t\,C'$. In fact, $HC=0$ does not imply $H\,t\,C=0$, i.e., t does not transform exact sequences into exact sequences (Exerc. 4). However,

2.8† Proposition. *If $t\colon R\text{-}\mathcal{M}od\to\mathcal{AG}$ transforms short exact sequences $0\to M'\to M\to M''\to 0$ into short exact sequences then $H\,t\,C\cong t\,HC$ for all complexes C in $R\text{-}\mathcal{M}od$. In particular, t transforms arbitrary exact sequences (=acyclic complexes) into exact sequences.*

Proof of 2.8. In the diagram

$$\begin{array}{ccccccccc} & & & & & & 0 & & \\ & & & & & & \downarrow & & \\ 0 & \longrightarrow & ZC & \longrightarrow & C & \longrightarrow & BC & \longrightarrow & 0 \\ & & & & & \searrow^{\partial} & \downarrow & & \\ & & & & & & C & & \end{array}$$

row and column are exact. Applying t we get

$$\begin{array}{ccccccccc} & & & & & & 0 & & \\ & & & & & & \downarrow & & \\ 0 & \longrightarrow & tZC & \longrightarrow & tC & \longrightarrow & tBC & \longrightarrow & 0 \\ & & & & & {\scriptstyle t\partial}\searrow & \downarrow & & \\ & & & & & & tC & & \end{array}$$

with exact row and column. It follows that $ZtC = \ker(t\partial) = tZC$ and $BtC = \operatorname{im}(t\partial) = tBC$. Now apply t to the exact sequence

$$0 \to BC \to ZC \to HC \to 0$$

and get an exact sequence

$$0 \to BtC \to ZtC \to tHC \to 0,$$

hence $tHC = HtC$. ∎

Another special case where HC determines HtC is the following.

2.9⁊ Proposition. *If R is hereditary and C, C' are free R-complexes such that $HC \cong HC'$ then $HtC \cong HtC'$ for all additive functors t.*

Indeed, $HC \cong HC' \Rightarrow C \simeq C'$ by II, 4.8, hence $tC \simeq tC'$ by 2.6, hence $HtC \cong HtC'$. ∎

Note, however, that this proof does not express HtC in terms of HC. In fact, much of the following section will be devoted to this problem, a problem, by the way, which, historically, was one of the main motives for homological algebra.

Because additive functors do not, in general, preserve exactness it makes sense to classify them according to their behaviour on (short) exact sequences. For the convenience of the reader we list the usual notations although we shall only use some of them. If for all short exact sequences $0 \to M' \to M \to M'' \to 0$ in R-$\mathcal{M}od$ the portion of $0 \to tM' \to tM \to tM'' \to 0$ which is listed in the second column below is exact then the functor t gets the name which is listed in the first column.

(2.10)⁊	*exact*	$0 \to tM' \to tM \to tM'' \to 0$
	left exact	$0 \to tM' \to tM \to tM''$
	right exact	$tM' \to tM \to tM'' \to 0$
	half exact	$tM' \to tM \to tM''$
	mono-functor	$0 \to tM' \to tM$
	epi-functor	$tM \to tM'' \to 0.$

In some of these cases one can get the exact sequence on the right under weaker assumptions, e.g.,

2.11*7* **Proposition.** *If* t: *R-$\mathcal{M}od \to \mathcal{AG}$ is covariant right exact and* $M' \xrightarrow{j} M \to M'' \to 0$ *is exact then* $tM' \xrightarrow{tj} tM \to tM'' \to 0$ *is exact,* i.e. it is not necessary to assume j monomorphic. This implies, for instance, that *compositions of covariant right exact functors are right exact.*

Proof. We have the following exact sequences:

$$0 \to \ker(j) \to M' \to \operatorname{im}(j) \to 0, \quad 0 \to \operatorname{im}(j) \to M \to M'' \to 0,$$

hence

$$t(\ker(j)) \to tM' \to t(\operatorname{im}(j)) \to 0, \quad t(\operatorname{im}(j)) \to tM \to tM'' \to 0,$$

hence, by splicing the last two sequences, $tM' \to tM \to tM'' \to 0$. ∎

2.12 Exercises. *1.* If (X, A) is a pair of spaces and t: $\mathcal{AG} \to \mathcal{AG}$ an additive functor we can apply t to the singular complex $S(X, A)$ and then take homology. The resulting sequence of groups $HtS(X, A)$ is called the *t-homology of* (X, A) and is denoted by $H(X, A; t)$. Study the formal properties of $H(X, A; t)$ in analogy to the treatment of $H(X, A) = H(X, A; \mathrm{Id})$ in Chapter III. Prove $H_i(\mathbb{S}^n; t) \cong t\mathbb{Z}$ for $i = 0, n$, and $= 0$ otherwise ($n > 0$).—We shall come back to these functors $H(X, A; t)$ in § 7.

2. Prove: If t: R-$\mathcal{M}od \to \mathcal{AG}$ is a functor which takes direct sum representations $\{M_i \to M\}_{i=1,2}$ into direct sum representations then t is additive (this is the converse of Prop. 2.3).

3. Construct an abelian group A such that the functor $tX = \operatorname{Hom}_{\mathbb{Z}}(A, X)$ is not strongly additive.

4. The complex

$$C\colon \cdots \xleftarrow{2} \mathbb{Z}_4 \xleftarrow{2} \mathbb{Z}_4 \xleftarrow{2} \mathbb{Z}_4 \leftarrow \cdots$$

is acyclic, $HC = 0$, but tC is not acyclic if $tX = \operatorname{Hom}_{\mathbb{Z}}(\mathbb{Z}_2, X)$.

5. If t: R-$\mathcal{M}od \to \mathcal{AG}$ is additive, and C is a free R-complex such that HC is free and $C_i = 0$ for $i < 0$ then $HtC \cong tHC$ (hint: use 1.12, Exerc. 4).

6. Verify: For every abelian group A the functor $tX = \operatorname{Hom}(X, A)$, $X \in \mathcal{AG}$, is contravariant, strongly additive, left exact; and $tX = X \otimes A$ (= tensor product; cf. § 5) is covariant, strongly additive, right exact. If A is finitely generated then $tX = \operatorname{Hom}(A, X)$ is covariant, strongly additive, left exact, and Hom(Hom(A, X); $\mathbb{Q}$) is contravariant, strongly additive, right exact. The functor t which assigns to every abelian group its torsion-free part, $tX = X/\text{torsion}(X)$, is not half-exact.

3. Derived Functors

3.1 Let $R\text{-}\mathcal{M}od^f$ denote the category of *free* (left) R-modules, and t: $R\text{-}\mathcal{M}od^f \to \mathcal{AG}$ an additive functor. If possible, we want to express HtC in terms of HC, where C is a free R-complex. The simplest non-trivial complexes C are perhaps those with *one* non-vanishing homology module, say $HC = (A, 0)$. This leads to the definition of a resolution: A (free) *resolution* is a free R-complex P such that $P_j = 0$ for $j < 0$, and $H_j P = 0$ for $j \neq 0$; *a resolution of* $M \in R\text{-}\mathcal{M}od$ is a resolution P together with an isomorphism $H_0 P \cong M$. If P, P' are resolutions we denote by $\pi(P, P')$ the abelian group of homotopy classes of chain maps $P \to P'$. Resolutions and homotopy classes of chain maps form a category, denoted by $R\text{-}\mathcal{R}es$, and 0-homology is a functor, H_0: $R\text{-}\mathcal{R}es \to R\text{-}\mathcal{M}od$.

3.2 Proposition. *H_0 is an equivalence of categories, i.e., there exists a functor F: $R\text{-}\mathcal{M}od \to R\text{-}\mathcal{R}es$ such that $H_0 F$ and FH_0 are equivalent with the respective identity functors.*

3.3⌝[4] Corollary and Definition. *There exist functors t_j: $R\text{-}\mathcal{M}od \to \mathcal{AG}$, $j = 0, 1, \ldots$, unique up to equivalence, such that*

$$H_j t P \cong t_j H_0 P, \quad j = 0, 1, \ldots, \tag{3.4}$$

naturally in $P \in R\text{-}\mathcal{R}es$. These functors are called the derived functors of t. *If φ: $t \to t'$ is a natural transformation, then there are unique natural transformations φ_j: $t_j \to t'_j$, called the* derived transformations *such that the following diagram commutes*

$$\begin{array}{ccc} H_j t P & \xrightarrow{H_j(\varphi)} & H_j t' P \\ \| \wr & & \| \wr \\ t_j H_0 P & \xrightarrow{\varphi_j} & t'_j H_0 P, \end{array} \qquad P \in R\text{-}\mathcal{R}es.$$

Proof of 3.3. Put $t_j = H_j t F$ where F is as in 3.2. Then $t_j H_0 P = H_j t F H_0 P \cong H_j t(\text{id}) P = H_j t P$, as required. If $\tilde{t}_j$ also satisfies 3.4 then $\tilde{t}_j M \cong \tilde{t}_j (H_0 F) M = \tilde{t}_j H_0 (FM) \cong H_j t(FM) \cong t_j M$. Similarly for φ_j. ∎

[4] We retain the ⌝-convention of §2 with the additional rule that for cofunctors t one replaces t_j by t^j and $H_{-j} t C$ by $H^j t C$. This applies, for instance to 3.3 which, as it stands, is formulated for covariant t.

3.5 Lemma. *If Q is an R-complex such that $H_jQ=0$ for $j\neq 0$, and P is a free R-complex such that $P_j=0$ for $j<0$ then*

$$H_0\colon\ \pi(P,Q)\cong \operatorname{Hom}_R(H_0P, H_0Q),$$

where π denotes homotopy classes of chain maps. In particular, this applies if P and Q are resolutions.

Proof. We can assume that $Q_j=0$ for $j<0$; if this is not already the case we replace Q_0 by Z_0Q and Q_j by 0 for $j<0$ without changing either side of the asserted isomorphism.

In order to show that H_0 is epimorphic we have to fill the diagram

$$\begin{array}{ccccccccc}
\cdots \xrightarrow{\partial} & P_2 & \xrightarrow{\partial} & P_1 & \xrightarrow{\partial} & P_0 & \xrightarrow{[\,]} & H_0P & \to 0 \\
 & \Big\downarrow{\scriptstyle f_2} & & \Big\downarrow{\scriptstyle f_1} & & \Big\downarrow{\scriptstyle f_0} & & \Big\downarrow{\scriptstyle \alpha} & \\
\cdots \xrightarrow{\partial} & Q_2 & \xrightarrow{\partial} & Q_1 & \xrightarrow{\partial} & Q_0 & \xrightarrow{[\,]} & H_0Q & \to 0
\end{array}$$

for any given α. According to II, 4.7, this can be done step by step.

Suppose now $f\colon P\to Q$ is a chain map such that $H_0f=0$. We have to show that $f\simeq 0$, i.e. we have to construct $s=(s_k\colon P_k\to Q_{k+1})$ such that $\partial s_k+s_{k-1}\partial=f_k$. Proceed by induction on k starting with $s_{-1}=0$. The inductive step from $k-1$ to $k\geq 0$ consists in filling the diagram

$$\begin{array}{ccccc}
P_k & \xrightarrow{(\mathrm{id},\,\mathrm{id})} & P_k\oplus P_k & \longrightarrow & 0 \\
\Big\downarrow{\scriptstyle s_k} & & \Big\downarrow{\scriptstyle (f_k,\,-s_{k-1}\partial)} & & \Big\downarrow \\
Q_{k+1} & \xrightarrow{\partial} & Q_k & \xrightarrow{\partial} & Q_{k-1},
\end{array}$$

where for $k=0$ one replaces Q_{-1} by H_0Q. By II, 4.7 again, the filling s_k exists. ∎

3.6 Corollary. *If P, P' are resolutions and $f\colon P\to P'$ is a chain map such that $H_0f\colon H_0P\cong H_0P'$ then f is a homotopy equivalence* (this is a special case of 1.12 Exerc. 4).

Proof. By 3.5, there is a chain map $g\colon P'\to P$ such that $H_0g=(H_0f)^{-1}$, hence $H_0(fg)=\mathrm{id}$, $H_0(gf)=\mathrm{id}$, hence $fg\simeq\mathrm{id}$, $gf\simeq\mathrm{id}$ by 3.5. ∎

Proof of 3.2. By induction on k we define module-homomorphisms $\partial_k\colon F_kM\to F_{k-1}M$ as follows: $F_{-2}M=0$, $F_{-1}M=M$, F_kM for $k\geq 0$ is the *free* R-module generated by the elements x of $\ker(\partial_{k-1})$, and $\partial_k(x)=x$.

If $\alpha\colon M \to M'$ is an R-homomorphism then we define R-homomorphism $F_k\alpha\colon F_kM \to F_kM'$ such that $F_{-1}\alpha=\alpha$ and $F_k\alpha$ for $k\geq 0$ takes a free generator x of F_kM into the generator $(F_{k-1}\alpha)(x)$ of F_kM'. Then F_k is a functor $R\text{-}\mathcal{M}od \to R\text{-}\mathcal{M}od$, and $\partial_k\colon F_k \to F_{k-1}$ is a natural transformation. Moreover, the sequence

$$(3.7)\qquad 0 \leftarrow M \xleftarrow{\partial_0} F_0M \xleftarrow{\partial_1} F_1M \leftarrow \cdots$$

is obviously exact. Hence $FM=(F_kM, \partial_k)_{k\geq 0}$ is a resolution which depends functorially on M, and ∂_0 induces $H_0FM\cong M$. In other words, we have a functor $F\colon R\text{-}\mathcal{M}od \to R\text{-}\mathcal{R}es$, and an equivalence $H_0F\sim \mathrm{Id}$. In particular, we have a natural isomorphism $\rho P\colon H_0(FH_0P)=(H_0F)(H_0P)\cong H_0P$ for $P\in R\text{-}\mathcal{R}es$. By 3.5, we can define $H_0^{-1}(\rho P)\in\pi(FH_0P, P)$; this is a natural transformation $H_0^{-1}\rho\colon FH_0 \to \mathrm{Id}$. But $H_0^{-1}(\rho P)$ is also a homotopy equivalence, by 3.6, hence $H_0^{-1}\rho\colon FH_0\sim \mathrm{Id}$. ∎

3.8 Proposition. (i) *For free modules M we have $t_jM=0$ if $j>0$, and a natural isomorphism $\iota\colon t_0|R\text{-}\mathcal{M}od^f\cong t$.*

(ii) *For any additive functor $T\colon R\text{-}\mathcal{M}od \to \mathcal{AG}$ and any natural transformation $\varphi\colon t\to T|R\text{-}\mathcal{M}od^f$ there is a unique natural transformation $\Phi\colon t_0\to T$ such that $\Phi|R\text{-}\mathcal{M}od^f=\varphi\iota$.*

For instance, if R is a (skew) field then every module M is free, hence $t_0\cong t$ and $t_j=0$ for $j>0$. For general R again, part (ii) of 3.8 is a characterization of t_0 by a *universal property* (cf. Mitchell, VI.5). The functors t_j for $j>0$ can be characterized as being the *satellites* of t_0 (cf. Cartan-Eilenberg, V.6).

Proof. (i) If M is free then $(M,0)$ is a resolution of M, hence $t_jM\cong H_jt(M,0)=H_j(tM,0)$.

(ii) Consider the diagram

$$(3.9)\qquad \begin{array}{ccccc} 0\leftarrow t_0M & \longleftarrow & t_0F_0M & \longleftarrow & t_0F_1M \\ \downarrow{\scriptstyle\Phi} & & \downarrow{\scriptstyle\varphi\iota} & & \downarrow{\scriptstyle\varphi\iota} \\ 0\leftarrow TM & \longleftarrow & TF_0M & \longleftarrow & TF_1M \end{array}$$

whose rows are obtained from 3.7 by applying t_0 resp. T. The first row is exact because $t_0M=H_0tFM\cong H_0t_0FM$, the last isomorphism by part (i). Therefore, 3.9 admits a unique filler $\Phi\colon t_0M\to TM$. If M is free then $\varphi\iota\colon t_0M\to TM$ is a filler, hence $\Phi=\varphi\iota$. ∎

3.10 Proposition. *If* $0 \to M' \xrightarrow{j} M \xrightarrow{q} M'' \to 0$ *is an exact sequence in R-*$\mathcal{M}od$ *then there is an exact sequence*

$$\cdots \to t_2 M'' \to t_1 M' \xrightarrow{t_1 j} t_1 M \xrightarrow{t_1 q} t_1 M'' \to t_0 M' \xrightarrow{t_0 j} t_0 M \xrightarrow{t_0 q} t_0 M'' \to 0.$$

In particular, t_0 *is right exact.*

Proof. Take a resolution P' of M' and consider the complex

$$Q\colon 0 \leftarrow M \xleftarrow{\varepsilon} P'_0 \xleftarrow{\partial} P'_1 \xleftarrow{\partial} \cdots$$

where ε is the composite $P'_0 \xrightarrow{[\,]} H_0 P' \cong M' \xrightarrow{j} M$. Its homology is concentrated in dimension -1 and agrees with $\operatorname{coker}(j) = M''$ there. If P'' is a resolution of M'' then, by 3.5, there is a chain map $f\colon P''^{-} \to Q$ such that $H_{-1} f\colon H_0 P'' \cong H_{-1} Q$ (in other words, f is a chain map $P'' \to Q$ of degree -1). The mapping-cone Cf has the following form (cf. II, 1.6)

$$0 \leftarrow M \leftarrow P'_0 \oplus P''_0 \leftarrow P'_1 \oplus P''_1 \leftarrow P'_2 \oplus P''_2 \leftarrow \cdots, \tag{3.11}$$

and it is exact because Hf is isomorphic (cf. II, 2.14). The terms to the right of M therefore constitute a resolution P of M (with $P_i = P'_i \oplus P''_i$). It contains P', and $H_0 P' \to H_0 P$ is clearly isomorphic with $j\colon M' \to M$. Further, $P/P' \cong P''$. Altogether, we have an exact sequence

$$0 \to P' \to P \to P'' \to 0 \tag{3.12}$$

of resolutions whose homology sequence $0 \to H_0 P' \to H_0 P \to H_0 P'' \to 0$ is isomorphic with $0 \to M' \xrightarrow{j} M \xrightarrow{q} M'' \to 0$. If we apply t we get an exact (because $P_i = P'_i \oplus P''_i$) sequence

$$0 \to tP' \to tP \to tP'' \to 0; \tag{3.13}$$

its homology sequence has the form which 3.10 asserts. ∎

3.14 Corollary. *Every additive functor* $t\colon R\text{-}\mathcal{M}od^f \to \mathcal{AG}$ *admits a unique (up to equivalence) right-exact extension* $R\text{-}\mathcal{M}od \to \mathcal{AG}$, namely t_0. If t_0 is exact then $t_j = 0$ for $j > 0$.

Indeed, if T is another extension then there is a natural homomorphism $\Phi\colon t_0 M \to TM$, defined by diagram 3.9 with $\varphi = \mathrm{id}$. In this diagram the rows are exact (T being right-exact), and the two vertical arrows on the right are isomorphic, hence Φ is isomorphic. If t_0 is exact then $H_j t_0 C \cong t_0 H_j C$ for every complex C (cf. 2.8). In particular, if P is a resolution of M then $t_j M \cong H_j t_0 P \cong t_0 H_j P = 0$ for $j > 0$. ∎

3.15 Proposition. *If* $t\colon R\text{-}\mathcal{M}od^f \to \mathcal{AG}$ *is strongly additive then the derived functors* $t_j\colon R\text{-}\mathcal{M}od \to \mathcal{AG}$, $j \geq 0$, *are also strongly additive.*

Proof. If $(M_\gamma)_{\gamma\in\Gamma}$ is a family of R-modules, choose resolutions P^γ, $\gamma\in\Gamma$. Then $P=\oplus_\gamma P^\gamma$ is a resolution of $\oplus_\gamma M_\gamma$, hence

$$t_j(\oplus_\gamma M_\gamma)\cong H_j t(\oplus_\gamma P^\gamma)\cong H_j(\oplus_\gamma t P^\gamma)\cong \oplus_\gamma H_j t P^\gamma \cong \oplus_\gamma t_j M_\gamma. \quad ▮$$

3.16 Proposition. *If R is a hereditary ring, and t: R-$\mathcal{M}od^f \to \mathcal{AG}$ is any additive functor then $t_j=0$ for $j>1$, and t_1 is left exact.*

Proof. Given $M\in R$-$\mathcal{M}od$ choose an epimorphism ε: $P_0\to M$ whose domain P_0 is free (e.g. $\varepsilon=\partial_0$ in the proof of 3.2). Then $P_1=\ker(\varepsilon)$ is free, hence $P=(P_0\leftarrow P_1\leftarrow 0\leftarrow\cdots)$ is a resolution of M such that $P_j=0$ for $j>1$, hence $t_j M\cong H_j(tP)=0$ for $j>1$. Proposition 3.10 then shows that t_1 is left exact. ▮

3.17 Exercises. *1.* If H: $\mathcal{K}\to\mathcal{K}'$ is a functor between arbitrary categories such that (i) H: $[X,Y]\to[HX,HY]$ is bijective for all $X,Y\in\mathcal{K}$, and (ii) every $X'\in\mathcal{K}'$ is equivalent with an object of the form HX, $X\in\mathcal{K}$, then H is an equivalence of categories, i.e. there exists a functor F: $\mathcal{K}'\to\mathcal{K}$ such that $FH\sim\mathrm{Id}$, $HF\sim\mathrm{Id}$. Compare this with the proof of 3.2.

2. If t: R-$\mathcal{M}od^f\to\mathcal{AG}$ is an additive functor show that t_j is left exact if and only if $t_{j+1}=0$.

3. If R is hereditary and t: R-$\mathcal{M}od^f\to\mathcal{AG}$ is a monofunctor then $t_1=0$ and t_0 is exact.

4. Prove that the *connecting homomorphism* $t_{j+1}M''\to t_j M'$ which occurs in 3.10 is natural with respect to mappings of short exact sequences.

5. If $R=\mathbb{Z}/p^2\mathbb{Z}$ where p is a prime then $t_{j+1}M\cong t_j M$ for all $j>0$, $M\in R$-$\mathcal{M}od$, t: R-$\mathcal{M}od^f\to\mathcal{AG}$.

4. Universal Coefficient Formula

As before we consider additive functors t: R-$\mathcal{M}od^f\to\mathcal{AG}$ which we extend, as in 2.6, to complexes C of free modules. Assuming R to be *hereditary* we prove the universal coefficient formula,

$$H_n tC\cong t_0 H_n C\oplus t_1 H_{n-1} C;$$

the name is motivated by some important special cases (see § 7).

We retain the 7-convention of the preceeding sections which permits us to concentrate on *covariant* functors.

4.1 Let C be a free R-complex. Consider the inclusions $BC \xrightarrow{\iota} ZC \xrightarrow{i} C$ and the boundary map $\partial: C \to BC$. They can be viewed as chain maps provided, in the second case, we shift dimension indices by one, i.e. replace BC by its suspension BC^+ (recall that $C_n^+ = C_{n-1}$, $\partial^{C^+} = -\partial^C$; see II, 1.3, Example 4). In particular, we can and shall apply $H \circ t$ to these maps. If R is hereditary then $B_n C$ is free hence $B_n C \xrightarrow{\iota_n} Z_n C$ is a resolution of $H_n C$ so that $\operatorname{coker}(t\,\iota_n) \cong t_0 H_n C$, $\ker(t\,\iota_n) \cong t_1 H_n C$.

4.2 Universal Coefficient Theorem. *If R is hereditary and C is a free R-complex then there are unique maps α, β which make the following diagram commutative.*

$$\begin{array}{ccccccccc}
 & & \operatorname{coker}(t\,\iota) & \cong & t_0 HC & & & & \\
 & & \uparrow & & \vdots\,\alpha & & & & \\
tBC & \xrightarrow{t\iota} & tZC & \xrightarrow{H(ti)} & HtC & \xrightarrow{H(t\partial)} & tBC^+ & \xrightarrow{t\iota} & tZC \\
 & & & & \vdots\,\beta & & \uparrow & & \\
 & & & & t_1 HC^+ & \cong & \ker(t\,\iota) & &
\end{array} \tag{4.3}$$

The maps α, β are natural in C (i.e. commute with chain maps). *The sequence*

$$0 \to t_0 H_n C \xrightarrow{\alpha=\alpha_n} H_n t\,C \xrightarrow{\beta=\beta_n} t_1 H_{n-1} C \to 0 \tag{4.4}$$

is exact, and splits. (Universal Coefficient Sequence.)

4.5 Remark. Because the sequence splits $H_n t\,C \cong t_0 H_n C \oplus t_1 H_{n-1} C$; however, the splitting is not natural, for general t and R (see Exerc. 1).

4.6 Remark. In terms of elements and representatives the maps α, β are as follows: Let $x \in tZC$ and $\bar{x} \in t_0 HC$ its coset. Then $(t\,i)(x) \in Zt\,C$ and $\alpha(\bar{x})$ is its homology class, $\alpha(\bar{x}) = [(t\,i)\,x]$. As to β, let $y \in ZtC = \ker(t\partial C\colon tC \to tC)$. Then $d = (t\partial\colon tC \to tBC^+)$ maps y into $\ker(t\,\iota) = t_1 HC^+$, and $\beta[y] = d(y)$.

Proof of 4.2. We first show that the middle row of 4.3 is exact; this implies existence and uniqueness of α, β, and exactness of 4.4. Consider the exact sequence

$$0 \to ZC \xrightarrow{i} C \xrightarrow{\partial} BC^+ \to 0. \tag{4.7}$$

Because BC^+ is free 4.7 splits in every dimension; therefore

$$0 \to tZC \xrightarrow{ti} t\,C \xrightarrow{t\partial} tBC^+ \to 0 \tag{4.8}$$

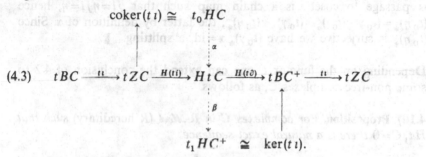

is also exact. The following is a portion of its homology sequence

$$(4.9)\qquad tBC^{+} \xrightarrow{d_*} tZC \xrightarrow{H(ti)} HtC \xrightarrow{H(t\partial)} tBC^{+} \xrightarrow{d_*} tZC.$$

We want to show $d_* = t\iota$, i.e., 4.9 is the middle row of 4.3 which is therefore exact. As remarked, 4.7 splits in every dimension: we find $q: BC^{+} \to C$, $j: C \to ZC$ with $\partial q = \mathrm{id}$, $ji = \mathrm{id}$, $jq = 0$, $ij + q\partial = \mathrm{id}$. Then tq and tj split the sequence 4.8, hence (II, 2.12) $d_* = (tj)(t\partial^C)(tq) = t(j\partial^C q)$. But $\partial q = \mathrm{id}$, $ji = \mathrm{id}$, clearly imply $j\partial^C q = \iota$, hence $d_* = t(\iota)$, as required.

It remains to split 4.4. As before let $j: C \to ZC$ be a retraction onto the cycles ($ji = \mathrm{id}$). Then the composition $\gamma: C \xrightarrow{j} ZC \xrightarrow{\eta} HC$, where η is passage to cosets, is a chain map such that $\gamma i = \eta j i = \eta$, hence $(t_0\eta)_* = (t_0\gamma)_*(t_0 i)_* = (t_0\gamma)_*\,\alpha(t_0\eta)_*$, the latter by definition of α. Since $(t_0\eta)_*$ is surjective we have $(t_0\gamma)_*\,\alpha = \mathrm{id}$, a splitting. ∎

Depending on the functor t, one can extend the conclusion of 4.2 to some non-free complexes C, as follows.

4.10 Proposition. *For complexes C in R-$\mathcal{M}od$* (R hereditary) *such that $Ht_1C = 0$ there is a natural exact sequence*

$$(4.11)\qquad 0 \to t_0 H_n C \xrightarrow{\alpha_n} H_n t_0 C \xrightarrow{\beta_n} t_1 H_{n-1} C \to 0,$$

and this sequence splits. (Note that $t_0 C = tC$, $t_1 C = 0$ if C is free (see 3.8).)

Proof. Consider the commutative diagram

$$(4.12)\qquad \begin{array}{ccccccccc}
 & & 0 & & 0 & & 0 & & \\
 & & \downarrow & & \downarrow & & \downarrow & & \\
\cdots \longrightarrow & & K_{n+1} & \longrightarrow & K_n & \longrightarrow & K_{n-1} & \longrightarrow \cdots & \\
 & & \downarrow{\scriptstyle i_{n+1}} & & \downarrow{\scriptstyle i_n} & & \downarrow{\scriptstyle i_{n-1}} & & \\
\cdots \longrightarrow & & C^f_{n+2} \oplus C^f_{n+1} & \xrightarrow{d_{n+1}} & C^f_{n+1} \oplus C^f_n & \xrightarrow{d_n} & C^f_n \oplus C^f_{n-1} & \longrightarrow \cdots & \\
 & & \downarrow{\scriptstyle \pi_{n+1}} & & \downarrow{\scriptstyle \pi_n} & & \downarrow{\scriptstyle \pi_{n-1}} & & \\
\cdots \longrightarrow & & C_{n+1} & \xrightarrow{\partial_{n+1}} & C_n & \xrightarrow{\partial_n} & C_{n-1} & \longrightarrow \cdots & \\
 & & \downarrow & & \downarrow & & \downarrow & & \\
 & & 0 & & 0 & & 0 & &
\end{array}$$

where C_n^f is the free R-module generated by the elements of C_n, $\pi_n = (\partial_{n+1} p_{n+1}, p_n)$ and $p_n\colon C_n^f \to C_n$ is the homomorphism which associates with every generator of C_n^f the corresponding element of C_n (this was $\partial_0\colon F_0 C_n \to C_n$ in the proof of 3.2), $K_n = \ker(\pi_n)$ and ι_n = inclusion, and the components of d_n are zero or inclusion ($d_n | C_{n+1}^f = 0$, $d_n | C_n^f = (\mathrm{id}, 0)$). We view the rows as complexes so that (4.12) is a short exact sequence of chain maps

(4.13) $$0 \to K \xrightarrow{\iota} \hat{C} \xrightarrow{\pi} C \to 0.$$

Further, $\hat{C}$ is free, hence also K, so that $K_n \to \hat{C}_n$ is a resolution of C_n. Applying $H \circ t$ therefore gives $t_0 C, t_1 C$, i.e., we get an exact sequence of chain maps

$$0 \to t_1 C \to tK \xrightarrow{t\iota} t\hat{C} \xrightarrow{t_0\pi} t_0 C \to 0,$$

or

(4.14) $$0 \to tK/t_1 C \xrightarrow{t\iota} t\hat{C} \xrightarrow{t_0\pi} t_0 C \to 0.$$

Now $\hat{C}$ is nullhomotopic; in fact, $\hat{C}$ is the *cone* of the complex $\{C_{n+1}^f, \partial = 0\}$ (it is clearly acyclic, and the cycles, $Z_n \hat{C} = C_{n+1}^f$, are direct summands; use II, 3.6), hence $t\hat{C} \simeq 0$, hence $Ht\hat{C} = 0$, hence $H_n C \cong H_{n-1} K$ and $H_n t_0 C \cong H_{n-1}(tK/t_1 C)$ from the homology sequences of 4.13, 4.14. Further, $H t_1 C = 0$ by assumption, hence $H_{n-1}(tK/t_1 C) \cong H_{n-1} tK$ from the homology sequence. By 4.2 we have a natural exact sequence (which splits)

$$0 \to t_0 H_{n-1} K \to H_{n-1} tK \to t_1 H_{n-2} K \to 0.$$

Inserting $H_j K = H_{j+1} C$, $H_{n-1} tK = H_n t_0 C$ gives the result. ∎

4.15 Exercises. *1.* Consider the functor $t\colon \mathscr{AG} \to \mathscr{AG}$, $tA = A/2A$, i.e. divide A by $\{a + a \mid a \in A\}$. Show that no non-zero *natural* homomorphism $\varphi\colon H_n t C \to t_0 H_n C$ exists (for free complexes C). In particular, *there is no natural isomorphism* $H_n t C \cong t_0 H_n C \oplus t_1 H_{n-1} C$. Hint: Show first that $\varphi = 0$ if C is the following complex: $C_i = 0$ for $i \neq n, n-1$, $C_n = C_{n-1} = \mathbb{Z}$, $\partial_n = 2$. For any other complex C', and $y \in H_n t C'$ there exists a chain map $f\colon C \to C'$ such that $y \in \mathrm{im}(tf)_*$; then apply naturality.

2. If $t\colon R\text{-}\mathscr{Mod}^f \to \mathscr{AG}$ is an additive functor (R hereditary) and $0 \to C' \xrightarrow{i} C \xrightarrow{p} C'' \to 0$ is an exact sequence of free R-complexes there results a diagram involving the maps i_*, p_*, ∂_* of the homology sequence and the maps α, β of the universal coefficient sequence. Check for commutativity.

3. Show that the universal coefficient sequence (4.11) commutes with natural transformations $t \to t'$ of additive functors.

5. Tensor and Torsion Products

We discuss strongly additive covariant functors $t\colon R\text{-}\mathcal{M}od^f \to \mathcal{AG}$ of free modules and show that they are completely characterized by tR, the value of t on the coefficient ring. The derived functors are called *torsion products;* in symbols, $t_j M = \operatorname{Tor}_j^R(tR, M)$. The functor t_0 is better known as the *tensor product;* its value on M is denoted by $(tR)\otimes_R M$, or simply $(tR)\otimes M$ when there is no danger of confusion. If R is hereditary then one also writes $(tR) *_R M$ or $(tR) * M$ for $t_1 M$ (while $t_j = 0$ for $j > 1$ in this case).—Dual results are discussed in the next §, and relations between the two cases will be established thereafter.

5.1 Definition. Let $t\colon R\text{-}\mathcal{M}od^f \to \mathcal{AG}$ be covariant, additive. The ring R is itself a (left) R-module, (via the ordinary product rx), and the *right translations*

$$\rho_r\colon R \to R, \quad \rho_r(x) = xr, \quad r \in R,$$

are module homomorphisms. We can therefore apply t and get $t(\rho_r)\colon tR \to tR$. Since $t(\rho_{rr'}) = t(\rho_{r'} \circ \rho_r) = t(\rho_{r'}) \circ t(\rho_r)$, and $t(\rho_1) = t(\mathrm{id}) = \mathrm{id}$ we can define a *right R-structure* on tR by $yr = (t\rho_r)(y)$, $r \in R$, $y \in tR$. We always have this structure in mind when we refer to tR as a (right) R-module.

If $\Phi\colon t \to t'$ is a natural transformation then the naturality condition applied to $\rho_r\colon R \to R$ says precisely that $\Phi_R\colon tR \to t'R$ is an R-module homomorphism. Let $[t, t']$ denote the class of all natural transformations and let $e\colon [t, t'] \to \operatorname{Hom}_R(tR, t'R)$ denote the map which to each $\Phi\colon t \to t'$ assigns its value Φ_R on R.

5.2 Proposition. *If t is strongly additive then $e\colon [t, t'] \cong \operatorname{Hom}_R(tR, t'R)$, $e(\Phi) = \Phi_R$. I.e., a natural transformation $\Phi\colon t \to t'$ is completely determined by its value $\Phi_R\colon tR \to t'R$ on the coefficient ring, and this value can be prescribed.*

5.3 Corollary. *If both $t, t'\colon R\text{-}\mathcal{M}od^f \to \mathcal{AG}$ are strongly additive and $tR \cong t'R$ (as R-modules) then $t \sim t'$.*

Proof. Let $tR \xrightarrow{\varphi} t'R \xrightarrow{\varphi'} tR$ be reciprocal isomorphisms. By 5.2 natural transformations $t \xrightarrow{\Phi} t' \xrightarrow{\Phi'} t$ exist with $\Phi_R = \varphi$, $\Phi'_R = \varphi'$, hence $(\Phi'\Phi)_R = \varphi'\varphi = \mathrm{id}$, hence $\Phi'\Phi = \mathrm{id}$ by the uniqueness part of 5.2. Similarly $\Phi\Phi' = \mathrm{id}$.

5.4 Corollary. *Let $T, T'\colon R\text{-}\mathcal{M}od \to \mathcal{AG}$ be covariant additive functors and assume T is strongly additive and right exact. Then*

$$e\colon [T, T'] \to \operatorname{Hom}_R(TR, T'R), \quad e(\Phi) = \Phi_R,$$

is bijective. If also T' is strongly additive and right exact, and if Φ_R: $TR \to T'R$ is an isomorphism then Φ: $T \to T'$ is an equivalence.

Proof. Put $t = T|R\text{-}\mathcal{M}od^f$, $t' = T'|R\text{-}\mathcal{M}od^f$. Then $T \cong t_0$ by 3.14, and $[T, T'] \cong [t_0, T'] \cong [t, t']$ by 3.8 (ii). Our first assertion, $[T, T'] \cong \mathrm{Hom}_R(tR, t'R)$, now follows from 5.2; and the second follows from the first as 5.3 does from 5.2.

Proof of 5.2. Assume $\Phi_R = 0$. Let M be a free R-module and ι: $R \to M$ an R-homomorphism. The commutative diagram

$$\begin{array}{ccc} tR & \xrightarrow{\;t\iota\;} & tM \\ \Big\downarrow{\scriptstyle \Phi_R = 0} & & \Big\downarrow{\scriptstyle \Phi_M} \\ t'R & \xrightarrow[\;t'\iota\;]{} & t'M \end{array}$$

shows $\mathrm{im}(t\iota) \subset \ker(\Phi_M)$. Because t is strongly additive (and M is free) the modules $\mathrm{im}(t\iota)$ generate tM, as ι varies. Hence $\Phi_M = 0$. Since e is clearly additive, this proves that e is injective.

To prove surjectivity, let φ: $tR \to t'R$ be an R-module homomorphism. Let M be a free R-module and $i = \{i_\gamma : R \to M\}_{\gamma \in \Gamma}$ a direct sum representation (equivalently: a base). By assumption $\{t i_\gamma : tR \to tM\}_{\gamma \in \Gamma}$ is also a direct sum representation. We can therefore define

(5.5) $\qquad \Phi^i$: $tM \to t'M \quad$ by $\quad \Phi^i \circ (t i_\gamma) = (t' i_\gamma) \circ \varphi$.

We claim: *$\Phi = \Phi^i$ depends only on M* (not on the base i), *and is a natural transformation with $e(\Phi) = \varphi$.*

Let g: $R \to M$ be any R-homomorphism. Then $g(1)$ is a finite linear combination of base-elements, $g(1) = \sum_k r_k i_k(1) = \sum_k i_k(r_k)$, hence $g = \sum_k i_k \circ \rho_k$, where $r_k \in R$ and ρ_k: $R \to R$ denotes right translation by r_k. Therefore

$$\text{(5.6)} \qquad \begin{aligned} \Phi^i \circ (tg) &= \Phi^i \circ \left(\textstyle\sum_k t i_k \circ t\rho_k\right) = \textstyle\sum_k \Phi^i \circ t i_k \circ t\rho_k = \sum_k t' i_k \circ \varphi \circ t\rho_k \\ &= \textstyle\sum_k t' i_k \circ t' \rho_k \circ \varphi = t'\left(\sum i_k \circ \rho_k\right) \circ \varphi = (t' g) \circ \varphi, \end{aligned}$$

the 3rd equality by 5.5, the 4th because φ is an R-homomorphism.

Let now M, N be two free R-modules with basis i, j, and let f: $M \to N$ an R-homomorphism. We shall show

(5.7) $\qquad \Phi^j \circ (tf) = (t'f) \circ \Phi^i$.

Taking $M = N$, $f = \mathrm{id}$, this gives $\Phi^j = \Phi^i$, i.e., $\Phi_M = \Phi^i$ depends only on M; taking f arbitrary again, it shows that $\{\Phi_M\}$ is natural. Clearly $\Phi_R = \varphi$, so it remains to prove 5.7. Because $\{t\,i_\gamma\colon tR \to tM\}$ is a direct sum representation it suffices to show that 5.7 holds after composition with $t\,i_\gamma$. But

$$\Phi^j \circ (tf) \circ (t i_\gamma) = \Phi^j \circ t(f i_\gamma) = t'(f i_\gamma) \circ \varphi = (t'f) \circ (t' i_\gamma) \circ \varphi = (t'f) \circ \Phi^i \circ t i_\gamma$$

(the 2nd equality uses 5.6, the last 5.5). ∎

We have seen that a strongly additive functor $t\colon R\text{-}\mathcal{M}od^f \to \mathcal{AG}$ (respectively a strongly additive right exact functor $t\colon R\text{-}\mathcal{M}od \to \mathcal{AG}$) is entirely determined by the R-module tR. We now show that this module can be prescribed.

5.8 Proposition and Definition (compare Eilenberg, and Watts). *For every right R-module L there exists a unique (up to equivalence) covariant strongly additive functor* $t\colon R\text{-}\mathcal{M}od^f \to \mathcal{AG}$ *[resp. strongly additive right-exact* $t\colon R\text{-}\mathcal{M}od \to \mathcal{AG}$*] such that* $tR \cong L$. It is called *the tensorproduct with L*, in symbols $tM = L \otimes_R M$. The derived functors $t_j\colon R\text{-}\mathcal{M}od \to \mathcal{AG}$ are called *torsion products;* in symbols, $t_j M = \operatorname{Tor}_j^R(L, M)$. In particular, $\operatorname{Tor}_0^R(L, M) = L \otimes_R M$. If R is hereditary, we also write $L *_R M$ instead of $\operatorname{Tor}_1^R(L, M)$.

Proof. Only the existence of $t\colon R\text{-}\mathcal{M}od^f \to \mathcal{AG}$ has to be shown (see 3.14 for the extension to $R\text{-}\mathcal{M}od$, and 5.3, 5.4 for uniqueness). In every free R-module X we pick a basis $BX \subset X$; for $X = R$ we choose $BX = \{1\}$. Let tX be the set of all functions $\omega\colon BX \to L$ which vanish almost everywhere. In analogy to singular chains (III, 2) we think of ω as a finite linear combination of elements $b \in BX$ with coefficients $\omega_b = \omega(b)$ taken in L, i.e. we write $\omega = \sum_{b \in BX} \omega_b \cdot b$. These linear combinations can be added by adding coefficients, and thereby form an abelian group. If $\alpha\colon X \to X'$ is an R-homomorphism then for every $b \in BX$ we have $\alpha(b) = \sum_{b' \in BX'} \alpha_{b'}^b \cdot b'$, a finite linear combination with coefficients $\alpha_{b'}^b \in R$ (this is the *matrix of* α). We define

$$(5.9) \quad t\alpha\colon tX \to tX', \qquad (t\alpha)\left(\textstyle\sum_{b \in BX} \omega_b \cdot b\right) = \textstyle\sum_{b' \in BX'} \left(\sum_{b \in BX} \omega_b\, \alpha_{b'}^b\right) \cdot b'.$$

Then $t(\mathrm{id}) = \mathrm{id}$ is clear, and $t(\alpha \circ \alpha') = (t\alpha) \circ (t\alpha')$ is the usual multiplication rule for the matrix of a composite map. Thus, $t\colon R\text{-}\mathcal{M}od^f \to \mathcal{AG}$ is a covariant functor.

If $\alpha, \beta\colon X \to X'$ are two R-homomorphisms then clearly $(\alpha + \beta)_{b'}^b = \alpha_{b'}^b + \beta_{b'}^b$, hence $t(\alpha + \beta) = t\alpha + t\beta$, i.e., t is additive. Obviously $tR = L$ (as R-modules); it remains to establish *strong* additivity.

By 2.5, it suffices to show that $\{t i_\gamma\}\colon \bigoplus_{\gamma \in \Gamma} tX_\gamma \to t(\bigoplus_{\gamma \in \Gamma} X_\gamma)$ is *epimorphic* for every family $\{X_\gamma\}$ of free modules (i_γ = inclusion of the γ-th summand), i.e. we have to show that every $y \in t(\bigoplus_{\gamma \in \Gamma} X_\gamma)$ is contained in some partial sum $t(\oplus_{k\in K} X_k)=\oplus_{k\in K} tX_k$ with finite $K\subset\Gamma$. Now y is a finite linear combination $y=\sum\omega_b\cdot b$, and every $b\in B(\oplus_{\gamma\in\Gamma} X_\gamma)$ is contained in a finite partial sum of $\oplus_{\gamma\in\Gamma} X_\gamma$, hence a finite set $K\subset\Gamma$ exists such that $\omega_b\neq 0\Rightarrow b\in\oplus_{k\in K} X_k$. Let

$$\oplus_{\gamma\in\Gamma} X_\gamma \xrightarrow{p} \oplus_{k\in K} X_k \xrightarrow{j} \oplus_{\gamma\in\Gamma} X_\gamma$$

denote projection and inclusion. Then $\omega_b\neq 0\Rightarrow (jp)\,b=b$, hence

$$(tj)(tp)\,y=t(jp)\sum\omega_b\cdot b=\sum\omega_b\cdot(jp)\,b=\sum\omega_b\cdot b=y,$$

hence $y\in\operatorname{im}(tj)=t(\oplus_{k\in K} X_k)$. ∎

5.10 Definition. If L, L' are right R-modules and $f\colon L\to L'$ is an R-homomorphism then, by 5.4, there is a unique natural transformation $f\otimes_R M\colon L\otimes_R M\to L'\otimes_R M$ such that $f\otimes_R R\colon L\to L'$ agrees with f. If $g\colon M\to M'$ is an R-homomorphism then $(L'\otimes_R g)\circ(f\otimes_R M)=(f\otimes_R M')\circ(L\otimes_R g)$, by naturality of $f\otimes_R$. We denote this homomorphism by

$$f\otimes_R g\colon L\otimes_R M\to L'\otimes_R M';$$

in particular, $f\otimes_R M=f\otimes\mathrm{id}_M$, $L\otimes_R g=\mathrm{id}_L\otimes_R g$. It follows immediately from the definitions that

$$(f\otimes_R g)\circ(f'\otimes_R g')=(f\circ f')\otimes_R(g\circ g'),\qquad \mathrm{id}_L\otimes_R\mathrm{id}_M=\mathrm{id}_{L\otimes M},$$

whenever the compositions are defined. These formulas assert that $\otimes_R$ *is a functor of two variables* $(L,M)\in(\mathcal{Mod}\text{-}R)\times(R\text{-}\mathcal{Mod})$. Moreover,

$$(5.11)\qquad \begin{aligned}(f_1+f_2)\otimes_R g&=f_1\otimes_R g+f_2\otimes_R g,\\ f\otimes_R(g_1+g_2)&=f\otimes_R g_1+f\otimes_R g_2,\end{aligned}$$

the first equation because $(f_1+f_2)\otimes_R$ and $(f_1\otimes_R)+(f_2\otimes_R)$ agree on R, the second equation because $L\otimes_R$ is additive.

5.12 Example. We want to compute $L\otimes_R M$, $L*_R M$ if $R=\mathbb{Z}$ and M is a cyclic group. If M is free cyclic, $M\cong\mathbb{Z}$, then $L\otimes M\cong L$ by definition, and $L*\mathbb{Z}=0$ as for any derived functor t_1. If M is finite cyclic, say of order n, $M=\mathbb{Z}_n$, then we apply $L\otimes$ to the exact sequence

$$0\to\mathbb{Z}\xrightarrow{n\cdot\mathrm{id}}\mathbb{Z}\to\mathbb{Z}_n\to 0$$

and get (by 3.10) an exact sequence

$$(5.13)\qquad 0\to L*\mathbb{Z}_n\to L\xrightarrow{n\cdot\mathrm{id}} L\to L\otimes\mathbb{Z}_n\to 0,$$

hence

$$(5.14)\qquad L\otimes\mathbb{Z}_n\cong L/nL,\quad L*\mathbb{Z}_n\cong\{y\in L\mid n\cdot y=0\}.$$

5.15 Corollary. *If L is a finitely generated abelian group and p a prime number then* $\dim(L \otimes \mathbb{Z}_p) = \operatorname{rank}(L) + \dim(L * \mathbb{Z}_p)$, *where* dim *denotes the vector space dimension over the field* $\mathbb{Z}_p$.

This formula is useful in connection with the Euler characteristic (cf. 7.21). If L is cyclic then the formula is immediate from 5.14. In the general case, L is a direct sum of cyclic groups, and the formula follows because both sides are additive in L. ∎

As an interesting exercise the reader might prove the same result for non-finitely-generated L provided every element in $\bigcap_{n=1}^{\infty} p^n L$ has finite order prime to p.

5.16 Example. An R-module L is called *flat* if $L\otimes$ is an exact functor. We want to determine all flat abelian groups ($=\mathbb{Z}$-modules). We claim: *The functor* $L\otimes\colon \mathcal{AG} \to \mathcal{AG}$ *is exact if and only if L is torsion-free*, i.e., L has no (non-zero) elements of finite order, i.e., the map $n\colon L \to L$ is injective for all integers $n \neq 0$.

Proof. If $y \in L$ is not zero but of finite order, say $n \cdot y = 0$, then 5.12 shows that

$$0 \to L \otimes \mathbb{Z} \xrightarrow{\mathrm{id} \otimes n} L \otimes \mathbb{Z} \to L \otimes \mathbb{Z}_n \to 0$$

is *not exact*, hence $L\otimes$ is not exact.

If $L = \mathbb{Z}$ then $L\otimes = \mathrm{id}$ is obviously exact. If L is (finitely generated and) free then $L\otimes$ is a (finite) direct sum of identity functors and therefore exact. Now take any torsionfree abelian group L, let $F \subset L$ be a finitely generated subgroup, and let $0 \to X_1 \xrightarrow{j} X_0 \to M \to 0$ be an exact sequence with free X_1, X_0. We have a commutative diagram

$$\begin{array}{ccc} L \otimes X_1 & \xrightarrow{\mathrm{id} \otimes j} & L \otimes X_0 \\ \cup\big\uparrow & & \cup\big\uparrow \\ F \otimes X_1 & \longrightarrow & F \otimes X_0 \end{array}$$

in which the vertical arrows are monomorphic by the very construction of the tensor-product (as a group of functions with values in F respectively L). The lower horizontal map is monomorphic because F is free (see above), hence the restriction of $\mathrm{id} \otimes j$ to $F \otimes X_1$ is monomorphic. But every $\omega \in L \otimes X_1$ has the form $\sum \omega_b \cdot b$, hence $\omega \in F \otimes X$ where F is generated by $\{\omega_b\}$, hence the whole map $\mathrm{id} \otimes j$ is monomorphic. Since $X_1 \to X_0$ is a resolution of M this proves $L * M = \ker(\mathrm{id} \otimes j) = 0$, and the exact sequence 3.10 shows that $L\otimes$ is exact. ∎

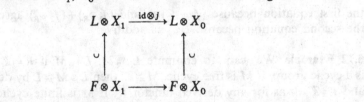

Concluding this section we make a few comments on $L \otimes_R M$ as a *functor of L*. The notation $L \otimes M$ already suggests some symmetry between L and M; this will be fully justified in § 8 (see also Exerc. 1c). Here we only show that $L \otimes_R$ and $\otimes_R M$ have analogous exactness properties.

5.17 Proposition. *For every exact sequence* $0 \to L' \to L \to L'' \to 0$ *in* $\mathcal{M}od$-R *and every* $M \in R$-$\mathcal{M}od$ *there is an exact sequence*

$$\begin{aligned}(5.18)\qquad &\cdots \to \operatorname{Tor}^R_{j+1}(L'', M) \to \operatorname{Tor}^R_j(L', M) \to \operatorname{Tor}^R_j(L, M) \to \operatorname{Tor}^R_j(L'', M)\\ &\to \operatorname{Tor}^R_{j-1}(L', M) \to \cdots \to L \otimes_R M \to L'' \otimes_R M \to 0.\end{aligned}$$

In particular, $\otimes_R M$ is right exact (and $*_R M$ is left exact if R is hereditary).

Proof. If F is a resolution of M then F_i is a direct sum of terms R, hence $L \otimes F_i$ is a direct sum of terms L, hence $0 \to L' \otimes F_i \to L \otimes F_i \to L'' \otimes F_i \to 0$ is a direct sum of sequences $0 \to L' \to L \to L'' \to 0$; in particular, it is exact. Therefore $0 \to L' \otimes_R F \to L \otimes_R F \to L'' \otimes_R F \to 0$ is an exact sequence of complexes whose homology sequence has the required form 5.18. ∎

5.19 Proposition. *If* $\cdots \to E_{j+1} \to E_j \to E_{j-1} \to \cdots \to E_0$ *is a resolution of* $L \in \mathcal{M}od$-R *then* $H_j(E \otimes_R M) \cong \operatorname{Tor}^R_j(L, M)$; *i.e. in order to compute* $\operatorname{Tor}(L, M)$ *one can resolve either variable.*

Proof. Note first that $E_j \otimes_R$ is an exact functor ($E_j = \bigoplus R$ implies that $E_j \otimes_R$ is a direct sum of identity functors), hence $\operatorname{Tor}^R_n(E_j, M) = 0$ for $n > 0$, by 3.14. Now consider the modules $L_j = \operatorname{coker}(E_{j+1} \to E_j)$; we have $L_0 \cong L$, $L_j \cong B_{j-1}\, E$ for $j > 0$, and for every j an exact sequence $0 \to L_{j+1} \to E_j \to L_j \to 0$. The corresponding long exact sequence 5.18 shows $\operatorname{Tor}^R_{n+1}(L_j, M) \cong \operatorname{Tor}^R_n(L_{j+1}, M)$ for $n > 0$ (because $\operatorname{Tor}^R_n(E_j, M) = 0$), hence by iteration, $\operatorname{Tor}^R_j(L, M) \cong \operatorname{Tor}^R_1(L_{j-1}, M)$ for $j > 0$. The last term occurs in the following commutative diagram

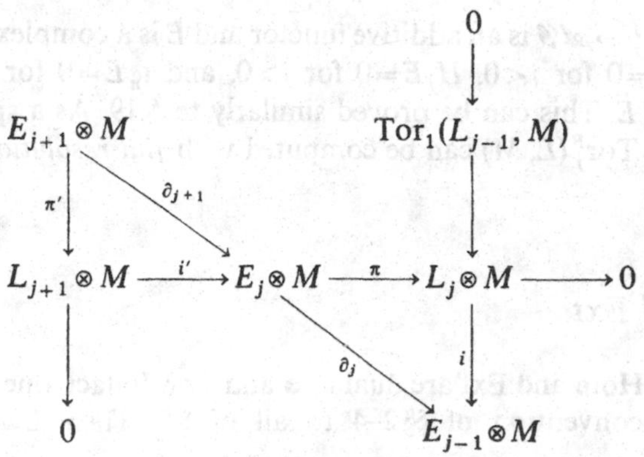

whose rows and columns are bits of exact sequences 5.18. We get $\mathrm{Tor}_1(L_{j-1}, M) \cong \ker(i) \cong \ker(i\,\pi)/\ker(\pi) = \ker(\partial_j)/\mathrm{im}(i') = \ker(\partial_j)/\mathrm{im}(i'\,\pi') = \ker(\partial_j)/\mathrm{im}(\partial_{j+1}) = H_j(E \otimes M)$. This proves 5.19 for $j>0$. As to $j=0$, we have an exact sequence $E_1 \to E_0 \to L \to 0$, hence (2.11) an exact sequence $E_1 \otimes M \to E_0 \otimes M \to L \otimes M \to 0$, hence $H_0(E \otimes M) \cong L \otimes M$. ∎

5.21 Remark. There is an obvious analogy between the preceding proof and the proof of V, 1.3. Both are "degenerating-spectral-sequence arguments" (cf. Godement, I.4.4).

5.22 Exercises. *1.* (a) Any direct sum $t = \oplus t_\gamma$ of (strongly additive) right-exact functors $\mathcal{Mod}\text{-}R \to \mathcal{AG}$ is (strongly additive) right-exact.

(b) If $\tau_1 \to \tau_0 \to t \to 0$ is an exact sequence of natural transformations and if τ_1, τ_0 are (strongly additive and) right exact then so is t.

(c) The tensor-product $L \otimes_R M$, as a *functor of* L (M fixed), is strongly additive and right exact. Consequently, $L \otimes_R M \cong M \otimes_R L$ if R is commutative.

2. The reader is urged to study also the usual existence proof for 5.8; cf. for instance, MacLane V.1. Still another possibility to construct $tM = L \otimes_R M$ (for free M) is as follows: Put $tM = \mathrm{Hom}_R(\mathrm{Hom}_R(M, R), L)$ for finitely generated-, and $tM = \varinjlim(tM_\alpha)$ for arbitrary free M, where the direct limit (see VIII, 4) is taken over all finitely generated submodules M_α of M.

3. If A is a finite abelian group then $L *_{\mathbb{Z}} A \cong \mathrm{Hom}_{\mathbb{Z}}(A, L)$, naturally in $L \in \mathcal{AG}$.

4. (a) If $j\colon \mathbb{Z} \to \mathbb{Q}$ is the inclusion then the kernel of $L = L \otimes \mathbb{Z} \xrightarrow{\mathrm{id} \otimes j} L \otimes \mathbb{Q}$ coincides with torsion(L), the subgroup of elements of finite order.

(b) $L * (\mathbb{Q}/\mathbb{Z}) \cong$ torsion(L).

5. If $t\colon R\text{-}\mathcal{Mod}^f \to \mathcal{AG}$ is an additive functor and E is a complex in $R\text{-}\mathcal{Mod}$ such that $E_j = 0$ for $j<0$, $H_j E = 0$ for $j>0$, and $t_n E = 0$ for $n>0$ then $H_j t_0 E \cong t_j H_0 E$. This can be proved similarly to 5.19. As a special case, it asserts that $\mathrm{Tor}_j^R(L, M)$ can be computed with *flat resolutions* (instead of free ones).

6. Hom and Ext

The functors Hom and Ext are dual to ⊗ and Tor. In fact, one can simply apply the ⁊-convention of §§2–4 to all of §5. Then $L \otimes_R$ becomes

$\mathrm{Hom}_R(-,L)$, and $\mathrm{Tor}_j^R(L,-)$ becomes $\mathrm{Ext}_R^j(-,L)$. However, because of the importance of Hom and Ext and because too many 7-s might confuse the reader we give a separate—if somewhat repetitious—treatment. Proofs will be abbreviated or omitted, and notations are taken over from § 5. The section numbers are chosen to correspond with § 5; thus 6.1 is dual to 5.1 etc.

6.1 Definition. Let $t\colon R\text{-}\mathcal{M}od^f \to \mathcal{A}\mathcal{G}$ be an additive cofunctor. Define a *left R-structure* on tR by $ry = t(\rho_r)\,y$, $r \in R$, $y \in tR$, where $\rho_r\colon R \to R$ is the right translation by r. If $\Phi\colon t' \to t$ is a natural transformation then $\Phi_R\colon t'R \to tR$ is an R-module homomorphism. Let $[t', t]$ denote the class of all natural transformations $t' \to t$.

6.2 Proposition. *If t is strongly additive (contravariant) then*

$$e\colon\ [t', t] \to \mathrm{Hom}_R(t'R, tR), \qquad e(\Phi) = \Phi_R,$$

is bijective.

6.3 Corollary. *If both $t, t'\colon R\text{-}\mathcal{M}od^f \to \mathcal{A}\mathcal{G}$ are strongly additive, and $tR \cong t'R$ (as R modules) then $t \sim t'$.* ▮

6.4 Proposition. *Let $T, T'\colon R\text{-}\mathcal{M}od \to \mathcal{A}\mathcal{G}$ be additive cofunctors and assume T is strongly additive and left exact. Then*

$$e\colon\ [T', T] \to \mathrm{Hom}_R(T'R, TR), \qquad e(\Phi) = \Phi_R,$$

is bijective. If also T' is strongly additive and left exact, and if $\Phi_R\colon T'R \to TR$ is an isomorphism then $\Phi\colon T' \to T$ is an equivalence.

Proof of 6.2. Assume $\Phi_R = 0$. Let M be a free R-module and $\iota\colon R \to M$ an R-homomorphism. The commutative diagram

$$\begin{array}{ccc} tR & \xleftarrow{\ t\iota\ } & tM \\ \Big\uparrow{\scriptstyle \Phi_R = 0} & & \Big\uparrow{\scriptstyle \Phi_M} \\ t'R & \xleftarrow[\ t'\iota\ ]{} & t'M \end{array}$$

shows $\mathrm{im}(\Phi_M) \subset \ker(t\iota)$. Because t is strongly additive we have

$$\bigcap_{\iota} \ker(t\iota) = \{0\},$$

hence $\Phi_M = 0$. This proves that e is injective.

To prove surjectivity let $\varphi\colon t'R \to tR$ be an R-homomorphism. Let M be a free R-module and $i = \{i_\gamma\colon R \to M\}_{\gamma \in \Gamma}$ a direct sum representation.

By assumption $\{t\,i_\gamma\colon t\,M \to t\,R\}_{\gamma\in\Gamma}$ is a direct product representation. We can therefore define

$$(6.5) \qquad \Phi^i\colon t'M \to t\,M \quad \text{by} \quad (t\,i_\gamma)\circ\Phi^i = \varphi\circ(t'\,i_\gamma).$$

The reader will have no difficulty in dualizing the rest of the proof of 5.2, i.e. to show that $\Phi_M = \Phi^i$ is independent of the base i, and is a natural transformation Φ with $e(\Phi)=\varphi$. ▮

6.8 Proposition and Definition. *For every left R-module L there exists a unique (up to equivalence) strongly additive cofunctor t:* $R\text{-}\mathcal{M}od^f \to \mathcal{AG}$ *[resp. strongly additive left-exact t:* $R\text{-}\mathcal{M}od \to \mathcal{AG}$*] such that* $t\,R \cong L$. *An example of such a cofunctor is* $t\,M = \mathrm{Hom}_R(M,L)$. Its derived functors t^j are denoted by $t^jM = \mathrm{Ext}^j_R(M,L)$; in particular, $\mathrm{Ext}^0_R(M,L) \cong \mathrm{Hom}_R(M,L)$.

If R is *hereditary* then we also write $\mathrm{Ext}_R(M,L)$ or $\mathrm{Ext}(M,L)$ instead of $\mathrm{Ext}^1_R(M,L)$ (while $\mathrm{Ext}^j_R = 0$ for $j>1$ in this case).

In order to prove 6.8 one has only to verify that $\mathrm{Hom}_R(-,L)$ is indeed strongly additive (for left-exactness and $\mathrm{Hom}_R(R,L)=L$ see 1.6 and 1.9). But $\mathrm{Hom}_R(\oplus_\gamma M_\gamma, L) \cong \prod_\gamma \mathrm{Hom}_R(M_\gamma, L)$ holds by the very definition of the direct sum (I, 2.13, 2.14). ▮

6.12 Example. We want to compute $\mathrm{Hom}_R(M,L)$, $\mathrm{Ext}_R(M,L)$ if $R=\mathbb{Z}$ and M is a cyclic group. Clearly $\mathrm{Hom}_{\mathbb{Z}}(\mathbb{Z},L)=L$, $\mathrm{Ext}_{\mathbb{Z}}(\mathbb{Z},L)=0$. If M is finite cyclic, say $M=\mathbb{Z}_n$, we apply $\mathrm{Hom}(-,L)$ to the exact sequence $0\to\mathbb{Z}\xrightarrow{n}\mathbb{Z}\to\mathbb{Z}_n\to 0$ and get (3.10) an exact sequence

$$(6.13) \qquad 0\to\mathrm{Hom}(\mathbb{Z}_n,L)\to L\xrightarrow{n} L\to\mathrm{Ext}(\mathbb{Z}_n,L)\to 0,$$

hence

$$(6.14) \qquad \begin{aligned} \mathrm{Hom}(\mathbb{Z}_n,L) &\cong \{y\in L \mid n\,y=0\} \cong L*\mathbb{Z}_n,\\ \mathrm{Ext}(\mathbb{Z}_n,L) &\cong L/nL \cong L\otimes\mathbb{Z}_n. \end{aligned}$$

6.16 Example. An R-module L is called *injective* if $\mathrm{Hom}_R(-,L)$ is an exact functor. Which abelian groups ($\mathbb{Z}$-modules) are injective? We claim: *The cofunctor* $\mathrm{Hom}(-,L)\colon \mathcal{AG}\to\mathcal{AG}$ *is exact if and only if L is divisible, i.e. the map* $n\colon L\to L$ *is surjective for all integers* $n\neq 0$.

While this result is dual to 5.16 its proof is more difficult; we shall only give some indications and refer to Mitchell, II.15.4 for more detail. Firstly, if $n\colon L\to L$ is not surjective, $n\neq 0$, then the sequence 6.13 shows that $\mathrm{Hom}(-,L)$ is not exact. Conversely, assume L is divisible. One has to prove that $\mathrm{Hom}(M,L)\to\mathrm{Hom}(M',L)$ is surjective for every group M and subgroup M', i.e. one has to show that every homomorphism $\alpha\colon M'\to L$ extends to M. Let $y\in M-M'$; if $m\,y\in M'$ for some integer

$m \neq 0$ let $n \in \mathbb{Z}$ generate the ideal of all such m, choose $z \in L$ such that $nz = \alpha(ny)$, and define an extension β of α to $\{M', y\}$, the subgroup generated by M' and y, by $\beta|M' = \alpha$, $\beta(y) = z$. If $m \neq 0 \Rightarrow my \notin M'$ one extends by $\beta(y) = 0$. By iteration (transfinite if M/M' is not finitely generated) this procedure leads to an extension $M \to L$. ∎

6.21 An R-module M is called *projective* if $\mathrm{Hom}_R(M, -)$ is an exact functor. We claim, *a module M is projective if and only if M is a direct summand of a free module.*

Proof. Let $\{M_\gamma\}_{\gamma \in \Gamma}$ be any family of modules. Clearly $\mathrm{Hom}_R(\oplus_\gamma M_\gamma, -) \cong \prod_\gamma \mathrm{Hom}_R(M_\gamma, -)$ is exact if and only if each functor $\mathrm{Hom}_R(M_\gamma, -)$ is exact. Since $\mathrm{Hom}_R(R, -) = \mathrm{id}$, this proves first that every free module $F = \oplus_\gamma R$ is projective, and then that every direct summand of a free module is projective.

Assume now $\mathrm{Hom}_R(M, -)$ is exact. Choose an exact sequence

$$0 \to G \to F \xrightarrow{p} M \to 0$$

such that F is free, apply $\mathrm{Hom}_R(M, -)$ and get an exact sequence $\mathrm{Hom}_R(M, F) \xrightarrow{p} \mathrm{Hom}_R(M, M) \to 0$. In particular, there exists

$$\beta \in \mathrm{Hom}_R(M, F)$$

such that $\mathrm{id}_M = p(\beta) = p \circ \beta$, hence β maps M isomorphically onto a direct summand of F. ∎

6.22 Exercises. *1.* For every $M \in R\text{-}\mathcal{M}od$ and every exact sequence $0 \to L' \to L \to L'' \to 0$ in $R\text{-}\mathcal{M}od$ there is an exact sequence

$$0 \to \mathrm{Hom}_R(M, L') \to \mathrm{Hom}_R(M, L) \to \cdots \to \mathrm{Ext}_R^{j-1}(M, L'') \to$$
$$\mathrm{Ext}_R^j(M, L') \to \mathrm{Ext}_R^j(M, L) \to \mathrm{Ext}_R^j(M, L'') \to \mathrm{Ext}_R^{j+1}(M, L') \to \cdots.$$

This is 5.17 dualized. Show that M is projective if and only if

$$\mathrm{Ext}_R^1(M, -) = 0.$$

2. If R is hereditary and $L \in R\text{-}\mathcal{M}od$ admits a resolution $F_1 \to F_0$ by *finitely generated* free modules ("L is finitely resolvable") then

$$\mathrm{Ext}_R(M, L) \cong \mathrm{Ext}_R(M, R) \otimes_R L.$$

Further, L is projective if and only if $\mathrm{Ext}_R(L, R) = \{0\}$.

3. If A is a finite abelian group then $\mathrm{Ext}_{\mathbb{Z}}(A, L) \cong L \otimes_{\mathbb{Z}} A$, naturally in $L \in \mathcal{AG}$.

4. If P is a projective R-module then there exists a *free* module F such that $P \oplus F$ is free (Eilenberg). Hint: By 6.21 there exists some module P' such that $P \oplus P'$ is free. Consider the relation $(P \oplus P') \oplus (P \oplus P') \oplus \cdots \cong P \oplus (P' \oplus P) \oplus (P' \oplus P) \oplus \cdots$.

5. If C is a complex of free abelian groups such that $H_n C$ contains elements of infinite order then $H_{-n}\operatorname{Hom}(C, \mathbb{Q}) \neq \{0\}$ (because $H\operatorname{Hom}(C, \mathbb{Q}) \cong \operatorname{Hom}(HC, \mathbb{Q})$). If p is a prime such that $(H_{n-1}C) * \mathbb{Z}_p = \{x \in H_{n-1}C \mid px = 0\} \neq \{0\}$ then $H_{-n}\operatorname{Hom}(C, \mathbb{Z}_p) \neq \{0\}$ (because $H\operatorname{Hom}(C, \mathbb{Z}_p) \cong \operatorname{Hom}(H(C \otimes \mathbb{Z}_p), \mathbb{Z}_p) \supset \operatorname{Hom}(H_{n-1}C * \mathbb{Z}_p, \mathbb{Z}_p)$). Consequently, if $H\operatorname{Hom}(C, k) = \{0\}$ for every prime field k then $HC = \{0\}$, hence $C \simeq 0$.

7. *Singular Homology and Cohomology with General Coefficient Groups*

We apply additive functors t of abelian groups to singular complexes SX of spaces X and discuss the formal properties and the significance of the resulting homology groups $HtS(X)$.

7.1 Definition. The singular complex $S(X, A)$ of a pair of spaces consists of *free* abelian groups. Therefore, any additive functor $t\colon \mathscr{AG}^f \to \mathscr{AG}$, defined on free abelian groups, can be applied to $S(X, A)$ and yields a new complex $tS(X, A)$. Its homology groups are denoted by $H(X, A; t) = HtS(X, A)$, and are called the (singular) *t-homology groups of* (X, A).

Usually one considers *strongly additive* functors only, i.e. tensor products and Hom-functors, and one uses a special notation as follows. The complex $S(X, A) \otimes G$[5] respectively $\operatorname{Hom}(S(X, A), G)$ is called *singular (chain- resp. cochain-) complex of* (X, A) *with coefficients in* G (G an abelian group), and is denoted by $S(X, A; G)$ respectively $S^*(X, A; G)$. The elements of $S_n(X, A; G) = S_n(X, A) \otimes G$ respectively

$$S^n(X, A; G) = (S^*(X, A; G))^n = \operatorname{Hom}(S_n(X, A), G)$$

are called *singular n-chains* respectively *n-cochains of* (X, A) *with coefficients in* G.

By definition (cf. proof of 5.8) an n-chain $c \in S_n(X; G)$ is a *finite* linear combination $c = \sum_\sigma c_\sigma \cdot \sigma$ of singular n-simplices $\sigma\colon \Delta_n \to X$ with coefficients $c_\sigma \in G$; addition is given by $(c + c')_\sigma = c_\sigma + c'_\sigma$. The n-chains in $S_n(X, A; G) = S_n(X; G)/S_n(A; G)$ can be thought of as finite linear combinations $\sum_\sigma c_\sigma \cdot \sigma$ where $\sigma(\Delta_n) \not\subset A$. Dually, n-cochains $\varphi \in S^n(X, A; G)$

[5] Or rather $G \otimes S(X, A)$. However, $M \otimes N \cong N \otimes M$, by 8.13.

are functions $\varphi(\sigma)$ such that $\varphi(\sigma)=0$ if $\sigma(\Delta_n)\subset A$; these functions are added by adding values. As for ordinary (integral) chains, the boundary operator is given by an alternating sum, $\partial c=\sum_\sigma\sum_{i=0}^{n}(-1)^i c_\sigma\cdot(\sigma\,\varepsilon_n^i)$, respectively $[\partial(\varphi)](\tau)=\sum_{i=0}^{n+1}(-1)^i\varphi(\tau\,\varepsilon_{n+1}^i)$ where $\tau\colon \Delta_{n+1}\to X$. The boundary operator for cochains is usually denoted by δ; thus $\delta(\varphi)=\varphi\circ\partial$. Often it will be appropriate to replace δ by $(-1)^{n+1}\delta$ (cf. 10.28).

If in the preceding notation we replace S by Z, B, H we get singular *(co-)cycles, (co-)boundaries, (co-)homology with coefficients in G*. For example, $H^n(X,A;G)=H_{-n}\operatorname{Hom}(S(X,A);G)=H^n S^*(X,A;G)$ is called *n-th cohomology group of* (X,A) *with coefficients in* G, and $H^*(X,A;G)=\{H^n(X,A;G)\}_{n\in\mathbb{Z}}$.

If G is an R-module (R some ring) then $S(X,A;G)=S(X,A)\otimes_{\mathbb{Z}}G$ and $S^*(X,A;G)=\operatorname{Hom}_{\mathbb{Z}}(S(X,A),G)$ are complexes of modules. In particular, the homology of these complexes consists of R-modules, i.e. *the (co-) homology of* (X,A) *with coefficients in an R-module consists of R-modules.*

The formal properties of ordinary integral homology $H(X,A)=H(X,A;\mathbb{Z})$ carry over to arbitrary coefficients; in fact, they carry over without further complications to t-homology where $t\colon \mathcal{AG}^f\to\mathcal{AG}$ is any additive functor. We list the most important properties. As in §§ 2–4 we use the ⁊-convention, i.e. we formulate the results for covariant functors t only and we mark by ⁊ all sections which are also valid after replacing covariant by contravariant, reversing arrows in the range category of t, exchanging lower and upper indices and stars, etc.

7.2⁊ If $f\colon (X,A)\to(Y,B)$ is a map of pairs then $tSf\colon tS(X,A)\to tS(Y,B)$ is a chain map. The induced homomorphism of homology is denoted by

$$f_*=H(f;t)\colon\ H(X,A;t)\to H(Y,B;t).$$

It clearly satisfies $(fg)_*=f_*g_*$, $\mathrm{id}_*=\mathrm{id}$, i.e. *t-homology* $H(X,A;t)$ *is a covariant functor from pairs of spaces to graded abelian groups.*

7.3⁊ For any pair (X,A) the sequence $0\to SA\xrightarrow{i}SX\xrightarrow{j}S(X,A)\to 0$ is exact and splits in every dimension. Since t is applied dimension-wise the sequence $0\to tSA\xrightarrow{ti}tSX\xrightarrow{tj}tS(X,A)\to 0$ is also exact (and splits in every dimension). In particular, *there results* (cf. II, 2.9 and III, 3.2) *a connecting homomorphism* $\partial_*\colon H_{q+1}(X,A;t)\to H_q(A;t)$ *and a (natural) exact sequence*

$$\cdots\xrightarrow{\partial_*}H_{q+1}(A;t)\xrightarrow{i_*}H_{q+1}(X;t)\xrightarrow{j_*}H_{q+1}(X,A;t)\xrightarrow{\partial_*}H_q(A;t)$$
$$\xrightarrow{i_*}H_q(X;t)\xrightarrow{j_*}\cdots,$$

called (t-) homology sequence of (X,A).

7.4? If $f, g\colon (X, A)\to(Y, B)$ are homotopic maps then $Sf, Sg\colon S(X,A)\to S(Y,B)$ are homotopic by III, 5.1, hence $tSf\simeq tSg\colon tS(X,A)\to tS(Y,B)$ by 2.6, and therefore $f_*=g_*\colon H(X,A;t)\to H(Y,B;t)$. I.e., *t-homology is homotopy invariant; it can be viewed as a functor on the category whose morphisms are homotopy classes of continuous maps (of pairs).*

7.5 If (X, A) is a pair and $\mathcal{U}=\{U\}$ is a family of subsets $U\subset X$ such that every point of X is contained in the interior of A or in the interior of some U then $S(\mathcal{U}, \mathcal{U}\cap A)\to S(X,A)$ was shown to be a homotopy equivalence (III, 7.3), where $S\mathcal{U}\subset SX$ is the subcomplex generated by all SU, and $S(\mathcal{U}, \mathcal{U}\cap A)=S\mathcal{U}/S(\mathcal{U}\cap A)$. It follows (by 2.6) that $tS(\mathcal{U}, \mathcal{U}\cap A)\simeq tS(X,A)$.

As a corollary (III, 7.4) we obtained that the inclusion $j\colon (X-B, A-B)\to(X,A)$ induces a homotopy equivalence $S(X-B, A-B)\simeq S(X,A)$ for every subset $B\subset A$ whose closure $\bar{B}$ is contained in the interior $\mathring{A}$ of A. It follows (2.6) that $tS(X-B, A-B)\simeq tS(X,A)$, hence $H(X-B, A-B; t)\cong H(X,A;t)$ for all B such that $\bar{B}\subset\mathring{A}$. I.e., *t-homology satisfies the same excision property* III, 7.4 *as ordinary homology.*

7.6? The Mayer-Vietoris sequences (III, 8) were deduced from exact sequences of the type

$$0\to S(X_1\cap X_2)\to SX_1\oplus SX_2\to S\{X_1, X_2\}\to 0$$

together with the fact that $S\{X_1, X_2\}\simeq S(X_1\cup X_2)$ if $(X; X_1, X_2)$ is an excisive triad. Because the exact sequence splits in every dimension it remains exact after applying t, and by 2.6 we have $tS\{X_1,X_2\}\simeq tS(X_1\cup X_2)$ for excisive triads. Thus, the Mayer-Vietoris sequences generalize to t-homology, i.e., *for every excisive triad* $(X; X_1, X_2)$ *we have exact (Mayer-Vietoris)-sequences*

$$\cdots\to H_{n+1}(X_1\cup X_2; t)\xrightarrow{d_*} H_n(X_1\cap X_2; t)\xrightarrow{(j_{1*},\,-j_{2*})} H_n(X_1;t)\oplus H_n(X_2;t)$$
$$\xrightarrow{(i_{1*},\,i_{2*})} H_n(X_1\cup X_2; t)\to\cdots,$$

and

$$\cdots\to H_{n+1}(X, X_1\cup X_2; t)\xrightarrow{d_*} H_n(X, X_1\cap X_2; t)\xrightarrow{(j_{1*},\,-j_{2*})} H_n(X, X_1; t)$$
$$\oplus\, H_n(X, X_2; t)\xrightarrow{(i_{1*},\,i_{2*})} H_n(X, X_1\cup X_2; t)\to\cdots.$$

The proposition III, 8.11 which describes the boundary operator d_* generalizes similarly. In fact, the present section 7.6 can be deduced purely formally from the preceding sections 7.2–7.5; this is carried out in Eilenberg-Steenrod I, 14–15.

7.7⁊ If X is a contractible space (e.g. a point) then $\eta: SX \simeq (\mathbb{Z}, 0) \simeq S(\text{point})$, where η denotes augmentation (III, 4.5), hence $tSX \simeq (t\mathbb{Z}, 0)$; in particular, $H_i(X;t) = t\mathbb{Z}$ if $i=0$, and $=0$ otherwise. For any non-empty X the augmentation $\eta: SX \to (\mathbb{Z}, 0)$ has a right inverse, hence $SX = (\mathbb{Z}, 0) \oplus \ker(\eta)$, hence $H(X;t) = HtSX = H(t\mathbb{Z}, 0) \oplus Ht(\ker(\eta)) = (t\mathbb{Z}, 0) \oplus \tilde{H}(X;t)$, where $\tilde{H}(X;t) = Ht(\ker(\eta)) = \ker(H(X;t) \to H(\text{point};t))$. These groups, $\tilde{H}(X;t)$, constitute the *reduced* t-homology of X. They differ from $H(X;t)$ only in dimension 0, and they fit into a reduced t-homology sequence

$$\cdots \to \tilde{H}_{q+1}(A;t) \to \tilde{H}_{q+1}(X;t) \to H_{q+1}(X,A;t) \to \tilde{H}_q(A;t) \to \tilde{H}_q(X;t) \to \cdots,$$

just as ordinary reduced homology (III, 4.4).

7.8⁊ The t-homology of a sphere $\mathbb{S}^n$ can be computed as in IV, 2. More simply, one observes that $H(\mathbb{S}^n)$ is free, hence $S(\mathbb{S}^n) \simeq H(\mathbb{S}^n)$ by II, 4.9, hence $tS(\mathbb{S}^n) \simeq tH(\mathbb{S}^n)$ by 2.6, hence $H(\mathbb{S}^n;t) \cong t(H(\mathbb{S}^n)) = (t\mathbb{Z}, 0) \oplus (t\mathbb{Z}, n)$. Similarly, $H(\mathbb{R}^n, \mathbb{R}^n - 0; t) = (t\mathbb{Z}, n)$. In general, the t-homology groups can always be expressed in terms of integral homology groups; just apply the universal coefficient theorem 4.2 to the complex $C = S(X, A)$. In case $t = \otimes G$, or $= \operatorname{Hom}(-, G)$, the result asserts: *There are natural exact sequences*

$$(7.9) \qquad 0 \to H_n(X,A) \otimes G \to H_n(X,A;G) \to H_{n-1}(X,A) * G \to 0,$$

$$(7.10) \qquad 0 \to \operatorname{Ext}(H_{n-1}(X,A), G) \to H^n(X,A;G) \to \operatorname{Hom}(H_n(X,A), G) \to 0$$

which split (but not naturally). In particular, $H_n(X,A;G)$ *is determined by* $H(X,A)$. However, the same is *not* true for induced homomorphisms: $H(f;G)$ is not determined by $H(f;\mathbb{Z})$; cf. Exerc. 2.

7.11⁊ For cellular spaces X we have established (V, 1.3) an isomorphism $H(X, X^{-1}) \cong HWX$ where WX is the cellular complex of X (reminder: $W_n X = H_n(X^n, X^{n-1})$). In trying to generalize that result to t-homology one encounters difficulties. Let us assume, however, that WX is a *free complex* (e.g. if X is a CW-space). Then $HWX \cong H(X, X^{-1})$ implies $WX \simeq S(X, X^{-1})$, hence $tWX \simeq tS(X, X^{-1})$, hence $HtWX \cong H(X, X^{-1}; t)$. I.e., *if the cellular complex* WX *is free (as is always the case for* CW*-spaces) then* $H(X, X^{-1}; t) = HtWX$. *Moreover, in that case,* tWX *can be identified with the complex* $W(X;t)$ *which is defined as follows*

$$(7.12) \qquad W_n(X;t) = H_n(X^n, X^{n-1}; t), \qquad \partial_n = i_* \partial_*,$$

where

$$H_n(X^n, X^{n-1}; t) \xrightarrow{\partial_*} H_{n-1}(X^{n-1}; t) \xrightarrow{i_*} H_{n-1}(X^{n-1}, X^{n-2}; t).$$

Proof. The maps $\alpha=\alpha_n\colon tH_n(X^n,X^{n-1})\to H_n(X^n,X^{n-1};t)$ of the universal coefficient theorem 4.2 (applied to $C=S(X^n,X^{n-1})$) define a homomorphism $\alpha\colon tWX\to W(X;t)$ of graded groups, which is isomorphic because $H(X^n,X^{n-1})$ is free. The only question is whether α is a chain map, i.e., whether the composite diagram

$$\begin{array}{ccccc} tH_n(X^n,X^{n-1}) & \xrightarrow{t_0(\partial_*)} & t_0H_{n-1}X^{n-1} & \longrightarrow & tH_{n-1}(X^{n-1},X^{n-2}) \\ \downarrow{\scriptstyle\alpha} & & \downarrow{\scriptstyle\alpha} & & \downarrow{\scriptstyle\alpha} \\ H_n(X^n,X^{n-1};t) & \xrightarrow{\partial_*} & H_{n-1}(X^{n-1};t) & \longrightarrow & H_{n-1}(X^{n-1},X^{n-2};t) \end{array}$$

is commutative. The square on the right commutes because α is compatible with chain maps (is natural). But the boundary operator ∂_* is also induced by a chain map (II, 2.12; note that SX^{n-1} is a direct summand of SX^n), hence the left square also commutes. ∎

7.13 The results of IV, 6 on (ordinary) homology of open subsets of $\mathbb{S}^n$ generalize almost verbatim to homology with arbitrary coefficients G (whereas difficulties arise for t-homology). In some detail, let $B\subset A\subset\mathbb{S}^n$, $n>0$, be arbitrary sets, let $P\in A$, and

$$H_n(\mathbb{S}^n-B,\mathbb{S}^n-A;G)\xrightarrow{j_P}H_n(\mathbb{S}^n,\mathbb{S}^n-P;G)\xleftarrow[\cong]{i_P}H_n(\mathbb{S}^n;G)$$

the homomorphisms induced by inclusions. Then (cf. IV, 6.1) for every $y\in H_n(\mathbb{S}^n-B,\mathbb{S}^n-A;G)$ the map

$$Jy\colon A\to H_n(\mathbb{S}^n;G),\qquad (Jy)(P)=i_P^{-1}j_P(y),$$

is locally constant, $(Jy)|B=0$, and (cf. IV, 6.2)

$$J=J(A,B)\colon H_n(\mathbb{S}^n-B,\mathbb{S}^n-A;G)\to\Gamma(A,B;G)$$

is a homomorphism into the group $\Gamma(A,B;G)$, whose elements are locally constant functions $A\to H_n(\mathbb{S}^n;G)$ which vanish on B. *If $X\subset Y\subset\mathbb{S}^n$ are neighborhood retracts then* (cf. IV, 6.4)

(7.14a) $$H_i(Y,X;G)=0\quad\text{for } i>n,$$

(7.14b) $$J\colon H_n(Y,X;G)\cong\Gamma(\mathbb{S}^n-X,\mathbb{S}^n-Y;G).$$

The proofs are the same as in IV, 6.—The reader may find it interesting to look at the corollaries and applications of IV, 6.4 again (IV, 6 and IV, 7) and to generalize them to arbitrary coefficients; the case $G=\mathbb{Z}_2$ is especially instructive.

7.15 So far, we have viewed (co-)homology with coefficients in G as a functor of (X, A) alone; the group G was fixed. However, a homomorphism $\varphi\colon G \to G'$ of (coefficient) groups induces chain maps $\mathrm{id} \otimes \varphi\colon S(X, A) \otimes G \to S(X, A) \otimes G'$, resp. $\mathrm{Hom}(\mathrm{id}, \varphi)\colon \mathrm{Hom}(S(X, A), G) \to \mathrm{Hom}(S(X, A), G')$, and by passage to homology, $\varphi_* = H(X, A; \varphi)\colon H(X, A; G) \to H(X, A; G')$ resp. $\varphi^* = H^*(X, A; \varphi)\colon H^*(X, A; G) \to H^*(X, A; G')$. *This turns (co-)homology into a functor of the coefficients G.* For fixed $\varphi\colon G \to G'$ the maps $\varphi_* = H(X, A; \varphi)$, $\varphi^* = H^*(X, A; \varphi)$ *are natural* with respect to the variable (X, A); they are the simplest examples of *(co-)homology operations.* (= natural transformations between (co-)homology groups).

If $0 \to G' \xrightarrow{\iota} G \xrightarrow{\pi} G'' \to 0$ is an exact sequence of abelian groups then the sequences

$$0 \to S(X, A) \otimes G' \xrightarrow{\mathrm{id} \otimes \iota} S(X, A) \otimes G \xrightarrow{\mathrm{id} \otimes \pi} S(X, A) \otimes G'' \to 0,$$

$$0 \to \mathrm{Hom}(S(X, A), G') \xrightarrow{\iota \circ} \mathrm{Hom}(S(X, A), G) \xrightarrow{\pi \circ} \mathrm{Hom}(S(X, A), G'') \to 0,$$

are also exact (because $S(X, A)$ is free; see 6.21). The connecting homomorphisms

$$(7.16) \qquad \begin{aligned} \beta\colon & H_{n+1}(X, A; G'') \to H_n(X, A; G'), \\ \beta\colon & H^n(X, A; G'') \to H^{n+1}(X, A; G) \end{aligned}$$

which are associated with these sequences (II, 2.7) are usually called *Bockstein-homomorphisms* (of the coefficient sequence ι, π). They are *natural* with respect to the variable (X, A), thus providing another example of (co-)homology operations (this one between groups of *different* dimensions). *The sequences*

$$(7.17) \qquad \begin{aligned} \cdots \xrightarrow{\pi_*} & H_{n+1}(X, A; G'') \xrightarrow{\beta} H_n(X, A; G') \xrightarrow{\iota_*} H_n(X, A; G) \\ \xrightarrow{\pi_*} & H_n(X, A; G'') \xrightarrow{\beta} \cdots, \end{aligned}$$

$$(7.18) \qquad \begin{aligned} \cdots \xrightarrow{\pi^*} & H^{n-1}(X, A; G'') \xrightarrow{\beta} H^n(X, A; G') \xrightarrow{\iota^*} H^n(X, A; G) \\ \xrightarrow{\pi^*} & H^n(X, A; G'') \xrightarrow{\beta} \cdots \end{aligned}$$

are exact and natural; they are called the *(co-)homology sequences of the coefficient sequence* (ι, π).

7.19 In V, 5 the *rank* of abelian groups was used to define the Euler characteristic of graded groups resp. of spaces. For many rings R, a rank-function ρ_R can be defined on finitely generated R-modules (compare Swan II, 4.6, and also Cohn 2.4), and can be used to define an Euler-characteristic χ_R on finitely generated graded R-modules, or spaces. For simplicity, we consider the case of a field R only (besides $\mathbb{Z}$)

with $\rho_R = \dim_R =$ vector space dimension. If the characteristic of R is zero, $\text{char}(R)=0$, and X is a space, then

$$\dim_R H_i(X;R) = \dim_R(H_i(X;\mathbb{Q})\otimes R) = \dim_{\mathbb{Q}}(H_i(X;\mathbb{Q})) = \text{rank}(H_i X)$$

which brings us back to V, 5. If $\text{char}(R)=p>0$ then $\dim_R(H_i(X;R)) = \dim_R(H_i(X;\mathbb{Z}_p)\otimes R) = \dim_{\mathbb{Z}_p}(H_i(X;\mathbb{Z}_p))$ which reduces the problem to prime fields $\mathbb{Z}_p$.

Assume then $G=\{G_i\}_{i\in\mathbb{Z}}$ is a finitely generated $(\sum_i \dim(G_i)<\infty)$ graded vector space over $\mathbb{Z}_p$, and define $\chi_p G = \sum_{i\in\mathbb{Z}}(-1)^i \dim(G_i)$; if (X,A) is a pair of spaces put $\chi_p(X,A) = \chi_p[H(X,A;\mathbb{Z}_p)]$ if the latter is defined. Call this the *$\mathbb{Z}_p$-characteristic* of G resp. (X,A). Just as in V, 5.2 one proves:

If K is a complex of $\mathbb{Z}_p$-vector-spaces such that $\chi_p K$ is defined then $\chi_p(HK)$ is defined and equals $\chi_p K$. It follows (cf. V, 5.7) that

$$\chi_p(X) = \chi_p(A) + \chi_p(X,A) \tag{7.20}$$

for pairs (X,A) of spaces such that two of the numbers in 7.20 are defined. In general, the $\mathbb{Z}_p$-characteristic differs from the Euler-characteristic (choose X such that $\tilde{H}X = (\mathbb{Q},1)$; then $\chi(X)=0$ and $\chi_p(X)=1$). However,

7.21 Proposition. *If (X,A) is a pair of spaces with finitely generated homology $H(X,A)$ then $\chi_p(X,A)=\chi(X,A)$.*

Proof. By 7.9 we have $H_i(X,A;\mathbb{Z}_p) \cong H_i(X,A)\otimes\mathbb{Z}_p \oplus H_{i-1}(X,A) * \mathbb{Z}_p$, hence $\chi_p(X,A) = \chi_p[H(X,A)\otimes\mathbb{Z}_p] - \chi_p[H(X,A)*\mathbb{Z}_p]$. On the other hand, 5.15 implies

$$\chi(X,A) = \chi[H(X,A)] = \chi_p[H(X,A)\otimes\mathbb{Z}_p] - \chi_p[H(X,A)*\mathbb{Z}_p]. \quad \blacksquare$$

For instance, 7.21 applies to compact CW-pairs (X,A)—but then the result also follows directly (cf. V, 5.9 or V, 5.10 Exerc. 3).

7.22 Exercises. *1.* Show that $H^1(X;G) \cong \text{Hom}(H_1 X, G)$ for all spaces X and abelian groups G. In particular, $H^1(X;\mathbb{Z})$ is always torsionfree (whereas $H_1(X;\mathbb{Z})$ can be any abelian group; see V, 6 Exerc. 2).

2. The space $P_2\mathbb{R}/P_1\mathbb{R}$ which is obtained from real projective plane by shrinking a projective line to a point is homeomorphic with the 2-sphere. Show that the identification map $P_2\mathbb{R}\to P_2\mathbb{R}/P_1\mathbb{R}$ induces trivial homomorphisms in (reduced) integral homology but not so with coefficients $\mathbb{Z}_2$. Compare this with 4.15 Exerc. 1.

3. Let $\sigma\colon (\Delta_n, \dot{\Delta}_n) \to (\mathbb{R}^n, \mathbb{R}^n - 0)$ be a singular simplex whose homology class generates $H_n(\mathbb{R}^n, \mathbb{R}^n - 0) \cong \mathbb{Z}$ Show that $G \to H_n(\mathbb{R}^n, \mathbb{R}^n - 0; G)$, $g \mapsto [g\cdot\sigma]$ is an isomorphism (G = abelian group).

4. If F is a *free* abelian group then $H(X;F)\cong H(X;\mathbb{Z})\otimes F$. If G is *any* abelian group then there exists an exact sequence $0\to F_1\xrightarrow{\iota} F_0\xrightarrow{\pi} G\to 0$ in which F_1, F_0 are free abelian groups (cf. proof of 3.16). Its homology sequence contains the portion

$$H_n(X;F_1)\xrightarrow{\iota_*} H_n(X;F_0)\xrightarrow{\pi_*} H_n(X;G)\xrightarrow{\beta} H_{n-1}(X;F_1)\xrightarrow{\iota_*} H_n(X;F_0),$$

hence an exact sequence $0\to \operatorname{coker}(\iota_*)\to H_n(X;G)\to \ker(\iota_*)\to 0$. Show that this sequence is isomorphic with the universal coefficient sequence $0\to (H_n X)\otimes G\to H_n(X;G)\to (H_{n-1}X)*G\to 0$.

5. If (X,A) is a pair of spaces such that $H^*(X,A;k)=0$ for every prime field k (equivalently: $H^*(X;k)\cong H^*(A;k)$) then $H(X,A;G)=0$ (equivalently: $H(A;G)\cong H(X;G)$) for all abelian groups G. This follows from 6.22 Exerc. 5.

6*. If X is a space such that $\chi(X)$ and $\chi_p(X)$ arc defined, and if every element in $\bigcap_{i>0} p^i HX$ has finite order prime to p then $\chi(X)=\chi_p(X)$. Compare remark after 5.15.

8. *Tensorproduct and Bilinearity*

We define bilinear maps and show (8.11, 8.19) how the tensorproduct can be used to reduce them to homomorphisms of abelian groups. Conversely, bilinear maps can be used to deduce properties of the tensor-product-functor (8.13, 8.17).

8.1. Definition. For every $M\in R\text{-}\mathcal{M}od$ and $y\in M$ we have an R-homomorphism $\hat{y}\colon R\to M$, $\hat{y}(r)=ry$; similarly, for right R-Modules $L\in \mathcal{M}od\text{-}R$ and $x\in L$ we have $\hat{x}\colon R\to L$, $\hat{x}(r)=xr$. We then define $x\otimes_R y\in L\otimes_R M$ to be the image of $1\in R=R\otimes_R R$ under $\hat{x}\otimes_R\hat{y}\colon R\otimes_R R\to L\otimes_R M$ (cf. 5.10); in formulas,

$$x\otimes_R y=(\hat{x}\otimes_R\hat{y})(1),\qquad x\in L,\qquad y\in M. \tag{8.2}$$

In particular, if $L=R$ resp. $M=R$ then $\hat{r}$ is the left resp. right translation with $r\in R$. It follows that

$$r\otimes_R y=ry,\qquad x\otimes_R r=xr,\qquad x\in L,\qquad y\in M,\qquad r\in R. \tag{8.3}$$

We have the following equations:

$$\begin{aligned}(x_1+x_2)\otimes_R y&=x_1\otimes_R y+x_2\otimes_R y,\\ x\otimes_R(y_1+y_2)&=x\otimes_R y_1+x\otimes_R y_2,\end{aligned} \tag{8.4}$$

$$(xr)\otimes_R y=x\otimes_R(ry),\qquad r\in R. \tag{8.5}$$

The former, 8.4, are quite obvious; they are special cases of 5.11. As to 8.5,

$$(x\,r)\otimes y=(\hat{x}\otimes\hat{y})(\hat{r}\otimes\hat{1})(1)=(\hat{x}\otimes\hat{y})(r)=(\hat{x}\otimes\hat{y})(\hat{1}\otimes\hat{r})(1)=x\otimes(r\,y).$$

8.6 Definition. Given modules $L\in\mathcal{M}od\text{-}R$, $M\in R\text{-}\mathcal{M}od$, and an abelian group $N\in\mathcal{AG}$, then a mapping $\zeta: L\times M\to N$ is called *R-biadditive* if

$$\begin{aligned}\zeta(x_1+x_2,y)&=\zeta(x_1,y)+\zeta(x_2,y),\\ \zeta(x,y_1+y_2)&=\zeta(x,y_1)+\zeta(x,y_2),\end{aligned}\tag{8.7}$$

$$\zeta(x\,r,y)=\zeta(x,r\,y)\quad\text{for all } x\in L,\quad y\in M,\quad r\in R.\tag{8.8}$$

For instance, the structure maps $R\times M\to M$, $(r,y)\mapsto r\,y$, resp. $L\times R\to L$, $(x,r)\mapsto x\,r$, are R-biadditive. Formulas 8.4, 8.5 assert that

$$\pi=\pi_{LM}\colon L\times M\to L\otimes_R M,\quad \pi(x,y)=x\otimes_R y,\tag{8.9}$$

is R-biadditive.

If $\zeta,\eta: L\times M\to N$ are R-biadditive then $\zeta\pm\eta: L\times M\to N$, $(\zeta\pm\eta)(x,y)=\zeta(x,y)\pm\eta(x,y)$ is also R-biadditive. The set $\operatorname{Biad}_R(L\times M,N)$ of all R-biadditive maps is thereby an abelian group. If $f: L'\to L$, $g: M'\to M$, $h: N\to N'$ are (R-) homomorphisms and $\zeta: L\times M\to N$ is R-biadditive then $L'\times M'\to N'$, $(x',y')\mapsto h\,\zeta(f x', g y')$ is also R-biadditive. This defines a homomorphism $\operatorname{Biad}_R(L\times M,N)\to\operatorname{Biad}_R(L'\times M',N')$ and turns Biad_R into a group-valued functor of L, M (contravariant), and N (covariant). For the moment being, only the functorial dependence on M will play a role.

A function of two variables can always be viewed as a function of one (the second) variable whose values are functions of the first variable. In the case of Biad_R this becomes

8.10 Proposition. *The homomorphisms*

$$\Phi\colon \operatorname{Biad}_R(L\times M,N)\rightleftarrows\operatorname{Hom}_R(M,\operatorname{Hom}_{\mathbb{Z}}(L,N))\colon\Psi,$$

$[(\Phi\zeta)\,y]\,x=\zeta(x,y)$, $(\Psi\eta)(x,y)=[\eta(y)]\,x$, $x\in L$, $y\in M$, *are reciprocal natural isomorphisms, where the group* $\operatorname{Hom}_{\mathbb{Z}}(L,N)$ *on the right is viewed as a left R-module via* $(r\,\alpha)\,x=\alpha(x\,r)$, $\alpha\in\operatorname{Hom}(L,N)$, $r\in R$, $x\in L$.

Proof. If we neglect the R-structure (i.e., take $R=\mathbb{Z}$) then it is quite obvious from the definitions that Φ, Ψ are reciprocal isomorphisms. As to the R-structure, the formulas $[\Phi(\zeta)(r\,y)]\,x=\zeta(x,r\,y)$, $(r[\Phi(\zeta)\,y])\,x=[\Phi(\zeta)\,y](x\,r)=\zeta(x\,r,y)$ show that $\Phi(\zeta)$ is an R-homomorphism if and only if ζ satisfies 8.8.

It remains to prove naturality of Φ resp. Ψ; this is left to the reader. ∎

The following proposition reduces R-biadditive maps to additive maps. It can also serve as an axiomatic description of the tensorproduct by R-biadditive maps (cf. Bourbaki, 1948).

8.11 Proposition. *If* $\zeta\colon L\times M\to N$ *is an* R*-biadditive map then there exists a unique homomorphism* $\xi\colon L\otimes_R M\to N$ *such that* $\xi(x\otimes_R y)=\zeta(x,y)$, $x\in L$, $y\in M$. *In other words, composition by* π_{LM} *(see 8.9) is an isomorphism,* $\circ\pi_{LM}\colon \mathrm{Hom}_{\mathbb{Z}}(L\otimes_R M,N)\cong \mathrm{Biad}_R(L\times M,N)$.

Proof. Clearly $\circ\pi_{LM}$ is a natural transformation between functors of $M\in R\text{-}\mathcal{M}od$. Both functors are contravariant, strongly additive, left exact: the first, by 2.11, because it is the composition of $L\otimes_R$ and $\mathrm{Hom}_{\mathbb{Z}}(-,N)$, the second because $\mathrm{Biad}_R(L\times M,N)\cong \mathrm{Hom}_R(M,\mathrm{Hom}_{\mathbb{Z}}(L,N))$, by 8.10. Therefore, by 6.4, it suffices to show that $\circ\pi_{LR}\colon \mathrm{Hom}_{\mathbb{Z}}(L\otimes_R R,N)\cong \mathrm{Biad}_R(L\times R,N)$. But this agrees with the composition

$$\mathrm{Hom}_{\mathbb{Z}}(L\otimes R,N)\cong \mathrm{Hom}_{\mathbb{Z}}(L,N)\cong \mathrm{Hom}_R(R,\mathrm{Hom}_{\mathbb{Z}}(L,N)) \overset{(8.10)}{\cong} \mathrm{Biad}_R(L\times R,N);$$

one has only to insert the definitions. ∎

8.12 Corollary. *The elements* $x\otimes_R y$, $x\in L$, $y\in M$, *generate the abelian group* $L\otimes_R M$.

Proof. Let K be the subgroup generated by all $x\otimes_R y$, put $N=(L\otimes_R M)/K$, and let $\xi\colon L\otimes_R M\to N$ be the projection. Then $\zeta(x,y)=\xi(x\otimes_R y)=0$ hence $\xi=0$ by the uniqueness part of 8.11, hence $K=L\otimes_R M$. ∎

8.13 Proposition. *For* $L\in\mathcal{M}od\text{-}R$, $M\in R\text{-}\mathcal{M}od$ *we have a natural isomorphism* $L\otimes_R M=M\otimes_{R^{\mathrm{op}}}L$, $x\otimes_R y\mapsto y\otimes_{R^{\mathrm{op}}}x$, *where* R^{op} *denotes the opposite ring* (recall that $R\text{-}\mathcal{M}od=\mathcal{M}od\text{-}R^{\mathrm{op}}$, $\mathcal{M}od\text{-}R=R^{\mathrm{op}}\text{-}\mathcal{M}od$; see 1.1).

Proof. Clearly $L\times M\to M\otimes_{R^{\mathrm{op}}}L$, $(x,y)\mapsto y\otimes_{R^{\mathrm{op}}}x$, is R-biadditive, hence a homomorphism $L\otimes_R M\to M\otimes_{R^{\mathrm{op}}}L$ with $x\otimes_R y\mapsto y\otimes_{R^{\mathrm{op}}}x$. Similarly in the other direction, and the two composites are identity maps. ∎

The symmetry $L\otimes M\cong M\otimes L$ shows that the tensorproduct is right exact in each variable. This implies

8.14 Proposition. *If* $L'\xrightarrow{i}L\xrightarrow{p}L''\to 0$, $M'\xrightarrow{j}M\xrightarrow{q}M''\to 0$, *are exact sequences in* $\mathcal{M}od\text{-}R$ *resp.* $R\text{-}\mathcal{M}od$ *then*

(8.15) $(L'\otimes_R M)\oplus(L\otimes_R M')\xrightarrow{(i\otimes \mathrm{id},\,\mathrm{id}\otimes j)}L\otimes_R M\xrightarrow{p\otimes q}L''\otimes_R M''\to 0$

is also exact.

Proof. Clearly 8.15 is a complex, and $p\otimes q=(\mathrm{id}\otimes q)\circ(p\otimes \mathrm{id})$ is surjective. It remains to prove $(p\otimes q)\,z=0 \Rightarrow z\in \operatorname{im}(i\otimes \mathrm{id}, \mathrm{id}\otimes j)$. Consider then the commutative diagram

$$\begin{array}{ccccc} & & L\otimes M' & \xrightarrow{p\otimes \mathrm{id}} & L''\otimes M' \to 0 \\ & & \big\downarrow{\scriptstyle \mathrm{id}\otimes j} & & \big\downarrow{\scriptstyle \mathrm{id}\otimes j} \\ L'\otimes M & \xrightarrow{i\otimes \mathrm{id}} & L\otimes M & \xrightarrow{p\otimes \mathrm{id}} & L''\otimes M \\ & & & & \big\downarrow{\scriptstyle \mathrm{id}\otimes q} \\ & & & & L''\otimes M''. \end{array}$$

Because the right column is exact we can find $t\in L''\otimes M'$ such that $(\mathrm{id}\otimes j)\,t=(p\otimes \mathrm{id})\,z$. Pick $y\in(p\otimes \mathrm{id})^{-1}\,t$, then $(p\otimes \mathrm{id})(\mathrm{id}\otimes j)\,y=(p\otimes \mathrm{id})\,z$, hence (because the second row is exact) we can find $x\in L'\otimes M$ such that $(i\otimes \mathrm{id})\,x=z-(\mathrm{id}\otimes j)\,y$, hence $z\in\operatorname{im}(i\otimes \mathrm{id}, \mathrm{id}\otimes j)$. ∎

8.16 Definition. If R, S are two rings and if M is both an R- and an S-module such that the two operations commute then M is called a bimodule (say R-left, S-right). That the operations commute means that multiplication with $r\in R$ (i.e., the map $\Theta_r\colon M\to M$ of 2.2) is an S-homomorphism, or multiplication with $s\in S$ is an R-homomorphism Θ_s. We can therefore apply functors $L\otimes_R -$, $L\in\mathcal{M}od$-R, to Θ_s and we can turn $L\otimes_R M$ into a right S-module by $x\,s=(\mathrm{id}\otimes\Theta_s)\,x$, $x\in L\otimes_R M$, $s\in S$. Similarly $M\otimes_S N$ is a left R-module for every $N\in S$-$\mathcal{M}od$. For instance, if R is commutative we can always take $S=R$ and let the two structures coincide.

8.17 Proposition. *If L, M, N are modules as in* 8.16 *then we have a natural isomorphism*

$$(L\otimes_R M)\otimes_S N\cong L\otimes_R(M\otimes_S N), \qquad (x\otimes_R y)\otimes_S z\mapsto x\otimes_R(y\otimes_S z).$$

Proof. For every $z\in N$ define an R-biadditive map $L\times M\to L\otimes_R(M\otimes_S N)$, $(x,y)\mapsto x\otimes(y\otimes z)$. By 8.11, it induces a homomorphism $L\otimes_R M\to L\otimes_R(M\otimes_S N)$, and hence an S-biadditive map $(L\otimes_R M)\times N\to L\otimes_R(M\otimes_S N)$ such that $(x\otimes y, z)\mapsto x\otimes(y\otimes z)$. Once more 8.11 applies and gives a homomorphism $(x\otimes_R y)\otimes_S z\mapsto x\otimes_R(y\otimes_S z)$. Similarly in the other direction, and the two composites are identity maps. ∎

8.18 *R* Commutative. In this case $L\otimes_R M$, as any other additive functor of M (cf. 2.2), has a natural R-structure; in formulas, $r(x\otimes_R y)=x\otimes_R r\,y$.

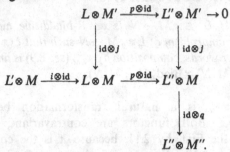

As a functor of L its R-structure is given by $r(x\otimes_R y)=(x\,r\otimes_R y)$; by 8.5 both structures agree. The isomorphisms $L\otimes_R M\cong M\otimes_R L$ (8.13) and $(L\otimes_R M)\otimes_R N\cong L\otimes_R(M\otimes_R N)$ are clearly R-maps. Thus $\otimes_R$: R-$\mathcal{M}od\times R$-$\mathcal{M}od\to R$-$\mathcal{M}od$ is a (strongly) additive (covariant, right exact) functor which is *associative*, *commutative* and has a *unit* object ($R\otimes_R=\mathrm{id}$).

If L, M, N are R-modules then $\mathrm{Hom}_R(L\otimes_R M, N)$ is a subgroup of $\mathrm{Hom}_{\mathbb{Z}}(L\otimes_R M, N)$. What is the corresponding subgroup of $\mathrm{Biad}_R(L\times M, N)$? The formula $\xi(x\otimes_R y)=\zeta(x,y)$ of 8.11 shows that $\xi\in\mathrm{Hom}_R(L\otimes_R M, N)$ if and only if ζ satisfies $\zeta(x\,r,y)=r\zeta(x,y)=\zeta(x,r\,y)$. Such an R-biadditive map is called *R-bilinear* (or simply *bilinear*).

8.19 Proposition. *If R is commutative then the equation $\xi(x\otimes_R y)=\zeta(x,y)$, $x\in L, y\in M$, defines a one-to-one correspondence between R-homomorphisms ξ: $L\otimes_R M\to N$ and R-bilinear maps ζ: $L\times M\to N$.* ▮

8.20 Exercises. *1.* If R is commutative then $\hat{x}\otimes_R\hat{y}$: $R\to L\otimes M$ is an R-homomorphism ($x\in L$, $y\in M$), hence $(\hat{x}\otimes_R\hat{y})\,r=r(\hat{x}\otimes_R\hat{y})(1)=r(x\otimes_R y)$; in particular, $\hat{x}\otimes_R\hat{y}$ is determined by $x\otimes_R y$. Conversely, if R is not commutative then $\hat{x}\otimes_R\hat{y}$ is not determined by $x\otimes_R y$ (hint: take $L=R$, $M=R$).

2. Show that the universal property 8.11 resp. 8.19 characterises the tensorproduct $L\otimes_R M\in\mathcal{AG}$ resp. $L\otimes_R M\in R$-$\mathcal{M}od$.

3. If R is commutative then $\mathrm{Hom}_R(L\otimes_R M, N)\cong\mathrm{Hom}_R(M, \mathrm{Hom}_R(L, N))$ for all R-modules L, M, N (compare 8.10, 8.11). This natural isomorphism expresses the fact that $L\otimes_R$- and $\mathrm{Hom}_R(L, -)$ are *adjoint* functors (compare K a n).

9. *Tensorproduct of Complexes. Künneth Formula*

We extend the definition of tensor products $C\otimes_R D$ to the case where both variables are complexes. Generalizing the universal coefficient formula (cf. § 4) we express $H(C\otimes_R D)$ in terms of HC, HD, at least if C or D is free (9.13); as before the ground ring R is assumed to be *hereditary*. Later on (cf. § 12) we shall see that for topological spaces X, Y one has $S(X\times Y)\simeq(SX)\otimes(SY)$; thus we can express $H(X\times Y)$ in terms of HX, HY.

9.1 Definition. Let C, D be complexes of right resp. left R-modules. Define a new complex $C\otimes_R D$ as follows

(9.2) $$(C\otimes_R D)_n=\oplus_{i+j=n}\,C_i\otimes_R D_j,$$

(9.3) $$\partial=\partial^{C\otimes D}\colon (C\otimes_R D)_n\to(C\otimes_R D)_{n-1},$$
$$\partial^{C\otimes D}|\,C_i\otimes D_j=\partial^C\otimes\mathrm{id}+(-1)^i\,\mathrm{id}\otimes\partial^D.$$

Since $\partial\partial|C_i\otimes D_j=(-1)^{i-1}\partial^C\otimes\partial^D+(-1)^i\partial^C\otimes\partial^D=0$, this defines indeed a complex.

If $f\colon C\to C'$, $g\colon D\to D'$ are chain maps then

(9.4) $\quad f\otimes_R g\colon C\otimes_R D\to C'\otimes_R D', \quad (f\otimes_R g)_n=\oplus_{i+j=n} f_i\otimes_R g_j,$

satisfies

$$\begin{aligned}\partial(f\otimes g)|C_i\otimes D_j&=(\partial f_i)\otimes g_j+(-1)^i f_i\otimes(\partial g_j)\\&=(f_{i-1}\,\partial)\otimes g_j+(-1)^i f_i\otimes(g_{j-1}\,\partial)=(f\otimes g)\,\partial\,|C_i\otimes D_j,\end{aligned}$$

i.e., $f\otimes_R g$ is a chain map. Thus, the tensorproduct is a covariant functor $\partial\,\mathcal{M}od\text{-}R\times\partial R\text{-}\mathcal{M}od\to\partial\mathcal{AG}$, resp. $\partial R\text{-}\mathcal{M}od\times\partial R\text{-}\mathcal{M}od\to\partial R\text{-}\mathcal{M}od$ if R is commutative.

Similarly one can define torsion products $C*D$ of complexes by $(C*D)_n=\oplus_{i+j=n}C_i*D_j$ etc.; in fact, $\otimes$ could be replaced by any additive functor $\mathcal{M}od\text{-}R\times R\text{-}\mathcal{M}od\to\mathcal{AG}$. We omit the details because we shall not really use these complexes.

9.5 Proposition. *If C_R, ${}_RD_S$, ${}_SE$ are complexes of modules on which the ground rings R, S act as indicated by the indices then*

(9.6) $\quad \tau\colon C\otimes_R D\cong D\otimes_{R^{op}} C, \quad \tau(x\otimes y)=(-1)^{|x|\,|y|}\,y\otimes x,$

(where $|\;|$ denotes dimensions; if $x\in C_n$ then $|x|=n$), and

(9.7) $\quad a\colon (C\otimes_R D)\otimes_S E\cong C\otimes_R(D\otimes_S E), \quad a[(x\otimes y)\otimes z]=x\otimes(y\otimes z).$

Proof. It is clear (8.13, 8.17) that τ and a are well-defined isomorphism of graded groups; the only question is whether they commute with ∂. Now

$$\begin{aligned}\tau\,\partial(x\otimes y)&=(-1)^{|\partial x|\,|y|}\,y\otimes\partial x+(-1)^{|x|+|x|\,|\partial y|}\,\partial y\otimes x\\&=(-1)^{|x|\,|y|}(\partial y\otimes x+(-1)^{|y|}\,y\otimes\partial x)=\partial\tau(x\otimes y),\end{aligned}$$

$$\begin{aligned}a\,\partial[(x\otimes y)\otimes z]&=a[(\partial x\otimes y)\otimes z+(-1)^{|x|}(x\otimes\partial y)\otimes z+(-1)^{|x|+|y|}(x\otimes y)\otimes\partial z]\\&=\partial x\otimes(y\otimes z)+(-1)^{|x|}\,x\otimes(\partial y\otimes z)+(-1)^{|x|+|y|}\,x\otimes(y\otimes\partial z)\\&=\partial[x\otimes(y\otimes z)]=\partial a[(x\otimes y)\otimes z].\quad\blacksquare\end{aligned}$$

9.8 Remark. A useful rule for memorizing signs is that whenever two objects u, v are permuted to which degrees $|u|, |v|$ are attached then a sign $(-1)^{|u|\,|v|}$ should be introduced. Examples are 9.6 and 9.3; the latter because $|\partial|=-1$.

9.9 Proposition. *If* $f^0 \simeq f^1\colon C \to C'$, $g^0 \simeq g^1\colon D \to D'$ *then*

$$f^0 \otimes g^0 \simeq f^1 \otimes g^1\colon C \otimes D \to C' \otimes D',$$

i.e., *the functor* $\otimes$ *is compatible with homotopies.*

9.10 Corollary. *If* $f\colon C \simeq C'$ *and* $g\colon D \simeq D'$ *are homotopy equivalences then also* $f \otimes g\colon C \otimes D \simeq C' \otimes D'$.

Proof of 9.9. Let $s\colon f^0 \simeq f^1$, i.e., $\partial s + s\partial = f^1 - f^0$. Then in $C_i \otimes D_j$ we have

$$\begin{aligned}\partial(s\otimes g^0)+(s\otimes g^0)\,\partial &= (\partial s\otimes g^0+(-1)^{i+1}\, s\otimes \partial g^0)+(s\partial\otimes g^0+(-1)^i\, s\otimes g^0\,\partial)\\ &= (f^1-f^0)\otimes g^0,\end{aligned}$$

hence $s \otimes \mathrm{id}\colon f^0 \otimes g^0 \simeq f^1 \otimes g^0$. By symmetry (9.6), $f^1 \otimes g^0 \simeq f^1 \otimes g^1$. ∎

Other properties of $\otimes$ like right exactness or strong additivity follow immediately from the module case and will not be formulated. An abstract characterization of the tensorproduct of complexes is sketched in § 10 Exerc. 3.

We want to express $H(C \otimes_R D)$ in terms of HC, HD, and we begin by generalizing the map α of 4.7.

9.11 Proposition. *Given complexes* C_R, ${}_RD$ *(not necessarily free) there exists a unique homomorphism* $\alpha\colon HC \otimes_R HD \to H(C \otimes_R D)$ *such that* $\alpha([x] \otimes [y]) = [x \otimes y]$ *for* $x \in ZC$, $y \in ZD$ ([ ] denotes homology classes). *The map* α *is natural in* (C, D).

Proof. Uniqueness is obvious. To prove existence define $a\colon ZC \times ZD \to H(C \otimes D)$ by $a(x, y) = [x \otimes y]$. If $[x] = [x']$, $[y] = [y']$ then $x = x' + \partial c$, $y = y' + \partial d$, hence

$$\begin{aligned}a(x,y) &= [x'\otimes y' + x'\otimes \partial d + \partial c\otimes y] = [x'\otimes y' \pm \partial(x'\otimes d) + \partial(c\otimes y)]\\ &= [x'\otimes y'] = a(x',y'),\end{aligned}$$

hence a induces $\bar{a}\colon HC \times HD \to H(C \otimes D)$, and this, in turn, $\alpha\colon HC \otimes HD \to H(C \otimes D)$ because a, and hence $\bar{a}$, is clearly R-biadditive (cf. 8.11).

If $f\colon C \to C'$, $g\colon D \to D'$ are chain maps then $(f \otimes g)_* \alpha([x] \otimes [y]) = [fx \otimes gy] = \alpha(f_* \otimes g_*)([x] \otimes [y])$, which proves naturality. ∎

9.12 Lemma. *If* C *is a free complex and* $\partial^C = 0$ *(hence* $C = HC$*) then* α *is an isomorphism.*

Proof. If $C=(R,n)$, i.e., $C_k=0$ for $k\neq n$, $C_n=R$, then 9.12 is obvious. In general, C is a direct sum of such complexes (by assumption), and both $HC\otimes HD$, $H(C\otimes D)$ commute with direct sums. ∎

9.13 Künneth Theorem. *If R is a hereditary ring, and C, D are R-complexes such that $H(C*D)=0$*[6] *then there is a natural exact sequence*

$$(9.14)\qquad 0\to(HC\otimes HD)_n\xrightarrow{\alpha}H_n(C\otimes D)\xrightarrow{\beta}(HC*HD)_{n-1}\to 0$$

which splits (but the splitting is not natural; cf. 4.15 Exerc. 1).

Proof. With minor modifications this proof is the same as for 4.2, 4.10: one replaces tC by $C\otimes D$. Assume C is free, first. Then

$$(9.15)\qquad 0\to ZC\xrightarrow{\iota}C\xrightarrow{\partial}BC^+\to 0$$

is an exact sequence of free complexes ($C_n^+=C_{n-1}$), hence

$$(9.16)\qquad 0\to ZC\otimes D\xrightarrow{\iota\otimes\mathrm{id}}C\otimes D\xrightarrow{\partial\otimes\mathrm{id}}BC^+\otimes D\to 0$$

is also exact. The following is a portion of its homology sequence,

$$(9.17)\qquad \begin{aligned}&H(BC^+\otimes D)\xrightarrow{d_*}H(ZC\otimes D)\xrightarrow{(\iota\otimes\mathrm{id})_*}H(C\otimes D)\\&\xrightarrow{(\partial\otimes\mathrm{id})_*}H(BC^+\otimes D)\xrightarrow{d_*}H(ZC\otimes D).\end{aligned}$$

Let $q\colon BC^+\to C$, $j\colon C\to ZC$ be maps which split 9.15 (as after 4.9), then $q\otimes\mathrm{id}$, $j\otimes\mathrm{id}$ split 9.16, hence (II, 2.12) $d_*=[(j\otimes\mathrm{id})\circ\partial^{C\otimes D}\circ(q\otimes\mathrm{id})]_*$. But

$$\begin{aligned}(j\otimes\mathrm{id})\,\partial^{C\otimes D}(q\otimes\mathrm{id})\,(x\otimes y)&=(j\otimes\mathrm{id})\,(\partial\,qx\otimes y\pm qx\otimes\partial y)\\&=j\,\partial qx\otimes y\pm(jq)\,x\otimes\partial y=x\otimes y\end{aligned}$$

(using $j\,\partial q=\iota$, $jq=0$), hence $d_*=(\iota\otimes\mathrm{id})_*$, where $\iota\colon BC\to ZC$ is the inclusion.

Applying naturality of α to (ι,id) now gives a commutative diagram

$$\begin{array}{ccc}H(BC^+\otimes D)&\xrightarrow{d_*=(\iota\otimes\mathrm{id})_*}&H(ZC\otimes D)\\ \uparrow{\scriptstyle\alpha\,\cong}&&\uparrow{\scriptstyle\alpha\,\cong}\\ BC^+\otimes HD&\xrightarrow[\iota\otimes\mathrm{id}]{}&ZC\otimes HD\end{array}$$

[6] In most applications C or D will be flat or even free so that $C*D=0$. However, the assumption $H(C*D)=0$ is more appropriate because it is homotopy invariant.

in which α is isomorphic by 9.12. Hence

$$\operatorname{coker}(d_*) \cong \operatorname{coker}(\iota \otimes \mathrm{id}) \cong HC \otimes HD, \quad \ker(d_*) \cong \ker(\iota \otimes \mathrm{id}) \cong HC * HD,$$

the second equation because $BC^+ \to ZC$ is a resolution of HC. Inserting this in 9.17 we obtain a natural exact sequence

$$(9.18) \qquad 0 \to HC \otimes HD \xrightarrow{\alpha'} H(C \otimes D) \xrightarrow{\beta'} HC^+ * HD \to 0.$$

Consider now the general case $H(C * D) = 0$. We reduce it to the free case exactly as in § 4 (proof of 4.10): There is a natural exact sequence (4.13)

$$0 \to K \to \hat{C} \to C \to 0$$

such that $\hat{C}$, K are free, and $\hat{C} \simeq 0$, hence exact sequences

$$0 \to C * D \to K \otimes D \to \hat{C} \otimes D \to C \otimes D \to 0,$$

$$0 \to \frac{K \otimes D}{C * D} \to \hat{C} \otimes D \to C \otimes D \to 0.$$

Now, $\quad \hat{C} \simeq 0 \Rightarrow HK^+ \cong HC,$

$$H(C * D) = 0 \Rightarrow H(K \otimes D) \cong H\left(\frac{K \otimes D}{C * D}\right),$$

$$\hat{C} \simeq 0 \Rightarrow \hat{C} \otimes D \simeq 0 \Rightarrow H\left(\frac{K \otimes D}{C * D}\right)^+ \cong H(C \otimes D),$$

hence $H(K \otimes D)^+ = H(C \otimes D)$. Inserting this in the sequence 9.18 for (K^+, D) gives a natural exact sequence

$$(9.19) \qquad 0 \to HC \otimes HD \xrightarrow{\alpha''} H(C \otimes D) \xrightarrow{\beta} HC^+ * HD \to 0.$$

It remains to show $\alpha'' = \alpha$, and to split the sequence. Consider first the case $C = (R, n)$. Then $C \otimes = HC \otimes$ is essentially the identity functor (except for a shift of indices), and it is immediate from the definitions that $\alpha'' = \mathrm{id} = \alpha$.

In the general case, pick $x \in Z_n C$ and define a chain map f: $(R, n) \to C$ by $f(1) = x$. Apply naturality of α'' to (f, id) and get

$$\begin{aligned} \alpha''([x] \otimes [y]) &= \alpha''(f_* \otimes \mathrm{id}_*)(1 \otimes [y]) = (f \otimes \mathrm{id})_* \alpha''(1 \otimes [y]) \\ &= (f \otimes \mathrm{id})_* [1 \otimes y] = [x \otimes y] = \alpha([x] \otimes [y]), \end{aligned}$$

hence $\alpha'' = \alpha$.

If C, D are free then chain maps γ: $C \to HC$, ε: $D \to HD$ exist (cf. II, 4.6) such that $\gamma x = [x]$, $\varepsilon y = [y]$ for $x \in ZX$, $y \in ZD$, hence

$$[(\gamma \otimes \varepsilon)_* \alpha]([x] \otimes [y]) = (\gamma \otimes \varepsilon)_* [x \otimes y] = [x] \otimes [y],$$

or $(\gamma \otimes \varepsilon)_* \alpha = \mathrm{id}$; hence $(\gamma \otimes \varepsilon)_*$ splits the Künneth sequence 9.14. In the general case, pick free complexes C', D' and chain maps $f\colon C' \to C$, $g\colon D' \to D$ such that f_*, g_* are isomorphisms (cf. II, 4.6). By naturality we get a commutative diagram

$$\begin{array}{ccccccccc} 0 \to & HC' \otimes HD' & \xrightarrow{\alpha} & H(C' \otimes D') & \xrightarrow{\beta} & HC'^{+} * HD'^{+} & \to 0 \\ (9.20) \quad & \downarrow {\scriptstyle Hf \otimes Hg} \,\cong & & \downarrow {\scriptstyle H(f \otimes g)} & & \downarrow {\scriptstyle Hf * Hg} \,\cong & \\ 0 \to & HC \otimes HD & \xrightarrow{\alpha} & H(C \otimes D) & \xrightarrow{\beta} & HC^{+} * HD^{+} & \to 0. \end{array}$$

By the five lemma, $H(f \otimes g)$ is isomorphic, hence the second row is isomorphic to the first row which was already shown to split. ∎

9.21 Exercises. *1.* Use the Künneth theorem to prove: If P, Q are *flat* complexes in $\mathcal{M}od$-R resp. R-$\mathcal{M}od$ (R hereditary) such that $H_i P = 0$, $H_i Q = 0$ for $i \neq 0$ then $H_j(H_0 P \otimes_R Q) \cong H_j(P \otimes_R Q) \cong H_j(P \otimes_R H_0 Q)$ for all j; and this group agrees with $H_0 P \otimes_R H_0 Q$ if $j = 0$, with $H_0 P *_R H_0 Q$ if $j = 1$, with zero for every other j. Compare this with Exerc. 5 in § 5.

2. Generalize the definition of the tensorproduct of complexes to arbitrary functors $t(L, M)$ of two variables (=modules). Use the γ-convention (§ 2) if t is (partly) contravariant. Try to generalize the Künneth theorem.

3. If $0 \to C' \xrightarrow{i} C \xrightarrow{p} C'' \to 0$ is an exact sequence of free (flat) complexes and D an arbitrary complex one can apply the Künneth theorem to the terms of $0 \to C' \otimes D \to C \otimes D \to C'' \otimes D \to 0$. There results a diagram involving the maps i_*, p_*, ∂_* of the homology sequences and the Künneth maps α, β. Check for commutativity.

4. The product of two finite CW-spaces X, Y is itself a CW-space whose cells are products $c \times d$ of cells of X resp. of Y. Show that there is a chain-isomorphism $(WX) \otimes_{\mathbb{Z}} (WY) \to W(X \times Y)$, $c \otimes d \mapsto c \times d$ (W=cellular chain complex; cf. V, 4.1). Use this and the Künneth theorem to compute the homology of $P_m \mathbb{R} \times P_n \mathbb{R}$.

5.* If C, D are finitely generated free $\mathbb{Z}$-complexes then

$$(C \otimes \mathbb{Z}_m) \otimes (D \otimes \mathbb{Z}_n) = C \otimes D \otimes \mathbb{Z}_{gcd(m,n)},$$

hence $\alpha\colon H(C \otimes \mathbb{Z}_m) \otimes H(D \otimes \mathbb{Z}_n) \to H(C \otimes D \otimes \mathbb{Z}_{gcd(m,n)})$. Show that every element in $H(C \otimes D \otimes \mathbb{Z}_k)$ can be obtained from elements of the form $x \in H(C \otimes \mathbb{Z}_m)$, $y \in H(D \otimes \mathbb{Z}_n)$ by applying combinations of α, coefficient homomorphisms, Bockstein homomorphisms, and addition. Are all of these operations needed?

10. Hom of Complexes. Homotopy Classification of Chain Maps

Dualizing §9, we define a functor $(C, D) \mapsto \operatorname{Hom}(C, D)$ from pairs of complexes to complexes. As to be expected, there are Künneth relations expressing $H \operatorname{Hom}(C, D)$ in terms of HC, HD if the ground ring R is hereditary (10.11). Because $H_0 \operatorname{Hom}(C, D)$ turns out to be the groups of homotopy classes of chain maps $C \to D$, we get as a corollary a simple expression for this group (10.13).

10.1 Definition. Let C, D be (left) R-complexes. Define a new complex $\operatorname{Hom}_R(C, D)$ as follows,

(10.2) $$\operatorname{Hom}_R(C, D)_n = \prod_{i\in\mathbb{Z}} \operatorname{Hom}_R(C_i, D_{i+n}),$$

(10.3) $$\partial\colon \operatorname{Hom}_R(C, D)_n \to \operatorname{Hom}_R(C, D)_{n-1},\ \partial\{f_i\} = \{\partial^D f_i\} - \{(-1)^n f_i \partial^C\},$$

for $\{f_i\} \in \prod_i \operatorname{Hom}_R(C_i, D_{i+n}) = \operatorname{Hom}_R(C, D)_n$. This defines a complex because

$$\partial\partial\{f_i\} = \{\partial^D \partial^D f_i\} - \{(-1)^{n-1} \partial^D f_i \partial^C\} - \{(-1)^n \partial^D f_i \partial^C\} + \{(-1)^{2n-1} f_i \partial^C \partial^C\} = 0.$$

If $g\colon C' \to C$, $h\colon D \to D'$ are chain maps then

(10.4) $$\operatorname{Hom}(g, h)\colon \operatorname{Hom}(C, D) \to \operatorname{Hom}(C', D'),\quad \operatorname{Hom}(g, h)_n = \prod_i \operatorname{Hom}(g_i, h_{i+n}),$$

satisfies

$$\begin{aligned}\partial \operatorname{Hom}(g, h)_n \{f_i\} &= \partial\{h_{i+n} f_i g_i\} = \{\partial^{D'} h_{i+n} f_i g_i\} - \{(-1)^n h_{i+n} f_i g_i \partial^{C'}\}\\ &= \{h_{i+n-1}(\partial^D f_i) g_i\} - \{(-1)^n h_{i+n}(f_i \partial^C) g_{i+1}\}\\ &= \operatorname{Hom}(g, h)_{n-1} \partial\{f_i\},\end{aligned}$$

i.e. $\operatorname{Hom}(g, h)$ *is a chain map.* Thus Hom_R is an additive functor $\partial R\text{-}\mathcal{Mod} \times \partial R\text{-}\mathcal{Mod} \to \partial\mathcal{AG}$ (resp. $\to \partial R\text{-}\mathcal{Mod}$ if R is commutative), contravariant in the first, covariant in the second variable. Similarly, we can define $\operatorname{Ext}_R(C, D)$ by $\operatorname{Ext}_R(C, D)_n = \prod_{j-i=n} \operatorname{Ext}_R(C_i, D_j)$, etc.

10.5 Remarks. The elements of $\operatorname{Hom}(C, D)_n$ are sequences $f_i\colon C_i \to D_{i+n}$, $i \in \mathbb{Z}$, of homomorphisms. Such a sequence is called a *map of degree n*. It is called a *chain* map of degree n if $\partial^D f = (-1)^n f \partial^C$. The boundary operator of $\operatorname{Hom}(C, D)$ therefore measures the deviation of f from being a chain map. In particular, $Z_0 \operatorname{Hom}(C, D)$ is the group of (ordinary) chain maps $C \to D$.

A chain map $f \in Z_n \operatorname{Hom}(C, D)$ of degree n is a boundary in $\operatorname{Hom}(C, D)$ if there exists a map $s = \{s_i\colon C_i \to D_{i+n+1}\}$ of degree $n+1$ such that $\partial^D s_i + (-1)^n s_{i-1} \partial^C = f_i$. Such an s is usually called a *homotopy of degree* $(n+1)$, and f is called nulhomotopic, $f \simeq 0$, if s exists. In particular, the boundary group $B_0 \operatorname{Hom}(C, D)$ consists precisely of all nulhomotopic chain maps, hence

(10.6)
$$H_0 \operatorname{Hom}(C, D) = \pi(C, D) = \textit{group of homotopy classes of chain maps } C \to D.$$

A chain map $f = \{f_i\colon C_i \to D_{i-n}\}$ of degree $-n$ can also be viewed as an ordinary chain map $f\colon C \to D^{(n)}$ of C into the n-fold suspension of D; similarly for homotopies, hence

(10.7)
$$H_{-n} \operatorname{Hom}(C, D) = \pi(C, D^{(n)}) = \pi(C^{(-n)}, D).$$

With every chain map $f\colon C \to D^{(n)}$ we can associate the induced map $f_*\colon HC \to (HD^{(n)}) = (HD)^{(n)}$. If $f \simeq 0$ then $f_* = 0$, hence a map

(10.8)
$$\alpha\colon H_n \operatorname{Hom}(C, D) \to \operatorname{Hom}(HC, HD)_n, \quad \alpha[f] = f_*, \quad f_*[z] = [fz],$$

for $f \in Z_n \operatorname{Hom}(C, D)$, $z \in ZC$. If $g\colon C' \to C$, $h\colon D \to D'$ are chain maps then the definitions show

(10.9)
$$\operatorname{Hom}(g, h)_*(f_*) = h_* f_* g_*, \quad f_* \in H \operatorname{Hom}(C, D);$$

in particular,

(10.10)
$$g^0 \simeq g^1,\ h^0 \simeq h^1 \Rightarrow \operatorname{Hom}(g^0, h^0)_* = \operatorname{Hom}(g^1, h^1)_*,$$

i.e. the functor Hom is compatible with homotopies.

10.11 Künneth Theorem. *Let* $_RC$, $_RD$ *be complexes over a hereditary ring* R *such that* $H[\operatorname{Ext}_R(C, D)] = 0$ (e.g., C free). *Then there are natural exact sequences*

(10.12)
$$0 \to \operatorname{Ext}_R(HC, HD)_{n+1} \xrightarrow{\beta} H_n \operatorname{Hom}_R(C, D) \xrightarrow{\alpha} \operatorname{Hom}_R(HC, HD)_n \to 0,$$

and these sequences split (unnaturally).

If $\partial^D = 0$ this reduces to the universal coefficient theorem 4.2 with $t = \mathrm{Hom}(-, D)$. For $n = 0$ we get the following

10.13 Corollary (homotopy classification). *Let ${}_R C$, ${}_R D$ be complexes over a hereditary ring R such that $H[\mathrm{Ext}_R(C, D)] = 0$. Then there is a natural exact sequence*

(10.14)

$$0 \to \prod_i \mathrm{Ext}(H_{i-1} C, H_i D) \xrightarrow{\beta} \pi(C, D) \xrightarrow{\alpha} \prod_i \mathrm{Hom}(H_i C, H_i D) \to 0,$$

and this sequence splits (unnaturally).

If C is free we knew already from II, 4.6 that α is epimorphic. 10.13 tells us, in addition, how many chain maps $C \to D$ induce the same homomorphism of homology.

The construction of the exact Künneth sequence 10.12 is dual to the construction of 9.14; the reader has only to apply the ⅂-convention (§2); in particular, Hom $= \otimes$⅂, Ext $= *$⅂. However, this procedure fails when it comes to split 10.12 (we have no "free⅂"). But exactness alone suffices to prove

10.15 Proposition. *If $f\colon C^2 \to C^1$, $g\colon D^1 \to D^2$ are chain maps which induce homology isomorphisms, $f_*\colon HC^2 \cong HC^1$, $g_*\colon HD^1 \cong HD^2$, and if $H\,\mathrm{Ext}_R(C^1, D^1) = 0$, $H\,\mathrm{Ext}_R(C^2, D^2) = 0$ then (f, g) induces isomorphisms between the Künneth sequences of (C^1, D^1) and (C^2, D^2). In particular,* $\mathrm{Hom}(f, g)_*\colon\ H\,\mathrm{Hom}(C^1, D^1) \cong H\,\mathrm{Hom}(C^2, D^2)$.—This follows immediately from the five lemma and naturality of α, β (compare with 9.20). ∎

Now, in order to split 10.12 we take free complexes C', D' and chain maps

$$C \xleftarrow{f} C' \xrightarrow{f'} HC', \qquad D \xleftarrow{g} D' \xrightarrow{g'} HD'$$

which induce homology isomorphisms, and such that $f'_* = \mathrm{id}$ (cf. 10.16 below, and II, 4.6). Then by 10.15 the maps

$$(C, D) \xleftarrow{(f, \mathrm{id})} (C', D) \xleftarrow{(\mathrm{id}, g)} (C', D') \xrightarrow{(\mathrm{id}, g')} (C', HD)$$

induce an isomorphism between the Künneth sequences 10.12 of (C, D) and (C', HD), so that it suffices to split the latter. But in that case

$$\mathrm{Hom}(f', \mathrm{id})_*\colon\ \mathrm{Hom}(HC', HD) \to H\,\mathrm{Hom}(C', HD)$$

is a right inverse of α.

It remains to show that f and g exist. This is contained in

10.16 Lemma. *Given any complex E over a hereditary ring R there is a chain map $h\colon \bar{E}\to E$ such that $\bar{E}$ is free, and $h_*\colon H\bar{E}\cong HE$. If $H_{n-1}E=H_nE=0$ for some n then we may take $\bar{E}_n=0$. If $H_{n-1}E$, H_nE are finitely generated, and R is noetherian (e.g. a principal ideal domain) then we may take $\bar{E}_n$ finitely generated. If E is free then h is a homotopy equivalence, by* II, 4.3.

Proof. Take a two-term resolution of H_nE (zero resp. finitely generated if H_nE is so), and place it in dimensions $n, n+1$. The resulting free complex $E(n)$ satisfies $H_nE(n)\cong H_nE$, $H_jE(n)=0$ for $j\neq n$. Put $\bar{E}=\oplus_n E(n)$. Then $H\bar{E}\cong HE$, and this isomorphism can be realized by a chain map (cf. II, 4.6). ∎

The relations between Hom and ⊗ of modules generalize to complexes. We discuss one instance (which will be needed later on) and indicate others in the exercises.

Assume R is a *principal ideal domain;* all modules, Hom, ⊗ are over R. If $f\colon L\to L'$, $g\colon M\to M'$ are R-homomorphisms then so is $f\otimes g\colon L\otimes M\to L'\otimes M'$. The assignment $(f,g)\mapsto f\otimes g$ is a natural bilinear map; by 8.19 it induces a natural R-homomorphism

$$(10.17)\qquad \gamma\colon \operatorname{Hom}(L,L')\otimes\operatorname{Hom}(M,M')\to\operatorname{Hom}(L\otimes M, L'\otimes M')$$

which is characterised by the confusing equation $\gamma(f\otimes g)=f\otimes g$. The confusion arises, of course, because $f\otimes g$ denotes two different things, and γ takes one into the other. In most cases the context will make it clear what is meant by $f\otimes g$; for the moment we think of it as an element of $\operatorname{Hom}(L,L')\otimes\operatorname{Hom}(M,M')$. Then 10.17 is characterised by

$$(\gamma(f\otimes g))(x\otimes y)=(fx)\otimes(gy).$$

10.18 Proposition. *If L, M are free modules, and if L, M or L, L' are finitely generated then γ is an isomorphism.*

Proof. If $L=M=R$ then both sides agree with $L'\otimes M'$, and $\gamma=\mathrm{id}$. If $L=\oplus R$, $M=\oplus R$, are finite sums, then γ is isomorphic because both sides are additive. Similarly, if $L=L'=R$ both sides agree, with $\operatorname{Hom}(M,M')$, and $\gamma=\mathrm{id}$. If $L=\oplus R$, $L'=\oplus R$ are finite sums, γ is isomorphic because both sides are additive. If $L=\oplus R$ is a finite sum, and L' is finitely generated then $L'\cong P_0/P_1$, where P_0, P_1 are finitely generated free modules. Now γ is isomorphic if L' is replaced by P_0 or P_1, and hence for L' itself because both sides, as functors of L', are right exact (L, M being free). ∎

10.19 Corollary. *If* L, M, L', M' *are as in* 10.18 *then*

$$\mathrm{Hom}(L, L') * \mathrm{Hom}(M, M') \cong \mathrm{Hom}(L \otimes M, L' * M').$$

Proof. As above, we choose an exact sequence $0 \to P_1 \to P_0 \to L' \to 0$ where P_0, P_1 are free (and finitely generated if L' is so). Then

$$(10.20) \qquad 0 \to L' * M' \to P_1 \otimes M' \to P_0 \otimes M'$$

is exact by definition of $*$. Further

$$(10.21) \qquad 0 \to \mathrm{Hom}(L, P_1) \to \mathrm{Hom}(L, P_0) \to \mathrm{Hom}(L, L') \to 0$$

is exact and $\mathrm{Hom}(L, P_j)$ is free because L is free and finitely generated. Consider the commutative diagram

$$(10.22) \qquad \begin{array}{ccccc} 0 \to \mathrm{Hom}(L,L') * \mathrm{Hom}(M,M') \to & \mathrm{Hom}(L,P_1) \otimes \mathrm{Hom}(M,M') & \to & \mathrm{Hom}(L,P_0) \otimes \mathrm{Hom}(M,M') \\ & \downarrow \gamma & & \downarrow \gamma \\ 0 \to \mathrm{Hom}(L \otimes M, L' * M') \to & \mathrm{Hom}(L \otimes M, P_1 \otimes M') & \to & \mathrm{Hom}(L \otimes M, P_0 \otimes M'). \end{array}$$

The first row is exact by definition of $*$ and 10.21. The second row is exact because $L \otimes M$ is free, and 10.20 is exact. The two vertical arrows are isomorphic by 10.18. Therefore, the left terms are isomorphic. ∎

Let now C, C', D, D' be R-complexes and define

$$(10.23) \qquad \gamma\colon \mathrm{Hom}(C, C') \otimes \mathrm{Hom}(D, D') \to \mathrm{Hom}(C \otimes D, C' \otimes D')$$

by $(\gamma(f \otimes g))(x \otimes y) = (-1)^{|g||x|}(fx) \otimes (gy)$. *This is a chain map.* Indeed, if we apply definitions 9.3, 10.3 we find

$$\begin{aligned}
[\gamma\partial(f \otimes g)](x \otimes y) &= (-1)^{|g||x|}(\partial f x - (-1)^{|f|} f \partial x) \otimes g y \\
&\quad + (-1)^{|f| + |x||\partial g|} f x \otimes (\partial g y - (-1)^{|g|} g \partial y), \\
[\partial\gamma(f \otimes g)](x \otimes y) &= (-1)^{|g||x|}(\partial f x \otimes g y + (-1)^{|f| + |x|} f x \otimes \partial g y) \\
&\quad - (-1)^{|f| + |g|}((-1)^{|g||\partial x|} f \partial x \otimes g y \\
&\quad + (-1)^{|g||x| + |x|} f x \otimes g \partial y).
\end{aligned}$$

The right sides agree, hence $\gamma\partial = \partial\gamma$.

10.24 Proposition. *If one of the following assumptions I–III holds then* γ *is a homotopy equivalence* (*R being a principal ideal domain*).

I. C and D are free, HC and HD are bounded and of finite type[7].

II. C and D are free, HC and HD are bounded from below, C' and D' are bounded from above, HC is of finite type, HD or C' is of finite type.

III. C and D are free, HC is bounded and of finite type, C' and D' are bounded, HD or C' is of finite type.

If one of I–III holds, and also $H(C' * D')=0$, then the Künneth theorem 9.13 applies to $\mathrm{Hom}(C,C')\otimes\mathrm{Hom}(D,D')$, hence a natural split-exact sequence

$$\begin{aligned} &0\to\oplus_{j+k=n}H_j\mathrm{Hom}(C,C')\otimes H_k\mathrm{Hom}(D,D')\\ (10.25)\qquad &\to H_n\mathrm{Hom}(C\otimes D, C'\otimes D')\\ &\to\oplus_{j+k=n-1}H_j\mathrm{Hom}(C,C')*H_k\mathrm{Hom}(D,D')\to 0.\end{aligned}$$

If we take $C=D'=(R,0)$, and $C'=(M,0)$ where M is an R-module then case III has the following

10.26 Corollary. *There is a natural split-exact sequence*

$$(10.27)\qquad 0\to M\otimes H^n(D;R)\to H^n(D;M)\to M*H^{n+1}(D;R)\to 0$$

for free R-complexes D, and R-modules M such that HD is of finite type or M is finitely generated. ▮

Here and later we use the notation $H^n(D;M)=H_{-n}\mathrm{Hom}(D,M)$.

Proof of 10.24. If HC is bounded and/or of finite type then C is homotopy equivalent to a free complex $\bar{C}$ which is bounded and/or of finite type (cf. 10.16). Similarly for D. Since γ is compatible with homotopies we can replace C, D by $\bar{C}$, $\bar{D}$, i.e. we can assume that C, D themselves satisfy the conditions which we required for HC, HD. Each one of the conditions I–III then implies that $\mathrm{Hom}(C,C')_i=\prod_p\mathrm{Hom}(C_p,C'_{p+i})$ is actually a finite product (=sum); similarly, for $\mathrm{Hom}(D,D')$. Therefore, the left side of 10.23 is, in dimension n, a direct sum of terms $\mathrm{Hom}(C_p,C'_r)\otimes\mathrm{Hom}(D_q,D'_s)$ with $p+q=r+s+n$. Similarly, each of I–III implies that the right side is the corresponding sum of terms $\mathrm{Hom}(C_p\otimes D_q, C'_r\otimes D'_s)$. By 10.18, γ maps each term isomorphically, and is therefore itself isomorphic.

It remains to justify the application of the Künneth theorem 9.13, i.e., we have to show that $\mathrm{Hom}(C,C')*\mathrm{Hom}(D,D')$ is acyclic. But $\mathrm{Hom}(C,C')*$

[7] A graded module G is said to be *bounded* (*from above, from below*) if $G_j=0=G_{-j}$ ($G_j=0$, $G_{-j}=0$) for large j. It is said to be of *finite type* if every G_j is finitely generated.

$\operatorname{Hom}(D, D') \simeq \operatorname{Hom}(C \otimes D, C' * D')$ by an easy extension of 10.19; one can also copy the proof of 10.19, replacing P_1, P_0 by free complexes. Finally, $\operatorname{Hom}(C \otimes D, C' * D')$ is acyclic because the Künneth theorem 10.11 for *Hom* applies, and $H(C' * D') = 0$. ∎

10.28 Remark. If $L \in R\text{-}\mathcal{M}od$ and C is a complex of left R-modules we can form $\operatorname{Hom}_R(C, L)$ in the sense of 2.6, i.e. we can apply the functor $\operatorname{Hom}_R(-, L)$ to the complex C; or we can view L as a complex, $L = (L, 0)$, and form $\operatorname{Hom}_R(C, (L, 0))$ in the sense of 10.1. These two complexes agree as graded groups but the boundary operators differ by a sign. In the first case $\partial(\varphi) = \varphi \circ \partial$, in the second $\partial(\varphi) = -(-1)^{|\varphi|} \varphi \circ \partial$. In most applications this difference does not matter—the complexes are isomorphic, after all. When it does matter we shall always take $\partial(\varphi) = -(-1)^{|\varphi|} \varphi \circ \partial$, this being preferable from a systematic point of view.

The homology of $\operatorname{Hom}(C, L)$ is often called *cohomology of C with coefficients in L*, and is denoted by $H^*(C; L)$; with indices, $H^q(C; L) = H_{-q} \operatorname{Hom}(C, L)$.

10.29 Exercises. *1. The composition map*

$$\operatorname{Hom}_R(D, D') \otimes_{\mathbf{Z}} \operatorname{Hom}_R(C, D) \otimes_{\mathbf{Z}} \operatorname{Hom}_R(C', C) \to \operatorname{Hom}_R(C', D'),$$
$$\{f_i\} \otimes \{g_j\} \otimes \{h_k\} \mapsto \{f_{l+|h|+|g|} \circ g_{l+|h|} \circ h_l\},$$

is a chain map. In particular, the *evaluation map* $\operatorname{Hom}_R(C, D) \otimes_{\mathbf{Z}} C \to D$, $\{g_j\} \otimes x \mapsto g_{|x|}(x)$, is a chain map. Study the maps which are obtained by passing to homology and composing with α (cf. 9.11).

2. Show that $\Phi\colon \operatorname{Hom}_{\mathbf{Z}}(C \otimes_R D, E) \to \operatorname{Hom}_R(C, \operatorname{Hom}_{\mathbf{Z}}(D, E))$, $[\Phi\{f_i\}\, x]\, y = f_{|x|+|y|}(x \otimes_R y)$, is a chain isomorphism.—Exercises 1 and 2 illustrate how useful the sign rule 9.8 is.

*3**. If $t\colon \partial R\text{-}\mathcal{M}od \to \partial\mathcal{AG}$ is a (covariant) functor between complexes then we define a *∂-structure on t* to be a natural chain-map $\tau\colon \operatorname{Hom}_R(D, D') \to \operatorname{Hom}_{\mathbf{Z}}(tD, tD')$ such that $Z_0\tau\colon Z_0 \operatorname{Hom}_R(D, D') \to Z_0 \operatorname{Hom}_{\mathbf{Z}}(tD, tD')$ agrees with $t\colon [D, D'] \to [tD, tD']$ where $[\,]$ denotes the set of chain maps. Show that $t = C \otimes_R -$ (C fixed) admits a ∂-structure. Prove that a strongly additive right exact functor t with ∂-structure is completely determined by its value on $R = (R, 0)$; in fact, $tD \cong t(R, 0) \otimes_R D$ (compare 5.4, 5.8). Formulate and prove the dual result for cofunctors (see 6.4, 6.8). Show that the functor "n-skeleton", defined by $(tX)_i = X_i$ for $i \leq n$, $(tX)_i = 0$ for $i > n$, $\partial^{tX} = \partial^X$ or 0, does not admit any ∂-structure (hint: it does not preserve homotopies).

*4**. If C, C', D, D' are complexes over a principal ideal domain take chain maps $\bar{C} \to C$, $\bar{C}' \to C'$, ..., as in lemma 10.16. They induce a commutative diagram

$$
\begin{array}{ccccc}
\operatorname{Hom}(C,C')\otimes\operatorname{Hom}(D,D') & \to & \operatorname{Hom}(\bar{C},C')\otimes\operatorname{Hom}(\bar{D},D') & \leftarrow & \operatorname{Hom}(\bar{C},\bar{C}')\otimes\operatorname{Hom}(\bar{D},\bar{D}') \\
\downarrow{\scriptstyle\gamma} & & \downarrow{\scriptstyle\gamma} & & \downarrow{\scriptstyle\gamma} \\
\operatorname{Hom}(C\otimes D, C'\otimes D') & \to & \operatorname{Hom}(\bar{C}\otimes\bar{D}, C'\otimes D') & \leftarrow & \operatorname{Hom}(\bar{C}\otimes\bar{D}, \bar{C}'\otimes\bar{D}').
\end{array}
\tag{10.30}
$$

Use this and the Künneth theorems 9.13, 10.11 to establish an exact sequence 10.25 under weaker assumptions than above. For instance, the complexes C, D need not be free if one assumes instead that

(i) The complexes $\operatorname{Ext}(C, C')$, $\operatorname{Ext}(D, D')$, $\operatorname{Ext}(C\otimes D, C'\otimes D')$ are acyclic.

(ii) The complexes $C * D$, $C' * D'$, $\operatorname{Hom}(C, C') * \operatorname{Hom}(D, D')$ are acyclic.

I don't know whether the assumptions on C', D' can be replaced by the corresponding assumptions on HC', HD'.

11. Acyclic Models

We have already used the method of acyclic models implicitly in proving homotopy invariance of singular homology (III, 5). Now we give it a general explicit formulation. We shall use it again in § 12 to prove the Eilenberg-Zilber theorem.

11.1 Definition. Let $\mathscr{K}$ be an arbitrary category, and $F\colon \mathscr{K}\to\mathscr{AG}$ a covariant functor to abelian groups. A *base of* F is a family of elements $\{m_j\}_{j\in J}$, such that $m_j\in FM_j$, $M_j\in\mathscr{K}$, and such that for every $X\in\mathscr{K}$ the abelian group FX is freely generated by $\{(F\sigma)m_j\}$, where $j\in J$, $\sigma\in\mathscr{K}(M_j,X)$. We say, F is *free* if it has a base.

If $\mathscr{M}\subset\operatorname{Ob}(\mathscr{K})$ is a class of objects containing all M_j, then one also says F has a *base in* $\mathscr{M}$, or F is free with *models* in $\mathscr{M}$. We shall often think of $\mathscr{M}$ as a subcategory of $\mathscr{K}$, having the same morphisms (between $M, M'\in\mathscr{M}$) as $\mathscr{K}$, i.e. as a *full* subcategory.

For instance, if $\mathscr{K}=\mathscr{Top}$ then $FX=S_nX$ is freely generated by $\{\sigma(\iota_n)\}$, where $\iota_n=\mathrm{id}\in S_n(\Delta_n)$ and $\sigma\colon\Delta_n\to X$, hence the element ι_n is a base for S_n. If $\mathscr{K}=\mathscr{Top}\times\mathscr{Top}$ then $F(X,Y)=S_n(X\times Y)$ is free with base $(\iota_n,\iota_n)\in S_n(\Delta_n\times\Delta_n)$, and $F(X,Y)=(SX\otimes SY)_n=\bigoplus_{p+q=n}S_pX\otimes S_qY$ has a base in $\mathscr{M}_n=\{(\Delta_p,\Delta_q)\}_{p+q=n}$, namely $\{\iota_p\otimes\iota_q\}_{p+q=n}$.

11.2 Proposition. *Let $F\colon\mathscr{K}\to\mathscr{AG}$ be a free functor with base $\{m_j\in FM_j\}_{j\in J}$, and let $W\colon\mathscr{K}\to\mathscr{AG}$ be any functor. If $\{w_j\in WM_j\}_{j\in J}$ is any family then there is a unique natural transformation $\Phi\colon F\to W$ such that $\Phi(m_j)=w_j$,* *for all* $j \in J$. In other words, *natural transformations* $F \to W$ *are completely determined by their values on a base, and these values can be prescribed.* This *universal property* justifies the adjective *free* (compare with I, 2.20).

Proof. If Φ: $F \to W$ is a natural transformation then $\Phi((F\sigma)m_j) = (W\sigma)(\Phi m_j)$, for every σ: $M_j \to X$. Since $\{(F\sigma)m_j\}$ is a base of FX this shows that Φ is indeed determined by its values on $\{m_j\}$. This also indicates how to construct Φ when $\{w_j\}$ is given, namely Φ: $FX \to WX$ takes a free generator $(F\sigma)m_j$ of FX (where σ: $M_j \to X$) into $(W\sigma)w_j$. One has to check naturality: If g: $X' \to X$ is a morphism then

$$(\Phi \circ F(g))((F\sigma)m_j) = \Phi(F(g\sigma)m_j) = W(g\sigma)w_j = W(g)W(\sigma)w_j$$
$$= (W(g) \circ \Phi)((F\sigma)m_j),$$

hence $\Phi \circ F(g) = W(g) \circ \Phi$. ∎

11.3 Corollary. *Let* $\mathcal{M} \subset \mathcal{K}$ *be a full subcategory, and assume* $F: \mathcal{K} \to \mathcal{AG}$ *has a base* $\{m_j \in FM_j\}_{j \in J}$ *such that* $M_j \in \mathcal{M}$ *for all* j *(*F *has a base in* $\mathcal{M}$*). Then every natural transformation* $F|\mathcal{M} \to W|\mathcal{M}$ *has a unique extension* $F \to W$ (where W: $\mathcal{K} \to \mathcal{AG}$ is any functor).—Indeed, both $F|\mathcal{M} \to W|\mathcal{M}$ and $F \to W$ are characterized by their values on $\{m_j\}$. ∎

This corollary admits a useful generalization to quotients of free functors, as follows.

11.4 Proposition. *Let* $F_1 \xrightarrow{\rho} F_0 \xrightarrow{\pi} G \to 0$ *be an exact sequence of natural transformations between functors* $\mathcal{K} \to \mathcal{AG}$ *(exact means:* exact on every $X \in \mathcal{K}$*). Assume* F_0 *has a base in* $\mathcal{M}_0 \subset \mathcal{K}$*, and* F_1 *a base in* $\mathcal{M}_1 \subset \mathcal{K}$*. Let* W: $\mathcal{K} \to \mathcal{AG}$ *be a functor such that for every non-zero* $w' \in WM'$*,* $M' \in \mathcal{M}_1$*, there is a morphism* g: $M' \to M$ *with* $M \in \mathcal{M}_0$ *and* $(Wg)w' \neq 0$ (this is always fulfilled if $\mathcal{M}_1 \subset \mathcal{M}_0$). *Then every natural transformation* ψ: $G|\mathcal{M}_0 \to W|\mathcal{M}_0$ *admits a unique extension* Ψ: $G \to W$ *to the whole category* $\mathcal{K}$.

Proof. If Ψ_1, Ψ_2: $G \to W$ agree on $\mathcal{M}_0$, then $\Psi_1\pi$, $\Psi_2\pi$ agree on $\mathcal{M}_0$, hence $\Psi_1\pi = \Psi_2\pi$ by 11.3, hence $\Psi_1 = \Psi_2$ because π is surjective. Assume now ψ: $G|\mathcal{M}_0 \to W|\mathcal{M}_0$ is given; then $\varphi = \psi(\pi|\mathcal{M}_0)$: $F_0|\mathcal{M}_0 \to W|\mathcal{M}_0$ admits an extension Φ: $F_0 \to W$, by 11.3. If we show that $\Phi\rho = 0$ then we can define Ψ by $\Psi\pi = \Phi$ (because $G \cong$ cokernel (ρ)). Let $m' \in F_1 M'$, $M' \in \mathcal{M}_1$, and g: $M' \to M$ a morphism, $M \in \mathcal{M}_0$. Then

$$(Wg)(\Phi\rho)m' = (\Phi\rho)(F_1 g)m' = ((\Phi|\mathcal{M}_0)(\rho|\mathcal{M}_0))(F_1 g)m'$$
$$= (\psi(\pi|\mathcal{M}_0)(\rho|\mathcal{M}_0))(F_1 g)m' = 0,$$

the latter because $\pi\rho=0$. Thus $w'=(\Phi\rho)\,m'$ is annihilated by all $g\colon M'\to M$, hence it is zero by assumption, hence $(\Phi\rho)|\mathcal{M}_1=0$, hence $\Phi\rho=0$ because F_1 has a base in $\mathcal{M}_1$ (cf. 11.3). ∎

11.5 Lemma (compare with II, 4.7). *Let*

$$\begin{array}{ccccc} F & \xrightarrow{\tau_1} & W_0 & \xrightarrow{\tau_0} & W_{-1} \\ \vdots\,\varphi & & \downarrow \varphi_0 & & \downarrow \varphi_{-1} \\ W_1' & \xrightarrow{\tau_1'} & W_0' & \xrightarrow{\tau_0'} & W_{-1}' \end{array} \tag{11.6}$$

be a commutative diagram (without φ as yet) of natural transformations between functors $\mathcal{K}\to\mathcal{AG}$. Suppose F has a base in $\mathcal{M}\subset\mathcal{K}$, $\tau_0\tau_1=0$, and the second row is exact on $\mathcal{M}$ (i.e. *$W_1'M\to W_0'M\to W_{-1}'M$ is exact for every $M\in\mathcal{M}$*). *Then* 11.6 *can be completed by a natural transformation φ.*

Proof. For every $m\in FM$ we have $\tau_0'(\varphi_0\tau_1(m))=\varphi_{-1}\tau_0\tau_1(m)=0$. If $M\in\mathcal{M}$ then $\varphi_0\tau_1(m)=\tau_1'(w)$ for some $w\in W_1'M$, because the second row is exact. In particular, there are elements $\{w_j\in W_1'M_j\}_{j\in J}$ such that $\tau_1'(w_j)=\varphi_0\tau_1(m_j)$, for every basic generator $m_j\in FM_j$ of F. By 11.2, there is a natural transformation $\varphi\colon F\to W_1'$ such that $\varphi(m_j)=w_j$. Then $\tau_1'\varphi$ and $\varphi_0\tau_1$ agree on $\{m_j\}$, hence they agree by 11.2. ∎

11.7 Proposition (Acyclic Model Theorem). *Let $F, V\colon \mathcal{K}\to\partial\mathcal{AG}$ be covariant functors from $\mathcal{K}$ to complexes such that $F_i=0=V_i$ for $i<0$. Assume there are $\mathcal{M}_k\subset\mathcal{K}$ for $k=0,1,\ldots$, such that F_k has a base in $\mathcal{M}_k$, and $H_{k+1}VM=0$ for $M\in\mathcal{M}_{k+1}$ or $M\in\mathcal{M}_{k+2}$. Then every natural transformation $\varphi\colon H_0F\to H_0V$ is induced by a unique (up to natural homotopy) natural chain map $f\colon F\to V$. In symbols,*

$$H_0\colon\ \pi[F,V]\cong[H_0F,H_0V],$$

where $\pi[\,]$ denotes the group of (*natural*) *homotopy classes of natural chain maps, and $[\,]$ the group of natural transformations.*

Proof. Given φ, we have to find f, i.e. we have to fill the diagram

$$\begin{array}{ccccccccc} \cdots \xrightarrow{\partial} & F_2 & \xrightarrow{\partial} & F_1 & \xrightarrow{\partial} & F_0 & \longrightarrow & H_0F & \to 0 \\ & \vdots\, f_2 & & \vdots\, f_1 & & \vdots\, f_0 & & \downarrow \varphi & \\ \cdots \xrightarrow{\partial} & V_2 & \xrightarrow{\partial} & V_1 & \xrightarrow{\partial} & V_0 & \longrightarrow & H_0V & \to 0. \end{array} \tag{11.8}$$

According to 11.5 this can be done step by step (using $H_{k+1} VM=0$ for $M\in\mathcal{M}_{k+2}$).

Suppose now $f\colon F\to V$ is a natural chain map with $H_0 f=0$. We have to construct $s=\{s_k\colon F_k\to V_{k+1}\}$ such that $\partial s_k+s_{k-1}\partial=f_k$. Proceed by induction on k starting with $s_{-1}=0$. The inductive step from $k-1$ to $k\geq 0$ consists in filling the diagram

$$(11.9)\qquad \begin{array}{ccccc} F_k & \xrightarrow{(\mathrm{id},\,\mathrm{id})} & F_k\oplus F_k & \longrightarrow & 0 \\ {\scriptstyle s_k}\vdots\downarrow & & \downarrow{\scriptstyle (f_k,\,-s_{k-1}\partial)} & & \downarrow \\ V_{k+1} & \xrightarrow[\partial]{} & V_k & \xrightarrow[\partial]{} & V_{k-1}, \end{array}$$

where for $k=0$ one replaces V_{-1} by $H_0 V$. By 11.5 again, this can be done (using $H_{k+1} VM=0$ for $M\in\mathcal{M}_{k+1}$). ∎

In 11.7 we make no assumption about $H_0 VM$; if we do then we can improve the theorem as follows.

11.10 Corollary. *In the situation* 11.7, *assume that for every non-zero* $v\in H_0 VM'$, $M'\in\mathcal{M}_1$, *there is a morphism* $g\colon M'\to M$ *such that* $M\in\mathcal{M}_0$ *and* $(H_0 Vg)\,v\neq 0$. *Then every natural transformation* $H_0 F|\mathcal{M}_0\to H_0 V|\mathcal{M}_0$ *is induced by a unique (up to natural homotopy) natural chain map* $F\to V$; *in symbols,* $\pi[F,V]\cong[H_0 F|\mathcal{M}_0, H_0 V|\mathcal{M}_0]$. Thus, natural chain maps $F\to V$ are characterized (up to $\simeq$) by what they do to $H_0 FM$, $M\in\mathcal{M}_0$.— This follows because 11.4 (with $G=H_0 F$, $W=H_0 V$) asserts $[H_0 F|\mathcal{M}_0, H_0 V|\mathcal{M}_0]=[H_0 F, H_0 V]$, and the latter equals $\pi[F,V]$ by 11.7. ∎

11.11 Exercises. *1.* Call a functor $P\colon\mathcal{K}\to\mathcal{AG}$ *pro-free* if it is a direct summand of a free functor $F\colon\mathcal{K}\to\mathcal{AG}$, i.e. if natural transformations $P\xrightarrow{\iota}F\xrightarrow{\rho}P$ exist such that $\rho\iota\sim\mathrm{Id}$. Generalize the preceding results from free to pro-free functors. If $0\to V'\to V\to V''\to 0$ is an exact sequence of natural transformations between functors $\mathcal{K}\to\mathcal{AG}$ and P is pro-free then $0\to[P,V']\to[P,V]\to[P,V'']\to 0$ is also exact, i.e. pro-free functors are *projective* in the sense of 6.21.

2. If $\mathcal{K}$ is a *small* category (objects form a set) and $V\colon\mathcal{K}\to\mathcal{AG}$ is any functor then there exists a free functor $F\colon\mathcal{K}\to\mathcal{AG}$ and a natural epimorphism $\Phi\colon F\to V$. (Hint: For every $K\in\mathcal{K}$, $v\in VK$, $X\in\mathcal{K}$, let $F_{K,v}(X)$ denote the free abelian group generated by $\mathcal{K}(K,X)$, and $\Phi_{K,v}\colon F_{K,v}(X)\to VX$ the natural homomorphism given by $\alpha\mapsto(V\alpha)\,v$, $\alpha\in\mathcal{K}(K,X)$. Put $F=\bigoplus_{(K,v)}F_{K,v}$, $\Phi=\{\Phi_{K,v}\}$.) Use this and Exercise 1 to show that every projective functor $\mathcal{K}\to\mathcal{AG}$ is pro-free. Compare with Dold-MacLane-Oberst.

3. Using 11.10, show that the group of (natural) homotopy classes of natural chain maps $SX \to SX$, $X \in \mathcal{T}\!op$, is freely generated by the identity map; in symbols, $\pi[SX, SX] \cong \mathbb{Z}$. More generally, if I is a (non-empty) acyclic space, $\tilde{H}I = 0$, then $\pi[SX, S(X \times I)]$ is a free cyclic group, generated by $S(i_P)$ where $P \in I$, and $i_P: X \to X \times I$, $i_P(x) = (x, P)$. Compare this with III, 5.7 and 6.6.

4. As in Exercise 3, show $\pi[SX, SX \otimes SX] \cong \mathbb{Z}$, where $X \in \mathcal{T}\!op$. If $\psi: SX \to SX \otimes SX$ is a natural chain map, then there is an integer n such that $\psi(\sigma) = n(\sigma \otimes \sigma)$ for all zero-simplexes $\sigma: \Delta_0 \to X$ (this follows from naturality, applied to σ); the assignment $\psi \mapsto n$ induces the above isomorphism. In particular, there is a unique (up to natural $\simeq$) natural chain map $D: SX \to SX \otimes SX$ such that $D(\sigma) = \sigma \otimes \sigma$ for $\sigma: \Delta_0 \to X$. This D is called the *natural diagonal of* SX.

5.* Let $F, V: \mathcal{K} \to \partial\mathcal{AG}$ be functors from $\mathcal{K}$ to complexes such that $F_i = 0 = V_i$ for $i < 0$. Assume $\mathcal{M} \subset \mathcal{K}$ exists such that every F_k has a base in $\mathcal{M}$, and $H_k VM = 0$ for $M \in \mathcal{M}$ and $k > 0$. Let $\mathrm{Hom}(F, V)$ denote the following complex: $\mathrm{Hom}(F, V)_n = 0$ for $n < 0$, $\mathrm{Hom}(F, V)_0 =$ group of natural chain maps $F \to V$, $\mathrm{Hom}(F, V)_n = \prod_k [F_k, V_{k+n}]$ for $n > 0$ ([ ] as in 11.7), and boundary operator $\partial\{f_k\} = \{\partial^V \circ f_k\} - \{(-1)^n f_k \circ \partial^F\}$, as in 10.3. Use 11.7 to prove $H_n \mathrm{Hom}(F, V) = 0$ for $n \neq 0$, $H_0 \mathrm{Hom}(F, V) \cong [H_0 F, H_0 V]$.

If $C \in \partial\mathcal{AG}$ is a free complex with $C_n = 0$ for $n < 0$, and $\varphi: H_0 C \to [H_0 F, H_0 V]$ is a homomorphism then there is a unique (up to $\simeq$) chain map $\Phi: C \to \mathrm{Hom}(F, V)$ which induces φ (cf. 3.5). Passing to adjoint homomorphisms (8.20, Exerc. 3) shows: *If* $\psi_M: H_0 C \otimes H_0 FM \to H_0 VM$, $M \in \mathcal{M}$, *is a family of homomorphisms which is natural on* $\mathcal{M} \subset \mathcal{K}$ *then there is a unique (up to* $\simeq$*) natural chain map* $\Psi_X: C \otimes FX \to VX$, $X \in \mathcal{K}$, *such that* $H_0 \Psi_M = \psi_M$ for $M \in \mathcal{M}$. This is the *acyclic model theorem with parameters C*. It is contained in 11.7 if $C = (\mathbb{Z}, 0)$. It extends to complexes C such that $H\,\mathrm{Ext}(C, \mathrm{Hom}(F, V)) = 0$, $\mathrm{Ext}(H_{-1} C, [H_0 F, H_0 V]) = 0$ (use 10.13).

12. The Eilenberg-Zilber Theorem. Künneth Formulas for Spaces

Using acyclic models we prove $S(X \times Y) \simeq SX \otimes SY$ for every couple of spaces $X, Y \in \mathcal{T}\!op$. Combining this with the Künneth theorem 9.13 we can express $H(X \times Y)$ in terms of HX, HY. If every $H_i X$ is finitely generated then there is a similar formula (12.18) expressing the cohomology $H^*(X \times Y)$ in terms of $H^* X, H^* Y$.

12.1 Eilenberg-Zilber Theorem. *The functors $(SX)\otimes(SY)$ and $S(X\times Y)$ from $\mathcal{Top}\times\mathcal{Top}$ (couples of spaces) to $\partial\mathcal{AG}$ (complexes) are homotopy equivalent. More precisely, there are unique (up to homotopy) natural chain maps*

$$\Phi\colon (SX)\otimes(SY) \rightleftarrows S(X\times Y)\colon \Psi$$

such that

$$\Phi_0(\sigma\otimes\tau)=(\sigma,\tau),\quad \Psi_0(\sigma,\tau)=\sigma\otimes\tau,\quad \textit{for 0-simplices } \sigma\colon \Delta_0\to X,\ \tau\colon \Delta_0\to Y.$$

Any such chain map is a homotopy equivalence; in fact, there are natural homotopies $\Phi\Psi\simeq \mathrm{id}$, $\Psi\Phi\simeq \mathrm{id}$. Any such chain map will be called an *Eilenberg-Zilber map*, and will be denoted by *EZ*.

Analogous results hold for three or more spaces (or for a single space!) *and functors like $SX\otimes SY\otimes SZ$, $S(X\times Y)\otimes SZ$, $S(X\times Y\times Z)$*

Proof. Write $F(X,Y)=SX\otimes SY$, $F'(X,Y)=S(X\times Y)$. Both F and F' are free (cf. 11.1); in fact, F_k has a base in $\{(\Delta_p,\Delta_q)\}_{p+q=k}$, and F'_k in (Δ_k,Δ_k), namely $\{(\iota_p\otimes\iota_q)\}_{p+q=k}$ resp. (ι_k,ι_k), where $\iota_p=\mathrm{id}(\Delta_p)$. Because Δ_p and $\Delta_p\times\Delta_q$ are convex we have (III, 4.6)

$$S(\Delta_p\times\Delta_q)\simeq(\mathbb{Z},0),\qquad (S\Delta_p)\otimes(S\Delta_q)\simeq(\mathbb{Z},0)\otimes(\mathbb{Z},0)=(\mathbb{Z},0),$$

hence $H_k F$, $H_k F'$ vanish on all models (Δ_p,Δ_q) for $k>0$, and $(\Delta_p,\Delta_q)\to(\Delta_0,\Delta_0)$ induces isomorphisms of H_0F, H_0F'. We can therefore apply 11.10 (with $V=F$ or F'); since $\mathcal{M}_0=(\iota_0,\iota_0)$ is a single object, and $H_0F(\iota_0,\iota_0)$ resp. $H_0F'(\iota_0,\iota_0)$ is freely generated by $\iota_0\otimes\iota_0$ resp. (ι_0,ι_0) we see that unique (up to natural $\simeq$) natural chain maps $\Phi\colon F\to F'$, $\Psi\colon F'\to F$ exist such that $\Phi(\iota_0\otimes\iota_0)=(\iota_0,\iota_0)$, $\Psi(\iota_0,\iota_0)=\iota_0\otimes\iota_0$. Then $\Psi\Phi(\iota_0\otimes\iota_0)=\iota_0\otimes\iota_0$, $\Phi\Psi(\iota_0,\iota_0)=(\iota_0,\iota_0)$, hence (by 11.10 again) $\Psi\Phi\simeq\mathrm{id}$, $\Phi\Psi\simeq\mathrm{id}$. Finally, $\Phi(\iota_0\otimes\iota_0)=(\iota_0,\iota_0)$ implies $\Phi_0(\sigma\otimes\tau)=(\sigma,\tau)$ by naturality of Φ applied to $\sigma\colon\Delta_0\to X$, $\tau\colon\Delta_0\to Y$; and $\Psi(\iota_0,\iota_0)=\iota_0\otimes\iota_0$ implies $\Psi_0(\sigma,\tau)=\sigma\otimes\tau$. The obvious generalization to three or more spaces is left to the reader. ∎

12.2 Corollary. *For arbitrary Eilenberg-Zilber maps the following diagrams are homotopy commutative.*

$$\text{(12.3)}\qquad \begin{array}{ccc} SX\otimes SY & \xrightarrow{EZ} & S(X\times Y) \\ \downarrow{\scriptstyle \tau} & & \downarrow{\scriptstyle S(t)} \\ SY\otimes SX & \xrightarrow{EZ} & S(Y\times X), \end{array}\qquad \begin{array}{ccc} SX\otimes SY & \xleftarrow{EZ} & S(X\times Y) \\ \downarrow{\scriptstyle \tau} & & \downarrow{\scriptstyle S(t)} \\ SY\otimes SX & \xleftarrow{EZ} & S(Y\times X), \end{array}$$

where $t(x, y)=(y, x)$, $\tau(u\otimes v)=(-1)^{|u||v|}\, v\otimes u$ ("commutativity of EZ-maps").

$$(12.4)\qquad \begin{array}{ccccccc} SX\otimes SY\otimes SZ & \xrightarrow{EZ\otimes \mathrm{id}} & S(X\times Y)\otimes SZ & & SX\otimes SY\otimes SZ & \xleftarrow{EZ\otimes \mathrm{id}} & S(X\times Y)\otimes SZ \\ {\scriptstyle \mathrm{id}\otimes EZ}\downarrow & & \downarrow{\scriptstyle EZ} & & {\scriptstyle \mathrm{id}\otimes EZ}\uparrow & & \uparrow{\scriptstyle EZ} \\ SX\otimes S(Y\times Z) & \xrightarrow[EZ]{} & S(X\times Y\times Z), & & SX\otimes S(Y\times Z) & \xleftarrow[EZ]{} & S(X\times Y\times Z) \end{array}$$

("associativity of EZ-maps").

$$(12.5)\qquad \begin{array}{ccccccc} SX\otimes SP & \xrightarrow{EZ} & S(X\times P) & & SX\otimes SP & \xleftarrow{EZ} & S(X\times P) \\ {\scriptstyle \mathrm{id}\otimes\eta}\downarrow & & \downarrow{\scriptstyle \mathrm{proj}} & & {\scriptstyle \mathrm{id}\otimes\eta}\downarrow & & \downarrow{\scriptstyle \mathrm{proj}} \\ SX\otimes(\mathbb{Z},0) & \xrightarrow{\mathrm{id}} & SX, & & SX\otimes(\mathbb{Z},0) & \xleftarrow{\mathrm{id}} & SX, \end{array}$$

where P is a point, η = augmentation ("EZ preserves units").

Indeed, in each case the two ways of going from one corner to the opposite one induce the identity in dimension 0 (or on H_0), hence are (naturally) homotopic. ∎

12.6 Corollary. *For arbitrary EZ-maps Φ, Ψ and arbitrary pairs of spaces (X, A), (Y, B) we have commutative diagrams with exact rows*

$$(12.7)\qquad \begin{array}{ccccccccc} 0\to & SA\otimes SY+SX\otimes SB & \xrightarrow{\subset} & SX\otimes SY & \longrightarrow & SX/SA\otimes SY/SB & \to 0 \\ & {\scriptstyle \Phi'}\downarrow\uparrow{\scriptstyle \Psi'} & & {\scriptstyle \Phi}\downarrow\uparrow{\scriptstyle \Psi} & & {\scriptstyle \Phi''}\downarrow\uparrow{\scriptstyle \Psi''} & \\ 0\to & S\{A\times Y, X\times B\} & \xrightarrow{\subset} & S(X\times Y) & \longrightarrow & \dfrac{S(X\times Y)}{S\{A\times Y, X\times B\}} & \to 0. \end{array}$$

The vertical maps are induced by Φ, Ψ, and

$$S\{A\times Y, X\times B\}=\mathrm{im}[S(A\times Y)\oplus S(X\times B)\xrightarrow{(j^1, j^2)} S(X\times Y)]$$

as in III, 7.1. *Moreover, there are natural homotopies $\Phi'\Psi'\simeq\mathrm{id}$, $\Psi'\Phi'\simeq\mathrm{id}$, $\Phi''\Psi''\simeq\mathrm{id}$, $\Psi''\Phi''\simeq\mathrm{id}$.*

Proof. Naturality of Φ applied to $j^1\colon A\xrightarrow{\subset} X$ and id_Y shows

$$\Phi(SA\otimes SY)\subset S(A\times Y);$$

similarly $\Phi(SX\otimes SB)\subset S(X\times B)$, and analoguously for Ψ. This gives the maps Φ', Ψ', Φ'', Ψ''. Since the homotopy $\Phi\Psi\simeq\mathrm{id}$ is natural it maps

$S\{A\times Y, X\times B\}$ into itself, hence induces $\Phi'\Psi'\simeq \mathrm{id}$, $\Phi''\Psi''\simeq \mathrm{id}$. Similarly for $\Psi'\Phi'\simeq \mathrm{id}$, $\Psi''\Phi''\simeq \mathrm{id}$. ∎

12.8 Corollary. *For pairs of spaces $(X, A), (Y, B)$ we have natural maps*

$$\frac{SX}{SA}\otimes\frac{SY}{SB}\xrightarrow[\simeq]{EZ}\frac{S(X\times Y)}{S\{A\times Y, X\times B\}}\to S(X\times Y, A\times Y\cup X\times B). \tag{12.9}$$

The second map is a homotopy equivalence if and only if $(X\times Y, A\times Y, X\times B)$ is an excisive triad (e.g. if A and B are open, or one of them is empty; cf. III, 8.1). ∎

Combining 12.8 with the Künneth theorem 9.13 we get

12.10 Corollary. *For pairs of spaces $(X, A), (Y, B)$ such that $(X\times Y;$ $A\times Y, X\times B)$ is an excisive triad there exist natural exact sequences*

$$0\to\bigoplus_{i+j=n}[H_i(X,A)\otimes H_j(Y,B)]\xrightarrow{(EZ)_*\alpha} H_n(X\times Y, A\times Y\cup X\times B)$$
$$\xrightarrow{\beta(EZ)_*^{-1}}\bigoplus_{i+j=n-1}[H_i(X,A)*H_j(Y,B)]\to 0,$$

and these sequences split (but not naturally). ∎

We can, of course, apply any additive functor $\mathcal{AG}\to\mathcal{AG}$ to 12.9 and still get a homotopy equivalence (if $(X\times Y; A\times Y, X\times B)$ is excisive). For instance, if L, M are R-modules we get

$$\begin{aligned}(SX/SA\otimes L)\otimes_R(SY/SB\otimes M)&\cong(SX/SA\otimes SY/SB)\otimes(L\otimes_R M)\\&\simeq S(X\times Y; A\times Y\cup X\times B)\otimes(L\otimes_R M),\end{aligned}\tag{12.11}$$

hence (by 9.13)

12.12 Corollary. *For pairs $(X, A), (Y, B)$ as in 12.10, and modules $L\in\mathcal{Mod}$-R, $M\in R$-$\mathcal{Mod}$ over a hereditary ring R such that $L*_R M=0$ there exist natural exact sequences*

$$0\to H(X,A;L)\otimes_R H(Y,B;M)\to H(X\times Y, A\times Y\cup X\times B; L\otimes_R M)$$
$$\to H(X,A;L)*_R H(Y,B;M)^+\to 0,$$

and these split (not naturally). *In particular, if R is a field then*

$$H(X\times Y, A\times Y\cup X\times B; R)\cong H(X,A;R)\otimes_R H(Y,B;R). \tag{12.13}$$ ∎

We now compare the *co*homology of X, Y and $X\times Y$. Remark first that

$$\mathrm{Hom}_{\mathbf{Z}}(SX, M)\cong \mathrm{Hom}_R(SX\otimes_{\mathbf{Z}} R, M) \tag{12.14}$$

for every ring R and R-module M; both sides, indeed, can be identified (in dimension n) with the set of all functions f, defined on the set of all singular n-simplices $\sigma\colon \Delta_n \to X$, and with values $f(\sigma)$ in M. Under this identification the chain map γ of 10.23 (with $C' = (L, 0)$, $D' = (M, 0)$; L, M modules) becomes

$$\begin{aligned}(12.15)\qquad &\gamma\colon \operatorname{Hom}_{\mathbf{Z}}(SX, L) \otimes_R \operatorname{Hom}(SY, M) \to \operatorname{Hom}_{\mathbf{Z}}(SX \otimes SY, L \otimes_R M),\\ &(\gamma(f \otimes_R g))(\sigma \otimes_{\mathbf{Z}} \tau) = (-1)^{|g||\sigma|}(f\sigma) \otimes_R (g\tau),\end{aligned}$$

and Proposition 10.24, case II, asserts that 12.15 is a homotopy equivalence if the graded R-modules $H(X; R)$, $H(Y; R)$ are of finite type, or if $H(X; R)$ is of finite type and L is finitely generated (R a principal ideal domain).

In this argument one can replace X, Y by pairs (X, A), (Y, B). Moreover, one may replace $SX/SA \otimes SY/SB$ by the homotopy equivalent complex $S(X \times Y, A \times Y \cup X \times B)$ if $(X \times Y; A \times Y, X \times B)$ is excisive (cf. 12.8). Proposition 10.24, case II, then implies

12.16 Proposition. *Let L, M be modules over a principal ideal domain R, and let (X, A), (Y, B) be pairs of spaces such that $(X \times Y; A \times Y, X \times B)$ is excisive. If the graded modules $H(X, A; R)$, $H(Y, B; R)$ are of finite type, or if $H(X, A; R)$ is of finite type and L is finitely generated then*

$$\begin{aligned}(12.17)\qquad &\operatorname{Hom}_{\mathbf{Z}}(S(X, A), L) \otimes_R \operatorname{Hom}_{\mathbf{Z}}(S(Y, B), M)\\ &\quad \to \operatorname{Hom}_{\mathbf{Z}}(S(X \times Y, A \times Y \cup X \times B), L \otimes_R M)\end{aligned}$$

is a homotopy equivalence. If, moreover, $L *_R M = 0$ then the Künneth theorem 9.13 applies and yields natural split-exact sequences

$$\begin{aligned}(12.18)\qquad &0 \to \textstyle\bigoplus_{i+j=n} H^i(X, A; L) \otimes_R H^j(Y, B; M)\\ &\to H^n(X \times Y, A \times Y \cup X \times B; L \otimes_R M)\\ &\to \textstyle\bigoplus_{i+j=n+1} H^i(X, A; L) *_R H^j(Y, B; M) \to 0.\end{aligned}$$

In particular, if R is a field, and (X, A), (Y, B) are pairs of spaces such that $(X \times Y; A \times Y, X \times B)$ is excisive and $H(X, A; R)$ of finite type then

$$(12.19)\qquad H^*(X \times Y, A \times Y \cup X \times B; M) \cong H^*(X, A; R) \otimes_R H^*(Y, B; M)$$

for all vector spaces M over R. ∎

We conclude this chapter by some remarks on diagonal chain maps $SX \to SX \otimes SX$. For every space X we have the diagonal map $\Delta\colon X \to X \times X$, $\Delta x = (x, x)$; it induces a natural chain map $\Delta\colon SX \to S(X \times X)$. If

$EZ\colon S(X\times Y)\to SX\otimes SY$ is an Eilenberg-Zilber map then we can take $Y=X$ and compose EZ with Δ; the composite natural chain map

(12.20) $$D\colon SX \xrightarrow{\Delta} S(X\times X) \xrightarrow{EZ} SX\otimes SX$$

is called a *natural diagonal of SX. It depends on the choice of EZ but its homotopy class doesn't.*

If A_1, A_2 are subspace of X then D maps $S\{A_1, A_2\}$—the subcomplex of SX which is generated by SA_1, SA_2—into $SA_1\otimes SX+SX\otimes SA_2$; this follows from 12.6 or directly from naturality of D. Passing to quotients it induces therefore a (relative) *diagonal*

(12.21) $$D\colon SX/S\{A_1, A_2\}\to SX/SA_1\otimes SX/SA_2$$

which is still unique up to (natural) homotopy. Even more generally, we have $D\colon SX/S\{\mathscr{A}_1, \mathscr{A}_2\}\to SX/S\mathscr{A}_1\otimes SX/S\mathscr{A}_2$ where $\mathscr{A}_1, \mathscr{A}_2$ are arbitrary families of subsets of X, and $\{\mathscr{A}_1, \mathscr{A}_2\}$ is their union.

The properties of Eilenberg-Zilber maps carry over to diagonals. In particular, 12.3, 12.4, 12.5 become

(12.22) $$\tau D\simeq D \quad (\textit{commutativity}),$$

where $\tau\colon SX\otimes SX\to SX\otimes SX$ permutes factors, $\tau(u\otimes v)=(-1)^{|u||v|}v\otimes u$. ∎

(12.23) $$(\mathrm{id}\otimes D)\circ D\simeq(D\otimes \mathrm{id})\circ D \quad (\textit{associativity}),$$

both sides being maps $SX\to SX\otimes SX\otimes SX$. ∎

(12.24) $$(\mathrm{id}\otimes\eta)\circ D\simeq \mathrm{id}\simeq(\eta\otimes\mathrm{id})\circ D \quad (\textit{units}),$$

where $\eta\colon SX\to(\mathbb{Z},0)$ is the augmentation, and

$$SX\otimes(\mathbb{Z},0)=SX=(\mathbb{Z},0)\otimes SX. \ ∎$$

These relations still make sense, and are true, in the relative case discussed above.

The map $EZ\colon S(X\times Y)\to SX\otimes SY$ which enters into the definition of the diagonal D can be recaptured from D; more precisely

(12.25) $$EZ=(p\otimes q)\circ D,$$

where $D=D_{X\times Y}\colon S(X\times Y)\to S(X\times Y)\otimes S(X\times Y)$, and $p\colon X\times Y\to X$, $q\colon X\times Y\to Y$ are projections. Indeed, if we apply naturality of EZ to (p,q) we get $EZ\circ(p\times q)=(p\otimes q)\circ EZ$, and if we compose this (on the right) with $\Delta=\Delta_{X\times Y}\colon S(X\times Y)\to S((X\times Y)\times(X\times Y))$ we get 12.25 because $(p\times q)\circ\Delta=\mathrm{id}$, $EZ\circ\Delta=D$.

Natural diagonals can be defined, and their properties derived, without referring to Eilenberg-Zilber maps (but using acyclic modules; cf. 11.11 Exerc. 4). In fact, natural diagonals $SX \to SX \otimes SX$ and Eilenberg-Zilber maps $S(X \times Y) \to SX \otimes SY$ are formally equivalent notions (cf. Exerc. 5).

12.26 Exercises. *1**. For every $0 \le j \le n$ define linear maps $\xi_j^n, \bar{\xi}_j^n \colon \Delta_j \to \Delta_n$, $\xi_j^n(e_i) = e_i$, $\bar{\xi}_j^n(e_i) = e_{i+n-j}$, $i = 0, 1, \ldots, j$, where $\{e_i\}$ are the vertices of Δ. Show that the following sequence AW of homomorphisms

(12.27) $\quad AW \colon S_n(X \times Y) \to (SX \otimes SY)_n, \quad (AW)(\sigma, \tau) = \sum_{0 \le j \le n} (\sigma\, \xi_j^n) \otimes (\tau\, \bar{\xi}_{n-j}^n)$

(where $(\sigma, \tau) \colon \Delta_n \to X \times Y$) is an Eilenberg-Zilber map; in particular, $\partial(AW) = (AW)\,\partial$. Show that AW is strictly associative (not only up to homotopy) in the sense of 12.4 but *not strictly commutative* (12.3).—The notation AW stands for Alexander-Whitney who, implicitly, used this map in their definition of cup-products.

*2**. If p, q are non-negative integers then a (p, q)-*shuffle* (μ, ν) is a pair of disjoint sets of integers

$$1 \le \mu_1 < \mu_2 < \cdots < \mu_p \le p+q, \qquad 1 \le \nu_1 < \nu_2 < \cdots < \nu_q \le p+q$$

between 1 and $p+q$. Let *sign*(μ, ν) be the sign of the permutation $(\mu_1, \mu_2, \ldots, \mu_p, \nu_1, \ldots, \nu_q)$ (of the integers $1, \ldots, p+q$). Define a linear map

$$\eta^\mu \colon \Delta_{p+q} \to \Delta_p, \quad \text{by} \quad \eta^\mu(e^i) = e^j \quad \text{if} \quad \mu_j \le i < \mu_{j+1},$$

where e^i are the vertices of Δ, and $\mu_0 = 0$, $\mu_{p+1} = p+q+1$. Define homomorphisms

$$(12.28) \qquad \begin{aligned} &\nabla_{pq} \colon S_p X \otimes S_q Y \to S_{p+q}(X \times Y), \\ &\nabla_{pq}(\sigma \otimes \tau) = \sum \operatorname{sign}(\mu, \nu)(\sigma \circ \eta^\mu, \tau \circ \eta^\nu), \end{aligned}$$

where $\sigma \colon \Delta_p \to X$, $\tau \colon \Delta_q \to Y$, and the sum ranges over all (p, q) shuffles (μ, ν). Show that the following sequence ∇ of maps

(12.29) $\quad \nabla_n = \{\nabla_{pq}\}_{p+q=n} \colon (SX \otimes SY)_n = \bigoplus_{p+q=n} (S_p X \otimes S_q Y) \to S(X \times Y)_n$

is an Eilenberg-Zilber map; in particular, $\partial \nabla = \nabla \partial$. Show that the "shuffle map" ∇ is strictly associative and commutative in the sense of 12.3, 12.4.

3. By 1.12 Exerc. 4, if C is a free R-complex such that $C_i = 0$ for $i < 0$ and HC is also free then $C \simeq HC$. Use this and the Eilenberg-Zilber theorem to show that $H(X \times Y; M) \cong H(X; R) \otimes_R H(Y; M)$ if X is a space such that $H(X; R)$ is *free* (as a right R-module; $M \in R$-$\mathcal{M}od$). If $H(X; R)$ is a free right R-module of *finite type* then one also finds $H^*(X \times Y; M) \cong H^*(X; R) \otimes_R H^*(Y; M)$. Similarly for pairs (X, A), (Y, B) of spaces. Compare with 12.13 and 12.19.

4. If $X = Y = \mathbb{S}^n \vee \mathbb{S}^n \vee \mathbb{S}^n \vee \cdots$, is an infinite wedge of spheres then the map $\mathrm{Hom}_{\mathbf{Z}}(SX, R) \otimes_R \mathrm{Hom}_{\mathbf{Z}}(SY, R) \to \mathrm{Hom}_{\mathbf{Z}}(S(X \times Y), R)$ of 12.17 is not a homotopy equivalence (does not induce homology isomorphisms). A more general result (and a hint) can be found in VII, 7 Exerc. 1.

5. Let $\mathcal{K}$ be a category with products $\sqcap: \mathcal{K} \times \mathcal{K} \to \mathcal{K}$ (cf. I, 1.15), and let $\Theta: \mathcal{K} \to \mathcal{K} \times \mathcal{K}$ denote the diagonal functor, $\Theta X = (X, X)$. Show that for arbitrary functors $S: \mathcal{K} \to \mathcal{L}$, $T: \mathcal{K} \times \mathcal{K} \to \mathcal{L}$ there is a 1-1 correspondence between natural transformations $D: S \to T \circ \Theta$ and natural transformations $E: S \circ \sqcap \to T$, given by $D_X = E_{\Theta X} \circ S\Delta$, or $E_{XY} = T(p, q) \circ D_{X \sqcap Y}$, where $\Delta = (\mathrm{id}, \mathrm{id}): X \to X \sqcap X$ is the diagonal morphism, and $p, q: X \sqcap Y \to X, Y$ are the projections.

If $\mathcal{K} = \mathcal{T}\!op$ is the category of topological spaces, $\mathcal{L} = \partial\mathcal{A}\mathcal{G}$ the category of complexes, S the singular complex, $T(X, Y) = SX \otimes SY$, then 12.20 (or 12.25) shows that natural diagonals $D: SX \to SX \otimes SX$ correspond to Eilenberg-Zilber maps $E: S(X \times Y) \to SX \otimes SY$. Verify that a natural chain map $D: SX \to SX \otimes SX$ corresponds to an Eilenberg-Zilber map (is a natural diagonal) if and only if $D\sigma = \sigma \otimes \sigma$ for every 0-simplex σ.

Chapter VII

Products

There are many products in (co-)homology theory of spaces; we shall treat about eight here. All of them are combinations of the following ingredients: (i) Relations between $\otimes$ and Hom which are familiar from (multi-)linear algebra; (ii) the mappings $\alpha\colon HC\otimes HD \to H(C\otimes D)$ and $\alpha\colon H\,\mathrm{Hom}(C,D) \to \mathrm{Hom}(HC,HD)$ of VI, 9.11, 10.8; (iii) the Eilenberg-Zilber mappings VI, 12.1—plus, of course, the standard functorial properties of (co-)homology. The significance of products lies in the extra structure which they introduce in (co-)homology. The $\smile$-product, for instance, turns $H^*(X;R)$ into a graded ring (*cohomology ring*) and makes $H^*(-,R)$ a functor from $\mathcal{T}\!op$ to the category $\mathcal{GR}g$ of graded rings (R a ring with unit). This functor provides a much more accurate picture of $\mathcal{T}\!op$ than the mere cohomology *group* which is obtained by composing $H^*(-,R)$ with the forget-functor $F\colon \mathcal{GR}g \to \mathcal{GAG}$ (F assigns to every ring its additive group).

In the whole Chapter VII the following rule applies: If a is a (co-)chain resp. (co-)homology class with coefficients in $L\in\mathcal{M}od\text{-}R$ and b is with coefficients in $M\in R\text{-}\mathcal{M}od$ then any one of the products $a\perp b$ which we consider has coefficients in $L\otimes_R M$. Sometimes this will be explicitely stated but in other cases we shall not write the coefficients (in order to simplify the notations), and then it is implicitely understood. If C is an R-complex and M an R-module we use the following abbreviations: $H\,\mathrm{Hom}(C,M)=H^*(C,M)$, $Z\,\mathrm{Hom}(C,M)=Z^*(C,M)$, $B\,\mathrm{Hom}(C,M)=B^*(C,M)$; with indices, $H_{-q}\mathrm{Hom}(C,M)=H^q(C,M)$ etc. The elements of these groups are called cohomology classes (cocycles, coboundaries) of C with coefficients in M. If $f\colon C\to D$ is a chain map then we write $f^*=H\,\mathrm{Hom}(f,M)\colon H^*(D,M)\to H^*(C,M)$ for the induced homomorphism. The analogous notations VI, 7.1 for singular cohomology will also be used.

With minor exceptions there are only the following logical dependencies between the various §§ of this chapter:

$$\begin{array}{c} 3\leftarrow 2\rightarrow 5\rightarrow 6, \\ \downarrow \\ 4 \end{array} \qquad 7\rightarrow 8 \begin{array}{l}\nearrow 9\\ \searrow 10\end{array}, \qquad 11\rightarrow 12.$$

Thus, the reader can study cap-products (§12) without reading §§1–10 first—although he will find the going easier if he knows §§7–8.

For simplicity, we assume from §2 on that the ground ring R is *commutative*—although at the cost of some notational inconvenience this restriction could easily be avoided.

1. The Scalar Product

1.1 Definition. For every (right) R-complex C and R-module M we define a map

$$\operatorname{Hom}_R(C,M)\times C\to M,\quad (\varphi,c)\mapsto\varphi(c).$$

This is clearly biadditive, and R-linear in the second variable $c\in C$; if R is commutative then it is R-bilinear. It induces therefore (cf. VI, 8.11 and 8.19) an *R-homomorphism*

$$\text{(1.2)}\qquad \begin{aligned} &e\colon \operatorname{Hom}_R(C,M)\otimes_{\mathbb{Z}} C\to M,\quad e(\varphi\otimes c)=\varphi(c);\\ \text{resp.}\quad &e\colon \operatorname{Hom}_R(C,M)\otimes_R C\to M\qquad \text{if } R \text{ is commutative}.\end{aligned}$$

This is a chain map.

$$\begin{aligned} e\partial(\varphi\otimes c)&=e(\partial(\varphi)\otimes c+(-1)^{|\varphi|}\varphi\otimes\partial c)=(\partial(\varphi))c+(-1)^{|\varphi|}\varphi(\partial c)\\ &=(\partial\circ\varphi-(-1)^{|\varphi|}\varphi\circ\partial)c+(-1)^{|\varphi|}\varphi(\partial c)=0=\partial e(\varphi\otimes c).\end{aligned}$$

We can therefore pass to homology and compose with α (cf. VI, 9.11),

$$\text{(1.3)}\qquad H^*(C,M)\otimes HC\xrightarrow{\alpha}H(\operatorname{Hom}(C,D)\otimes C)\xrightarrow{e_*}M.$$

The composite map 1.3, or the corresponding biadditive (resp. bilinear) map $H^*(C,M)\times HC\to M$, is called the *scalar product*, and the image of $x\otimes\xi$ is called the scalar product of x and ξ. We write

$$\text{(1.4)}\qquad \langle x,\xi\rangle=e_*\alpha(x\otimes\xi),\qquad x\in H^*(C,M),\qquad \xi\in HC.$$

With representative (co-)chains 1.4 becomes

$$\text{(1.5)}\qquad \langle[\varphi],[z]\rangle=\varphi(z),\quad \text{for } \varphi\in Z^*(C,M),\qquad z\in ZC.$$

This shows that $\langle\,,\rangle$ can also be expressed in terms of the map $\alpha\colon H\operatorname{Hom}_R(C,M)\to\operatorname{Hom}_R(HC,M)$ of VI, 10.8, namely

$$\text{(1.6)}\qquad \langle x,\xi\rangle=(\alpha(x))(\xi).$$

Therefore the universal coefficient sequence VI, 4.4 gives

1.7 Proposition. *If R is hereditary and $H\,\mathrm{Ext}_R(C,M)=0$* (e.g. if C is free) *then*

$$H^n(C,M)\to \mathrm{Hom}_R(H_n C, M), \quad x\mapsto\langle x,-\rangle$$

is epimorphic, and its kernel is isomorphic with $\mathrm{Ext}_R(H_{n-1}C,M)$.

For instance, if R is a field (hence $\mathrm{Ext}_R=0$) and $H_n C$ has finite vector-space dimension then $\langle\,,\rangle\colon H^n(C,R)\times H_n C\to R$ is a *dual pairing* in the sense of linear algebra.

If $f\colon C\to D$ is a chain map and $\psi\in\mathrm{Hom}(D,M)$ then $(\psi f)c=\psi(fc)$, hence from 1.5,

$$(1.8)\qquad \langle f^*y,\xi\rangle=\langle y,f_*\xi\rangle, \quad y\in H^*(D,M), \quad \xi\in HC;$$

i.e. *f^* and f_* are transposed maps* (in the sense of linear algebra) *with respect to the scalar product* $\langle\,,\rangle$.

Similarly, *δ^* and ∂_* are transposed maps.* More precisely, if $0\to C'\xrightarrow{i} C\xrightarrow{p} C''\to 0$ is an exact sequence of chain maps (over R) such that $0\leftarrow\mathrm{Hom}(C',M)\leftarrow\mathrm{Hom}(C,M)\leftarrow\mathrm{Hom}(C'',M)\leftarrow 0$ is also exact then

$$(1.9)\qquad \begin{gathered}\langle\delta^*x',\xi''\rangle=-(-1)^{|x'|}\langle x',\partial_*\xi''\rangle,\\ \text{for } x'\in H^*(C',M), \quad \xi''\in HC''.\end{gathered}$$

Indeed, $\xi''=[pc]$ for some $c\in C$, and $\partial_*\xi''=[z']$ where $iz'=\partial c$; similarly, $x'=[\varphi\circ i]$ for some $\varphi\in\mathrm{Hom}(C,M)$, and $\delta^*x'=[\varphi'']$ where $\varphi''\circ p=\delta\varphi=-(-1)^{|\varphi|}\varphi\circ\partial$. Hence $\langle\delta^*x',\xi''\rangle=\varphi''(pc)=(\varphi''\circ p)c=-(-1)^{|\varphi|}\varphi\partial c$ $=-(-1)^{|\varphi|}(\varphi\circ i')z'=-(-1)^{|\varphi|}\langle x',\partial_*\xi''\rangle$. ∎

Note that 1.9 would take the simpler form $\langle\delta^*x',\xi''\rangle=\langle x',\partial_*\xi''\rangle$ if we defined $\delta\varphi=\varphi\circ\partial$ (compare VI, 10.28); in later §§ of this chapter, however, $\delta\varphi=-(-1)^{|\varphi|}\varphi\circ\partial$ is far more convenient.

More generally than above, we can tensor the map 1.2 with a left module M' and get

$$(1.10)\qquad \begin{gathered} e\otimes_R\mathrm{id}\colon \mathrm{Hom}_R(C,M)\otimes(C\otimes_R M')\to M\otimes_R M';\\ (e\otimes_R\mathrm{id})_*\alpha\colon H^*(C,M)\otimes H(C\otimes_R M')\to M\otimes_R M';\\ \langle x,\xi\rangle=(e\otimes_R\mathrm{id})_*\alpha(x\otimes\xi)\in M\otimes_R M'\\ \text{for } x\in H^*(C,M), \quad \xi\in H(C\otimes_R M').\end{gathered}$$

If, moreover, a homomorphism $\pi\colon M\otimes_R M'\to N$ is given one can compose with π. Sometimes $\pi\langle x,\xi\rangle$ is still denoted by $\langle x,\xi\rangle$ and called the *scalar product of x, ξ with respect to the pairing π.*

1.11 Example. For topological spaces X one easily proves (cf. VI, 12.14)

$$\mathrm{Hom}_R(SX \otimes_{\mathbb{Z}} R, M) \cong \mathrm{Hom}_{\mathbb{Z}}(SX, M).$$

Therefore, with $C = SX \otimes_{\mathbb{Z}} R$, the scalar product 1.4 becomes

$$\langle\, ,\rangle\colon H^n(X; M) \times H_n(X; R) \to M, \tag{1.12}$$

and *this is a dual pairing provided $M = R$ is a field and $H_n(X; R)$ has finite vector-space dimension.* Similarly, if X is replaced by a pair (X, A) of spaces, or/and a second module M' is used as in 1.10, one gets

$$\langle\, ,\rangle\colon H^n(X; A; M) \times H_n(X, A; M') \to M \otimes_R M'. \tag{1.13}$$

1.14 Exercises. *1.* If R is any ring (not necessarily hereditary), and X is a space such that $H(X; R)$ is R-free then

$$H^*(X; M) \to \mathrm{Hom}_R(H(X; R), M), \quad x \mapsto \langle x, -\rangle$$

is isomorphic (hint: use VI, 2.12 Exerc. 5).

2. Use 1.8 to show: If $f\colon \mathbb{S}^n \to \mathbb{S}^n$, $n > 0$, has degree k then $f^*(x) = kx$ for every $x \in H^*(\mathbb{S}^n; M)$.

3. If X is a space such that $H_n(X; \mathbb{Q})$ has finite vector-space dimension, and $f\colon X \to X$ is a continuous map then the endomorphisms f_* of $H_n(X; \mathbb{Q})$ and f^* of $H^n(X; \mathbb{Q})$ have the same *trace* (in fact, the same characteristic polynomial). This (rather trivial) remark can be useful in computing fixed-point indices (cf. 9.12 Exerc. 3).

2. *The Exterior Homology Product*

From now on the ground ring R is assumed to be *commutative.*

2.1 Definition. The exterior homology product $HX \times HY \to H(X \times Y)$ is obtained from the Eilenberg-Zilber map $SX \otimes SY \to S(X \times Y)$ by passing to homology and composing with α. More generally, let (X, A), (Y, B) be arbitrary pairs of spaces, L and M R-modules, and consider the composite chain map

$$\begin{aligned} \left(\frac{SX}{SA} \otimes L\right) \otimes_R \left(\frac{SY}{SB} \otimes M\right) &\xrightarrow{EZ} \frac{S(X \times Y)}{S\{A \times Y, X \times B\}} \otimes (L \otimes_R M) \\ &\xrightarrow{j} \frac{S(X \times Y)}{S(A \times Y \cup X \times B)} \otimes (L \otimes_R M), \end{aligned} \tag{2.2}$$

where $S\{A\times Y, X\times B\}\subset S(X\times Y)$, as in III, 7.1, is generated by simplices in $A\times Y$ or $X\times B$, and j is induced by inclusion. Passage to homology and composition with

$$\alpha\colon H\left(\frac{SX}{SA}\otimes L\right)\otimes_R H\left(\frac{SY}{SB}\otimes M\right)\to H\left(\frac{SX}{SA}\otimes L\otimes_R\frac{SY}{SB}\otimes M\right)$$

gives

$$(2.3)\qquad j_*(EZ)_*\,\alpha\colon H(X,A;L)\otimes_R H(Y,B;M) \to H(X\times Y, A\times Y\cup X\times B; L\otimes_R M),$$

or with indices,

$$(2.3')\qquad H_i(X,A;L)\otimes_R H_k(Y,B;M)\to H_{i+k}(X\times Y, A\times Y\cup X\times B; L\otimes_R M).$$

This map or the corresponding bilinear map is called the *exterior homology product*. We write

$$(2.4)\qquad \xi\times\eta=j_*(EZ)_*\,\alpha(\xi\otimes\eta)\in H(X\times Y, A\times Y\cup X\times B; L\otimes_R M),$$

where $\xi\in H(X,A;L)$, $\eta\in H(Y,B;M)$.

In terms of representative relative cycles this reads

$$(2.5)\qquad [a]\times[b]=[EZ(a\otimes_R b)],$$

where $a\in(SX)\otimes L$, $\partial a\in(SA)\otimes L$, $b\in(SY)\otimes M$, $\partial b\in(SB)\otimes M$.

The Eilenberg-Zilber map EZ is a homotopy equivalence (VI, 12.1), and the map j is a homotopy equivalence if $(X\times Y; A\times Y, X\times B)$ is an excisive triad (III, 8.1). Therefore, the Künneth theorem VI, 9.13 implies (cf. VI, 12.12)

2.6 Proposition. *If (X,A), (Y,B) are pairs of spaces such that $(X\times Y; A\times Y, X\times B)$ is an excisive triad (e.g., A, B open, or $B=\emptyset$), and if $L*_R M=0$ (R being hereditary) then*

$$\oplus_{i+k=n} H_i(X,A;L)\otimes_R H_k(Y,B;M)\to H_n(X\times Y; A\times Y\cup X\times B; L\otimes_R M),$$
$$\xi\otimes\eta\mapsto\xi\times\eta$$

is a split-monomorphism whose cokernel is naturally isomorphic with $\oplus_{i+k=n-1} H_i(X,A;L)*_R H_k(Y,B;M)$. ∎

We now list some properties of $\times$. If $f\colon(X,A)\to(X',A')$, $g\colon(Y,B)\to(Y',B')$ are maps then naturality of EZ says $(f\times g)\,EZ(a\otimes b)=EZ(fa\otimes gb)$, hence by 2.5 (with $\xi=[a]$, $\eta=[b]$)

$$(2.7)\qquad (f\times g)_*(\xi\times\eta)=(f_*\xi)\times(g_*\eta)\qquad \textbf{(naturality).}$$

Commutativity (VI, 12.3) and associativity (VI, 12.4) of EZ imply

(2.8) $$t_*(\xi \times \eta) = (-1)^{|\xi||\eta|}\, \eta \times \xi \qquad \textbf{(commutativity)},$$

and

(2.9) $$(\xi \times \eta) \times \zeta = \xi \times (\eta \times \zeta) \qquad \textbf{(associativity)},$$

where $\xi \in H(X, A)$, $\eta \in H(Y, B)$, $\zeta \in H(Z, C)$ (with appropriate coefficients) and $t\colon X \times Y \to Y \times X$ is given by $t(x, y) = (y, x)$.

If $Y = P$ is a point, $B = \emptyset$, and $1^P = 1 \in R = H_0(Y; R)$ then $(X \times Y, A \times Y \cup X \times B) = (X, A)$ and

(2.10) $$1^P \times \xi = \xi \times 1^P = \xi \qquad \textbf{(unit element)}.$$

This follows from VI, 12.5.

Compatibility of $\times$ and ∂_* is expressed by the following commutative diagram (coefficients omitted)

(2.11) $$\begin{array}{ccc} H(X,A) \otimes H(Y,B) & \xrightarrow{\ \times\ } & H(X \times Y, A \times Y \cup X \times B) \\ & & \big\downarrow{\scriptstyle \partial_*} \\ \big\downarrow{\scriptstyle (\partial_* \otimes \mathrm{id},\, (-1)^{\dim} \mathrm{id} \otimes \partial_*)} & & H(A \times Y \cup X \times B, A \times B) \\ & & \big\uparrow{\scriptstyle (i_{1*},\, i_{2*})} \\ [HA \otimes H(Y,B)] \oplus [H(X,A) \otimes HB] & \xrightarrow{\ \times \oplus \times\ } & H(A \times Y, A \times B) \oplus H(X \times B, A \times B), \end{array}$$

where i_1, i_2 are inclusions; i.e. we claim

(2.12) $$\partial_*(\xi \times \eta) = i_{1*}[(\partial_* \xi) \times \eta] + i_{2*}[(-1)^{|\xi|}\, \xi \times \partial_* \eta] \qquad \textbf{(stability)}.$$

In the important special case $B = \emptyset$ we have $i_1 = \mathrm{id}$, $i_{2*} = 0$, and stability reduces to

(2.13) $$\partial_*(\xi \times \eta) = (\partial_* \xi) \times \eta, \qquad \xi \in H(X, A), \qquad \eta \in HY.$$

Proof of 2.12. Let $a \in SX$, $b \in SY$ be representatives of ξ, η; in particular, $\partial a \in SA$, $\partial b \in SB$. Then

$$EZ(\partial a \otimes b) \in S(A \times Y) \subset S(A \times Y \cup X \times B) \quad \text{represents } i_{1*}[(\partial_* \xi) \times \eta],$$

$$EZ(a \otimes \partial b) \in S(X \times B) \subset S(A \times Y \cup X \times B) \quad \text{represents } i_{2*}[\xi \times \partial_* \eta],$$

and

$$\partial(EZ)(a \otimes b) = (EZ)\,\partial(a \otimes b) = EZ(\partial a \otimes b) + (-1)^{|\xi|} EZ(a \otimes \partial b)$$

represents $\partial_*(\xi \times \eta)$. ∎

2.14 Example. If we identify $\mathbb{R}^m \times \mathbb{R}^n = \mathbb{R}^{m+n}$ then

$$((\mathbb{R}^m - \{0\}) \times \mathbb{R}^n) \cup (\mathbb{R}^m \times (\mathbb{R}^n - \{0\})) = \mathbb{R}^{n+m} - \{0\},$$

and we get

$$H_m(\mathbb{R}^m, \mathbb{R}^m - \{0\}) \otimes H_n(\mathbb{R}^n, \mathbb{R}^n - \{0\}) \xrightarrow{\times} H_{m+n}(\mathbb{R}^{m+n}, \mathbb{R}^{m+n} - \{0\}).$$

If all coefficients are taken in $\mathbb{Z}$ then each of these groups is isomorphic with $\mathbb{Z}$ and the map is an isomorphism, by 2.6. *In other words, if o^i is a generator of $H_i(\mathbb{R}^i, \mathbb{R}^i - \{0\})$ then $o^m \times o^n = \pm o^{m+n}$.*

More generally, we consider pairs (V, K) where V is open in $\mathbb{R}^m \subset \mathbb{S}^m = \mathbb{R}^m \cup \{\infty\}$ and $K \subset V$ is compact. By IV, 6.4b, there is a unique element $o_K \in H_m(V, V-K)$ such that for every $P \in K$ the image of o_K under

$$H_m(V, V-K) \to H_m(V, V-P) \cong H_m(\mathbb{S}^m, \mathbb{S}^m - P) \cong \tilde{H}_m \mathbb{S}^m$$

is a fixed generator of $\tilde{H}_m \mathbb{S}^m \cong \mathbb{Z}$. Each of the two classes o_K which correspond to the two generators of $\tilde{H}_m S^m$ is called a *fundamental class around* K. If $V \subset V'$ then the inclusion clearly takes fundamental classes of $H_m(V, V-K)$ into fundamental classes of $H_m(V', V'-K)$; this justifies the expression "around K" and the notation o_K in which V does not appear. Generalizing the formula $o^m \times o^n = \pm o^{m+n}$ we have

2.15 Proposition. *If $o_K \in H_m(V, V-K)$ and $o_{K'} \in H_n(V', V'-K')$ are fundamental classes ($K' \subset V' \subset \mathbb{R}^n$) then*

$$o_K \times o_{K'} \in H_{m+n}(V \times V', V \times V' - K \times K') \quad \textit{is also fundamental.}$$

The proof follows by moving $o_K \otimes o_{K'}$ around the diagram

$$\begin{array}{ccc} H_n(V, V-K) \otimes H_n(V', V'-K') & \to H_m(V, V-P) \otimes H_n(V', V'-P') & \cong \tilde{H}_m \mathbb{S}^m \otimes \tilde{H}_n S^n \\ \downarrow \times & \cong \downarrow \times & \\ H_{m+n}(V \times V', V \times V' - K \times K') & \to H_{m+n}(V \times V', V \times V' - P \times P') & \cong \tilde{H}_{m+n} \mathbb{S}^{m+n}, \end{array}$$

where $P \in K$, $P' \in K'$. The diagram commutes by 2.7, and the second vertical arrow is isomorphic by 2.6. ∎

2.16 Exercises. *1.* Show that for every topological space Y and every $y \in Y$ one has a commutative diagram of isomorphisms

$$\begin{array}{ccc} H_n(Y, y) & \xrightarrow{[\iota] \times} & H_{n+1}(I \times Y, \dot{I} \times Y \cup I \times \{y\}) \\ \downarrow \pm\sigma & & \downarrow p_* \\ H_{n+1}(\Sigma Y, \{1\} \times \{y\}) & \xrightarrow{i_*} & H_{n+1}(\Sigma Y, I \times \{y\}), \end{array}$$

where Σ denotes suspension (III, 8.16 example 3), $I=[0,1]$, $\dot{I}=\{0\}\cup\{1\}$, σ is the isomorphism III, 8.18, i=inclusion, p=identification map, and $[\iota]\in H_1(I,\dot{I})$ is the homology class of the linear map $\iota\colon \Delta_1\to I$, $\iota(e^j)=j$.

2. Use 2.15 to prove $\deg(f\times g)=\deg(f)\deg(g)$ for proper maps $f\colon V\to \mathbb{R}^m$, $g\colon V'\to\mathbb{R}^n$ of open subsets V resp. V' of $\mathbb{R}^m$ resp. $\mathbb{R}^n$.

3*. If $c_n z^n + c_{n-1} z^{n-1} + \cdots + c_1 z + c_0$ is a non-zero complex polynomial of degree $\le n$, $(c_i\in\mathbb{C})$, then $[c_0, c_1, \ldots, c_n]$ is a point in $P_n\mathbb{C}$. Every point in projective space $P_n\mathbb{C}$ is of this form and two polynomials define the same point in $P_n\mathbb{C}$ if and only if they are proportional. Thus $P_n\mathbb{C}$ can be identified with the set of all non-zero complex polynomials provided one identifies polynomials if they differ only by a scalar $\lambda\in\mathbb{C}, \lambda\neq 0$ (equivalently: if the polynomials have the same roots).

(i) Multiplication of polynomials defines a mapping $\mu_{i,k}\colon P_i\mathbb{C}\times P_k\mathbb{C}\to P_{i+k}\mathbb{C}$. Check for continuity and prove

$$(\mu_{ik})_*(v_i\times v_k)=\pm\frac{(i+k)!}{i!\,k!}\,v_{i+k},$$

where v_j is a generator of $H_j(P_j\mathbb{C};\mathbb{Z})\cong\mathbb{Z}$ (hint: Pick a polynomial $w\in P_{i+k}\mathbb{C}$ with $i+k$ distinct roots. Then $\mu_{ik}^{-1}(w)$ consists of $\dfrac{(i+k)!}{i!\,k!}$ points w_ν. Use $H_j(P_j\mathbb{C})\cong H_j(P_j\mathbb{C}, P_j\mathbb{C}-w)$, $w\in P_j C$, and compute $\times$ and μ_* in terms of these local groups).

(ii) Let $SP^n(P_1\mathbb{C})$ denote the n-th symmetric power of $P_1\mathbb{C}\approx\mathbb{S}^2$, i.e. the space which is obtained from the ordinary n-th power $\times^n P_1\mathbb{C}_2$ by identifying points which differ only by a permutation of coordinates. Show that

$$\times^n P_1\mathbb{C}\to P_n\mathbb{C},\quad (a_1z+b_1, a_2z+b_2, \ldots, a_nz+b_n)\mapsto \textstyle\prod_{\nu=1}^n (a_\nu z+b_\nu)$$

induces a homeomorphism $SP^n(P_1\mathbb{C})\approx P_n\mathbb{C}$.

(iii) Define and investigate the analoguous notions for real projective spaces (coefficients $\mathbb{Z}$ or $\mathbb{Z}_2$).

3. The Interior Homology Product (Pontrjagin Product)

If $X\times X\xrightarrow{\mu} X$ is a multiplication (see below) then the composite map $HX\times HX\xrightarrow{\times} H(X\times X)\xrightarrow{\mu_*} HX$ is called the interior homology product with respect to μ. In more detail:

3.1 Definition. A continuous map $\mu\colon X\times X\to X$ is called a *multiplication* (on X); we write $\mu(x_1,x_2)=x_1x_2$ if there is no danger of confusion. An element $e \in X$ is called a *homotopy unit* of μ if the maps

$$X \to X, \quad x \mapsto ex, \quad x \mapsto xe,$$

are homotopic to the identity map. Further, μ is *homotopy-associative* resp. *homotopy-commutative* if the two maps

$$X \times X \times X \to X, \quad (x_1, x_2, x_3) \mapsto x_1(x_2 x_3),\ (x_1 x_2) x_3,$$

resp. the two maps

$$X \times X \to X, \quad (x_1, x_2) \mapsto x_1 x_2,\ x_2 x_1,$$

are homotopic.

If $(X, \mu)(X', \mu')$ are spaces with multiplications then $h\colon X \to X'$ is a *homotopy-homomorphism* if the two maps

$$X \times X \to X', \quad (x_1, x_2) \mapsto h(x_1 x_2),\ h(x_1) h(x_2),$$

are homotopic. A space X with a multiplication with homotopy-unit e is called an *h-space;* we use the notation *H-space* if the multiplication is also homotopy associative. A homotopy homomorphism $h\colon X \to X'$ between h-spaces (H-spaces) is called an *h-map* (H-map), provided $h(e)$ lies in the path component of e'. Not every X admits an h-space structure; for instance, $\mathbb{S}^{2k}$ does not as we shall see in 10.1.

3.2 Definition. If (X, μ) is a space with multiplication then the composite

$$(3.3) \quad H(X; L) \otimes_R H(X; M) \xrightarrow{\times} H(X \times X; L \otimes_R M) \xrightarrow{\mu_*} H(X; L \otimes_R M)$$

or the corresponding bilinear map is called the *Pontrjagin product* with respect to μ. We write

$$(3.4) \quad \mu_*(\xi_1 \times \xi_2) = \xi_1 \cdot \xi_2, \quad \xi_1 \in H(X; L), \quad \xi_2 \in H(X; M).$$

The properties 2.7–2.10 of the exterior product imply

(3.5) *If $h\colon X \to X'$ is a homotopy homomorphism then*

$$h_*(\xi_1 \cdot \xi_2) = h_*(\xi_1) \cdot h_*(\xi_2),$$

i.e. h_* *is a homomorphism with respect to* $\cdot$

(3.6) If $e \in X$ is a homotopy unit, and $[e] \in H_0(X; R)$ is the homology class of $e \in Z_0 SX$ then

$$[e] \cdot \xi = \xi \cdot [e] = \xi \quad \text{for all } \xi \in H(X; M).$$

(3.7) *If μ is homotopy associative* then

$$\xi_1 \cdot (\xi_2 \cdot \xi_3) = (\xi_1 \cdot \xi_2) \cdot \xi_3.$$

(3.8) *If μ is homotopy commutative* then

$$\xi_1 \cdot \xi_2 = (-1)^{|\xi_1||\xi_2|} \xi_2 \cdot \xi_1.$$

The proofs are immediate. As an illustration we give it for 3.8. By assumption the diagram

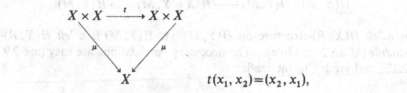

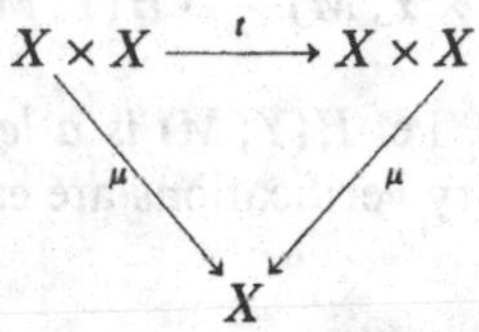

$$t(x_1, x_2) = (x_2, x_1),$$

is homotopy commutative, i.e., $\mu \simeq \mu t$, hence $\mu_* = \mu_* t_*$. Apply this to $\xi_1 \times \xi_2$, use 2.8, and get

$$\xi_1 \cdot \xi_2 = (-1)^{|\xi_1||\xi_2|} \xi_2 \cdot \xi_1. \quad \blacksquare$$

The formal properties of the Pontrjagin product suggest the following

3.9 Definition. Let $A = \{A_i\}_{i \in \mathbb{Z}}$ be a graded abelian group. A *multiplication* in A is a homomorphism $\nu\colon A \otimes A \to A$ of graded groups; with indices this reads, $\nu_{ik}\colon A_i \otimes A_k \to A_{i+k}$; $i, k \in \mathbb{Z}$. We write $\nu(a \otimes b) = a \cdot b$. A *unit* for ν is an element $1 \in A_0$ such that $1 \cdot a = a \cdot 1 = a$ for all $a \in A$. The multiplication is called *associative* resp. *commutative* if $a \cdot (b \cdot c) = (a \cdot b) \cdot c$, resp. $a \cdot b = (-1)^{|a||b|} b \cdot a$, for all $a, b, c \in A$. The pair (A, ν)—or simply A—is called a *graded ring* if ν is associative and has a unit; if it is also commutative then (A, ν) is a *commutative* graded ring.

Note that $\bar{A} = \bigoplus_{i \in \mathbb{Z}} A_i$ is an ordinary ring with respect to the induced multiplication (defined by $[\{a_i\} \cdot \{b_j\}]_n = \sum_{i+j=n} a_i \cdot b_j$, where $[\ ]_n$ denotes the component in A_n). However, $\bar{A}$ will, in general, *not* be *commutative* (in the ordinary sense) if A is commutative (in the graded sense).—If $A_i = 0$ for $i < 0$ then $\hat{A} = \prod_{i \in \mathbb{Z}} A_i$ is also a ring via $[\{a_i\} \cdot \{b_j\}]_n = \sum_{i+j=n} a_i \cdot b_j$.

If A is a graded ring and G a graded abelian group then a *left A-structure* on M is a homomorphism $\vartheta\colon A \otimes G \to G$ of graded groups such that $\vartheta(1 \otimes g) = g$ and $\vartheta(a \otimes \vartheta(b \otimes g)) = \vartheta((a \cdot b) \otimes g)$, for all $g \in G$; $a, b \in A$. If we write $\vartheta(a \otimes g) = a \cdot g$ this takes the familiar form $1 \cdot g = g, a \cdot (b \cdot g) = (a \cdot b) \cdot g$. The pair (G, ϑ)—or simply G—is called a *left A-module.*

3.10 Definition. If (X, μ) is an H-space then 3.6 and 3.7 assert that $H(X; R)$ under Pontrjagin multiplication is a graded ring. It is called the *Pontrjagin ring* of (X, μ). If $h: X \to X'$ is an H-map then $h_*: H(X; R) \to H(X'; R)$ is a homomorphism of graded rings (cf. 3.5); thus, the Pontrjagin ring is a functor from H-maps to homomorphisms of graded rings.

If X is an H-space and Y is any space then a *left-operation* of X on Y is a map $\eta: X \times Y \to Y$ (we write $\eta(x, y) = x \cdot y$) such that $y \mapsto e \cdot y$ is homotopic to id_Y, and the two maps $(x_1, x_2, y) \mapsto x_1 \cdot (x_2 \cdot y), (x_1 \cdot x_2) \cdot y$ are homotopic. In this situation the composite map

$$H(X; R) \otimes H(Y; M) \xrightarrow{\times} H(X \times Y; M) \xrightarrow{\eta_*} H(Y; M)$$

is a left $H(X; R)$-structure on $H(Y; M)$, i.e. *$H(Y; M)$ is a left $H(X; R)$-module* (M an R-module).—The necessary verifications are easy (use 2.9, 2.10), and are left to the reader.

3.11 Examples. If $\mu: X \times X \to X$ is a multiplication such that X is a group with respect to μ and if moreover $x \mapsto x^{-1}$ is continuous then (X, μ) is called a *topological group*. For instance, the space of all invertible $n \times n$ matrices (over $\mathbb{R}, \mathbb{C}, \mathbb{H}$) is a topological group under ordinary multiplication of matrices; it is called the *general linear group* $Gl(n; F)$ where $F = \mathbb{R}, \mathbb{C}, \mathbb{H}$. Matrices of determinant $+1$, orthogonal matrices, unitary matrices a.o. form subgroups and are also topological groups. The Pontrjagin rings of these and other groups have been computed by A. Borel 1954.

Other examples of H-spaces are provided by the loop spaces ΩY; they play an important role in homotopy theory. If Y is a space and $y_0 \in Y$ then $\Omega Y = \Omega(Y, y_0)$, as a set, consists of all paths $w: [0, 1] \to Y$ such that $w(0) = w(1) = y_0$ (so-called *loops*).

Any two loops v, w can be composed:

$$(v \cdot w)(t) = \begin{cases} v(2t) & \text{for } 0 \le 2t \le 1 \\ w(2t-1) & \text{for } 1 \le 2t \le 2. \end{cases}$$

This defines a mapping $\mu: \Omega Y \times \Omega Y \to \Omega Y, \mu(v, w) = v \cdot w$. If ΩY is equipped with the compact-open topology then μ is continuous, in fact, $(\Omega Y, \mu)$ is an H-space (cf. tomDieck-Kamps-Puppe, § 11). In many cases the Pontrjagin ring of ΩY can be computed in terms of homological properties of Y (see Adams 1956).

3.12 Exercises. *1.* Generalize the Pontrjagin product to the relative case, $H(X, A) \otimes H(X, B) \to H(X, A \cdot X \cup X \cdot B)$, and study its properties.

2. Let $P_\infty \mathbb{C} = \bigcup_{k=1}^{\infty} P_k \mathbb{C}$ denote the infinite dimensional complex projective space, with the weak topology (A is closed $\Leftrightarrow$ every $A \cap P_k \mathbb{C}$ is closed). We can think of $P_\infty \mathbb{C}$ as the set of *all* non-zero complex polynomials where two of them are identified if they are proportional (cf. 2.16, Exerc. 3). Show that ordinary multiplication of polynomials turns $P_\infty \mathbb{C}$ into a strictly (not just up to homotopy) associative H-space. Use 2.16, Exerc. 3(i) to determine the Pontrjagin ring $H(P_\infty \mathbb{C}; \mathbb{Z})$. Show that $\hat{H}(P_\infty \mathbb{C}; \mathbb{Q}) \cong \mathbb{Q}[[v]]$ = ring of formal power series over $\mathbb{Q}$ in one indeterminate v.

3. If G is a graded abelian group then $\operatorname{Hom}(G, G)$, as defined in VI, 10.2, is also graded abelian. Under composition of endomorphisms it is even a graded ring, and $\operatorname{Hom}(G, G) \otimes G \to G$, $\{\varphi_i\} \otimes g \mapsto \varphi_{|g|}(g)$, turns G into a left $\operatorname{Hom}(G, G)$-module. There is a natural 1-1-correspondence between left A-structures ϑ on G and homomorphisms Θ: $A \to \operatorname{Hom}(G, G)$ of graded rings (compare 3.9 and VI, 1.1).

4. Intersection Numbers in $\mathbb{R}^n$

Intuitively and vaguely one might expect that compact subsets X, Y of $\mathbb{R}^n$ whose dimensions add up to n intersect in a finite number of points, at least if they are in "general position". Moreover, if no intersection points lie on the boundary of X or Y then the total number of intersection points should be invariant under small deformations of X and Y. The following can be viewed as an approximation of this program, with compact sets being replaced by singular chains.

For simplicity, all homology groups will have coefficients in a fixed commutative ring R (which will usually not appear in the notation). In practice, $R = \mathbb{Z}$ or $\mathbb{Z}_2$.

4.1 Definition. Let $A \subset X \subset \mathbb{R}^n$, $B \subset Y \subset \mathbb{R}^n$ be such that $A \cap Y = \emptyset$, $X \cap B = \emptyset$, and consider the map d: $(X \times Y, A \times Y \cup X \times B) \to (\mathbb{R}^n, \mathbb{R}^n - 0)$, $d(x, y) = x - y$. The composition

$$(4.2) \quad H_{n-i}(X, A) \times H_i(Y, B) \xrightarrow{\times} H_n(X \times Y, A \times Y \cup X \times B) \xrightarrow{(-1)^i d_*} H_n(\mathbb{R}^n, \mathbb{R}^n - 0)$$

is called the *intersection pairing*. We write

$$(4.3) \quad \xi \circ \eta = (-1)^i d_*(\xi \times \eta), \quad \text{for } \xi \in H_{n-i}(X, A), \quad \eta \in H_i(Y, B),$$

and call this element of $H_n(\mathbb{R}^n, \mathbb{R}^n - 0) \cong R$ the *intersection number* of ξ and η. We shall see that 4.2 does indeed provide an algebraic measure of the geometric situation near $X \cap Y$ (cf. 4.6, 4.8, 4.11).

4.4 Remark. Classically (see Seifert-Threlfall, §73) one defines intersection numbers of singular chains $c \in S_{n-i}\mathbb{R}^n$, $c' \in S_i\mathbb{R}^n$ whenever $\mathrm{Carr}(c) \cap \mathrm{Carr}(\partial c') = \emptyset = \mathrm{Carr}(\partial c) \cap \mathrm{Carr}(c')$, where the *carrier*, $\mathrm{Carr}(c)$, is the smallest subset X of $\mathbb{R}^n$ such that $c \in SX$. But this condition just means that (X, A), (Y, B) exist such that $X \cap B = \emptyset = A \cap Y$ and $c \in SX$, $\partial c \in SA$, $c' \in SY$, $\partial c' \in SB$; therefore we can take homology classes $[c] \in H_{n-i}(X, A)$, $[c'] \in H_i(Y, B)$ and form the intersection number $[c] \circ [d]$. The following proposition shows that this number does not depend on the choice of (X, A), (Y, B).

4.5 Proposition. *If* $f\colon (X, A) \xrightarrow{\subset} (X', A')$, $g\colon (Y, B) \xrightarrow{\subset} (Y', B')$ *are inclusion maps in* $\mathbb{R}^n$ *and* $A' \cap Y' = \emptyset = X' \cap B'$ *then*

$$\xi \circ \eta = (f_* \xi) \circ (g_* \eta), \quad \text{for } \xi \in H_{n-i}(X, A),\ \eta \in H_i(Y, B).$$

This is obvious from naturality 2.7 of ×-products. ∎

For instance, we can always take $X' = X$, $A' = X - Y$, $Y' = Y$, $B' = Y - X$, and thus factor the intersection pairing 4.2 through $H(X, X - Y) \times H(Y, Y - X)$. This in turn is isomorphic (by excision III, 7.4) with $H(X \cap V, (X - Y) \cap V) \times H(Y \cap V, (Y - X) \cap V)$ where V is an arbitrary neighborhood of $\overline{X \cap Y}$. Roughly speaking then, *the intersection number* $\xi \circ \eta$ *depends only on the parts of* ξ, η *in* V, *where* V *is an arbitrary neighborhood of* $\overline{X \cap Y}$. In particular, if $X \cap Y = \emptyset$ we can take $V = \emptyset$ and get

4.6 Proposition. *If* $X \cap Y = \emptyset$ *then all intersection pairings* $H_{n-i}(X, A) \times H_i(Y, B) \to H_n(\mathbb{R}^n, \mathbb{R}^n - 0)$ *are zero.*

In fact, this is obvious because $X \cap Y = \emptyset$ implies $d(X \times Y) \subset (\mathbb{R}^n - 0)$. ∎

If $X \cap Y$ decomposes into several parts which do not touch each other, more precisely, if $\{V_l\}_{l=1,2,\ldots}$ are mutually disjoint open sets such that $\overline{X \cap Y} \subset V = \bigcup_l V_l$ then

$$(4.7) \quad H(X, X - Y) \cong H(X \cap V, (X - Y) \cap V) \cong \oplus_l H(X \cap V_l, (X - Y) \cap V_l),$$

and

$$(4.8) \quad \xi \circ \eta = \textstyle\sum_l \xi_l \circ \eta,$$

where $\xi = \{\xi_l\}$ is the decomposition of $\xi \in H(X, X - Y)$ corresponding to (4.7). The number $\xi_l \circ \eta = \xi_l \circ \eta_l$ is called the *intersection of* ξ *and* η *in* V_l. It may be thought of as a "local" intersection number; formula 4.8 says: *the global intersection of* ξ *and* η *is the sum of their local intersections* (the proof is easy, and left to the reader. ∎ To 4.5 we have the

4.9 Corollary. *All intersection pairings*

$$H_{n-i}(X,\emptyset)\times H_i(Y,\emptyset)\to H_n(\mathbb{R}^n,\mathbb{R}^n-0)$$

are zero—because if $A=\emptyset=B$ we can factor through $H_{n-i}(\mathbb{R}^n,\emptyset)\times H_i(\mathbb{R}^n,\emptyset)$. ∎

The following is an important example of a non-zero intersection number.

4.10 Example. Let X, Y be sub-vectorspaces of $\mathbb{R}^n$ of complementary dimensions $n-i$, i. Assume they are in general position, i.e., $X\cap Y=\{0\}$. *If* $\xi\in H_{n-i}(X,X-0;\mathbb{Z})\cong\mathbb{Z}$ *and* $\eta\in H_i(Y,Y-0;\mathbb{Z})\cong\mathbb{Z}$ *are generators then* $\xi\circ\eta\in H_n(\mathbb{R}^n,\mathbb{R}^n-0,\mathbb{Z})$ *is also a generator. In fact, if* $\varphi:\mathbb{R}^{n-i}\cong X,\psi:\mathbb{R}^i\cong Y$ *are linear isomorphisms and* $o_k\in H_k(\mathbb{R}^k,\mathbb{R}^k-0;\mathbb{Z})$ *is a generator then*

$$\varphi_*(o_{n-i})\circ\psi_*(o_i)=(\varphi,\psi)_*(o_{n-i}\times o_i), \tag{4.11}$$

where

$$(\varphi,\psi)\colon\ \mathbb{R}^{n-i}\times\mathbb{R}^i\to\mathbb{R}^n,\qquad (\varphi,\psi)(a,b)=\varphi(a)+\psi(b).$$

Proof. The diagram

$$\begin{array}{ccccc}
H_{n-i}(\mathbb{R}^{n-i},\mathbb{R}^{n-i}-0)\otimes H_i(\mathbb{R}^i,\mathbb{R}^i-0) & \xrightarrow{\times} & H_n(\mathbb{R}^{n-i}\times\mathbb{R}^i,\mathbb{R}^{n-i}\times\mathbb{R}^i-0) & & \\
\Big\downarrow{\scriptstyle \varphi_*\otimes\psi_*} & & \swarrow{\scriptstyle (\varphi\times\psi)_*} & \searrow{\scriptstyle (\varphi,-\psi)_*} & \\
H_{n-i}(X,X-0)\otimes H_i(Y,Y-0) & \xrightarrow{\times} & H_n(X\times Y,X\times Y-0) & \xrightarrow{d_*} & H_n(\mathbb{R}^n,\mathbb{R}^n-0)
\end{array}$$

is commutative. Following $o_{n-i}\times o_i$ along the lower way $\downarrow\to\to$ gives $(-1)^i\varphi_*(o_{n-i})\circ\psi_*(o_i)$, whereas the upper way $\to\searrow$ leads to

$$(\varphi,-\psi)_*(o_{n-i}\times o_i)=(-1)^i(\varphi,\psi)_*(o_{n-i}\times o_i),$$

the latter because $(-\mathrm{id})_*(o_i)=(-1)^i o_i$ by IV, 4.3. ∎

Formula 4.11 is one of the reasons why a sign $(-1)^i$ was introduced in defining intersection numbers. Another reason is the following (recall VI, 9.8).

4.12 Proposition. *In the notation of* 4.3, *we have* $\xi\circ\eta=(-1)^{i(n-i)}\eta\circ\xi$.

Proof.

$$\begin{aligned}\xi\circ\eta&=(-1)^i d_*(\xi\times\eta)=(-1)^{i+i(n-i)}d_*t_*(\eta\times\xi)\\&=(-1)^{i+i(n-i)+n}d_*(\eta\times\xi)=(-1)^{i(n-i)}\eta\circ\xi,\end{aligned}$$

where $t\colon Y\times X\to X\times Y$, $t(y,x)=(x,y)$. The second equality stems from 2.8, the third from $dt(y,x)=x-y=-d(y,x)$. ∎

Our definition of intersection numbers used the group structure of $\mathbb{R}^n$. We now give a characterisation in purely topological terms.

4.13 Lemma. *Let $D=\{(x,y)\in\mathbb{R}^n\times\mathbb{R}^n\,|\,x=y\}$, the diagonal. Then for every $P\in\mathbb{R}^n$ the maps*

$$i^P\colon (\mathbb{R}^n,\mathbb{R}^n-0)\to(\mathbb{R}^n\times\mathbb{R}^n,\mathbb{R}^n\times\mathbb{R}^n-D),\qquad i^P(x)=(x+P,P),$$
$$d\colon (\mathbb{R}^n\times\mathbb{R}^n,\mathbb{R}^n\times\mathbb{R}^n-D)\to(\mathbb{R}^n,\mathbb{R}^n-0),\qquad d(x,y)=x-y,$$

are reciprocal homotopy equivalences.

Proof. Clearly $di^P=\mathrm{id}$; a homotopy $\Theta\colon \mathrm{id}\simeq i^P d$ is given by $\Theta_t(x,y)=[x+t(P-y),y+t(P-y)]$. ∎

4.14 Corollary. *Up to the sign $(-1)^i$ the intersection pairing 4.2 coincides with the composite*

$$H_{n-i}(X,A)\times H_i(Y,B)\xrightarrow{\times}H_n(X\times Y,A\times Y\cup X\times B)$$
$$\xrightarrow{j_*}H_n(\mathbb{R}^n\times\mathbb{R}^n,\mathbb{R}^n\times\mathbb{R}^n-D)\xrightarrow{(i^P_*)^{-1}}H_n(\mathbb{R}^n,\mathbb{R}^n-0)$$

(where j = inclusion).

Indeed, $(i^P_*)^{-1}=d_*$ by 4.13, hence

$$(i^P_*)^{-1}j_*(\xi\times\eta)=(dj)_*(\xi\times\eta)=d_*(\xi\times\eta)=(-1)^i\,\xi\circ\eta.\quad ∎$$

Essentially then, the intersection number of (ξ,η) agrees with $j_*(\xi\times\eta)$— and no group structure in $\mathbb{R}^n$ is needed to define this element.

4.15 Proposition (topological invariance). *Under an injective map $h\colon\mathbb{R}^n\to\mathbb{R}^n$ all intersection numbers are multiplied with the degree of h, i.e. $(h_*\xi)\circ(h_*\eta)=\deg(h)(\xi\circ\eta)$.*

By definition (IV, 5.1), $\deg(h)=\deg_Q(h)$ is the composite map

$$H_n\mathbb{S}^n\cong H_n(\mathbb{R}^n,\mathbb{R}^n-h^{-1}Q)\xrightarrow{h_*}H_n(\mathbb{R}^n,\mathbb{R}^n-Q)\cong H_n\mathbb{S}^n,$$

where $Q\in h\mathbb{R}^n$. Since $h(\mathbb{R}^n)$ is open and $h\colon\mathbb{R}^n\approx h(\mathbb{R}^n)$ (cf. IV, 7.4) the number $\deg_Q(h)$ does not depend on $Q\in h\mathbb{R}^n$, and equals ±1 (IV, 5.4 and 5.12); according to these two cases h is called *orientation preserving* or *orientation reversing*. Thus, *intersection numbers remain invariant or change sign depending on whether h preserves or reverses orientation.*

Proof. Consider the diagram

$$\begin{array}{ccc}
H(X,A)\times H(Y,B) & \xrightarrow{\times} & H(X\times Y, A\times Y\cup X\times B) \\
\downarrow{\scriptstyle h_*\times h_*} & & \downarrow{\scriptstyle (h\times h)_*} \\
H(hX,hA)\times H(hY,hB) & \xrightarrow{\times} & H(hX\times hY, hA\times hY\cup hX\times hB)
\end{array}$$

$$\begin{array}{ccccc}
\xrightarrow{j_*} & H(\mathbb{R}^n\times\mathbb{R}^n, \mathbb{R}^n\times\mathbb{R}^n - D) & \xleftarrow[\cong]{i_*^0} & H(\mathbb{R}^n,\mathbb{R}^n-0)=H_n\mathbb{S}^n \\
& \downarrow{\scriptstyle (h\times h)_*} & & \downarrow{\scriptstyle (h-Q)_*} \\
\xrightarrow{j_*} & H(\mathbb{R}^n\times\mathbb{R}^n, \mathbb{R}^n\times\mathbb{R}^n - D) & \xleftarrow[\cong]{i_*^Q} & H(\mathbb{R}^n,\mathbb{R}^n-0)=H_n\mathbb{S}^n,
\end{array}$$

where $Q=h(0)$ and $(h-Q)(x)=h(x)-Q$. The first square is commutative by naturality of $\times$-products, the 2nd and 3rd square are commutative even before applying H. By 4.14 the rows of the diagram coincide (up to sign) with the intersection pairings; further $(h-Q)_*=\deg_Q(h)=\deg(h)$. Therefore, starting with $(\xi,\eta)\in H_{n-i}(X,A)\times H_i(Y,B)$ in the upper left corner, and chasing it to the lower right along the edges of the diagram gives $(-1)^i\deg(h)(\xi\circ\eta)=(-1)^i(h_*\xi)\circ(h_*\eta)$. ∎

4.16 Remark. One can generalize 4.15 to injective maps h which are only defined in a *neighborhood of* $\overline{X\cap Y}$ (compare remark after 4.5). If this neighborhood is itself homeomorphic with $\mathbb{R}^n$ (as is often the case) statement and proof remain virtually unchanged. The general case is more complicated; it will be dealt with in VIII, 13 where we treat intersections in general manifolds.

4.17 Exercises. *1.* Let $A\subset X\subset\mathbb{R}^n$, $B\subset Y\subset\mathbb{R}^n$, $A'\subset X'\subset\mathbb{R}^{n'}$, $B'\subset Y'\subset\mathbb{R}^{n'}$, $\xi\in H(X,A)$, $\eta\in H(Y,B)$, $\xi'\in H(X',A')$, $\eta'\in H(Y',B')$ be such that $\xi\circ\eta$ and $\xi'\circ\eta'$ are defined. Then $(\xi\times\xi')\circ(\eta\times\eta')\in H_{n+n'}(\mathbb{R}^{n+n'},\mathbb{R}^{n+n'}-0)$ is defined and equals $(-1)^{|\xi'||\eta|}(\xi\circ\eta)\times(\xi'\circ\eta')$.

2*. Let $P\in\mathbb{S}^p$, $Q\in\mathbb{S}^q$, and let $W\subset\mathbb{S}^p\times\mathbb{S}^q$ be a subset which contains $\mathbb{S}^p\times Q\cup P\times\mathbb{S}^q=\mathbb{S}^p\vee\mathbb{S}^q$ and also contains a neighborhood V of $(P,Q)\in\mathbb{S}^p\times\mathbb{S}^q$. Prove: *No injective map* $J\colon W\to\mathbb{R}^{p+q}$ *exists.* (Hint: $J(\mathbb{S}^p\times Q)$ and $J(P\times\mathbb{S}^q)$ intersect in just one point $J(P,Q)$. The intersection number of the generators of H_p, H_q can be determined within $J(V)$, and by 4.10 and 4.15 it turns out to be ± 1. This is impossible by 4.9).
For $q=1$ this result is closely related to the Jordan theorem (IV, 7.2); how? Draw pictures for $p=q=1$.

*3**. If $A, B \subset \mathbb{R}^n$ are disjoint sets we define the *linking product* to be the composite

$$H_{n-i}A \times \tilde{H}_{i-1}B \overset{\mathrm{id}\times\partial_*^{-1}}{\cong} H_{n-i}A \times H_i(\mathbb{R}^n, B) \xrightarrow{\circ} H_n(\mathbb{R}^n, \mathbb{R}^n - 0) \cong R.$$

For $\xi \in H_{n-i}A$, $\zeta \in \tilde{H}_{i-1}B$ we write $L(\xi, \zeta) = \xi \circ \partial_*^{-1}(\zeta)$ and call this the *linking number* of ξ and ζ.

(a) Study the properties of L which correspond to those of the intersection product $\circ$. In particular, compare $L(\xi, \zeta)$ and $L(\zeta, \xi)$.

(b) Let $f: \mathbb{S}^{n-1} \to \mathbb{R}^n$ be a map, and $s \in \tilde{H}_{n-1}\mathbb{S}^{n-1}$ a generator. For every $P \in (\mathbb{R}^n - f(\mathbb{S}^{n-1}))$ define $w(P, f) = L([P], f_* s)$. *This is called the winding number of f around P.* Prove: If f is injective then $w(P, f)$ assumes exactly two values, namely 0 and ± 1 (Hint: compare with the proof of IV, 7.2).

5. The Fixed Point Index

If V is an open set of $\mathbb{R}^n$ and $g: V \to \mathbb{R}^n$ is a mapping then the degree of g over $Q \in \mathbb{R}^n$ was interpreted (IV, 5) as being the "number" of points in $g^{-1}(Q)$, assuming this set is finite or at least compact. The fixed point set F_g of g agrees with $(\iota - g)^{-1}(0)$ where ι = inclusion; therefore the "number" of fixed points should be measured by the degree of $(\iota - g)$ over 0, provided F_g is compact. This degree is called the fixed point index I_g of g. We establish some elementary properties of I, in particular (using $\times$-products) an invariance property (5.9) which allows to extend the definition of I_g to maps g of ENRs (=euclidian neighbourhood retracts; cf. 5.10).—All homology groups will be taken with integral coefficients $\mathbb{Z}$.

5.1 Recall first (2.14) that for every generator o of $H_n\mathbb{S}^n \cong \mathbb{Z}$ (where $\mathbb{S}^n = \mathbb{R}^n \cup \{\infty\}$, $n > 0$) and every pair $K \subset V$ (where $V \subset \mathbb{R}^n$ is open, K compact) there is a *fundamental class* $o_K \in H_n(V, V-K)$ *around* K. This class o_K is the image of o under $H_n\mathbb{S}^n \to H_n(\mathbb{S}^n, \mathbb{S}^n - K) \cong H_n(V, V-K)$, and it is characterized by the property (IV, 6.4) that its image under $H_n(V, V-K) \to H_n(V, V-P) \cong \mathbb{Z}$ agrees with o_P for every $P \in K$. Clearly $(-o)_K = -(o_K)$.

5.2 Definition. Let $V \subset \mathbb{R}^n$ be an open set, and $g: V \to \mathbb{R}^n$ a map. Assume $F = F_g = \{x \in V \mid g(x) = x\}$, *the fixed point set of* g, *is compact* (n.b. F is always *closed in* V). Consider the map

$$(\iota - g)_*: H_n(V, V-F) \to H_n(\mathbb{R}^n, \mathbb{R}^n - 0) \cong \mathbb{Z}$$

(where $(\iota - g)\,x = x - g(x)$), and define the *fixed point index* $I_g \in \mathbb{Z}$ of g by

$$(5.3) \qquad (\iota - g)_*(o_F) = I_g \cdot o_0$$

(recall that o_0 generates $H_n(\mathbb{R}^n, \mathbb{R}^n - 0)$). This definition does not depend on the choice of the generator $o \in H_n \mathbb{S}^n$ because $(-o)_F = -(o_F)$ and $(-o)_0 = -(o_0)$.

5.4 Proposition. *Given* $g\colon V \to \mathbb{R}^n$ *as in* 5.2, *let* W *be an open set,* K *a compact set such that* $F_g \subset K \subset W \subset V$. *Then* $(\iota - g)$ *maps* $(W, W - K)$ *into* $(\mathbb{R}^n, \mathbb{R}^n - 0)$, *and* $(\iota - g)_*(o_K) = I_g\, o_0$.

Thus, I_g depends only on $g|W$ where W is any neighborhood of the fixed point set F, and in order to compute I_g we may replace F by any larger compact set $K \subset W$.—The proof is obvious because the inclusion $(W, W - K) \to (V, V - F)$ takes o_K into o_F. ∎

5.5 Units. *A constant map* $g\colon V \to \mathbb{R}^n$ *has index* 1 *if* $g(V) \in V$, *and index* 0 *if* $g(V) \notin V$.

Proof. If $g\,V \notin V$ then $F = \emptyset$ hence $o_F = 0$. If $g\,V = P \in V$ then

$$\iota - g\colon (V, V - P) \to (\mathbb{R}^n, \mathbb{R}^n - 0)$$

takes o_P into o_0. ∎

5.6 Additivity. *Given* $g\colon V \to \mathbb{R}^n$ *as in* 5.2, *assume* V *is represented as a finite union of open sets* V_i, $i = 1, \ldots, r$, *such that every* $F^i = \{x \in V_i | g(x) = x\}$ *is compact and* $F^i \cap F^j = \emptyset$ *for* $i \neq j$. *Then* $F_g = \bigcup_i F^i$, *and* $I_g = \sum_i (I_{g|V_i})$.

This expresses the local nature of I; it asserts that the "global" index I_g is the sum of the "local" indices $I_{g|V_i}$.

Proof. We can surround each F^i by an open neighborhood W_i such that $W_i \subset V_i$ and $W_i \cap W_j = \emptyset$ for $i \neq j$; put $W = \bigcup_i W_i$. Then $I_g = I_{g|W}$, and $I_{g|V_i} = I_{g|W_i}$ by 5.4. But $H(W, W - F) \cong \bigoplus_i H(W_i, W_i - F^i)$ (because the W_i are disjoint), and $o_F = \{o_{F^i}\}$, hence

$$I_{g|W}\, o_0 = (\iota - g)_*(o_F) = \textstyle\sum_i (\iota - g)_*(o_{F^i}) = (\sum_i I_{g|W_i})\, o_0. \quad ∎$$

5.7 Multiplicativity. *Let* $g\colon V \to \mathbb{R}^n$, $g'\colon V' \to \mathbb{R}^{n'}$ *be maps as in* 5.2. *Then the fixed point set of* $g \times g'\colon V \times V' \to \mathbb{R}^n \times \mathbb{R}^{n'} = \mathbb{R}^{n+n'}$ *is* $F_{g \times g'} = F_g \times F_{g'}$, *and* $I_{g \times g'} = I_g\, I_{g'}$.

Proof. Put $F=F_g$, $F'=F_{g'}$. By 2.15, $o_F\times o'_{F'}$ resp. $o_0\times o'_{0'}$ is fundamental around $F\times F'=F_{g\times g'}$ resp. $0\times 0'\in\mathbb{R}^{n+n'}$; hence

$$\begin{aligned}I_{g\times g'}(o_0\times o'_{0'})&=(\imath\times\imath'-g\times g')_*(o_F\times o'_{F'})=[(\imath-g)\times(\imath'-g')]_*(o_F\times o'_{F'})\\&=[(\imath-g)_*o_F]\times[(\imath'-g')_*o'_{F'}]=(I_gI_{g'})(o_0\times o'_{0'});\end{aligned}$$

the third equality by 2.7. ∎

5.8 Homotopy Invariance. *If* $g_t\colon V\to\mathbb{R}^n$, $0\le t\le 1$, *is a deformation such that* $K=\{x\in V\,|\,g_t(x)=x$ *for some* $t\}=\bigcup_t F_{g_t}$ *is compact then* $I_{g_0}=I_{g_1}$ (n.b. $\bigcup_t F_{g_t}$ is always *closed* in V).

This means: If during a deformation the fixed points stay away from the boundary of V (including ∞) then their "total number" remains unchanged. An example where a fixed point disappears at ∞ is the following: $g_t\colon\mathbb{R}\to\mathbb{R}$, $g_t(x)=1+tx$; clearly $I_{g_0}=1$, $I_{g_1}=0$.

Proof of 5.8. By 5.4, we have $I_{g_t}o_0=(\imath-g_t)_*(o_K)$. But $(\imath-g_t)\colon(V,V-K)\to(\mathbb{R}^n,\mathbb{R}^n-0)$ is a deformation, hence $(\imath-g_0)_*=(\imath-g_1)_*$ by III, 5.2. ∎

5.9 Commutativity. *If* $U\subset\mathbb{R}^n$, $U'\subset\mathbb{R}^{n'}$ *are open sets and* $f\colon U\to\mathbb{R}^{n'}$, $g\colon U'\to\mathbb{R}^n$ *are maps then the two composites*

$$gf\colon V=f^{-1}U'\to\mathbb{R}^n,\qquad fg\colon V'=g^{-1}U\to\mathbb{R}^{n'}$$

have homeomorphic fixed point sets, $F_{gf}\approx F_{fg}$. *If these sets are compact then* $I_{gf}=I_{fg}$.

Proof. The first assertion is clear: the restrictions of f, g define reciprocal homeomorphisms $F_{gf}\approx F_{fg}$. Assume then these sets are compact and define

$$\gamma\colon V\times V'\to\mathbb{R}^n\times\mathbb{R}^{n'},\qquad \gamma(x,y)=(gy,fx).$$

Using homotopy invariance we shall show $I_\gamma=I_{gf}$, $I_\gamma=I_{fg}$, and thus prove the proposition. We first use the deformation

$$\gamma_t(x,y)=[t\,gf(x)+(1-t)\,gy,fx],\qquad x\in V,\ y\in V',\ 0\le t\le 1.$$

A fixed point of γ_t satisfies $y=fx$ and $x=t\,gf(x)+(1-t)\,gf(x)=gf(x)$, i.e., the fixed point set of γ_t is $F_{\gamma_t}=\{(x,y)\,|\,x\in F_{gf},\ y=fx\}$. This is clearly compact and independent of t, hence (5.8) $I_\gamma=I_{\gamma_0}=I_{\gamma_1}$. The map γ_1 is a restriction of $\delta\colon V\times\mathbb{R}^{n'}\to\mathbb{R}^n\times\mathbb{R}^{n'}$, $\delta(x,y)=(gfx,fx)$, hence $I_{\gamma_1}=I_\delta$ by 5.4. Now we deform δ by $\delta_t(x,y)=[gfx,(1-t)fx]$. A fixed point (x,y) of δ_t satisfies $x=gfx$, $y=(1-t)fx$, hence $\bigcup_t F_{\delta_t}$ coincides with the image of

$$F_{gf}\times[0,1]\to V\times\mathbb{R}^{n'},\qquad (x,t)\mapsto(x,(1-t)fx).$$

This image being compact, we can apply 5.8 once more and get $I_\delta = I_{\delta_0} = I_{\delta_1}$ where $\delta_1(x,y) = (gfx, 0)$. But δ_1 is a product map, therefore $I_{\delta_1} = I_{gf}\, I_{\text{constant}} = I_{gf}$, by 5.7 and 5.5. Altogether $I_\gamma = I_{gf}$. By symmetry of γ we also find $I_\gamma = I_{fg}$; explicitely this uses the deformations

$$[g\,y,\, t\,f\,g(y) + (1-t)\,f\,x] \quad \text{and} \quad [(1-t)\,g\,y,\, f\,g\,y]. \quad ∎$$

Property 5.9 suggests the following generalization of the fixed point index. Suppose Y is any topological space, $U \subset Y$ is an open set, and $h\colon U \to Y$ a map which factors through some open set V of $\mathbb{R}^n$, i.e. $h = \beta\alpha$ where $U \xrightarrow{\alpha} V \xrightarrow{\beta} Y$. Then the index of h (if it can be defined at all) should coincide with the index of $\alpha\beta\colon \beta^{-1}U \to V \subset \mathbb{R}^n$. The question is, of course, whether this index is independent of the decomposition $h = \beta\alpha$. I don't know the answer in general, however, it is affirmative if U is an ENR (= euclidian neighborhood retract; cf. IV, 8).

5.10 Proposition and Definition. *If Y is any topological space, and $U \subset Y$ is an open set which is also an ENR then every mapping $h\colon U \to Y$ admits a decomposition $h = \beta\alpha$ where $U \xrightarrow{\alpha} V \xrightarrow{\beta} Y$ and V is open in some euclidian space $\mathbb{R}^n$. If $F_h = \{y \in U \mid h\,y = y\}$ is compact then the fixed point index $I_{\alpha\beta}$ of $\alpha\beta\colon \beta^{-1}U \to V \subset \mathbb{R}^n$ is defined and is independent of the decomposition $h = \beta\alpha$ (i.e. depends only on h).* This number is then, by definition, *the fixed point index of h*; in symbols $I_h = I_{\alpha\beta}$.

If $Y = \mathbb{R}^n$ we can take $V = U$, $\alpha = \mathrm{id}$, $\beta = h$, and we see that the definition agrees with 5.2 in this case. Also note that every open set $U \subset Y$ is an ENR if Y is an ENR; in IV, 8 we showed that the class of ENRs is fairly large.

Proof. By assumption there is a euclidian neighborhood retraction $U \xrightarrow{i} V' \xrightarrow{r} U$, $r\,i = \mathrm{id}$, where V' is open in some $\mathbb{R}^{n'}$. Then $U \xrightarrow{i} V' \xrightarrow{hr} Y$ is a euclidian decomposition, as required. If $U \xrightarrow{\alpha} V \xrightarrow{\beta} Y$ is any euclidian decomposition then $F_{\alpha\beta} \approx F_{\beta\alpha} = F_h$; assuming this to be compact we have to show that $I_{\alpha\beta}$ depends only on h. Consider the maps

$$\alpha\,r\colon V' \to V \subset \mathbb{R}^n, \qquad i\,\beta\colon \beta^{-1}U \to V' \subset \mathbb{R}^{n'}.$$

The two composites $(\alpha\,r)(i\,\beta) = \alpha\,\beta$ and $(i\,\beta)(\alpha\,r) = i\,h\,r$ have the same index by 5.9, in symbols $I_{\alpha\beta} = I_{ihr}$; clearly the right side I_{ihr} is independent of the decomposition α, β. ∎

The properties 5.4–5.9 of I carry over to the more general situation 5.10. We formulate the generalizations but omit some of the proofs; they consist of rather obvious reductions to 5.4–5.9. The notation is as in 5.10, with compact fixed point set F_h.

(5.11) *If W is an open set such that $F_h \subset W \subset U$ then $I_h = I_{h|W}$.* ∎

(5.12) *If h is constant then $I_h = 1$ if $h(U) \in U$, and $I_h = 0$ if $h(U) \notin U$.* ∎

(5.13) *If U is represented as a finite union of open sets U_i, $i = 1, \ldots, r$, such that $U_i \cap U_j \cap F_h = \emptyset$ for $i \neq j$ then $I_h = \sum_{i=1}^{r} (I_{h|U_i})$.* This is reduced to 5.6 by putting $V_i = \beta^{-1} U_i$. ∎

(5.14) *If $h: U \to Y$, $h': U' \to Y'$ are as in 5.10 with compact fixed point sets then $I_{h \times h'} = I_h I_{h'}$ where $h \times h': U \times U' \to Y \times Y'$.* ∎

(5.15) *If $h_t: U \to Y$ is a deformation, $0 \leq t \leq 1$, and $\bigcup_t F_{h_t}$ is compact then $I_{h_0} = I_{h_1}$.*

Proof. Choose a euclidian neighborhood retraction $U \xrightarrow{i} V \xrightarrow{r} U$. Then $I_{h_t} = I_{ih_t r}$ by Definition 5.10, and the right side does not depend on t by 5.8. ∎

(5.16) *If $U \subset Y$, $U' \subset Y'$ are open subsets (and ENRs), and $k: U \to Y'$, $k': U' \to Y$ are maps then $k'k: k^{-1} U' \to Y$, $kk': k'^{-1} U \to Y'$ have homeomorphic fixed point sets, $F_{k'k} \approx F_{kk'}$. If these sets are compact, then $I_{k'k} = I_{kk'}$.*

Proof. Choose euclidian neighborhood retractions $U \xrightarrow{i} V \xrightarrow{r} U$, $U' \xrightarrow{i'} V' \xrightarrow{r'} U'$. Then $k'k|(k'k)^{-1} U = r(ik'k)$ and $kk' = (kk'r')i'$ are euclidian factorizations, hence $I_{k'k} = I_{ik'kr}$, $I_{kk'} = I_{i'kk'r'}$ by (5.11 and) Definition 5.10. But $ik'kr = (ik'r')(i'kr)$ and $i'kk'r' = (i'kr)(ik'r')$ have the same index by 5.9. ∎

5.17 Exercises. *1.* If $g: \mathbb{R} \to \mathbb{R}$ has a compact fixed point set then $I_g = 0$ or ± 1.

2. Construct maps $g: \mathbb{R}^2 \to \mathbb{R}^2$ with prescribed fixed point index whose only fixed point is the origin 0. Draw pictures.

3. If $\varphi: \mathbb{R}^n \to \mathbb{R}^n$ is a linear map then F_φ is compact if and only if $+1$ is not an eigenvalue of φ. In that case, $(\mathrm{id} - \varphi)$ is an isomorphism, 0 is the only fixed point of φ, and $I_\varphi = (-1)^\eta$ where η is the number of real eigenvalues λ such that $\lambda > 1$.

4. If $V \subset \mathbb{R}^n$ is open, $0 \in V$, and $g: V \to \mathbb{R}^n$ is a continuous map such that $gx \neq \lambda x$ for all non-zero $x \in V$ and all real numbers $\lambda > 1$ then $g(0) = 0$. If, moreover, F_g is compact then $I_g = 1$. (Hint: consider the deformation $g_t(x) = t(gx)$.)

5*. Let $\mathscr{G}$ denote the class of all continuous maps $g\colon U \to Y$ such that Y is an ENR, U is an open subset of Y, and $F_g = \{x \in U \mid g\,x = x\}$ is compact. *Theorem. If $I\colon \mathscr{G} \to \mathbb{Z}$ is a function with the properties* 5.11–5.16 (actually, 5.14 follows from the others) *then I is the fixed point index*, $I(g) = I_g$. Program for a proof (compare R.F. Brown Pac. J. 35 (1970) 544–558): Use the proof of 5.10 to reduce to the case $Y = \mathbb{R}^n$. Use differentiable (or simplicial) approximation and make the graph of g transverse to the diagonal; this reduces to the case where F_g is finite, or (with 5.13) even $F_g = \{0\}$, and $Dg(0)$ has no eigenvalue $+1$. Approximate g by $Dg(0)$, reducing the problem to linear maps. Use eigenvalues to reduce to the case $n = 1$.

6*. Let Y, Y' be spaces, $U \subset Y$, $U' \subset Y'$ open subsets and $\varphi\colon U \to Y'$, $\varphi'\colon U' \to Y$ maps such that the fixed point sets $F_{\varphi\varphi'} \approx F_{\varphi'\varphi}$ are compact. If φ, φ' admit euclidian decompositions

$$\varphi\colon U \xrightarrow{\gamma} V \xrightarrow{\delta} Y', \qquad \varphi'\colon U' \xrightarrow{\gamma'} V' \xrightarrow{\delta'} Y,$$

where $V \subset \mathbb{R}^n$, $V' \subset \mathbb{R}^{n'}$ are open, then the two composites of the mappings $\gamma'\delta$ and $\gamma\delta'$ have equal indices (5.9), and these indices $I_{\gamma'\varphi\delta'} = I_{\gamma\varphi'\delta}$ do not depend on the decompositions of φ, φ'. Call this number the *fixed point index of the pair* φ, φ', in symbols $I_{\varphi,\varphi'} = I_{\varphi',\varphi}$. If U, U' are ENRs then $I_{\varphi,\varphi'} = I_{\varphi\varphi'} = I_{\varphi'\varphi}$. If $Y = Y'$, $U = U'$ and $\varphi' =$ inclusion then γ', δ' is a euclidian neighborhood retraction and $I_{\varphi,\varphi'} = I_\varphi$, by Proposition 5.10.

6. *The Lefschetz-Hopf Fixed Point Theorem*

This famous theorem expresses the fixed point index of $g\colon Y \to Y$, Y a compact ENR, in terms of the induced endomorphism $g_*\colon H(Y;\mathbb{Q}) \to H(Y;\mathbb{Q})$. We start with some algebraic preliminaries on endomorphisms of graded modules. R denotes a fixed commutative ring with unit; all modules, $\otimes$-products, and Hom are over R. The application will be to $R = \mathbb{Q}$.

6.1 Definition. Let $M = \{M_i\}_{i \in \mathbb{Z}}$ be a graded R-module, and let M^* denote the dual (graded) module, $M^*_{-i} = \operatorname{Hom}(M_i, R)$. For every graded R-module N define

(6.2) $\Theta = \Theta_{MN}\colon M^* \otimes N \to \operatorname{Hom}(M, N)$, $[\Theta(\varphi \otimes n)](m) = (-1)^{|m||n|}\varphi(m)\,n$,

(cf. VI, 10.1 for Hom). Clearly Θ is a homomorphism of graded modules, and is natural in M and N. (It is a special case of the map γ in VI, 10.23; take $C = M$, $C' = R = D$, $D' = N$.)

6.3 Proposition. *The image of Θ consists of those homomorphisms $\beta: M \to N$ which factor through a finitely generated free (graded) module, i.e. of composites $\beta: M \to F \to N$, where $F_i = R \oplus R \oplus \cdots \oplus R$ for all i, and $F_i = 0$ for almost all i.* These homomorphisms are called of *finite rank.*

If N is free then Θ is monomorphic. Hence in this case Θ maps $M^* \otimes N$ isomorphically onto $\operatorname{im}(\Theta) = \{\beta: M \to N \mid \beta$ *of finite rank*$\}$.

Proof. If $\varphi \in M^*$, $n \in N = \operatorname{Hom}(R, N)$ then, up to sign, $\Theta(\varphi \otimes n)$ agrees with the composite homomorphism $M \xrightarrow{\varphi} R \xrightarrow{n} N$. This proves the first part because elements of $M^* \otimes N$ are finite sums of terms $\varphi \otimes n$, and homomorphisms $M \to N$ of finite rank are finite sums of compositions $M \to R \to N$.

If N is free let $\{i_\gamma: R \to N\}_{\gamma \in \Gamma}$ be a direct sum representation. (N.b. the i_γ may have various degrees.) Every $a \in M^* \otimes N$ is then of the form $a = \sum_{\gamma \in \Gamma} \varphi_\gamma \otimes i_\gamma(1)$. If $p_\mu: N \to R$ is the μ-th projection ($p_\mu i_\mu = \mathrm{id}$, $p_\mu i_\gamma = 0$ for $\gamma \neq \mu$) then $(\mathrm{id} \otimes p_\mu)\, a = \sum_\gamma \varphi_\gamma \otimes p_\mu i_\gamma(1) = \varphi_\mu \otimes 1$, hence

$$\Theta_{MR}(\mathrm{id} \otimes p_\mu)\, a = \pm \varphi_\mu .$$

But $\Theta_{MR}(\mathrm{id} \otimes p_\mu)\, a = p_\mu \Theta_{MN}(a)$ by naturality of Θ (applied to p_μ). Therefore, $\Theta_{MN}(a) = 0$ implies $\varphi_\mu = \pm p_\mu \Theta_{MN}(a) = 0$ for all $\mu \in \Gamma$, hence $a = 0$. ∎

6.4 Definition. Let N be a graded R-module and let $e: N^* \otimes N \to R$ denote the evaluation map, $e(\varphi \otimes n) = \varphi(n)$. If N is free and $\beta: N \to N$ is an endomorphism of finite rank then $\Theta^{-1}(\beta) \in N^* \otimes N$ by 6.3, and $\Lambda(f) = e\Theta^{-1}(\beta) \in R$ is called the *trace* or *Lefschetz number* of β.

Since e annihilates all elements of dimension $\neq 0$ the Lefschetz number of β is zero unless $|\beta| = 0$, i.e. unless β is a sequence of endomorphisms $\beta_i: N_i \to N_i$, $i \in \mathbb{Z}$. In order to compute $\Lambda(\beta)$ in this case we pick a base Γ_i for each N_i; then $\beta(\gamma) = \sum_{\mu \in \Gamma_i} \beta^\gamma_\mu \cdot \mu$, for $\gamma \in \Gamma_i$, with matrix coefficients $\beta^\gamma_\mu \in R$. For every $j \in \mathbb{Z}$ and $\mu \in \Gamma_j$ define $\varphi^\mu \in N^*_{-j} = \operatorname{Hom}(N, R)_{-j}$ by $\varphi^\mu(\gamma) = \beta^\gamma_\mu$, $\gamma \in \Gamma_j$. If β is of finite rank then almost all φ^μ are zero, hence

$$a = \sum_{\mu \in \Gamma_j,\, j \in \mathbb{Z}} (-1)^j \varphi^\mu \otimes \mu \in N^* \otimes N$$

is defined, and $[\Theta(a)](\gamma) = \sum_{\mu, j} (-1)^{j + j|\gamma|} \varphi^\mu(\gamma) \cdot \mu = \sum_\mu \beta^\gamma_\mu \cdot \mu = \beta(\gamma)$, i.e., $\Theta(a) = \beta$. Therefore

$$\Lambda(\beta) = e(a) = \sum_{\mu, j} (-1)^j \varphi^\mu(\mu) = \sum_{j \in \mathbb{Z}} (-1)^j \sum_{\mu \in \Gamma_j} \beta^\mu_\mu . \tag{6.5}$$

In particular, we see that the last expression (which is often used to define $\Lambda(\beta)$) is independent of the choice of the bases Γ_j.

The Lefschetz-Hopf fixed point theorem now reads as follows.

6.6 Proposition. *Let Y be an ENR, K a compact subset of Y, and $f\colon Y \to K \subset Y$ a mapping. Then f has compact fixed point set, $(f|K)_*\colon H(K;\mathbb{Q}) \to H(K;\mathbb{Q})$ has finite rank, and $I_f = \Lambda(f|K)_*$.*

Proof. The fixed point set of f is closed in K and therefore compact. Let $Y \xrightarrow{j} V \xrightarrow{r} Y$ be a euclidian neighborhood retraction ($rj = \mathrm{id}$; V open in $\mathbb{R}^n$); then $jK \approx K$, the index of f equals the index of $g = jfr\colon V \to K \approx jK \subset \mathbb{R}^n$ by 5.10, and $f|K \approx g|K$. We have to show therefore, that $I_g = \Lambda(g|K)_*$. We use rational homology throughout (and omit the coefficients $\mathbb{Q}$) so that $H(X \times X') = (HX) \otimes (HX')$. The image of the fundamental class o_K under $H(V, V-K;\mathbb{Z}) \to H(V, V-K;\mathbb{Q})$ is still denoted by o_K.

Consider first the diagram

$$
(6.7)\qquad
\begin{array}{ccccc}
H(V,V-K)\otimes HV & \xrightarrow{\mathrm{id}\otimes g_*} & H(V,V-K)\otimes(HK) & \xrightarrow{\hat{d}\otimes \mathrm{id}} & (HK)^*\otimes(HK) \\
\uparrow{\scriptstyle \Delta_*} & & \downarrow{\scriptstyle d_*} & & \downarrow{\scriptstyle e} \\
H(V,V-K) & \xrightarrow{(\iota-g)_*} & H(\mathbb{R}^n,\mathbb{R}^n-0) & \xleftarrow[\cong]{o_0} & \mathbb{Q}
\end{array}
$$

where $\Delta\colon (V, V-K) \to (V, V-K) \times V$, $\Delta(x) = (x, x)$, is the diagonal map, $d\colon (V, V-K) \times K \to (\mathbb{R}^n, \mathbb{R}^n - 0)$, $d(x, y) = x - y$, is the difference as in 4.1, e is the evaluation of 6.4, and $\hat{d}\colon H(V, V-K) \to (HK)^* = \mathrm{Hom}(HK, \mathbb{Q})$ is so defined as to make the right square commutative, $([\hat{d}(v)]\,k)\,o_0 = d_*(v \otimes k)$. The left square is commutative because $d(\mathrm{id} \times g)\Delta(x) = x - gx = (\iota - g)\,x$. By Definition 5.3, the lower row of 6.7 takes o_K into I_g. Going along the upper row must give the same, i.e.,

$$
(6.8)\qquad I_g = e(a_g), \quad \text{where} \quad a_g = (\hat{d} \otimes g_*)\,\Delta_*(o_K).
$$

By Definition 6.4 we can also write

$$
(6.9)\qquad I_g = \Lambda\big(\Theta(a_g)\big), \quad a_g = (\hat{d} \otimes g_*)\,\Delta_*(o_K).
$$

We shall see that $\Theta(a_g) = (g|K)_*$, and thus prove the theorem.

Consider the diagram

$$
(6.10)\qquad
\begin{array}{ccccc}
H(V,V-K)\otimes HV\otimes HK & \xrightarrow{\mathrm{id}\otimes t_*} & H(V,V-K)\otimes HK\otimes HV & \xrightarrow{d_*\otimes \mathrm{id}} & H(\mathbb{R}^n,\mathbb{R}^n-0)\otimes HV \cong HV \\
\downarrow{\scriptstyle \hat{d}\otimes g_*\otimes \mathrm{id}} & & \downarrow{\scriptstyle \hat{d}\otimes \mathrm{id}\otimes g_*} & & \downarrow{\scriptstyle g_*} \\
(HK)^*\otimes HK\otimes HK & \xrightarrow[\mathrm{id}\otimes t_*]{} & (HK)^*\otimes HK\otimes HK & \xrightarrow[e\otimes \mathrm{id}]{} & \mathbb{Q}\otimes HK \cong HK
\end{array}
$$

where $t(x,y)=(y,x)$. The right square is obtained from the right square of 6.7 by tensoring with g_* and is therefore commutative; commutativity of the left square is obvious. If we follow $\Delta_*(o_K)\otimes k\in H(V,V-K)\otimes HV\otimes HK$ along the lower way $\downarrow\rightarrow\rightarrow$ we get $[\Theta(a_g)]\,k$ by Definition 6.2 (recall that $t_*(\xi\otimes\eta)=(-1)^{|\xi||\eta|}\eta\otimes\xi$). Using the upper way $\rightarrow\rightarrow\downarrow$ instead must give the same, i.e.

$$\Theta(a_g)=g_*\circ\Phi(K,V), \tag{6.11}$$

where $\Phi(K,V)=\{\Phi_\lambda(K,V)\}_{\lambda\in\mathbb{Z}}$ is the following composition.

$$\begin{aligned}&\Phi_\lambda(K,V)\colon\ H_\lambda K\xrightarrow{o_K\times}H_{\lambda+n}[(V,V-K)\times K]\\ (6.12)\quad&\xrightarrow{(\Delta\times\mathrm{id})_*}H_{\lambda+n}[(V,V-K)\times V\times K]\xrightarrow{(\mathrm{id}\times t)_*}H_{\lambda+n}[(V,V-K)\times K\times V]\\ &\xrightarrow{(d\times\mathrm{id})_*}H_{\lambda+n}[(\mathbb{R}^n,\mathbb{R}^n-0)\times V]\xrightarrow{(o_0\times)^{-1}}H_\lambda V.\end{aligned}$$

We shall prove below that

(6.13) $\Phi_\lambda(K,V)=i_*(K,V)$, *where* $i(K,V)\colon K\to V$ *is the inclusion map.*

This gives $\Theta(a_g)=g_*\,i_*(K,V)=(g|K)_*$ by 6.11, hence (6.9) $I_g=\Lambda\Theta(a_g)=\Lambda(g|K)_*$, as required.

Proof of 6.13. It consists of several steps, and is based on the following naturality properties of Φ: If $K'\subset K\subset V\subset V'\subset\mathbb{R}^n$ with V' open, K' compact then

$$i_*(V,V')\,\Phi(K,V)=\Phi(K,V'), \tag{6.14}$$

$$\Phi(K,V)\,i_*(K',K)=\Phi(K',V), \tag{6.15}$$

where i_* is induced by inclusion. Property 6.14 is obvious because $i_*(o_K)=o_K$. For 6.15, one has to remark that in the definition of $\Phi(K',V)$ one can replace $(V,V-K')$ everywhere by $(V,V-K)$ and $o_{K'}$ by o_K. We use these properties as follows.

(6.16) $\quad\Phi(K,V)=i_*(K,V)\Rightarrow\Phi(K,V')=i_*(K,V')$,

(6.17) $\quad\Phi(K,V)=i_*(K,V)\Rightarrow\Phi(K',V)=i_*(K',V)$,

(6.18) $\quad\Phi(K,V')=i_*(K,V')\Rightarrow i_*(V,V')[\Phi(K,V)-i_*(K,V)]=0$,

(6.19) $\quad\Phi(K',V)=i_*(K',V)\Rightarrow[\Phi(K,V)-i_*(K,V)]\,i_*(K',K)=0$.

Step 1: $K=P=$ a point. By 6.16 we can replace V by a little open ball around P. Then the generator $1_P\in H_0P=H_0V$ maps as follows under Φ

$$1_P\mapsto o_P\times 1_P\mapsto o_P\times 1_P\times 1_P\mapsto o_P\times 1_P\times 1_P\mapsto o_0\times 1_P\mapsto 1_P.$$

Step 2: $\Phi_0\colon H_0 K \to H_0 V$. Since $H_0 K$ is generated by groups $H_0 P, P \in K$, this reduces to step 1 using 6.19.

Step 3: $K \approx \mathbb{S}^\lambda = \lambda$-sphere, $\lambda > 0$, V a neighborhood of which K is a deformation retract. Since $i_*\colon H_\lambda K \cong H_\lambda V \cong \mathbb{Q}$ there exists a number $q = q(K, V) \in \mathbb{Q}$ such that $\Phi_\lambda(K, V) = q\, i_*(K, V)$. Let A be the antipodal map of $K \approx \mathbb{S}^\lambda$, and let $r\colon V \to \mathbb{S}^\lambda$ be a retraction. Then $g = Ar$ is fixed point free, hence $I_g = 0$. Further $g_* \Phi_\lambda(K, V) = q g_* i_*(K, V) = q A_* = q(-1)^{\lambda+1}\,\mathrm{id}$ because A has degree $(-1)^{\lambda+1}$ (cf. IV, 4.3). The Lefschetz number of $\Theta(a_g) = g_* \Phi(K, V)$ (cf. 6.11) therefore equals

$$1 + (-1)^\lambda q(-1)^{\lambda+1} = 1 - q$$

by 6.5, and 6.9 becomes $0 = 1 - q$, hence $\Phi_\lambda(K, V) = i_*(K, V)$ as asserted.

Step 4: $K = C^\lambda = \lambda$-skeleton of a cubical lattice decomposition (cf. V, 3.4) of a cube C, $\lambda > 0$, V arbitrary. Using cellular homology (V, 4) we see that $H_\lambda C^\lambda$ is generated by groups $i_*(S, K)(H_\lambda S)$ where $S \approx \mathbb{S}^\lambda$ is the boundary of an $(\lambda+1)$-cube in C. Clearly S has an open neighborhood W in V of which it is a deformation retract (just thicken S a little within V; cf. 6.20 below), hence $\Phi_\lambda(S, W) = i_*(S, W)$ by step 3, hence $\Phi_\lambda(S, V) = i_*(S, V)$ by 6.16, hence $[\Phi_\lambda(K, V) - i_*(K, V)]\, i_*(S, K) = 0$ by 6.19, hence $\Phi_\lambda(K, V) = i_*(K, V)$ because the images of $i_*(S, K)$ generate $H_\lambda K$.

Step 5: K a CW-subspace of C^λ (cf. step 4), V arbitrary. Choose an open subset U of which C^λ is a deformation retract (a little thickening; cf. 6.20), and an open set W inside $U \cap V$ of which K is a deformation retract (6.20). Using cellular homology (V, 4) again we see that $i_*(K, C^\lambda)\colon H_\lambda K \to H_\lambda C^\lambda$ is monomorphic (C^λ has no $(\lambda+1)$-cells!) hence $i_*(W, U)\colon H_\lambda W \to H_\lambda U$ is also monomorphic. Now $\Phi_\lambda(C^\lambda, U) = i_*(C^\lambda, U)$ by step 4, hence $\Phi_\lambda(K, U) = i_*(K, U)$ by 6.17, hence $i_*(W, U)[\Phi_\lambda(K, W) - i_*(K, W)] = 0$ by 6.18, hence $\Phi_\lambda(K, W) = i_*(K, W)$ because $i_*(W, U)$ is monomorphic, hence $\Phi_\lambda(K, V) = i_*(K, V)$ by 6.16.

Step 6: V, K arbitrary. Choose a cubical lattice so fine that every closed cell which meets K lies in V, and let M be the union of these cells. Then M is a CW-subspace of some cube C as in steps 4–5, and $K \subset M \subset V$. By 6.17 it suffices to prove $\Phi_\lambda(M, V) = i_*(M, V)$. Let $M^\lambda = M \cap C^\lambda$ be the λ-skeleton of M. Then $\Phi_\lambda(M^\lambda, V) = i_*(M^\lambda, V)$ by step 5, hence $[\Phi_\lambda(M, V) - i_*(M, V)]\, i_*(M^\lambda, M) = 0$ by 6.19, hence $\Phi_\lambda(M, V) = i_*(M, V)$ because $i_*(M^\lambda, M)\colon H_\lambda M^\lambda \to H_\lambda M$ is epimorphic (use cellular homology!).

This finishes the proof except for an easy deformation argument which was used in steps 4–5 and which we prove now.

6.20 Lemma (compare V, 2.14, Exerc. 3). *If $X \subset Y$ are CW-subspaces of a (cubical) lattice decomposition* (V, 3.4) *of $\mathbb{R}^n$ and V is an open set containing X then X has an open neighborhood W in $Y \cap V$ of which it is a strong deformation retract.*

Proof by induction on $k = \dim(Y - X)$. If e is a (closed) k-cube in $Y - X$, and c_e its center then we can deform $(e - c_e)$ radially into its boundary. Doing this for all e defines a strong deformation retraction

$$\rho\colon Y - \bigcup_e \{c_e\} \to Y',$$

where Y' is obtained from Y by removing the interiors of all k-cubes e. In fact, ρ is such that its restriction $\rho^{-1}(U) \to U$ is a strong deformation retraction for every subset U of Y'. Moreover, if U is open in $Y' \cap V$ then

$$W = \{y \in \rho^{-1} U \mid \text{the deformation path of } y \text{ lies in } V\}$$

is open in $Y \cap V$, and the restriction of ρ is a strong deformation retraction $W \xrightarrow{\simeq} U$. By induction hypothesis X has an open neighborhood U in $Y' \cap V$ of which it is a strong deformation retract. Composing the two deformations gives $W \simeq U \simeq X$, as required. ∎

6.21 Example. If N is a free graded R-module then the identity map $\mathrm{id}\colon N \to N$ is of finite rank (6.3) if and only if all N_i have finite bases and almost all N_i vanish. In this case, $\Lambda(\mathrm{id}) = \sum_j (-1)^j \beta_j$ (cf. 6.5) where β_j is the number of base elements in N_j. Thus the Lefschetz number Λ generalizes the Euler-Poincaré characteristic (V, 5.1): $\chi(N) = \Lambda(\mathrm{id}_N)$ for free graded abelian groups N (if N is not free, $\chi(N) = \chi(N \otimes \mathbb{Q})$, and $N \otimes \mathbb{Q}$ is always free over $R = \mathbb{Q}$). This implies (cf. 6.6):

6.22 Proposition. *If Y is a compact ENR, and $f\colon Y \to Y$ is a mapping such that $f_* = \mathrm{id}\colon H(Y; \mathbb{Q}) \to H(Y; \mathbb{Q})$ (for instance, if $f \simeq \mathrm{id}$) then $I_f = \chi(Y) =$ Euler-Poincaré characteristic of Y.* ∎

In particular, if Y is such that $\tilde{H}(Y; \mathbb{Q}) = 0$, i.e. if Y has the rational homology of a point, then $f_* = \mathrm{id}$ for all f, and $I_f = 1$ for all f. This applies to contractible spaces, or real projective spaces of even dimension, and others.

6.23 Remark. Formula 6.13 which was used to prove the fixed point theorem has independent interest; the reader is encouraged to study it and look for other applications. Note that $\Phi(K, V)$ as defined in 6.12 makes sense with any coefficients Γ, i.e. $\Phi(K, V)\colon H(K; \Gamma) \to H(V; \Gamma)$. For this, one can either take o_K with integral coefficients so that $o_K \times \eta$ has the same coefficients as η, or one can take ring coefficients Γ and replace o_K by its image under $H(V, V-K; \mathbb{Z}) \to H(V, V-K; \Gamma)$. In any case 6.13 holds, and the proof is practically the same. Only step 3 used the coefficients $\mathbb{Q}$; but for this step ($K \approx \mathbb{S}^\lambda$) one can take integral coefficients first (using $H(K; \mathbb{Z}) \subset H(K; \mathbb{Q})$) and from there pass to the general case. We now illustrate the significance of 6.13 by proving the following

6.24 Proposition. *Let Γ denote a field, $K \subset \mathbb{R}^n$ a compact set, and $\eta \in H(K; \Gamma)$ a homology class such that $i_*(\eta) \neq 0$ for some open neighborhood V of K (i = inclusion: $K \to V$). Then there exists a class $\xi \in H(\mathbb{R}^n, \mathbb{R}^n - K; \Gamma)$ such that the intersection number $\xi \circ \eta$ is not zero* (n.b. if K is a retract of a neighborhood V then $\eta \neq 0 \Rightarrow i_* \eta \neq 0$).

Proof. We take o_K with coefficients in Γ; then

$$\Delta_*(o_K) \in H((V, V-K) \times V; \Gamma) \cong H(V, V-K; \Gamma) \otimes_\Gamma H(V; \Gamma)$$

is of the form

$$\Delta_*(o_K) = \sum_k \xi_k \otimes \zeta_k$$

with

$$\zeta_k \in H(V; \Gamma), \qquad \xi_k \in H(V, V-K; \Gamma) \cong H(\mathbb{R}^n, \mathbb{R}^n - K; \Gamma).$$

Therefore, by 6.13,

$$\begin{aligned} 0 \neq o_0 \otimes i_*(\eta) &= o_0 \otimes (\Phi(K, V)\, \eta) = (d_* \otimes \mathrm{id})(\mathrm{id} \otimes \iota_*)(\Delta_*(o_K) \otimes \eta) \\ &= (d_* \otimes \mathrm{id})\left(\sum \pm \xi_k \otimes \eta \otimes \zeta_k\right) = \sum \pm d_*(\xi_k \otimes \eta) \otimes \zeta_k = \sum \pm (\xi_k \circ \eta) \otimes \zeta_k, \end{aligned}$$

hence $\xi_k \circ \eta \neq 0$ for at least one k. ∎

Proposition 6.24 is an instance of *Alexander duality* (which will be treated in more generality in VIII, 8.15).

6.25 Exercises. *1.* If M, N are complexes of R-modules then the map Θ of 6.1 is a chain map.

2. If

$$\begin{array}{ccccccccc} 0 & \to & N' & \to & N & \to & N'' & \to & 0 \\ & & \downarrow{\scriptstyle \beta'} & & \downarrow{\scriptstyle \beta} & & \downarrow{\scriptstyle \beta''} & & \\ 0 & \to & N' & \to & N & \to & N'' & \to & 0 \end{array}$$

is a commutative diagram of finitely generated free graded R-modules with exact rows then $\Lambda(\beta) = \Lambda(\beta') + \Lambda(\beta'')$. In analogy with V, 5.7 this implies

$$\Lambda(f_*^X) = \Lambda(f_*^A) + \Lambda(f_*),$$

for maps $f:(X,A)\to(X,A)$ of compact ENRs (f_*^X resp. f_*^A is the induced endomorphism of HX resp. HA; coefficients $\mathbb{Q}$). If $\bar{f}$: $X/A\to X/A$ is the induced map then $1+I_f=I_{\bar{f}}+I_{f|A}$ (hint: use IV, 8 Exercises 5 and 6).

3. If Y is an ENR and $f: Y\to Y$ is a map such that $\overline{fY}$ is compact and such that every $\zeta\in\tilde{H}(Y;\mathbb{Q})$ is annihilated by some power of f (i.e. $\bigcup_k \ker(f_*^k)=\tilde{H}(Y;\mathbb{Q})$, where f^k is the k-fold iterate of f) then $I_f=1$. Hint: The trace of a nilpotent endomorphism is zero.

4. If Y is a compact ENR and f: $Y\to Y$ is a map whose index $I_f=0$, can we then deform f into a fixed point free map? The answer is yes if Y is a manifold (Fadell), but no in general. For instance, if A is a compact ENR whose Euler-characteristic $\chi(A)$ is -1 (e.g. the non-orientable surface of genus 3; V, 3.11, Exerc. 2) then the wedge $Y=\mathbb{S}^2\vee A$ has $\chi(Y)=\chi(\mathbb{S}^2)+\chi(A)-1=0$ but every map f: $Y\to Y$ which induces the identity on $H(Y;\mathbb{Q})$ has a fixed point (hint: consider the compositions $A,\mathbb{S}^2\rightrightarrows A\vee\mathbb{S}^2\xrightarrow{g}A\vee\mathbb{S}^2\rightrightarrows A,\mathbb{S}^2$).

5. If f is a map as in 6.6 then $\Lambda(f|K)_*=\Lambda(f|K)^*$, i.e. Lefschetz numbers (or fixed point indices) can just as well be computed from *co*homology (compare 1.14, Exerc. 3).

7. The Exterior Cohomology Product

This product, $H^*X\times H^*Y\to H^*(X\times Y)$, is quite analogous to the exterior homology product of § 2.

7.1 Definition. Let (X,A), (Y,B) be pairs of spaces such that $(X\times Y;\ A\times Y, Y\times B)$ is an excisive triad, let L, M be R-modules and consider the composite chain map

$$\begin{aligned}
&\operatorname{Hom}\left(\frac{SX}{SA},L\right)\otimes_R\operatorname{Hom}\left(\frac{SY}{SB},M\right)\\
&\xrightarrow{\ \gamma\ }\operatorname{Hom}\left(\frac{SX}{SA}\otimes\frac{SY}{SB},L\otimes_R M\right)\\
(7.2)\qquad&\xrightarrow{\ \circ EZ\ }\operatorname{Hom}\left(\frac{S(X\times Y)}{S\{A\times Y,X\times B\}},L\otimes_R M\right)\\
&\xleftarrow{\ j\ }\operatorname{Hom}\left(\frac{S(X\times Y)}{S(A\times Y\cup X\times B)},L\otimes_R M\right),
\end{aligned}$$

where the chain map γ, as in VI, 10.23, is defined by

$$[\gamma(\varphi\otimes\psi)](c\otimes d)=(-1)^{|c||d|}\varphi(c)\otimes\psi(d),$$

EZ is an Eilenberg-Zilber map, and j is induced by the inclusion $S\{A\times Y, X\times B\}\subset S(A\times Y\cup X\times B)$ as in 2.2; the second and third arrow are homotopy equivalences. Passage to homology and composition with

$$\alpha\colon\ H^*(X,A;L)\otimes_R \operatorname{Hom}^*(Y,B;M)\to H[S^*(X,A;L)\otimes_R S^*(Y,B;M)]$$

gives

$$(7.3)\qquad \begin{aligned}(j^*)^{-1}(EZ)^*\gamma_*\alpha\colon\ & H^*(X,A;L)\otimes_R H^*(Y,B;M)\\ &\to H^*(X\times Y, A\times Y\cup X\times B; L\otimes_R M),\end{aligned}$$

or with indices

$$(7.3')\quad H^i(X,A;L)\otimes_R H^k(Y,B;M)\to H^{i+k}(X\times Y, A\times Y\cup X\times B; L\otimes_R M).$$

This map or the corresponding bilinear map is called the *exterior cohomology product*. We write

$$(7.4)\quad x\times y=(j^*)^{-1}(EZ)^*\gamma_*\alpha(x\otimes y)\in H^{i+k}(X\times Y, A\times Y\cup X\times B; L\otimes_R M),$$

for $x\in H^i(X,A;L)$, $y\in H^k(Y,B;M)$.

In terms of representative cocycles φ,ψ this reads

$$(7.5)\qquad [\varphi]\times[\psi]=[\gamma(\varphi\otimes\psi)\circ EZ],$$

where $\varphi\in S^*(X;L)$, $\varphi|SA=0$, $\varphi\circ\partial=0$, $\psi\in S^*(Y;M)$, $\psi|SB=0$, $\psi\circ\partial=0$.

N.B. One has to be careful in applying 7.5: $\gamma(\varphi\otimes\psi)\circ EZ$ vanishes on $S\{A\times Y, X\times B\}$ but not, in general, on $S(A\times Y\cup X\times B)$. However,

$$H\operatorname{Hom}\left(\frac{S(X\times Y)}{S\{A\times Y, X\times B\}}, L\otimes_R M\right)\cong H^*(X\times Y, A\times Y\cup X\times B; L\otimes_R M)$$

so that $[\gamma(\varphi\otimes\psi)\circ EZ]$ can be viewed as lying in the latter group. Of course, this little difficulty does not appear if one of A, B is empty.

In analogy to 2.6 we get from VI, 12.16

7.6 Proposition. *Let L, M be modules over a principal ideal domain R such that $L*_R M=0$. Let $(X,A),(Y,B)$ be pairs of spaces such that $(X\times Y; A\times Y, X\times B)$ is excisive and $H(X,A;R)$ of finite type. If, more-*

over, L is finitely generated or H(Y, B; R) of finite type then

$$\oplus_{i+j=n} H^i(X,A;L)\otimes_R H^j(Y,B;M)\to H^n(X\times Y, A\times Y\cup X\times B; L\otimes_R M),$$
$$x\otimes y\mapsto x\times y,$$

is a split-monomorphism whose cokernel is naturally isomorphic with

$$\oplus_{i+j=n+1} H^i(X,A;L)*H^j(Y,B;M). \blacksquare$$

The analogues (duals) of 2.7–2.13 are as follows.

7.7 Naturality. *If* $f\colon (X,A)\to(X',A')$, $g\colon (Y,B)\to(Y',B')$ *are maps of pairs as in 7.1 then*

$$(f\times g)^*(x'\times y')=(f^*x')\times(g^*y').$$

7.8 Commutativity. $t^*(x\times y)=(-1)^{|x||y|}y\times x$, *where* $t\colon X\times Y\to Y\times X$ *commutes factors.*

7.9 Associativity. $(x\times y)\times z=x\times(y\times z)$.

7.10 Units. *If* $Y=P$ *is a point,* $B=\emptyset$, *and* $1_P\in H^0(P;R)$ *is the cohomology class of the augmentation* $\eta\colon S_0P\to R$, $P\mapsto 1$, *then* $1_P\times x=x=x\times 1_P$ (*where* $P\times(X,A)=(X,A)=(X,A)\times P$). *If Y is an arbitrary space again, and* $\pi\colon Y\to P$ *then* $1_Y=\pi^*(1_P)\in H^0(Y;R)$ *is the class of the augmentation* $S_0Y\to R$, *and naturality 7.7 gives*

$$x\times 1_Y=(\mathrm{id}\times\pi)^*(x\times 1_P)=p^*(x),$$

where $p\colon (X,A)\times Y\to(X,A)=(X,A)\times P$ *is the projection.*

7.11 Stability (cf. also Exerc. 3). The following diagram (coefficients omitted) is commutative

$$\begin{array}{ccc}
H^*A\otimes H^*(Y,B) & \xrightarrow{\times} H^*(A\times Y, A\times B)\overset{i^*}{\cong} & H^*(A\times Y\cup X\times B, X\times B)\\
\downarrow{\scriptstyle \delta^*\otimes\mathrm{id}} & & \downarrow{\scriptstyle \delta^*}\\
H^*(X,A)\otimes H^*(Y,B) & \xrightarrow{\qquad\times\qquad} & H^*(X\times Y, A\times Y\cup X\times B),
\end{array}$$

where i = inclusion. In formulas,

(7.12) $\delta^*(i^*)^{-1}(a\times y)=(\delta^*a)\times y$, for $a\in H^*A$, $y\in H^*(Y,B)$.

In the important special case $B=\emptyset$ we have $i=\mathrm{id}$, and stability reduces to

(7.13) $\delta^*(a\times y)=(\delta^*a)\times y$, for $a\in H^*A$, $y\in H^*Y$.

7.14 Duality. This relates homology and cohomology cross-products: *If* $\xi \in H(X, A; R)$, $\eta \in H(Y, B; R)$, $x \in H^*(X, A; L)$, $y \in H^*(Y, B; M)$ *then* $\langle x \times y, \xi \times \eta \rangle = (-1)^{|y||\xi|} \langle x, \xi \rangle \otimes \langle y, \eta \rangle$.

Proofs of 7.7–7.14. Using representatives $\varphi, \varphi', \psi, \ldots$ for $x, x', y, \ldots$ we find, by 7.5, the following representatives for the terms of 7.7–7.10.

7.7 left: $\gamma(\varphi' \otimes \psi') \circ EZ \circ (f \times g)$;

7.7 right: $\gamma(\varphi' f \otimes \psi' g) \circ EZ = \gamma(\varphi' \otimes \psi') \circ (f \otimes g) \circ EZ$,

and these agree by naturality of EZ.

7.8 left: $\gamma(\varphi \otimes \psi) \circ EZ \circ t$;

7.8 right: $(-1)^{|x||y|} \gamma(\psi \otimes \varphi) \circ EZ = \gamma(\varphi \otimes \psi) \circ \tau \circ EZ$;

these agree by commutativity (VI, 12.3) of EZ.

7.9 left: $\gamma(\varphi \otimes \psi \otimes \rho) \circ (EZ \otimes \mathrm{id}) \circ EZ$;

7.9 right: $\gamma(\varphi \otimes \psi \otimes \rho) \circ (\mathrm{id} \otimes EZ) \circ EZ$;

these agree by associativity (VI, 12.4) of EZ.

7.10 left: $\gamma(\eta \otimes \varphi) \circ EZ$; 7.10 middle: φ;

these agree by VI, 12.5.

For 7.11 we choose a representative cocycle of $a \in H^* A$ first, and extend it to a cochain φ on X; in particular, $\delta \varphi | SA = 0$. As before, ψ denotes a representative cocycle of $y \in H^*(Y, B)$. The left side of 7.12 is represented by

$$\begin{aligned}\delta(\gamma(\varphi \otimes \psi) \circ EZ) &= (-1)^{|\varphi| + |\psi| + 1} \gamma(\varphi \otimes \psi) \circ EZ \circ \partial \\ &= (-1)^{|\varphi| + |\psi| + 1} \gamma(\varphi \otimes \psi) \circ \partial \circ EZ \\ &= (-1)^{|\varphi| + 1} \gamma(\varphi \circ \partial \otimes \psi) \circ EZ + (-1)^{|\varphi| + |\psi| + 1} \gamma(\varphi \otimes \psi \circ \partial) \circ EZ \\ &= \gamma(\delta \varphi \otimes \psi) \circ EZ,\end{aligned}$$

and the last expression also represents the right side of 7.12 (n.b. these cochains may not vanish on $S(A \times Y \cup X \times B)$, but only on $S\{A \times Y, X \times B\}$; by excision, that is enough).

For 7.14 we use representatives, too, and get

$$\langle [\varphi] \times [\psi], [a] \times [b] \rangle = \gamma(\varphi \otimes \psi) \circ \Psi \circ \Phi(a \otimes b),$$

where Ψ, Φ are EZ-maps going in opposite directions. Since $\Psi \circ \Phi \simeq \mathrm{id}$ the last term equals

$$\gamma(\varphi \otimes \psi)(a \otimes b) = (-1)^{|\psi||a|} \varphi(a) \otimes \psi(b) = (-1)^{|y||\xi|} \langle x, \xi \rangle \otimes \langle y, \eta \rangle. \quad \blacksquare$$

7.15 Exercises. 1*. *If R is a field then $\times: H^*(X;R)\otimes_R H^*(Y;R)\to H^*(X\times Y;R)$ is always injective. It is surjective if and only if $H(X;R)$ or $H(Y;R)$ is of finite type.* Because $H^i(X;R)\cong H_i(X;R)^*$, this reduces to the following algebraic assertion: If V, W are vector-spaces over R then $\gamma: V^*\otimes W^*\to(V\otimes W)^*$, $[\gamma(\varphi\otimes\psi)](v\otimes w)=\varphi(v)\psi(w)$ is always injective; it is surjective if and only if at least one of V, W is finite-dimensional. We indicate a proof. Let B, C be bases for V, W; then $V^*, W^*, (V\otimes W)^*$ may be identified with the function-sets $F(B,R)$, $F(C,R)$, $F(B\times C,R)$. If $\rho\in F(B\times C,R)$ and $c\in C$ let $\rho_c\in F(B,R)$ denote the partial function $\rho_c(b)=\rho(b,c)$. If $\rho=\gamma(\varphi\otimes\psi)$ then $\rho_c(b)=\varphi(b)\psi(c)$ hence all ρ_c are multiples of φ. Every $\rho\in\operatorname{im}(\gamma)$ is a finite linear combination of elements $\gamma(\varphi\otimes\psi)$, hence the set $\{\rho_c|c\in C\}$ contains no more than finitely many linearly independent elements (if $\rho\in\operatorname{im}(\gamma)$!). Let $F_e(B\times C,R)$ consist of all ρ such that $\{\rho_c|c\in C\}$ has finite rank. It is easy to see that $F_e(B\times C,R)=F(B\times C,R)$ if and only if at least one of B, C is finite. It remains to show that $\gamma: F(B,R)\otimes F(C,R)\to F_e(B\times C,R)$ is isomorphic. If $\rho\in F_e(B\times C,R)$, choose a maximal linearly independent set among the ρ_c, say $\varphi_1,\ldots,\varphi_r$. Then, for every $c\in C$ we have $\rho_c=\sum_{i=1}^r\psi_i(c)\,\varphi_i$; the coefficients $\psi_i(c)$ are uniquely determined functions of c, and the assignment $\rho\mapsto\sum_{i=1}^r\varphi_i\otimes_R\psi_i$ defines a map which is inverse to γ.—Generalize to free modules over principal ideals domains.

2. If R is a principal ideal domain and X is a space such that $H(X;R)$ is of finite type then there are two ways of expressing $H^*(X\times Y;R)$ in terms of $H(X;R)$, $H(Y;R)$. 1st way: Express $H(X\times Y;R)$ by the Künneth formula then apply the universal coefficient formula. 2nd way: Express $H^*(X;R)$, $H^*(Y;R)$ by the universal coefficient formula then apply 7.6. Compare the two results. Formulate and prove the underlying algebraic relations.

3. After 2.11, the reader might have expected the following diagram under the heading "stability".

$$
(7.16)\qquad
\begin{array}{ccc}
H^*A\otimes H^*B & \xrightarrow{\quad\times\quad} & H^*(A\times B)\\
 & & \big\downarrow{\scriptstyle\delta^*}\\
\Big\downarrow{\scriptstyle \delta^*\otimes\mathrm{id},\,(-1)^{\dim}\mathrm{id}\otimes\delta^*} & & H^*(X\times B\cup A\times Y, A\times B)\\
 & & \|\wr\\
[H^*(X,A)\otimes H^*B]\oplus[H^*A\otimes H^*(Y,B)] & \xrightarrow{\times\oplus\times} & H^*(X\times B,A\times B)\oplus H^*(A\times Y,A\times B).
\end{array}
$$

Because the group on the lower right is a direct product, 7.16 decomposes into two diagrams both of which can be seen to be special cases 7.12 of stability (up to naturality 7.7 and commutativity 7.8). In particular, 7.16 *is commutative.*

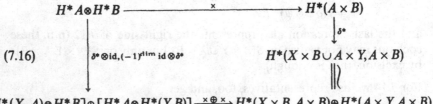

8. The Interior Cohomology Product ($\smile$-Product)

In VI, 12, Exerc. 5, it was indicated that Eilenberg-Zilber maps $EZ\colon S(X\times Y)\to SX\otimes SY$ and natural diagonals $D\colon SX\to SX\otimes SX$ are formally equivalent notions. When applied to the map EZ which occurs in the definition of exterior cohomology products $\times\colon H^*X\otimes H^*Y\to H^*(X\times Y)$ this equivalence gives the interior cohomology product $\smile\colon H^*X\otimes H^*X\to H^*X$. Although then $\times$ and $\smile$ are equivalent, it is convenient to have both of them. $\times$-products, for instance, may be easier to compute (compare proof of 9.4), $\smile$-products on the other hand provide a more familiar algebraic structure: they turn $H^*(X;R)$ into a (functorial) graded ring.

8.1 Definition. Let $(X;A_1,A_2)$ be an excisive triad, and let M_1,M_2 be R-modules. Consider the composite chain map

$$\begin{aligned}(8.2)\qquad &\operatorname{Hom}(SX/SA_1,M_1)\otimes_R\operatorname{Hom}(SX/SA_2,M_2)\\ &\xrightarrow{\gamma}\operatorname{Hom}(SX/SA_1\otimes SX/SA_2,M_1\otimes_R M_2)\\ &\xrightarrow{\circ D}\operatorname{Hom}\left(\frac{SX}{S\{A_1,A_2\}},M_1\otimes_R M_2\right)\\ &\xleftarrow{j}\operatorname{Hom}\left(\frac{SX}{S(A_1\cup A_2)},M_1\otimes_R M_2\right),\end{aligned}$$

where, as before, $(\gamma(\varphi_1\otimes\varphi_2))(a_1\otimes a_2)=(-1)^{|\varphi_2|\,|a_1|}(\varphi_1a_1)\otimes(\varphi_2a_2)$, and j is induced by inclusion. By assumption, j is a homotopy equivalence. Passage to homology and composition with α, as in VII, 7, gives

$$\begin{aligned}(8.3)\qquad (j^*)^{-1}D^*\gamma_*\alpha\colon\ &H^*(X,A_1;M_1)\otimes_R H^*(X,A_2;M_2)\\ &\to H^*(X,A_1\cup A_2;M_1\otimes_R M_2),\end{aligned}$$

or with dimension indices

$$(8.3')\qquad H^i(X,A_1;M_1)\otimes_R H^k(X,A_2;M_2)\to H^{i+k}(X,A_1\cup A_2;M_1\otimes_R M_2).$$

This map or the corresponding bilinear map is called the *interior cohomology product* or *cup-product* ($\smile$-product). We write

$$(8.4)\qquad x_1\smile x_2=(j^*)^{-1}D^*\gamma_*\alpha(x_1\otimes x_2),\qquad \text{for } x_\mu\in H^*(X,A_\mu;M_\mu).$$

In terms of representative cocycles φ_1,φ_2 this reads

$$(8.5)\qquad [\varphi_1]\smile[\varphi_2]=[\gamma(\varphi_1\otimes\varphi_2)\circ D],$$

where $\varphi_\mu \in S^*(X; M_\mu)$, $\varphi_\mu | SA_\mu = 0$, $\varphi_\mu \circ \partial = 0$. As in 7.5, one has to remember that $\gamma(\varphi_1 \otimes \varphi_2) \circ D$ vanishes on $S\{A_1, A_2\}$ but not necessarily on $S(A_1 \cup A_2)$.

The following properties 8.6–8.10 of $\smile$-products follow from properties of the natural diagonal just as 7.7–7.11 followed from properties of EZ.

8.6 Naturality. If $f\colon (X; A_1, A_2) \to (Y; B_1, B_2)$ is a map of excisive triads then

$$f^*(y_1 \smile y_2) = (f^* y_1) \smile (f^* y_2), \quad \text{for } y_\mu \in H^*(Y, B_\mu; M_\mu). \;\blacksquare$$

8.7 Commutativity. $x_1 \smile x_2 = (-1)^{|x_1||x_2|} x_2 \smile x_1$. ▮

8.8 Associativity. $x_1 \smile (x_2 \smile x_3) = (x_1 \smile x_2) \smile x_3$. This triple product lies in $H^*(X, A_1 \cup A_2 \cup A_3)$ if $x_\mu \in H^*(X, A_\mu)$. ▮

8.9 Units. $1_X \smile x = x = x \smile 1_X$, where $1_X \in H^0(X; R)$ is the class of the augmentation $S_0 X \to R$. ▮

8.10 Stability. The following diagram is commutative,

$$\begin{array}{ccccc}
H^*A_1 \otimes H^*(X, A_2) & \xrightarrow{\mathrm{id} \otimes i^*} & H^*A_1 \otimes H^*(A_1, A_1 \cap A_2) & \xrightarrow{\smile} & H^*(A_1, A_1 \cap A_2) \\
 & & & & \uparrow \cong\, j^* \\
\Big\downarrow \delta^* \otimes \mathrm{id} & & & & H^*(A_1 \cup A_2, A_2) \\
 & & & & \downarrow \delta^* \\
H^*(X, A_1) \otimes H^*(X, A_2) & & \xrightarrow{\qquad\smile\qquad} & & H^*(X; A_1 \cup A_2),
\end{array} \tag{8.11}$$

where i, j are inclusions. In formulas,

(8.12) $$\delta^*(j^*)^{-1}(a \smile i^* x) = (\delta^* a) \smile x, \quad a \in H^* A_1, \; x \in H^*(X, A_2).$$

In the important special case $A_2 = \emptyset$, this becomes

(8.13) $$\delta^*(a \smile i^* x) = (\delta^* a) \smile x,$$

where $i\colon A \to X$ is an inclusion map, $a \in H^* A$, $x \in H^* X$. ▮

The following two properties reflect the relation between Eilenberg-Zilber maps and natural diagonals.

(8.14) $$x_1 \smile x_2 = \Delta^*(x_1 \times x_2), \quad x_\mu \in H^*(X, A_\mu),$$

where $\Delta: (X, A_1 \cup A_2) \to (X \times X, A_1 \times X \cup X \times A_2)$ is the diagonal map, $\Delta P = (P, P)$; we have to assume that both $(X; A_1, A_2)$ and $(X \times X; A_1 \times X, X \times A_2)$ are excisive.

Proof. With representative cocycles φ_1, φ_2 the left side is $[\gamma(\varphi_1 \otimes \varphi_2) \circ D]$, the right is $[\gamma(\varphi_1 \otimes \varphi_2) \circ EZ \circ \Delta]$, and $D = EZ \circ \Delta$ by VI, 12.20. ∎

$$(8.15) \qquad x \times y = (p^* x) \smile (q^* y), \quad \text{if } x \in H^*(X, A),\ y \in H^*(Y, B),$$

and $p: (X \times Y, A \times Y) \to (X, A)$, $q: (X \times Y, X \times B) \to (Y, B)$ are the projections; we have to assume that $(X \times Y; A \times Y, X \times B)$ is excisive.

Proof. With representative cocycles φ, ψ the left side is $[(\varphi \otimes \psi) \circ EZ]$, the right is $[(\varphi \circ p) \otimes (\psi \circ q) \circ D] = [(\varphi \otimes \psi) \circ (p \otimes q) \circ D]$, and $EZ = (p \otimes q) \circ D$ by VI, 12.25. ∎

As a consequence of 8.15 we note

8.16 Multiplicativity. $(x_1 \times y_1) \smile (x_2 \times y_2) = (-1)^{|y_1||x_2|}(x_1 \smile x_2) \times (y_1 \smile y_2)$, if $x_\mu \in H^*(X, A_\mu)$, $y_\mu \in H^*(Y, B_\mu)$, and $(X; A_1, A_2)$, $(Y; B_1, B_2)$ are triads such that the products above are defined.

Proof.

$$\begin{aligned}(x_1 \times y_1) \smile (x_2 \times y_2) &= p^* x_1 \smile q^* y_1 \smile p^* x_2 \smile q^* y_2 \\ &= (-1)^{|y_1||x_2|} p^* x_1 \smile p^* x_2 \smile q^* y_1 \smile q^* y_2 \\ &= (-1)^{|y_1||x_2|} p^*(x_1 \smile x_2) \smile q^*(y_1 \smile y_2) \\ &= (-1)^{|y_1||x_2|} (x_1 \smile x_2) \times (y_1 \smile y_2). \quad ∎\end{aligned}$$

8.17 Remarks. If the coefficients are $M_1 = R = M_2$ then also $M_1 \otimes_R M_2 = R$. Properties 8.3′, 8.6–8.9 then assert that $H^*(X; R)$ is a *commutative graded ring* (in fact, an R-algebra), which depends *functorially* on X. It is called the *cohomology ring* (algebra) of X (with coefficients in R). Further, $H^*(X, A; M)$ is an $H^*(X; R)$-*module* with respect to $H^* X \otimes H^*(X, A) \xrightarrow{\smile} H^*(X, A)$. By restriction, $H^*(A; M)$ is also an $H^*(X; R)$-module, $H^* X \otimes H^* A \to H^* A \otimes H^* A \xrightarrow{\smile} H^* A$, and 8.13 asserts that $\delta^*: H^*(A; M) \to H^*(X, A; M)$ *is a homomorphism of* $H^*(X; R)$-*modules.*

If K, L are graded R-algebras then $K \otimes_R L$ is also a graded algebra with respect to the multiplication

$$(k_1 \otimes l_1) \cdot (k_2 \otimes l_2) = (-1)^{|l_1||k_2|}(k_1 k_2) \otimes (l_1 l_2).$$

This algebra is called the *tensor product* of the algebras K, L. Multiplicativity 8.16 then asserts that

$$H^*(X;R)\otimes_R H^*(Y;R) \xrightarrow{\times} H^*(X\times Y;R)$$

is an *algebra homomorphism*. Moreover, by 7.6,

8.18 Proposition. *If R is a principal ideal domain, and X, Y are spaces such that $H(X;R)$ is of finite type and all torsion products $H^i(X;R) * H^j(Y;R)$ vanish then $\times: H^*(X;R)\otimes H^*(Y;R)\to H^*(X\times Y;R)$ is an isomorphism of algebras.* ∎

8.19 Exercises. *1.* We can define cup-products of cochains by composing

$$S^*X\otimes S^*X \xrightarrow{\gamma} (SX\otimes SX)^* \xrightarrow{\circ D} S^*X, \tag{8.20}$$

where D is a natural diagonal. This is a chain map. It depends on the choice of D but its homotopy class does not. Show that for $D = AW\circ\Delta$, where AW is the Alexander-Whitney map VI, 12.27, this cup-product of cochains $\varphi_1\in S^pX$, $\varphi_2\in S^qX$ has the form

$$(\varphi_1\smile\varphi_2)\,\sigma = (-1)^{pq}\,\varphi_1(\sigma\,\overleftarrow{\varepsilon}_p^{\,p+q})\otimes\varphi_2(\sigma\,\overrightarrow{\varepsilon}_q^{\,p+q}), \tag{8.21}$$

where $\sigma\colon \Delta_{p+q}\to X$, and $\overleftarrow{\varepsilon}_p^{\,p+q}\colon \Delta_p\to\Delta_{p+q}$ resp. $\overrightarrow{\varepsilon}_q^{\,p+q}\colon \Delta_q\to\Delta_{p+q}$ cover the first $(p+1)$ resp. last $(q+1)$ vertices of Δ_{p+q} (cf. VI, 12.26 Exerc. 1). The formula 8.21 (up to sign) is often used to define cup-products directly, without referring to EZ-maps or natural diagonals. In particular, this was the procedure of Alexander and Whitney.—Show that the cup-product 8.21 is associative but not commutative.

2. Formulate the stability property of $\smile$-products which corresponds to 7.16.

9. $\smile$-Products in Projective Spaces. Hopf Maps and Hopf Invariant

9.1 We begin with some $\smile$-products in euclidian space. Coefficients are taken in a fixed commutative ring R; they will not appear in the notation. For $k\le n$ we consider $\mathbb{R}^k$ as subspace of $\mathbb{R}^n$, namely

$$\mathbb{R}^k = \{x = (x_1,\ldots,x_n)\in\mathbb{R}^n \mid x_i = 0 \text{ for } i>k\},$$

and we put

$$\hat{\mathbb{R}}^{n-k} = \{x\in\mathbb{R}^n \mid x_i = 0 \text{ for } i\le k\}\approx\mathbb{R}^{n-k};$$

clearly

$$(\mathbb{R}^n - \mathbb{R}^k) \cup (\mathbb{R}^n - \hat{\mathbb{R}}^{n-k}) = (\mathbb{R}^n - 0).$$

We claim

$$(9.2) \qquad \smile\colon H^k(\mathbb{R}^n, \mathbb{R}^n - \hat{\mathbb{R}}^{n-k}) \otimes_R H^{n-k}(\mathbb{R}^n, \mathbb{R}^n - \mathbb{R}^k) \cong H^n(\mathbb{R}^n, \mathbb{R}^n - 0).$$

Proof. Consider the diagram

$$\begin{array}{ccc} H^k(\mathbb{R}^k_0) \otimes_R H^{n-k}(\mathbb{R}^{n-k}_0) & \cong & H^k(\mathbb{R}^k_0 \times \mathbb{R}^{n-k}) \otimes_R H^{n-k}(\mathbb{R}^k \times \mathbb{R}^{n-k}_0) \\ & \searrow^{\times} \qquad \swarrow^{\smile} & \\ & H^n(\mathbb{R}^k \times \mathbb{R}^n, \mathbb{R}^k \times \mathbb{R}^{n-k} - 0), & \end{array}$$

where we use the abbreviation $\mathbb{R}^j_0 = (\mathbb{R}^j, \mathbb{R}^j - 0)$ to facilitate the printing. The diagram commutes by 8.15. The left arrow $\times$ is isomorphic by 7.6 (or by duality 7.14, using 2.14), hence the right arrow is isomorphic. But the right arrow agrees with (9.2) by naturality of $\smile$-products, applied to $\mathbb{R}^k \times \mathbb{R}^{n-k} = \mathbb{R}^n$. ∎

9.3 $\smile$-Products in Projective Spaces. We treat simultaneously the projective spaces P_n over the reals $\mathbb{R}$, the complex numbers $\mathbb{C}$, and the quaternions $\mathbb{H}$ (cf. V, 3.5). We recall that $H^i(P_n; R) \cong R$ if $i = 0, d, 2d, \ldots, nd$, and $H^i(P_n; R) = 0$ otherwise, where $d = 1, 2, 4$, and $R = \mathbb{Z}_2, \mathbb{Z}, \mathbb{Z}$ according to the cases $\mathbb{R}, \mathbb{C}, \mathbb{H}$. We shall prove

9.4 Proposition. $\smile\colon H^{id}(P_n; R) \otimes H^{jd}(P_n; R) \xrightarrow{\cong} H^{(i+j)d}(P_n; R)$, for $i, j \geq 0$, $i + j \leq n$.

In other words, if $x \in H^d(P_n; R) \cong R$ is a generator then $\{1, x, x^2, \ldots, x^n\}$ is a base of $H^*(P_n; R)$, and $x^{n+1} = 0$. Or again, $H^*(P_n; R) \cong R[x]/(x^{n+1}) =$ polynomial ring in x divided by the ideal (x^{n+1}).

Proof. Fix $k \leq n$, and consider P_k as subspace of P_n, namely

$$P_k = \{\zeta \in P_n \mid \zeta_{k+1} = 0 = \zeta_{k+2} = \cdots = \zeta_n\},$$

where ζ_ν are homogenuous coordinates. As we know (V, 4.10 and 6.13), $H^i P_n \cong H^i P_k$ for $i \leq dk$. Define $\hat{P}_{n-k} = \{\zeta \in P_n \mid \zeta_0 = 0 = \zeta_1 = \cdots = \zeta_{k-1}\}$. Then

$$(9.5) \qquad (P_n - \hat{P}_{n-k}) \simeq P_{k-1},$$

via the deformation retraction $\zeta \mapsto [\zeta_0, \ldots, \zeta_{k-1}, t\zeta_k, \ldots, t\zeta_n]$, $0 \leq t \leq 1$.

Further, we identify $\mathbb{R}^{dn}$ with $\{\zeta\in P_n|\zeta_k\neq 0\}$, $\mathbb{R}^{dk}=\mathbb{R}^{dn}\cap P_k$, $\hat{\mathbb{R}}^{d(n-k)}=R^{dn}\cap\hat{P}_{n-k}$. Consider the following diagram

$$(9.6)\qquad \begin{array}{ccccc} H^{dk}P_n & \longleftarrow & H^{dk}(P_n,P_n-\hat{P}_{n-k}) & \longrightarrow & H^{dk}(\mathbb{R}^{dn},\mathbb{R}^{dn}-\hat{\mathbb{R}}^{d(n-k)}) \\ \downarrow{\scriptstyle\rho} & & \downarrow{\scriptstyle\varphi} & & \downarrow{\scriptstyle\rho} \\ H^{dk}P_k & \xleftarrow{\rho} & H^{dk}(P_k,P_k-\hat{P}_0) & \xrightarrow{\rho} & H^{dk}(\mathbb{R}^{dk},\mathbb{R}^{dk}-0) \end{array}$$

in which all maps are induced by inclusions ($\hat{P}_0=P_k\cap\hat{P}_{n-k}$). We know that all maps marked ρ are isomorphisms. Further, φ is isomorphic because $H^iP_n\cong H^iP_k$ for $i\le dk$, and $P_n-\hat{P}_{n-k}\simeq P_{k-1}\simeq P_k-\hat{P}_0$. Therefore, all maps in 9.6 are isomorphisms.—We can also interchange the role of P_k, $\hat{P}_{n-k}$ and get a similar diagram $\widehat{9.6}$ of isomorphisms for $H^{d(n-k)}$. Consider then the diagram

$$(9.7)\qquad \begin{array}{ccc} H^{dk}P_n\otimes H^{d(n-k)}P_n & \longleftarrow & H^{dk}(P_n,P_n-\hat{P}_{n-k})\otimes H^{d(n-k)}(P_n,P_n-P_k) \\ \downarrow{\scriptstyle\smile} & & \downarrow{\scriptstyle\smile} \\ H^{dn}P_n & \xleftarrow{\rho} & H^{dn}(P_n,P_n-\hat{P}_0) \\ & \longrightarrow & H^{dk}(\mathbb{R}^{dn},R^{dn}-\hat{R}^{d(n-k)})\otimes H^{d(n-k)}(R^{dn},R^{dn}-R^{dk}) \\ & & \downarrow{\scriptstyle\smile} \\ & \xrightarrow{\rho} & H^{dn}(\mathbb{R}^{dn},\mathbb{R}^{dn}-0). \end{array}$$

The top row is obtained by tensoring the top rows of 9.6 and $\widehat{9.6}$, hence consists of isomorphisms. The lower row consists of isomorphisms as in 9.6. The right vertical is isomorphic by 9.2. Hence all maps are isomorphic, in particular the left vertical. This proves the theorem if $i+j=n$. The general case $i+j\le n$ follows because $H^*P_n\xrightarrow{\iota^*}H^*P_{i+j}$ is an isomorphism of rings up to dimension $d(i+j)$. ∎

9.8 Corollary. *If $0<k<n$ then P_k is not a retract of P_n.*

Proof. If $r: P_n\to P_k$ is a retraction, and $x_n\in H^d(P_n,R)$, $x_k\in H^d(P_k,R)$ are generators as in 9.4 then $r^*x_k=x_n$, hence $0=r^*(x_k^{k+1})=r^*(x_k)^{k+1}=x_n^{k+1}\neq 0$, a contradiction. ∎

9.9 Corollary. *The* Hopf *map* $h: \mathbb{S}^{dn-1} \to P_{n-1}$, $h(z_0, \ldots, z_{n-1}) = [z_0, \ldots, z_{n-1}]$, $n>1$, *is not nulhomotopic* (we use coordinates $z_\nu \in \mathbb{R}$ resp. $\mathbb{C}$ resp. $\mathbb{H}$ with $\sum_{\nu=0}^{n-1} \|z_\nu\|^2 = 1$ to describe the sphere $\mathbb{S}^{dn-1}$). In particular, for $n=2$, we get essential maps $\mathbb{S}^3 \to P_1\mathbb{C} \approx \mathbb{S}^2$, $\mathbb{S}^7 \to P_1\mathbb{H} \approx \mathbb{S}^4$.

Proof. If $h \simeq 0$ then there is a map $\Theta: \mathbb{B}^{dn} \to P_{n-1}$ with $\Theta|\mathbb{S}^{dn-1} = h$ (where $\mathbb{B}^{dn} = \{(z_0 \ldots z_{n-1}) | \sum_\nu \|z_\nu\|^2 \le 1\}$). But then we could define a retraction $r: P_n \to P_{n-1}$ as follows

$$r[\zeta_0, \ldots, \zeta_n] = \begin{cases} [\zeta_0, \ldots, \zeta_{n-1}] & \text{if } \sum_{\nu=0}^{n-1} \|\zeta_\nu\|^2 \ge \|\zeta_n\|^2 \\ \Theta\left(\dfrac{\zeta_0}{\zeta_n}, \ldots, \dfrac{\zeta_{n-1}}{\zeta_n}\right) & \text{if } \sum_{\nu=0}^{n-1} \|\zeta_\nu\|^2 \le \|\zeta_n\|^2. \end{cases}$$

This contradicts 9.8. ∎

The Hopf-Invariant. Analyzing the relation between P_n and the Hopf map $h: \mathbb{S}^{dn-1} \to P_{n-1}$ leads to the following

9.10 Definition. Let $f: X \to Y$ be a continuous map. Define the *mapping cone* Cf to be the space which is obtained from the topological sum $(X \times [0,1]) \oplus Y$ by shrinking $X \times \{0\}$ to a point, and identifying each $(x, 1) \in X \times [0,1]$ with $f(x) \in Y$. Alternatively, Cf is obtained from $CX \oplus Y$ by identifying each point x in the base X of the cone CX with $f(x) \in Y$ (see Fig. 10); one often writes $Cf = Y \cup_f CX$. We can view Y as a subspace of Cf (no identifications were made in Y), and we have

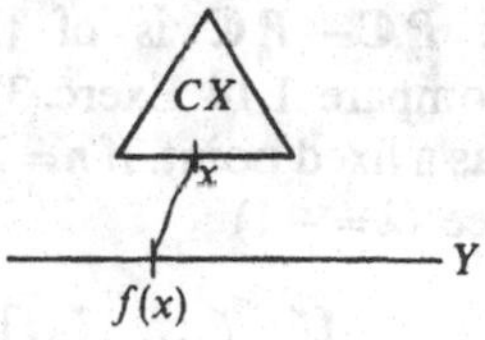

Fig. 10

9.11 Lemma. *The map* $f: X \to Y$ *is nulhomotopic if and only if* Y *is a retract of* Cf.

Proof. Let $((x,t)) \in Cf$ denote the equivalence class of $(x,t) \in X \times [0,1]$. The equation $r((x,t)) = \Theta(x,t)$ establishes a 1-1 correspondence between retractions $r: Cf \to Y$ and nulhomotopies $\Theta: f \simeq 0$. ∎

Every (co-)homological condition for the existence of a retraction $r: Cf \to Y$ therefore is also a condition for the existence of a nulhomotopy $\Theta: f \simeq 0$. For instance, if $f: \mathbb{S}^{2n-1} \to \mathbb{S}^n$, $n>1$, then $Cf = \mathbb{S}^n \cup_f \mathbb{B}^{2n}$. This is a CW-space having one cell in dimensions $0, n, 2n$, and no other cells. Hence $H^n(Cf; \mathbb{Z}) \cong \mathbb{Z} \cong H^{2n}(Cf; \mathbb{Z})$. If $y \in H^n(Cf; \mathbb{Z})$, $y' \in H^{2n}(Cf; \mathbb{Z})$ are generators then $y \smile y = \gamma(f)\, y'$, where $\gamma(f)$ is an integer. This number is easily seen to be an invariant of the homotopy class of f (in fact, $f \simeq f' \Rightarrow Cf \simeq Cf'$); it is called the *Hopf-invariant* of f. If $f \simeq 0$ and r

is the corresponding retraction ($ri=\mathrm{id}_{\mathbb{S}^n}$) then $0=r^*(i^*y\smile i^*y)=y\smile y=\gamma(f)y'$, hence $\gamma(f)=0$. The Hopf-invariant allows to exhibit essential maps $\mathbb{S}^{4k-1}\to\mathbb{S}^{2k}$ for all $k>0$ (see Steenrod-Epstein I, 5 for a simple proof, and for further properties of γ).

9.12 Exercises. *1**. We indicate another way of computing $\smile$-products in $P_n\mathbb{C}$ (coefficients $\mathbb{Z}$). Clearly $H^*(P_1\mathbb{C})$ has a base consisting of $1\in H^0$, $s\in H^2$. Therefore $\times^n(P_1\mathbb{C})=P_1\mathbb{C}\times P_1\mathbb{C}\times\cdots$ has as cohomology base

$$\{s_{i_1}\smile s_{i_2}\smile\cdots\smile s_{i_k}\in H^{2k}(\times^n P_1\mathbb{C})\},\quad 1\leq i_1<i_2<\cdots<i_k\leq n,$$

where $s_i=1\times 1\times\cdots\times s\times\cdots\times 1$, the s in i-th position (cf. 8.18). Let $x_j\in H^{2j}(P_n\mathbb{C})\cong\mathbb{Z}$ be a generator, $0\leq j\leq n$. Consider the map $\mu:\times^n P_1\mathbb{C}\to P_n\mathbb{C}$ of 2.16 Exerc. 3(ii), and prove $\pm\mu^*(x_1)=\sum_{i=1}^n s_i$, $\pm\mu^*(x_n)=n!(s_1\smile s_2\cdots\smile s_n)$ (hint: look at homology first, and use scalar products). This implies $\pm\mu^*(x_1\smile x_1\smile\cdots\smile x_1)=(\Sigma s_i)^n=n!\,s_1\smile s_2\smile\cdots\smile s_n=\mu^*(x_n)$ hence $(x_1)^n=\pm x_n$, and hence the multiplicative structure of $H^*(P_n\mathbb{C})$.

2. If $f\colon P_n\mathbb{C}\to P_n\mathbb{C}$, $n>0$, is a map then the induced endomorphism f_* of $H_n(P_n\mathbb{C};\mathbb{Z})\cong\mathbb{Z}$ is given by multiplication with an integer $\deg(f)\in\mathbb{Z}$, the *degree* of f. Show that $\deg(f)=\lambda^n$ for some $\lambda\in\mathbb{Z}$. In particular, n even $\Rightarrow\deg(f)\geq 0$. Hint: Study the ring endomorphism f^* of $H^*(P_n\mathbb{C};\mathbb{Z})$.

3. Using $\smile$-products show that the Lefschetz number of any map $f\colon P_n\mathbb{C}\to P_n\mathbb{C}$ is of the form $\Lambda(f_*)=1+\lambda+\lambda^2+\cdots+\lambda^n$ with $\lambda\in\mathbb{Z}$ (compare 1.14 Exerc. 3). If n is even, this is never zero, hence every f has a fixed point. If $n=2k-1$ is odd then the following map is fixed point free ($\lambda=-1$)

$$[\zeta_0,\zeta_1,\ldots,\zeta_{2k}]\mapsto[-\bar\zeta_1,\bar\zeta_0,-\bar\zeta_3,\bar\zeta_2,\ldots,-\bar\zeta_{2k},\bar\zeta_{2k-1}].$$

4. Show that $P_{n+1}F$, $F=\mathbb{R},\mathbb{C}$, or $\mathbb{H}$, is homeomorphic with the mapping cone of the Hopf map

$$\mathbb{S}^{d(n+1)-1}\to P_nF,\quad (z_0,\ldots,z_n)\mapsto[z_0,z_1,\ldots,z_n].$$

Deduce that the Hopf maps $\mathbb{S}^3\to P_1\mathbb{C}\approx\mathbb{S}^2$, $\mathbb{S}^7\to P_1\mathbb{H}\approx\mathbb{S}^4$ have Hopf invariant ± 1.

5. Using the commutative law for cup-products show that every map $\mathbb{S}^{4k+1}\to\mathbb{S}^{2k+1}$ has Hopf invariant zero.

Remark. If n,γ are even then there is an $\mathbb{S}^{2n-1}\to\mathbb{S}^n$ with Hopf invariant γ; cf. Steenrod-Epstein I, 5.2. Maps $\mathbb{S}^{2n-1}\to\mathbb{S}^n$ with odd Hopf invariants only occur for $n=2$, 4, 8; cf. Adams, 1960, and for a simpler proof Adams-Atiyah.

10. Hopf Algebras

In 3.1 we remarked without proof that even-dimensional spheres do not admit multiplications μ: $\mathbb{S}^{2n}\times\mathbb{S}^{2n}\to\mathbb{S}^{2n}$ with two-sided unit. Such a map would have bidegree (1, 1) (because of the unit), but

10.1 Proposition. *If* μ: $\mathbb{S}^{2n}\times\mathbb{S}^{2n}\to\mathbb{S}^{2n}$ *has bidegree* (α,β) *then* $\alpha\beta=0$.

Proof. Let $s\in H^{2n}(\mathbb{S}^{2n};\mathbb{Z})$ be a generator. Then $\mu^*(s)=\alpha(s\times 1)+\beta(1\times s)$ by definition of (α,β), hence

$$0=\mu^*(s\smile s)=\mu^*(s)^2=[\alpha(s\times 1)+\beta(1\times s)]^2$$
$$=\alpha^2(s^2\times 1)+\beta^2(1\times s^2)+2\alpha\beta(s\times s)=2\alpha\beta(s\times s). \quad ∎$$

Note that even-dimensionality of s was used in order to get $(1\times s)\smile(s\times 1)=s\times s$.

The proof of 10.1 is purely algebraic: The ring $H^*\mathbb{S}^{2n}$ admits no multiplicative homomorphism

$$H^*\mathbb{S}^{2n}\to H^*\mathbb{S}^{2n}\otimes H^*\mathbb{S}^{2n}\cong H^*(\mathbb{S}^{2n}\times\mathbb{S}^{2n})$$

such that $s\mapsto s\otimes 1+1\otimes s$. The question arises therefore which algebraic conditions on H^*X are imposed by the existence of a multiplication μ: $X\times X\to X$ with unit. Assume X is pathwise connected and $H^*(X\times X)\cong H^*X\otimes H^*X$ (cf. 8.18). Then μ^*: $H^*X\to H^*X\otimes H^*X$ is a homomorphism of algebras such that $\mu^*(x)=x\otimes 1+1\otimes x+r$, where

$$r\in\bigoplus_{i,j>0}H^iX\otimes H^jX, \qquad \text{for } |x|>0.$$

Which graded algebras A admit such maps $A\to A\otimes A$? For connected commutative graded algebras over a field R (perfect, if $\operatorname{char}(R)>0$) this question has been completely solved by Hopf-Leray-Borel (cf. Milnor-Moore, and 10.17 Exerc. 5). We now discuss the problem but give full details only if R is of characteristic zero (see 10.16).

10.2 Definition. We consider connected graded R-algebras A (R a commutative ring with unit 1). *A* graded algebra A is called *connected* if $A_i=0$ for $i<0$, $A_0\cong R$. For instance, the cohomology algebra $H^*(X;R)$ of a pathwise connected space X is connected.

Let μ: $A\otimes A\to A$ denote the multiplication, $\mu(a\otimes a')=aa'$. Define graded submodules $D^nA\subset A$, $n=0,1,\ldots$, as follows: $D^0A=A$; $(D^1A)_j=A_j$ if $j>0$, $(D^1A)_j=0$ if $j\le 0$; $D^{n+1}A=\operatorname{im}(D^nA\otimes D^1A\xrightarrow{\mu}A)$, $n\ge 0$. Clearly $D^nA\supset D^{n+1}A$, and $(D^nA)_k=0$ if $k<n$. The elements of D^2A are often

called *decomposable*. The elements of $D^n A$ might then be called $(n-1)$-*times decomposable*; they are linear combinations of elements $a_1 a_2 \dots a_n$ with $|a_i|>0$.

We shall also consider the modules $\Theta^n A = D^n A/D^{n+1} A$. Most of the time we shall assume that $D^{n+1} A$ is a direct summand of $D^n A$ for all n; in that case we say A is a *split-algebra*. (If R is a field then all connected R-algebras are split-algebras.) For split-algebras

$$D^n A \cong \Theta^n A \oplus D^{n+1} A \cong \oplus_{\nu \geq n} \Theta^\nu A. \tag{10.3}$$

D^n and Θ^n may be viewed as functors of connected algebras, i.e.

10.4 Proposition. *If $h: A \to A'$ is a homomorphism of connected algebras then $h(D^n A) \subset D^n A'$. Therefore we have induced homomorphisms*

$$D^n h: D^n A \to D^n A' \quad \text{and} \quad \Theta^n h: \Theta^n A \to \Theta^n A'. \quad \blacksquare$$

The module $\Theta^1 A = D^1 A/D^2 A$ will play a special role in the following. We first show that it can be thought of as "generating the algebra A".

10.5 Lemma. *If $M \subset D^1 A$ is a submodule which maps epimorphically onto $\Theta^1 A = D^1 A/D^2 A$ then M generates the algebra A.*

Proof. Fix an integer k. By decreasing induction on n we show that $(D^n A)_k$ is contained in the subalgebra $\{M\}$ which M generates. Since $(D^n A)_k = 0$ for $n > k$ we have a start. Consider a generator $a_1 a_2 \dots a_n$ of $(D^n A)_k$, where $a_i \in D^1 A$. By assumption, $a_i = m_i + b_i$ with $m_i \in M$, $b_i \in D^2 A$, hence $a_1 a_2 \dots a_n = m_1 m_2 \dots m_n + b$, where $b \in D^{n+1} A$. But $m_1 m_2 \dots m_n$ lies in the subalgebra which M generates, and so does b, by inductive hypothesis. ∎

10.6 Corollary. *If $h: A' \to A$ is a homomorphism of connected algebras such that $\Theta^1 h: \Theta^1 A' \to \Theta^1 A$ is epimorphic then h is epimorphic.*

Just apply 10.5 to $M = h D^1 A'$. ∎

10.7 Proposition. *If A, B are connected split-algebras then $A \otimes B = A \otimes_R B$ is also a connected split-algebra. Further*

$$D^n(A \otimes B) = \textstyle\sum_{0 \leq i \leq n} D^i A \otimes D^{n-i} B, \tag{10.8}$$

and

$$\Theta^n(A \otimes B) = \oplus_{0 \leq i \leq n} \Theta^i A \otimes \Theta^{n-i} B. \tag{10.9}$$

Proof. Connectedness is clear. If $|a\otimes b|>0$ then $|a|>0$ or $|b|>0$, hence $D^1(A\otimes B)=D^1 A\otimes B+A\otimes D^1 B$. Now proceed by induction on n. Suppose $n\geq 1$. Then

$$\begin{aligned} D^{n+1}(A\otimes B) &= \operatorname{im}[D^n(A\otimes B)\otimes D^1(A\otimes B)\to A\otimes B] \\ &= \operatorname{im}[(\textstyle\sum_i D^i A\otimes D^{n-i} B)\otimes(D^1 A\otimes B+A\otimes D^1 B)\to A\otimes B] \\ &= \textstyle\sum_i D^{i+1} A\otimes D^{n-i} B+\sum_i D^i A\otimes D^{n-i+1} B, \end{aligned}$$

which proves 10.8.

If A, B are split-algebras then (by 10.8 and 10.3)

$$\begin{aligned} D^n(A\otimes B) &= \textstyle\sum_i[(\bigoplus_{\nu\geq i}\Theta^\nu A)\otimes(\bigoplus_{\rho\geq n-i}\Theta^\rho B)] \\ &= \bigoplus_{\nu+\rho\geq n}(\Theta^\nu A\otimes\Theta^\rho B) \\ &= \bigoplus_{\nu+\rho=n}(\Theta^\nu A\otimes\Theta^\rho B)\oplus D^{n+1}(A\otimes B), \end{aligned}$$

which proves the rest of the proposition. ∎

10.10 Definition. Let A be a connected algebra. A *diagonal* is an algebra homomorphism $\psi\colon A\to A\otimes A$ such that

$$\psi(a)=a\otimes 1+1\otimes a+r \quad \text{with } r\in D^1 A\otimes D^1 A, \quad \text{for all } a\in D^1 A.$$

Our problem is (see text after 10.1): Which algebras admit a diagonal? For instance, if a topological space X admits a multiplication with unit (an h-space structure) then H^*X admits a diagonal (provided $H^*(X\times X)\cong H^*X\otimes H^*X$).

A connected algebra A together with a diagonal ψ is called a *hopf-algebra;* it is a *Hopf-algebra* if the diagonal is associative, i.e. if the two compositions

$$A\xrightarrow{\psi}A\otimes A\xrightarrow{\mathrm{id}\otimes\psi,\ \psi\otimes\mathrm{id}}A\otimes A\otimes A$$

agree. As remarked above, the cohomology algebra of a pathwise connected h- resp. H-space X is a hopf- resp. Hopf-algebra (provided $H^*(X\times X)\cong H^*X\otimes H^*X$). Also, the Pontrjagin-algebra H_*X of an H-space is a Hopf-algebra: the geometric diagonal $X\to X\times X$ induces an algebraic diagonal $H_*X\to H_*X\otimes H_*X$ (if the latter equals $H_*(X\times X)$). These two Hopf-algebras are related by duality (cf. Exerc. 3).

10.11 Lemma. *Let $h\colon A'\to A$ be a homomorphism of connected split-algebras. Assume A' is commutative and let $\psi\colon A\to A\otimes A$ be a diagonal* (note the symmetry of this assumption: $A'\otimes A'\xrightarrow{\mu'}A'$ and $A\xrightarrow{\psi}A\otimes A$ have to be algebra homomorphisms). *If $\Theta^i h\colon\Theta^i A'\cong\Theta^i A$, $\Theta^k h\colon\Theta^k A'\cong$*

$\Theta^k A$ for some i, k, then the composition ($n=i+k$)

$$\text{(10.12)}\quad \begin{aligned} &\Theta^n A' \xrightarrow{\Theta^n h} \Theta^n A \xrightarrow{\Theta^n \psi} \Theta^n(A\otimes A) \xrightarrow{\text{proj}} \Theta^i A \otimes \Theta^k A \\ &\xrightarrow{(\Theta^i h \otimes \Theta^k h)^{-1}} \Theta^i A' \otimes \Theta^k A' \xrightarrow{\text{inj}} \Theta^n(A'\otimes A') \xrightarrow{\Theta^n(\mu')} \Theta^n A' \end{aligned}$$

multiplies every element of $\Theta^n A'$ with the binominal coefficient $\binom{n}{i}$, i.e. 10.12 agrees with $\binom{n}{i}$ id.

The special case $h=\text{id}$ yields the following

10.13 Corollary. *If A is a connected commutative split-algebra and $\psi: A\to A\otimes A$ is a diagonal then the composition*

$$\Theta^n A \xrightarrow{\Theta^n\psi} \Theta^n(A\otimes A) \xrightarrow{\text{proj}} \Theta^i A\otimes \Theta^{n-i} A \xrightarrow{\text{inj}} \Theta^n(A\otimes A) \xrightarrow{\Theta^n \mu} \Theta^n A$$

equals $\binom{n}{i}$ id.

Proof of 10.11. It suffices to consider generators $a_1' a_2' \dots a_n' \in \Theta^n A'$, where $a_\nu' \in \Theta^1 A' \subset D^1 A'$. Let $a_\nu = h a_\nu'$. We have $\psi a_\nu = a_\nu \otimes 1 + 1\otimes a_\nu + r_\nu$ with $r_\nu \in D^2(A\otimes A)$, hence

$$\psi h(a_1' a_2' \dots a_n') = \psi(a_1 a_2 \dots a_n) = \textstyle\prod_\nu (a_\nu \otimes 1 + 1\otimes a_\nu) + r$$

with $r\in D^{n+1}(A\otimes A)$. The component of $\prod_\nu (a_\nu\otimes 1 + 1\otimes a_\nu)$ in $\Theta^i A\otimes \Theta^k A$ is $a=\sum \pm a_{\nu_1}\dots a_{\nu_i}\otimes a_{\rho_1}\dots a_{\rho_k}$; the sum extends over all i-tuples $\{\nu\}$ such that $1\le \nu_1<\nu_2<\dots<\nu_i\le n$, and $\{\rho_1,\dots,\rho_k\}$ is the complement of $\{\nu_1\dots\nu_i\}$ in $\{1,\dots,n\}$; the signs $\pm$ are caused by the commutation law $(1\otimes a_\rho)(a_\nu\otimes 1) = (-1)^{|a_\rho||a_\nu|} a_\nu \otimes a_\rho$.

Consider then the corresponding expression

$$a' = \textstyle\sum \pm a_{\nu_1}' \dots a_{\nu_i}' \otimes a_{\rho_1}' \dots a_{\rho_k}' \quad \text{in } \Theta^i A' \otimes \Theta^k A'.$$

Clearly $(\Theta^i h\otimes \Theta^k h) a' = a = \text{proj}\circ\psi\circ h(a_1'\dots a_n')$. But if we apply the multiplication ν' to a', each summand goes into $a_1'\dots a_n'$ (the signs disappear when we reverse the permutation), and the number of summands is $\binom{n}{i}$. ∎

10.14 Proposition. *Let $h: A'\to A$ be a homomorphism of connected split-algebras, such that $\Theta^1 h: \Theta^1 A' \cong \Theta^1 A$. If A' is torsionfree (as abelian group) and commutative (as graded algebra), and if A admits a diagonal then h is an isomorphism.*

Proof. By 10.6, h is epimorphic. It follows that $\Theta^n h$: $\Theta^n A' \to \Theta^n A$ is also epimorphic, for all n (just look at generators $a_1 \dots a_n \in \Theta^n A$ with $a_i \in \Theta^1 A$). We show that it is also monomorphic. By induction we can assume $\Theta^{n-1} h$: $\Theta^{n-1} A' \cong \Theta^{n-1} A$. Then 10.11, with $i=n-1$, $k=1$, asserts that the composition 10.12 is monomorphic ($\Theta^n A'$ is torsion-free!). In particular, the first factor of this composition, $\Theta^n h$, is monomorphic.

By decreasing induction on n we now show that $(D^n h)_j$: $(D^n A')_j \to (D^n A)_j$ is isomorphic (j fixed); for $n=0$ this is then the theorem. We have a start because $(D^n A')_j = 0 = (D^n A)_j$ for $n>j$. The inductive step follows from the exact sequence $0 \to D^{n+1} A' \to D^n A' \to \Theta^n A' \to 0$ and the five lemma. ∎

10.15 Example (compare Chevalley Chap. V). For every graded set $M=(M_1, M_2, \dots)$ there is a *free commutative graded R-algebra FM generated by M*. It contains M, and it is characterized by the following *universal property*: If A is any graded commutative R-algebra, and f: $M \to A$ is a map of graded sets then there exists a unique homomorphism h: $FM \to A$ of graded algebras such that $h|M=f$. In particular, *FM always admits a diagonal* ψ: $FM \to FM \otimes FM$, defined by $\psi(m) = m \otimes 1 + 1 \otimes m$, for $m \in M$.

If $M_1 = M_3 = \dots = \emptyset$ then FM is the *polynomial algebra* generated by M; if $M_2 = M_4 = \dots = \emptyset$ (and $\frac{1}{2} \in R$) then FM is the *exterior algebra* generated by M. A general construction is as follows (we shall use FM only if R is a field of characteristic zero).

Let ΦM denote the graded R-module which in dimension n is freely generated by the set of all finite sequences $(x_1, x_2, \dots, x_r)$ of elements in M such that $\sum |x_j| = n$; in particular, $(\Phi M)_0$ is free on one generator, namely the empty sequence, which we denote by 1. If $(x_1, \dots, x_r)$, $(y_1, \dots, y_r)$ are two sequences which differ only by a permutation σ then their odd-dimensional terms also differ by a permutation $\bar{\sigma}$ only, and we put $(x_1, \dots, x_r) = \operatorname{sign}(\bar{\sigma})(y_1, \dots, y_r)$; in particular, $2(x_1, \dots, x_r) = 0$ if some odd-dimensional element x_j occurs twice in the sequence. The quotient-module of ΦM by these relations is denoted by FM; if 2 is invertible in R then FM is also a *free* R-module (we just annihilated some base elements of ΦM, and identified some others, up to sign).

We define the *product* of two sequences by writing one after the other: $(x_1, \dots, x_r) \cdot (y_1, \dots, y_s) = (x_1, \dots, x_r, y_1, \dots, y_s)$. This turns ΦM into a connected graded R-algebra with unit 1 ("free graded R-algebra generated by M"). One easily verifies that the products pass to the quotient FM, and turn FM into a *commutative* connected graded R-algebra. It contains M, the set of one-term sequences, and the coset $[x_1, \dots, x_r] \in FM$ of the sequence $(x_1, \dots, x_r) \in \Phi M$ agrees with the product $x_1 x_2 \dots x_r$. In particular, $\Theta^1 FM = D^1 FM / D^2 FM$ is *freely generated by* M. If f maps M into a graded commutative R-algebra A then $h[x_1, \dots, x_r] = f(x_1) f(x_2) \dots f(x_r)$ is the unique homomorphic extension of f, hence FM does indeed have the required universal property.

10.16 Proposition (Hopf-Leray). *Let A be a connected graded commutative algebra over a field R of characteristic zero. If A admits a diagonal then A is free; in fact, if M is a base of $\Theta^1 A$ then $A \cong FM$.*

In other words, over a field of characteristic zero the only commutative connected algebras which admit diagonals are the free ones. In particular, this determines the multiplicative structure of $H^*(X; \mathbb{Q})$ if X is a connected h-space (and all $H^i(X; \mathbb{Q})$ are finitely generated so that $H^*(X \times X) = H^* X \otimes H^* X$). It shows again that even-dimensional spheres are not h-spaces but it also excludes many other spaces, like $P_k \mathbb{C}$. In the (important) finite dimensional case, i.e. if $H^i(X; \mathbb{Q}) = 0$ for large i, $H^*(X; \mathbb{Q})$ can have no generators of even dimension (their powers would have to be non-zero), hence $H^*(X; \mathbb{Q})$ is an exterior algebra on odd-dimensional generators. This is the classical result of Hopf.

Proof of 10.16. Lift $M \subset \Theta^1 A = D^1 A / D^2 A$ back to $D^1 A \subset A$ and let $h\colon FM \to A$ be the algebra homomorphism which extends the inclusion $M \to A$. Clearly, $\Theta^1 h\colon \Theta^1 FM \cong \Theta^1 A$, hence $h\colon FM \cong A$ by 10.14. ∎

10.17 Exercises. *1.* The multiplication map $\mu\colon A \otimes A \to A$ of any graded algebra induces homomorphisms $\mu^{ik}\colon \Theta^i A \otimes \Theta^k A \to \Theta^{i+k} A$ which turn $((\Theta^n A)_j)_{n, j \in \mathbb{Z}}$ into a *bigraded algebra*, the bigraded algebra *associated with A*. Ignoring the n-gradation one defines a (simply) graded algebra ΘA by $\Theta_r A = \oplus_n (\Theta^n A)_r$. Show that ΘA is a split-algebra, and $\Theta(\Theta A) \cong \Theta A$. Further: If A is commutative and ΘA is free-commutative then $A \cong \Theta A$.

2. If A is a split-algebra and $\psi\colon A \to A \otimes A$ is a diagonal then

$$\Theta \psi\colon \Theta A \to \Theta(A \otimes A) = \Theta A \otimes \Theta A$$

is also a diagonal, which does not depend on ψ.

3. If A is a graded R-module such that every A_i is free and finitely generated, then $A^* = \operatorname{Hom}_R(A; R)$ has the same property, and $(A \otimes_R A)^* \cong A^* \otimes_R A^*$, $A^{**} \cong A$; this is well-known linear algebra. Suppose now A is also a Hopf-algebra, with multiplication $\mu\colon A \otimes A \to A$ and diagonal $\psi\colon A \to A \otimes A$. Then $\mu^*\colon A^* \to A^* \otimes A^*$ and $\psi^*\colon A^* \otimes A^* \to A^*$ are the diagonal and the multiplication of a Hopf-algebra structure on A^*, and $A^{**} \cong A$ as Hopf-algebras. One says, A, A^* are *dual Hopf-algebras*. Show that the Pontrjagin algebra and the cohomology algebra of a connected H-space are dual Hopf-algebras (provided they are free and finitely generated in each dimension).

4. If $n>0$ is a natural number then

$$\gcd\left\{\binom{n}{i}\middle|\, 0<i<n\right\}=\begin{cases}1 \text{ if } n \text{ is not a power of a prime,}\\ p \text{ if } n=p^r,\ p \text{ prime, } r>0.\end{cases}$$

Use this and 10.11 to show that the conclusion of 10.14 holds if torsion-freeness is weakened to: $\Theta^{p^r} A$ has no p-torsion.

5. Over a field R of characteristic $p>0$ there are connected commutative algebras A which admit diagonals but are not free. For instance, if A is generated by one element x then $A \cong R[x]/(x^N)$ for some N with $0<N\leq\infty$ $(x^\infty=0)$, and there is one candidate for a diagonal ψ, namely $\psi(x)=x\otimes 1+1\otimes x$. Show that this does define a diagonal if (and only if) one of the following conditions hold: (i) $|x|$ is odd, $N\leq 2$; (ii) $p=2$, $N=2^r, 0\leq r\leq\infty$; (iii) $|x|$ is even, $N=p^r, 0\leq r\leq\infty$.

A theorem by A. Borel (cf. Milnor-Moore, 7.11) then asserts that over a *perfect* field R of characteristic $p>0$ the only connected commutative algebras A which admit diagonals (and satisfy $\dim_R(A_j)<\infty$) are multiple tensor-products of algebras on one generator as above.

6. Let R be a field of characteristic $p>0$ which is *not* perfect; pick $\rho\in R$ which is *not* of the form λ^p with $\lambda\in R$. Let x, y be two-dimensional indeterminates and put $A=R[x,y]/(x^p+\rho y^p)$. Show that A admits a diagonal but is not a tensor-product of algebras on one generator.

11. The Cohomology Slant Product

This product contains somewhat more information than the exterior cohomology product (VII, 7) but is often less convenient to deal with. Algebraically it is based (in the simplest case where all coefficients are in R) on the natural map $\mathrm{id}\otimes\colon D^*=\mathrm{Hom}(D,R)\to\mathrm{Hom}(C\otimes D, C)$, or rather on its adjoint $D^*\otimes(C\otimes D)\to C$, whereas the exterior cohomology product was based on $C^*\otimes D^* \xrightarrow{\gamma} (C\otimes D)^*$ which is the adjoint of the composite $D^* \xrightarrow{\mathrm{id}\otimes} \mathrm{Hom}(C\otimes D, C) \xrightarrow{*} \mathrm{Hom}(C^*, (C\otimes D)^*)$.

11.1 Definition. Let C, D be R-complexes, and L, M modules over R. Consider the composite chain map

$$E\colon \mathrm{Hom}(D,M)\otimes(C\otimes D\otimes L)\cong(C\otimes L)\otimes(\mathrm{Hom}(D,M)\otimes D)\xrightarrow{\mathrm{id}\otimes e} C\otimes L\otimes M,$$

where ω permutes factors and e is the evaluation map 1.2. Explicitely,

$$E(\psi \otimes c \otimes d \otimes l) = (-1)^{|\psi||c|}\, c \otimes l \otimes \psi(d). \tag{11.2}$$

Passage to homology and composition with α (cf. VI, 9.11) gives

$$H^*(D, M) \otimes H(C \otimes D \otimes L) \xrightarrow{E_*\alpha} H(C \otimes L \otimes M); \tag{11.3}$$

with dimension indices,

$$H^i(D, M) \otimes H_n(C \otimes D \otimes L) \to H_{n-i}(C \otimes L \otimes M). \tag{11.3'}$$

This map or the corresponding bilinear map is called the *cohomology slant product (for complexes)*. We write

(11.4) $y \diagdown \zeta = E_* \alpha(y \otimes \zeta) \in H_{n-i}(C \otimes L \otimes M)$, for $y \in H^i(D, M)$, $\zeta \in H_n(C \otimes D \otimes L)$.

The *cohomology slant product for spaces* (X, A), (Y, B) is obtained by taking

$$C = S(X, A; R) = \frac{S(X; R)}{S(A; R)}, \qquad D = S(Y, B; R),$$

and replacing $S(X, A; R) \otimes S(Y, B; R)$ by the homotopy equivalent complex

$$S(X \times Y, A \times Y \cup X \times B; R) \simeq \frac{S(X \times Y; R)}{S\{A \times Y, X \times B; R\}} \overset{EZ}{\simeq} S(X, A; R) \otimes S(Y, B; R).$$

We have, of course, to assume that $(X \times Y; A \times Y, X \times B)$ is an *excisive triad*. Under this assumption the cohomology slant product is then a homomorphism

$$H^i(Y, B; M) \otimes H_n(X \times Y, A \times Y \cup X \times B; L) \to H_{n-i}(X, A; L \otimes M). \tag{11.5}$$

As in the case of complexes, we write $(y \diagdown \zeta) \in H_{n-i}(X, A)$ for the slant product of $y \in H^i(Y, B)$ with $\zeta \in H_n(X \times Y, A \times Y \cup X \times B)$. In terms of representative relative (co-)cycles $\psi \in S^i(Y; M)$, $z \in S(X \times Y; L)$ we have

$$[\psi] \diagdown [z] = (-1)^{|\psi|(|z| - |\psi|)} \left[\textstyle\sum_\nu a_\nu \otimes \psi(b_\nu)\right], \tag{11.6}$$

where $(EZ)(z) = \sum_\nu a_\nu \otimes b_\nu$, $a_\nu \in S(X; L)$, $b_\nu \in S(Y; R)$. When applying this formula one has to be careful to choose the representative z in such a fashion that

$$\partial z \in S\{A \times Y, X \times B; L\} \text{ — and not only } \partial z \in S(A \times Y \cup X \times B; L).$$

If in 11.3, $C=(R,0)$, or *if in* 11.5, X *is a point and* $A=\emptyset$ *then the cohomology slant product reduces to the scalar product of* VII, 1.

The main formal properties of $\diagdown$ are as follows (coefficients omitted).

11.7 Naturality. If $f\colon (X,A)\to(X',A')$, $g\colon (Y,B)\to(Y',B')$ are maps of pairs as in 11.5 then

$$f_*(g^*y'\diagdown\zeta)=y'\diagdown(f\times g)_*\zeta,$$

for

$$y'\in H^*(Y',B'),\quad \zeta\in H(X\times Y, A\times Y\cup X\times B).$$

11.8 Associativity. $(x\times y)\diagdown\gamma=x\diagdown(y\diagdown\gamma)$, for $x\in H^*(X,A)$, $y\in H^*(Y,B)$, $\gamma\in H[(W,U)\times(X,A)\times(Y,B)]$. In particular, if W is a point, and $U=\emptyset$, this becomes

11.9 Duality. $\langle x\times y,\zeta\rangle=\langle x, y\diagdown\zeta\rangle$, for $x\in H^*(X,A)$, $y\in H^*(Y,B)$, $\zeta\in H(X\times Y, A\times Y\cup X\times B)$.

11.10 Units. $1_Y\diagdown\zeta=p_*\zeta$, where $1_Y\in H^0(Y;R)$, $\zeta\in H(X\times Y, A\times Y)$, and $p\colon (X\times Y, A\times Y)\to(X,A)$ is the projection.

11.11 Stability. *The following diagrams are commutative,*

$$\begin{array}{ccc}
H^*(Y,B)\otimes H(X\times Y, A\times Y\cup X\times B) & \xrightarrow{\quad\diagdown\quad} & H(X,A)\\
\Big\downarrow{\scriptstyle (-1)^{\dim}\,\mathrm{id}\otimes\partial_*} & & \Big\downarrow{\scriptstyle\partial_*}\\
H^*(Y,B)\otimes H(A\times Y\cup X\times B, X\times B) \overset{\mathrm{id}\otimes j_*}{\cong} H^*(Y,B)\otimes H(A\times Y, A\times B) & \longrightarrow & HA,
\end{array}\tag{11.12}$$

$$\begin{array}{ccc}
H^*B\otimes H(X\times Y, A\times Y\cup X\times B) & \xrightarrow{\delta^*\otimes\mathrm{id}} & H^*(Y,B)\otimes H(X\times Y, A\times Y\cup X\times B)\\
\Big\downarrow{\scriptstyle -(-1)^{\dim}\,\mathrm{id}\otimes\partial_*} & & \Big\downarrow{\scriptstyle\diagdown}\\
H^*B\otimes H(A\times Y\cup X\times B, A\times Y) \overset{\mathrm{id}\otimes j_*}{\cong} H^*B\otimes H(X\times B, A\times B) & \xrightarrow{\ \diagdown\ } & H(X,A).
\end{array}\tag{11.13}$$

where j denotes inclusion maps. In formulas,

$$\partial_*(y\diagdown\zeta)=(-1)^{|y|}\,y\diagdown j_*^{-1}\partial_*\zeta,\qquad \text{if } y\in H^*(Y,B),\ \zeta\in H(X\times Y, A\times Y\cup X\times B);\tag{11.14}$$

$$(\delta^*b)\diagdown\zeta+(-1)^{|b|}\,b\diagdown j_*^{-1}\partial_*\zeta=0,\qquad \text{if } b\in H^*B,\quad \zeta\in H(X\times Y, A\times Y\cup X\times B).\tag{11.15}$$

Note that $j_*=\mathrm{id}$ if $B=\emptyset$ in 11.14, or $A=\emptyset$ in 11.15.

11.16 Multiplicativity. $y \diagdown \omega \times \zeta = (-1)^{|y||\omega|} \omega \times (y \diagdown \zeta)$, if $y \in H^*(Y, B)$, $\omega \in H(W, U)$, $\zeta \in H(X \times Y, A \times Y \cup X \times B)$, and (W, U), (X, A), (Y, B) are pairs of spaces such that the products above are defined. In particular, if X is a point, and $A = \emptyset$, this becomes

$$y \diagdown \omega \times \eta = (-1)^{|y||\omega|} \omega \otimes \langle y, \eta \rangle, \tag{11.17}$$

for $\omega \in H(W, U)$, $y \in H^*(Y, B)$, $\eta \in H(Y, B)$.

Proof of 11.7. It is enough to consider complexes (instead of spaces) because EZ-maps are natural. Let then $f\colon C \to C'$, $g\colon D \to D'$ be chain maps. In the notation of 11.2 (L omitted) we have

$$fE(\psi' g \otimes c \otimes d) = (-1)^{|\psi'||c|} f(c) \otimes \psi' g(d) = E(\psi' \otimes fc \otimes gd),$$

i.e.

$$f \circ E \circ (g^+ \otimes \mathrm{id}) = E \circ [\mathrm{id} \otimes (f \otimes g)]\colon \ \mathrm{Hom}(D', M) \otimes (C \otimes D) \to C' \otimes M,$$

where $g^+ = \mathrm{Hom}(g, \mathrm{id}_{C \otimes D})$. Passage to homology and composition with $\alpha\colon H^*(D', M) \otimes H(C \otimes D) \to H[\mathrm{Hom}(D', M) \otimes C \otimes D]$ gives

$$f_* \circ (E_* \circ \alpha) \circ (g^* \otimes \mathrm{id}) = (E_* \circ \alpha) \circ (\mathrm{id} \otimes (f \otimes g)_*) \quad \text{(by naturality of } \alpha);$$

and applying this to $y' \otimes \zeta$ gives 11.7.

Proof of 11.8. As in the proof of 11.7 it is enough to consider complexes B, C, D (instead of spaces) because EZ-maps are (homotopy-) associative. Consider the diagram

$$\begin{array}{ccc} \mathrm{Hom}(C, M) \otimes \mathrm{Hom}(D, N) \otimes (B \otimes C \otimes D) & \xrightarrow{\mathrm{id} \otimes E} & \mathrm{Hom}(C, M) \otimes B \otimes C \otimes N \\ \downarrow{\scriptstyle \gamma \otimes \mathrm{id}} & & \downarrow{\scriptstyle E} \\ \mathrm{Hom}(C \otimes D, M \otimes N) \otimes (B \otimes (C \otimes D)) & \xrightarrow{E} & B \otimes M \otimes N, \end{array}$$

where γ is as in 7.2. A generator $\varphi \otimes \psi \otimes b \otimes c \otimes d$ in the upper left maps into $(-1)^{|\psi||b| + |\psi||c| + |\varphi||b|} b \otimes \varphi(c) \otimes \psi(d)$ on either way; the diagram is therefore commutative. Now pass to homology and apply the resulting equality to $\alpha(x \otimes y \otimes \zeta)$.

Proof of 11.10. Let $f = \mathrm{id}\colon (X, A) \to (X, A)$, and $g\colon Y \to P$ where P is a point. Then

$$1_Y \diagdown \zeta = f_*(g^*(1_P) \diagdown \zeta) = 1_P \diagdown (f \times g)_* \zeta = 1_P \diagdown (p_* \zeta \times 1^P) = p_* \zeta;$$

the 2nd equality by 11.7, the third by 2.10, the last by the very definition of slant-products (cf. 11.6).

Proof of 11.11. Consider 11.14 first. Choose representatives ψ of y and z of ζ as in 11.6. Then $\psi|SB=0$, $\delta\psi=0$, and $\partial z=\alpha+\beta$ where $\alpha\in S(A\times Y)$, $\beta\in S(X\times B)$. We have

$$\partial E(\psi\otimes EZ(z))=(-1)^{|\psi|}E(\psi\otimes EZ(\partial z))=(-1)^{|\psi|}E(\psi\otimes EZ(\alpha)),$$

the first equation because $\delta\psi=0$, the second because $\psi|SB=0$. But the outer terms of this represent the two sides of 11.14, or the two ways of moving $y\otimes z$ from the upper left corner of 11.12 to the lower right.

For 11.15 we choose a representative of b first and extend it to a cochain ψ in Y; then $\psi|SB$ represents b and $\delta\psi$ represents $\delta^* b$. We have

$$\partial E(\psi\otimes EZ(z))=E(\delta\psi\otimes EZ(z))+(-1)^{|\psi|}E(\psi\otimes EZ(\partial z)),$$

or

$$E(\delta\psi\otimes EZ(z))+(-1)^{|\psi|}E(\psi\otimes EZ(\beta))$$
$$=\partial E(\psi\otimes EZ(z))-(-1)^{|\psi|}E(\psi\otimes EZ(\alpha)).$$

The left side represents the sum in 11.15, the right side represents zero in $H(X,A)$ because $E(\psi\otimes EZ(\alpha))\in SA$.

Proof of 11.16. As before it is enough to consider complexes $B(=SW/SU)$, $C(=SX/SA)$, $D(=SY/SB)$. Let $\psi, u, z=\sum a_\nu\otimes b_\nu$ be representatives of y,ω,ζ. Then $\omega\times\zeta$, $y\diagdown(\omega\times\zeta)$, $\omega\times(y\diagdown\zeta)$ have the following representatives:

$$\sum(u\otimes a_\nu)\otimes b_\nu,\quad (-1)^{|\psi||u|}\sum(-1)^{|\psi||a_\nu|}(u\otimes a_\nu)\otimes\psi(b_\nu),$$
$$u\otimes\sum(-1)^{|\psi||a_\nu|}a_\nu\otimes\psi(b_\nu).\quad \blacksquare$$

11.17 Exercises. *1.* Show that $\diagdown\colon H^*(\mathbb{R}^n,\mathbb{R}^n-0)\otimes H(\mathbb{R}^{m+n},\mathbb{R}^{m+n}-0)\to H(\mathbb{R}^m,\mathbb{R}^m-0)$ is isomorphic—if one identifies

$$(\mathbb{R}^{m+n},\mathbb{R}^{m+n}-0)=(\mathbb{R}^m,\mathbb{R}^m-0)\times(\mathbb{R}^n,\mathbb{R}^n-0).$$

2. Define $\sigma\colon H(X\times Y)\to \operatorname{Hom}(H^*X,HX)$ by $(\sigma\zeta)y=y\diagdown\zeta$ (or better: $(-1)^{|y||\zeta|}y\diagdown\zeta$). Show that under suitable finiteness conditions on HX and HY (compare 7.6; assume R is a principal ideal domain) there is a split exact sequence

$$0\to\operatorname{Ext}(H^*Y,HX)^-\to H(X\times Y)\xrightarrow{\sigma}\operatorname{Hom}(H^*Y,HX)\to 0.$$

As in 7.16 Exerc. 2, this provides two possibilities of expressing $H(X\times Y)$ in terms of HX, HY, and yields algebraic relations between $\otimes$, $*$, Hom, and Ext.

3. What is the analogue of 7.16 for $\smallsetminus$-products?

4*. Let $K \subset V \subset \mathbb{R}^n$, where V is open, K compact; let $i\colon K \to V$ denote the inclusion-, and $\Delta\colon (V, V-K) \to (V, V-K) \times V$ the diagonal map. Show that

$$(y \smallsetminus \Delta_* o_K) \circ \zeta = \pm \langle y, i_* \zeta \rangle o_0$$

if $o_K \in H_n(V, V-K)$, $o_0 \in H_n(\mathbb{R}^n, \mathbb{R}^n - 0)$ are fundamental classes (cf. 2.14), $y \in H^* V$, $\zeta \in HK$, and $\circ$ is the intersection number of VII, 4. Hint: Use 6.13 as in the proof of 6.24.

12. The Cap-Product ($\frown$-Product)

This product, $\frown$, is related to the $\smallsetminus$-product as $\smile$ is to $\times$ (or to $\smile$, as $\smallsetminus$ is to $\times$). Roughly speaking then, § 12 is obtained from the preceding § 11 by putting $Y = X$, and replacing EZ by D (hence $X \times Y$ by X), $\smallsetminus$ by $\frown$. We shall perform this transcription for the definitions and propositions but we shall omit most of the proofs. An important property of $\frown$-products is that they make HX a graded $H^* X$-module; this extra structure in HX will be crucial in the study of manifolds (Chap. VIII).— As before, the ground ring is assumed to be commutative.

12.1 Definition. Let $(X; A_1, A_2)$ be an excisive triad, and let M_1, M_2 be R-modules. Consider the composite chain map

(12.2)
$$\begin{aligned}&\operatorname{Hom}\left(\frac{SX}{SA_2}, M_2\right) \otimes_R \left(\frac{SX}{S\{A_1, A_2\}} \otimes M_1\right) \xrightarrow{\mathrm{id} \otimes D} \\ &\operatorname{Hom}\left(\frac{SX}{SA_2}, M_2\right) \otimes_R \left(\frac{SX}{SA_1} \otimes \frac{SX}{SA_2} \otimes M_1\right) \xrightarrow{E} \frac{SX}{SA_1} \otimes M_1 \otimes_R M_2,\end{aligned}$$

where D is a natural diagonal (VI, 12.21), and E is essentially an evaluation (cf. 11.1). We pass to homology $\left(\text{using } H\dfrac{SX}{S\{A_1, A_2\}} \cong H(X, A_1 \cup A_2)\right)$, compose with α, and obtain

(12.3)
$$\begin{aligned}E_* (\mathrm{id} \otimes D)_* \alpha\colon H^k(X, A_2; M_2) \otimes H_n(X, A_1 \cup A_2; M_1) \\ \to H_{n-k}(X, A_1; M_1 \otimes_R M_2).\end{aligned}$$

This map or the corresponding bilinear map is called the *cap-product* (*$\frown$-product*). We write

$$(12.4) \qquad \begin{aligned} &x \frown \xi = E_*(\mathrm{id} \otimes D)_* \alpha(x \otimes \xi), \\ &\text{if } x \in H^*(X, A_2; M_2),\ \xi \in H(X, A_1 \cup A_2; M_1). \end{aligned}$$

In terms of representatives this reads

$$(12.5) \qquad [\varphi] \frown [c] = (-1)^{|\varphi|(|c| - |\varphi|)} \left[\textstyle\sum_v c_v^1 \otimes \varphi(c_v^2)\right],$$

where $Dc = \sum_v c_v^1 \otimes c_v^2$; the formula assumes $\varphi \in S^* X$, $\varphi | SA_2 = 0$, $\delta\varphi = 0$, $c \in SX$, $\partial c \in S\{A_1, A_2\}$.

The following properties 12.6–12.14 correspond to 11.7–11.15.

12.6 Naturality. $f_*((f^* x') \frown \xi) = x' \frown (f_* \xi)$, if $f\colon (X; A_1, A_2) \to (X'; A_1', A_2')$ is a map of excisive triads, $x' \in H^*(X', A_2')$, $\xi \in H(X, A_1 \cup A_2)$. ∎

12.7 Associativity. $(x_1 \smile x_2) \frown \xi = x_1 \frown (x_2 \frown \xi)$, if $x_i \in H^*(X, A_{i+1})$, $\xi \in H(X, A_1 \cup A_2 \cup A_3)$. ∎

12.8 Duality. $\langle x_1 \smile x_2, \xi \rangle = \langle x_1, x_2 \frown \xi \rangle$, if $x_i \in H^*(X, A_i)$, $\xi \in H(X, A_1 \cup A_2)$. ∎

In particular,

$$\langle 1, x \frown \xi \rangle = \langle x, \xi \rangle, \quad \text{for } x \in H^j(X, A),\ \xi \in H_j(X, A).$$

If X is path-connected, $x \frown \xi$ must be a multiple of $[P] \in H_0(X; R)$ where $P \in X$; the formula then implies $x \frown \xi = \langle x, \xi \rangle [P]$. More generally, this holds if only $X - A$ is contained in a path-component $\tilde{X}$ of X, and $P \in \tilde{X}$; it reduces to the connected case by excision $H(X, A) \cong H(\tilde{X}, \tilde{X} \cap A)$.

12.9 Units. $1 \frown \xi = \xi$, if $\xi \in H(X, A)$, and $1 \in H^0(X; R)$ is the augmentation class. ∎

12.10 Stability. The following diagrams are commutative,

$$(12.11) \qquad \begin{array}{ccccc} H^*(X, A_2) \otimes H(X, A_1 \cup A_2) & & \xrightarrow{\quad\frown\quad} & & H(X, A_1) \\ \Big\downarrow {\scriptstyle (-1)^{\dim} i^* \otimes \partial_*} & & & & \Big\downarrow {\scriptstyle \partial_*} \\ H^*(A_1, A_1 \cap A_2) \otimes H(A_1 \cup A_2, A_2) & \overset{\mathrm{id} \otimes j_*}{\cong} & H^*(A_1, A_1 \cap A_2) \otimes H(A_1, A_1 \cap A_2) & \xrightarrow{\frown} & H A_1, \end{array}$$

$$(12.12) \qquad \begin{array}{ccccc} H^* A_2 \otimes H(X, A_1 \cup A_2) & \xrightarrow{\delta^* \otimes \mathrm{id}} & H^*(X, A_2) \otimes H(X, A_1 \cup A_2) & \xrightarrow{\frown} & H(X, A_1) \\ \Big\downarrow {\scriptstyle -(-1)^{\dim} \mathrm{id} \otimes \partial_*} & & & & \Big\uparrow {\scriptstyle i_*} \\ H^* A_2 \otimes H(A_1 \cup A_2, A_1) & \overset{\mathrm{id} \otimes j_*}{\cong} & H^* A_2 \otimes H(A_2, A_1 \cap A_2) & \xrightarrow{\frown} & H(A_2, A_1 \cap A_2), \end{array}$$

where i, j denote inclusion maps. In formulas,

(12.13) $$\partial_*(x\frown\xi)=(-1)^{|x|}(i^*x)\frown(j_*^{-1}\partial_*\xi),$$
$$\text{if } x\in H^*(X,A_2),\ \xi\in H(X,A_1\cup A_2).$$

(12.14) $$(\delta^*a)\frown\xi+(-1)^{|a|}i_*(a\frown j_*^{-1}\partial_*\xi)=0,$$
$$\text{if } a\in H^*A_2,\ \xi\in H(X,A_1\cup A_2).\ \blacksquare$$

Note that $j_*=\mathrm{id}$ if $A_2=\emptyset$ in 12.13, or $A_1=\emptyset$ in 12.14.

The following two properties reflect the relation between Eilenberg-Zilber maps and natural diagonals.

(12.15) $$x\frown\xi=x\diagdown\Delta_*\xi,\quad \text{if } x\in H^*(X,A_2),\ \xi\in H(X,A_1\cup A_2),$$

and $\Delta\colon (X,A_1\cup A_2)\to(X\times X, A_1\times X\cup X\times A_2)$ is the diagonal map, $\Delta P=(P,P)$; we have to assume that both $(X;A_1,A_2)$ and $(X\times X;$ $A_1\times X, X\times A_2)$ are excisive. The proof is immediate from 12.5 and 11.6; if φ, c are representatives of x,ξ, and $Dc=\sum c_\nu^1\otimes c_\nu^2$, then both sides of 12.15 are represented by $(-1)^{|\varphi|(|c|-|\varphi|)}\sum_\nu c_\nu^1\otimes\varphi(c_\nu^2)$, the right side because $D=EZ\circ\Delta$. ∎

(12.16) $$y\diagdown\zeta=p_*(q^*y\frown\zeta),\quad \text{if } y\in H^*(Y,B),\ \zeta\in H(X\times Y, A\times Y\cup X\times B),$$

and $p\colon(X\times Y, A\times Y)\to(X,A)$, $q\colon(X\times Y, X\times B)\to(Y,B)$ are the projections; we have to assume that $(X\times Y; A\times Y, X\times B)$ is excisive. *Proof:* If ψ, z are representatives of y,ζ, and $Dz=\sum_\nu z_\nu^1\otimes z_\nu^2$, then the right side of 12.16 is $\pm p\sum z_\nu^1\otimes\psi\, q z_\nu^2$. By VI, 12.25, we have $EZ(z)=\sum(p z_\nu^1)\otimes(q z_\nu^2)$; therefore the left side of 12.16 is $\pm\sum(p z_\nu^1)\otimes\psi(q z_\nu^2)$. Clearly, the two expressions agree. ∎

As a consequence of 12.15 we find

12.17 Multiplicativity. $(x\times y)\frown(\xi\times\eta)=(-1)^{|y||\xi|}(x\frown\xi)\times(y\frown\eta)$, if $x\in H^*(X,A_2)$, $y\in H^*(Y,B_2)$, $\xi\in H(X,A_1\cup A_2)$, $\eta\in H(Y,B_1\cup B_2)$, and $(X;A_1,A_2)$, $(Y;B_1,B_2)$ are triads such that the products above are defined. Actually, our proof also assumes that products like $x\times y\diagdown\Delta_*(\xi\times\eta)$ are defined which (perhaps) requires further excision assumptions. We don't formulate these; they *are* satisfied if A_1, A_2, B_1, B_2 are open subsets, or if at most one of them is non-empty, or if $(X;A_1,A_2)$, $(Y;B_1,B_2)$ are CW-triads. Also, we indicate a general (and quite different) proof in Exerc. 4.

Proof of 12.17. Consider the diagram

$$(12.18)\qquad \begin{array}{ccccccc}
H(X\times X\times Y\times Y) & \xrightarrow{y\diagdown} & H(X\times X\times Y) & \xrightarrow{\tau_*} & H(Y\times X\times X) & \xrightarrow{x\diagdown} & H(Y\times X)\\
\downarrow{\scriptstyle (\mathrm{id}\times t\times \mathrm{id})_*} & & {\scriptstyle (\mathrm{id}\times t)_*}\searrow & & \swarrow{\scriptstyle (t'\times \mathrm{id})_*} & & \downarrow{\scriptstyle t'_*}\\
H(X\times Y\times X\times Y) & & \xrightarrow{y\diagdown} & H(X\times Y\times X) & \xrightarrow{x\diagdown} & & H(X\times Y),
\end{array}$$

where t, t', τ are maps which permute factors

$$\big(t(P,Q)=(P,Q),\ t'(Q,P)=(P,Q),\ \tau(P,P',Q)=(Q,P,P')\big);$$

for simplicity's sake we omitted all subspaces modulo which the homology groups have to be taken. The middle triangle of 12.18 is obviously commutative, the outside squares are commutative by naturality 11.7 of slant-products. Consider then the element $(\Delta^X\times\Delta^Y)_*(\xi\times\eta)=(\Delta^X_*\xi)\times(\Delta^Y_*\eta)$ in the upper left group $H(X\times X\times Y\times Y)$. Going down takes it into $\Delta^{X\times Y}_*(\xi\times\eta)$, going right then gives

$$x\diagdown y\diagdown\Delta_*(\xi\times\eta)\overset{11.8}{=}(x\times y)\diagdown\Delta_*(\xi\times\eta)\overset{12.15}{=}(x\times y)\frown(\xi\times\eta).$$

If we go first right and then down we get successively:

$$\begin{aligned}
y\diagdown(\Delta^X_*\xi)\times(\Delta^Y_*\eta)&\overset{11.16}{=}\pm(\Delta^X_*\xi)\times(y\diagdown\Delta^Y_*\eta)\overset{12.15}{=}\pm(\Delta^X_*\xi)\times(y\frown\eta)\\
&\overset{\tau_*}{\longmapsto}\pm(y\frown\eta)\times(\Delta^X_*\xi)\overset{x\diagdown}{\longmapsto}\pm x\diagdown(y\frown\eta)\times(\Delta^X_*\xi)\\
&\overset{11.16}{=}\pm(y\frown\eta)\times(x\diagdown\Delta^X_*\xi)\\
&\overset{12.15}{=}\pm(y\frown\eta)\times(x\frown\xi)\overset{t'_*}{\longmapsto}\pm(x\frown\xi)\times(y\frown\eta).
\end{aligned}$$

The sign which comes in is (-1) to the exponent

$$|y|\,|\xi|+|\xi|(|\eta|-|y|)+|x|(|\eta|-|y|)+(|\eta|-|y|)(|\xi|-|x|),$$

and this exponent is $\equiv|y|\,|\xi| \bmod 2$. ∎

12.19 Remark. If the coefficients for cohomology are $M_2=R$ then $M\otimes_R M_2=M$, and 12.3, 12.7, 12.9 assert that homology $H(X,A;M)$ is a graded $H^*(X;R)$-module. If $f\colon X\to X'$ is a map then $f^*\colon H^*(X';R)\to H^*(X;R)$ is a ring-homomorphism so that every $H^*(X;R)$-module becomes a $H^*(X';R)$-module; then 12.6 shows that f_* is a homomorphism of $H^*(X;R')$-modules. Similarly, 12.13 shows that $\partial_*\colon H(X,A;M)\to H(A;M)$ is an $H^*(X;R)$-homomorphism (of degree -1), and 12.17 asserts that the homology $\times$-product is a homomorphism of $H^*(X;R)\otimes_R H^*(Y;R)$-modules.

We conclude this section by a further (more difficult) stability formula. For simplicity's sake we make stronger assumptions than needed: we assume open subspaces whereas suitable excision conditions would suffice.

12.20 Proposition. *Let X_1, X_2, Y_1, Y_2 be open subsets of a space X such that $X_1 \cup Y_1 = X_2 \cup Y_2 = X_1 \cup X_2 = X$. Let*

$$x \in H^*(X_1 \cap X_2), \qquad \xi \in H(X, Y_1 \cap Y_2),$$

and let ξ' denote the image of ξ under the composite

$$H(X, Y_1 \cap Y_2) \to H(X, Y_1 \cup Y_2) \overset{j_*^{-1}}{\cong} H(X_1 \cap X_2, (X_1 \cap X_2) \cap (Y_1 \cup Y_2))$$

—coefficients omitted. Then

$$d_* j_* (x \frown \xi') = (d^* x) \frown \xi,$$

where

$$d_*\colon H(X, Y_1 \cup Y_2) \to H(X, Y_1 \cap Y_2)$$

and

$$d^*\colon H^*(X_1 \cap X_2) \to H^*(X_1 \cup X_2) = H^* X$$

are Mayer-Vietoris boundaries (cf. III, 8).

Proof. Consider the diagram

$$\begin{array}{ccccc}
H^*(X_1 \cap X_2) & \xrightarrow{\frown \xi'} & H(X_1 \cap X_2, (X_1 \cap X_2) \cap (Y_1 \cup Y_2)) & \xrightarrow{j_*} & H(Y, Y_1 \cup Y_2) \\
\downarrow{\scriptstyle \delta^*} & & & & \downarrow{\scriptstyle \partial_*} \\
H^*(X_1, X_1 \cap X_2) & & & & H(Y_1 \cup Y_2, Y_1) \\
\uparrow{\scriptstyle \cong} & & & & \uparrow{\scriptstyle \cong} \\
H^*(X_1 \cup X_2, X_2) & \longrightarrow & H^*(Y_2, X_2 \cap Y_2) & \xrightarrow{\frown \xi_1} & H(Y_2, Y_1 \cap Y_2) \\
\downarrow & & & & \downarrow \\
H^*(X_1 \cup X_2) & & \xrightarrow{\qquad \frown \xi \qquad} & & H(X, Y_1 \cap Y_2),
\end{array} \tag{12.21}$$

where all unmarked arrows are induced by inclusions. The composite columns are d^* resp. d_* (cf. III, 8.11); we have therefore to show that the outer diagram (without middle horizontal) is commutative. The element ξ_1 in the middle is the image of ξ under the composite

$$H(X, Y_1 \cap Y_2) \to H(X, (Y_1 \cap Y_2) \cup X_2) \overset{\text{exc}}{\cong} H(Y_2, (Y_1 \cap Y_2) \cup (X_2 \cap Y_2)).$$

Recall that $\frown$-products are induced by the chain map

$$S^* X \otimes SX \xrightarrow{\text{id} \otimes D} S^* X \otimes SX \otimes SX \xrightarrow{E} SX.$$

Let us denote this chain map by $\frown$, too, so that $[\varphi\frown z]=[\varphi]\frown[z]$ whenever φ, z are suitable relative (co-)cycles, and [ ] is passage to homology. Note further that $z\in SB$, where $B\subset X$, implies $\varphi\frown z\in SB$ for every $\varphi\in S^*X$; if, moreover, $\varphi|SB=0$ then $\varphi\frown z=0$. This will be used now.

We choose a representative cocycle φ of x and extend it to a cochain φ' on X; then $\delta\varphi'|SX_1$ represents $\delta^* x$. By excision, there exists a cocycle $\psi\in Z^*(X, X_2)$ such that $\psi|SX_1=\partial\varphi'|SX_1+\delta\psi'$, where $\psi'\in S^*(X_1, X_1\cap X_2)$; extend ψ' (by zero outside SX_1) to a cochain $\psi''\in S^*(X, X_2)$, and replace ψ by $\psi-\delta\psi''$; the new cocycle ψ then satisfies $\psi|SX_1=\delta\varphi'|SX_1$. Note that ψ represents the image of x in $H^*(X_1\cup X_2, X_2)$ and $H^*(X_1\cup X_2)$.

Because $X_1\cap Y_2$, $X_2\cap Y_1$, $X_1\cap X_2$ are open sets which cover X we can (cf. III, 7.3) find a representative μ of ξ such that $\mu=\mu_1+\mu_2+\mu'$ with $\mu_1\in S(X_1\cap Y_2)$, $\mu_2\in S(X_2\cap Y_1)$, $\mu'\in S(X_1\cap X_2)$ and, of course, $\partial\mu\in S(Y_1\cap Y_2)$. Then μ' represents ξ', and μ_1 represents ξ_1. It follows that the image of x in $H(Y_1\cup Y_2, Y_1)$ along the two ways of the upper part of 12.21 has the representatives $\psi\frown\mu_1$ resp. $\partial(\varphi\frown\mu')=(-1)^{|\varphi|}\varphi\frown\partial\mu'$. We show that these are in the same homology class, i.e. that the upper part of 12.21 commutes. Indeed, we have

$$\begin{aligned}\partial(\varphi'\frown\mu_1)&=\delta\varphi'\frown\mu_1+(-1)^{|\varphi|}\varphi'\frown\partial\mu_1\\&=\delta\varphi'\frown\mu_1+(-1)^{|\varphi|}\varphi'\frown\partial\mu-(-1)^{|\varphi|}\varphi'\frown\partial\mu_2-(-1)^{|\varphi|}\varphi'\frown\partial\mu',\end{aligned}$$

and

(i) $\delta\varphi'\frown\mu_1=\psi\frown\mu_1$ because $\mu_1\in SX_1$ and $\psi|SX_1=\delta\varphi'|SX_1$,
(ii) $\varphi'\frown\partial\mu'=\varphi\frown\partial\mu'$ because $\partial\mu'\in S(X_1\cap X_2)$ and $\varphi'|S(X_1\cap X_2)=\varphi$,
(iii) $\varphi'\frown\partial\mu-\varphi'\frown\partial\mu_2\in SY_1$ because $\partial\mu$ and μ_2 are in SY_1.

Hence, $\partial(\varphi'\frown\mu_1)=\psi\frown\mu_1-(-1)^{|\varphi|}\varphi\frown\partial\mu' \bmod SY_1$, as required.

It remains to prove commutativity in the lower part of 12.21. Let $\psi\in Z^*(X_1\cup X_2, X_2)$. Then $\psi\frown(\mu_2+\mu')=0$ because $\mu_2+\mu'\in SX_2$ and $\psi|SX_2=0$; hence, $\psi\frown\mu=\psi\frown\mu_1$. Passing to homology this gives $[\psi]\frown\xi=[\psi]\frown\xi'$, as required. ∎

12.22 Corollary. *Let V', W, W' be open subsets of a space X such that $V'\subset W'$ and $W\cup W'=X$. Then the following diagram*

$$\begin{array}{ccccc}H^*W & \xrightarrow{\delta^*} & H^*(X, W) & \longrightarrow & H^*(W', W\cap W')\\ \downarrow{\scriptstyle\frown\xi'} & & & & \downarrow{\scriptstyle\frown\xi_1}\\ H(W, W\cap W') & \longrightarrow & H(X, W') & \xrightarrow{\partial_*} & H(W', V')\end{array}\qquad(12.23)$$

commutes for every $\xi \in H(X, V')$, where ξ' resp. ξ_1 are the images of ξ in

$$H(X, W') \cong H(W, W \cap W') \text{ resp. } H(X, V' \cup W) \cong H(W', V' \cup (W \cap W')).$$

Indeed, this is just the upper part of 12.21 for $X_1 = X$, $X_2 = W$, $Y_1 = V'$, $Y_2 = W'$. ∎

But actually, an analysis of the proof shows *that we can take* $\xi \in H(X, A)$ *where A is any subset of* $V' \cup (W \cap W')$. For a better appreciation of this result, or of 12.22, the reader might consider the special case $V' = \emptyset$ (cf. also Exerc. 5).

12.24 Exercises. *1.* Show that $\langle x, \xi \rangle = \eta_*(x \frown \xi)$ for all $x \in H^*(X, A; M_2)$, $\xi \in H(X, A; M_1)$, where η is the augmentation map (tensored with $\mathrm{id}(M_1 \otimes M_2)$.

2. Show that $H(P_n\mathbb{C}; R)$ is a free $H^*(P_n\mathbb{C}; R)$-module with one generator.

3*. If $K \subset V \subset \mathbb{R}^n$ and $o_K \in H_n(V, V-K)$ are as in 11.17 Exerc. 4 then $(y \frown o_K) \circ \zeta = \pm \langle y, i_* \zeta \rangle\, o_0$, for $y \in H^* V$, $\zeta \in HK$.

4*. Let $(Y; B_1, B_2)$ be any triad, let $b \in SY$ be a chain such that $\partial b \in S\{B_1, B_2\}$, and $\psi \in Z^* Y$ a cocycle such that $\psi | SB_2 = 0$. With these data fixed, consider the following natural transformations $F, G\colon SX \to S(X \times Y)$,

$$F(c) = E(\psi\, q \otimes D(EZ)(c \otimes b)), \qquad G(c) = EZ(c \otimes E(\psi \otimes Db)),$$

where $q\colon S(X \times X) \to SY$ is the projection, and D, E, EZ are as above; cf. 12.1; VI, 12.20; VI, 12.25. (These two expressions are representatives for $q^*[\psi] \frown ([c] \times [b])$ resp. $[c] \times ([\psi] \frown [b])$; we want to show that they are homologous, up to sign.) Show that F and $(-1)^{|c||\psi|} G$ coincide if X is a point. Conclude by naturality that $qF = (-1)^{|c||\psi|} qG$, hence $(F - (-1)^{|c||\psi|} G)\colon SX \to \ker(q\colon S(X \times Y) \to SY)$. Show that F, G induce chain maps $\bar{F}, \bar{G}\colon SX \to S(X \times Y)/S(X \times B_1)$. Combining, get a natural chain map

$$(\bar{F} - (-1)^{|c||\psi|} \bar{G})\colon SX \to \ker(\bar{q}\colon S(X \times Y, X \times B_1) \to S(Y, B_1)).$$

The acyclic model theorem (cf. VI, 11) shows that it is nulhomotopic (S is a free functor, $\ker \bar{q}$ is acyclic on models $X = \Delta_p$), hence $\bar{F} \simeq (-1)^{|c||\psi|} \bar{G}$. By naturality of $\bar{F}$ and $\bar{G}$ (applied to inclusion maps $A \to X$), find induced chain maps $\bar{F}, \bar{G}\colon S(X, A) \to S(X \times Y)/S\{A \times Y, X \times B_1\}$, and an induced homotopy $\bar{F} \simeq (-1)^{|c||\psi|} \bar{G}$. Passage to homology then shows

$$(q^* y) \frown (\xi \times \eta) = (-1)^{|\xi||y|} \xi \times (y \frown \eta),$$

where $y=[\psi]\in H^*(Y,B_2)$, $\eta=[b]\in H(Y,B_1\cup B_2)$, $\xi\in H(X,A)$, provided $(Y;B_1,B_2)$ and $(X\times Y;A\times Y\cup X\times B_1,X\times B_2)$ are excisive. Similarly (or by applying $t\colon X\times Y\approx Y\times X$ to the above result) one finds $(p^*x)\frown(\xi\times\eta_1)=(x\frown\xi)\times\eta_1$, where $p\colon (X\times Y,A_2\times Y)\to(X,A_2)$ is the projection. Combining we find

$$\begin{aligned}(x\times y)\frown(\xi\times\eta)&=p^*x\smile q^*y\frown(\xi\times\eta)=(-1)^{|\xi||y|}p^*x\frown(\xi\times(y\frown\eta))\\&=(-1)^{|\xi||y|}(x\frown\xi)\times(y\frown\eta),\end{aligned}$$

proving 12.17.

5*. Let $V\subset W$, $V'\subset W'$ be open pairs in a space X and assume $W\cup W'=X$. Generalizing 12.22, show that the following diagram

$$\text{(12.25)}\qquad\begin{array}{ccccc}H^*(W,V)&\xrightarrow{\delta^*}&H^*(X,W)&\longrightarrow&H^*(W',W\cap W')\\ \downarrow{\scriptstyle\frown\xi_2}&&&&\downarrow{\scriptstyle\frown\xi_1}\\ H(W,W\cap W')&\longrightarrow&H(X,W')&\xrightarrow{\partial_*}&H(W',V')\end{array}$$

commutes for every $\xi\in H(X,(V'\cup W)\cap(V\cup W'))$, where ξ_1 resp. ξ_2 are the images of ξ in $H(X,V'\cup W)\cong H(W',V'\cup(W\cap W'))$ resp. $H(X,V\cup W')\cong H(W,V\cup(W\cap W'))$.—Instead of open pairs one can consider pairs of CW-subspaces of a CW-space X, or make the appropriate excision assumptions.

13. The Homology Slant Product, and the Pontrjagin Slant Product

The homology slant product has been used in connection with cohomology operations (Steenrod 1953), and S-duality (Spanier 1959). It will not be applied in this book and will therefore be treated very briefly. It is dual to the homology $\times$-product in the same sense as the cohomology-$\times$ is dual to the cohomology slant. The Pontrjagin-slant is obtained from the homology-slant by composition with a multiplication $\mu\colon X\times X\to X$; it turns the cohomology H^*X of an H-space into a module over the Pontrjagin ring HX.

13.1 Definition. Let C, D be complexes and L, M modules over R. Define

$$\vartheta\colon \mathrm{Hom}(C\otimes D,L)\otimes(C\otimes M)\to\mathrm{Hom}(D,L\otimes M)\quad\text{by}\quad[\vartheta(\rho\otimes c\otimes m)](d)=\rho(c\otimes d)\otimes m.$$

Verify that this is a chain map.

Let now (X,A), (Y,B) be pairs of spaces. Consider the composite chain map

$$\begin{aligned}&\mathrm{Hom}(S(X\times Y,A\times Y\cup X\times B),L)\otimes(S(X,A)\otimes M)\\&\quad\to\mathrm{Hom}\left(\frac{S(X\times Y)}{S\{A\times Y,X\times B\}},L\right)\otimes(S(X,A)\otimes M)\\&\quad\xrightarrow{(\circ EZ)\otimes\mathrm{id}}\mathrm{Hom}(S(X,A)\otimes S(Y,B),L)\otimes(S(X,A)\otimes M)\xrightarrow{\vartheta}\mathrm{Hom}(S(Y,B),L\otimes M).\end{aligned}$$

Passage to homology and composition with α (VI, 9.11) gives a homomorphism

$$H^*(X\times Y, A\times Y\cup X\times B; L)\otimes H(X, A; M)\to H^*(Y, B; L\otimes M). \tag{13.2}$$

This homomorphism or the corresponding bilinear map is called the ***homology slant product*** and denoted by /; more explicitly, we write $(z/\xi)\in H^{n-i}(Y, B; L\otimes M)$ for the image of $z\otimes\xi$, where $z\in H^n(X\times Y, A\times Y\cup X\times B; L)$, $\xi\in H_i(X, A; M)$. We leave it to the reader to establish the formal properties of this product. ***Duality*** with the homology cross-product is expressed by the formula

$$\langle z/\xi, \eta\rangle = \langle z, \xi\times\eta\rangle, \tag{13.3}$$

where $z\in H^*(X\times Y, A\times Y\cup X\times B, L)$, $\xi\in H(X, A; M)$, $\eta\in H(Y, B; N)$; both sides of 13.3 are elements of $L\otimes M\otimes N$.

13.4 Definition. Suppose now (X, μ) is a space with a multiplication $\mu\colon X\times X\to X$. Then the composite

$$H^n(X; L)\otimes H_i(X; M)\xrightarrow{\mu^*\otimes\mathrm{id}} H^n(X\times X; L)\otimes H_i(X; M)\xrightarrow{/} H^{n-i}(X; L\otimes M)$$

or the corresponding bilinear map is called the ***Pontrjagin slant product***. We write

$$(\mu^* x)/\xi = x\top\xi, \quad \text{for } x\in H^n(X; L),\ \xi\in H_i(X; M). \tag{13.5}$$

Again, we leave it to the reader to study the properties of $\top$, in particular, to establish the formulas for $\top$ which are implied by homotopy-associativity, -commutativity, -units of μ. For instance, if (X, μ) is an H-space then $\top$ makes $H^*(X; L)$ a graded module over the Pontrjagin ring $H(X; R)$.

Chapter VIII

Manifolds

A manifold is a space which is locally like euclidean space. Some of the most important topological spaces are manifolds: Lie groups and their homogeneous spaces are manifolds. If a (compact) Lie group operates on a manifold then the orbit of every point is a manifold; if the operation is sufficiently regular then the orbit space is also a manifold. The set of solutions $x=(x_1, \ldots, x_n)\in \mathbb{R}^n$ of a sufficiently regular system of equations $\alpha_\mu(x_1, \ldots, x_n)=0$, $\mu=1, \ldots, m$, is a manifold. These and other examples justify studying the special homology properties of manifolds.

1. Elementary Properties of Manifolds

1.1 Definition. A Hausdorff-space $M=M^n$ is said to be an *n-dimensional manifold*, or *n-manifold*, if every point of M has a neighborhood which is homeomorphic with an open set of $\mathbb{R}^n$. I.e., an n-manifold is a Hausdorff-space which is locally homeomorphic with $\mathbb{R}^n$. Because of invariance of dimension (IV, 3.8), if a manifold M is m-dimensional *and* n-dimensional then $m=n$ or $M=\emptyset$.

For instance, every open subset of $\mathbb{R}^n$ is an n-manifold. More generally, every open subset of an n-manifold is again an n-manifold. Spheres $\mathbb{S}^n$ and projective spaces $P_n\mathbb{R}$ resp. $P_n\mathbb{C}$ are manifolds of dimension n resp. $2n$. The solutions of systems of equations often form manifolds (cf. 1.7). The surfaces which we discussed in V, 3.11, Exerc. 1 and 2 are 2-manifolds.

1.2 Lemma. *Every point P of an n-manifold has an open neighborhood V which is homeomorphic with $\mathbb{R}^n$.* Any such V is called a *coordinate neighborhood* of P, and any homeomorphism $V \xrightarrow{\approx} \mathbb{R}^n$ is called a *chart* (around P).

Proof. Let $h\colon U \to W$ be a homeomorphism of a neighborhood U of P onto an open subset W of $\mathbb{R}^n$. The image of the interior, $h\mathring{U}$, is open in W, hence open in $\mathbb{R}^n$. Let W' be an open ball such that $hP \in W' \subset h\mathring{U}$. Then $V = h^{-1}W'$ is open in $\mathring{U}$, hence open in M, and $P \in V \approx W' \approx \mathbb{R}^n$. ∎

Obviously every coordinate neighborhood of a manifold is an ENR (= euclidean neighborhood retract). Therefore IV, 8.10 implies

1.3 Proposition. *If a manifold M is the union of finitely many coordinate neighborhoods (in particular, if M is compact) then M is an ENR.* ∎
More generally, manifolds with countable bases are ENR's (compare IV, 8.11 and 8.9).

By V, 4.11 this implies

1.4 Corollary. *If M is a compact manifold then $H_i(M; \mathbb{Z})$ is finitely generated for all i, and $H_i(M; \mathbb{Z}) = 0$ for sufficiently large i* (in fact, it vanishes for $i > \dim(M)$, as we shall see in 3.3). ∎

We now show that the solutions of a system of independent equations form a manifold.

1.5 Definition. Let W^m be an m-manifold, and let $g_1, g_2, \ldots, g_k$, $k \leq m$, be real valued continuous functions which are defined in a neighborhood of a point $P \in W$. We say, $g_1, \ldots, g_k$ are *topologically independent* at P if there are continuous functions $g_{k+1}, \ldots, g_m$, also defined in a neighborhood of P, such that $x \mapsto (g_1 x, \ldots, g_m x)$ maps some neighborhood of P homeomorphically onto an open subset of $\mathbb{R}^m$ (injectively would be enough by invariance of domaine; cf. IV, 7.4).
Clearly, if $g_1, \ldots, g_k$ are topologically independent at P then also at all points of a neighborhood of P. An important example is the following.

1.6 Proposition. *If $U \subset \mathbb{R}^m$ is an open set, and $g_1, \ldots, g_k$: $U \to \mathbb{R}$ are continuously differentiable functions whose differentials $dg_1(P), \ldots, dg_k(P)$: $\mathbb{R}^m \to \mathbb{R}$ are linearly independent at $P \in U$ then $g_1, \ldots, g_k$ are topologically independent at P.*

Proof. Let $g_{k+1}, \ldots, g_m$: $\mathbb{R}^m \to \mathbb{R}$ be linear maps such that $dg_1(P), \ldots, dg_k(P)$, $g_{k+1}, \ldots, g_m$ are linearly independent. Then the differential of g: $U \to \mathbb{R}^m$, $gx = (g_1 x, \ldots, g_m x)$ at P is isomorphic, $dg(P)$: $\mathbb{R}^m \cong \mathbb{R}^m$; therefore g is homeomorphic near P by the inverse function theorem (Dieudonne 10.2.5). ∎

In fact, g is even *diffeomorphic* near P, i.e. it has a differentiable inverse near P.

1.7 Proposition. *Let N be a subset of the manifold W^{n+k} which, locally, is the set of solutions of k independent equations. This means, for every $P\in N$ there exists a neighborhood V^P in W and functions $g_1^P, \ldots, g_k^P\colon V^P \to \mathbb{R}$ which are topologically independent at P and such that*

$$N\cap V^P = \{x\in V^P \mid g_1^P x = 0 = g_2^P x = \cdots = g_k^P x\}.$$

Then N is an n-manifold.

Proof. By assumption (of independence) there exists a homeomorphism $g^P\colon V^P \approx U$ such that U is open in $\mathbb{R}^{n+k} = \mathbb{R}^k \times \mathbb{R}^n$ and $N\cap V^P = (g^P)^{-1}[U\cap(\{0\}\times\mathbb{R}^n)]$ (in fact, it may be necessary to replace V^P by a smaller neighborhood first). Thus $N\cap V^P$ is a neighborhood of P in N which is homeomorphic to an open set in $\{0\}\times\mathbb{R}^n \approx \mathbb{R}^n$. ∎

1.8 Remark. A subset N of a manifold W as in 1.7 is called a *locally flat submanifold*. Not every subset N of W which is a manifold is locally flat: there are counterexamples even with $W=\mathbb{R}^3$, and $N\approx\mathbb{S}^1$ or $\mathbb{S}^2$ (cf. Artin-Fox). On the other hand it is not hard to see that every compact manifold (more generally: union of finitely many coordinate neighborhoods) is homeomorphic with a locally flat submanifold of some euclidean space (cf. Exerc. 5).

If we think of manifolds as solutions of systems of equations $g=0$ then we might also consider the solutions of combined equalities $g=0$ and inequalities $h\geq 0$. This leads to the following

1.9 Definition. Let $\mathbb{R}^n_+ = \{x\in\mathbb{R}^n \mid x_n \geq 0\}$, the "upper half" of $\mathbb{R}^n$. A Hausdorff-space $L=L^n$ is called an n-dimensional *∂-manifold* (or *n-manifold with boundary*) if every point $P\in L^n$ has a neighborhood U which is homeomorphic with an open set W of $\mathbb{R}^n_+$. Let $h\colon U\to W$ be such a homeomorphism. We say P is a *boundary point* of L if $h(P)\in\mathbb{R}^{n-1} = \{x\in\mathbb{R}^n \mid x_n = 0\}$; otherwise P is an *interior point* of L. The property of being a boundary (interior) point does not depend on the choice of $h\colon U\approx W$ (invariance of the boundary; cf. IV, 3.9). The set ∂L of all boundary points is an $(n-1)$-manifold (possibly empty), called the *boundary of L*; the set iL of all interior points is an n-manifold, called the *interior of L*. The interior is an open subset of L, the boundary is closed, and $iL\cup\partial L = L$, $iL\cap\partial L=\emptyset$.

As in 1.2 one shows that every point of iL resp. ∂L has an open neighborhood (called *coordinate neighborhood*) which is homeomorphic with $\mathbb{R}^n$ resp. $\mathbb{R}^n_+$; these homeomorphisms are still called *charts*.

1.10 Examples. Every manifold is a ∂-manifold (with empty boundary). $\mathbb{R}^n_+$ is a ∂-manifold, and every open subset of a ∂-manifold is also a

∂-manifold. The n-ball $\mathbb{B}^n$ is a ∂-manifold, with boundary $\partial\mathbb{B}^n = \mathbb{S}^{n-1}$. The product of a manifold M with a ∂-manifold L is a ∂-manifold, and $\partial(M \times L) = M \times \partial L$ (proof clear). More generally, the product of two ∂-manifolds L, L' is a ∂-manifold, and $\partial(L \times L') = \partial L \times L' \cup L \times \partial L'$ (proof left to the reader). Solutions of combined equalities and inequalities often form ∂-manifolds (Exerc. 4).

If two ∂-manifolds have homeomorphic boundaries then one can paste them together along the (common) boundary and get an ordinary manifold—just as $\mathbb{R}^n$ can be obtained by pasting two copies of $\mathbb{R}^n_+$. More precisely,

1.11 Proposition. *Let L, L' be n-dimensional ∂-manifolds and $\eta: \partial L \to \partial L'$ a homeomorphism. Then $M = (L \oplus L')/(y \sim \eta(y)$ for $y \in \partial L)$, the space which is obtained from the topological sum $L \oplus L'$ by identifying corresponding boundary points, is an n-manifold. It contains L, L' as subspaces via $L, L' \subset L \oplus L' \xrightarrow{\text{proj}} M$, and $L \cup L' = M$, $L \cap L' \approx \partial L \approx \partial L'$. We shall sometimes write $M = L \cup_\eta L'$.*

The proof is quite obvious: In order to show that $P \in L \cap L'$ has a coordinate neighborhood in M one uses coordinate neighborhoods of P in L and in L'. ∎

For example, if $L' = L$ and $\eta = \text{id}(\partial L)$ we say that $M = L \cup_{\text{id}} L$ is obtained by *doubling* L. If $L' = \partial L \times [0, 1)$ then $\partial L' = \partial L \times \{0\} \xleftarrow[j]{\approx} \partial L$ with $j(y) = (y, 0)$; in this case we say $M = L \cup_j (\partial L \times [0, 1))$ is obtained from L by *attaching a collar*.

1.12 Exercises. *1.* A space which is locally homeomorphic with $\mathbb{R}^n$ is always a T_1-space (points are closed) but not necessarily hausdorff. For instance, if X^n is obtained from two copies of $\mathbb{R}^n$ by identifying corresponding points *outside the origin* then X^n is the union of two coordinate neighborhoods $\approx \mathbb{R}^n$, but X^n is not hausdorff.

*2**. Let W be a well-ordered set which represents the first non-countable ordinal. Order $W \times [0, 1)$ lexicographically,

$$[(w, t) \leq (w', t')] \Leftrightarrow [w < w', \text{ or } (w = w' \text{ and } t \leq t')],$$

introduce the order topology in $W \times [0, 1)$, and call the resulting space LH ("long half-line"). Show that LH is a connected 1-dimensional ∂-manifold whose boundary $\partial(LH)$ consists of a single point and whose interior $i(LH)$ cannot be covered with countably many coordinate neighborhoods. By doubling LH one obtains the "long line" $LL = LH \cup_{\text{id}} LH$. Show that $LL \not\approx i(LH)$ (LL is "long at both ends", $i(LH)$ only at one). Compare Kneser-Kneser.

3. If $h: V \to \mathbb{R}^n$, $h': V' \to \mathbb{R}^n$ are charts of a manifold M^n then

$$h' h^{-1}: h(V \cap V') \to h'(V \cap V') \subset \mathbb{R}^n$$

is called *change of charts.* A set of charts $\mathscr{A} = \{h: V^h \approx \mathbb{R}^n\}$ is called *atlas* if $\bigcup_{h \in \mathscr{A}} V^h = M$. An atlas all of whose changes of charts are C^r (r-times continuously differentiable) is called C^r-*atlas.* Two C^r-atlasses $\mathscr{A}, \mathscr{A}'$ are C^r-*equivalent* if $\mathscr{A} \cup \mathscr{A}'$ is also a C^r-atlas. An equivalence class of C^r-atlasses is called a C^r-*structure* on M, and M together with a C^r-structure is a C^r-*manifold* (or *smooth* manifold if $r = \infty$). Not every manifold admits a C^1-structure (cf. Kervaire) but for $r > 0$ every C^r-atlas is C^r-equivalent with a C^∞-atlas (cf. Koch-Puppe).

(a) Define C^s-*maps* between C^r-manifolds, $s \leq r$, using the charts of the given C^r-structures.

(b) Show as in 1.7 that a subspace N of a C^r-manifold W^{n+k} which, locally, is the set of solutions of k C^s-functions with independent differentials ($1 \leq s \leq r$), inherits a C^s-structure. It is then called a C^s-submanifold of W.

(c) Adapt the proof of IV, 8.8 to show that every compact C^r-manifold is C^r-homeomorphic with a C^r-submanifold of some euclidean space.

4. Let N be a subset of a manifold W^{n+k} such that for every $P \in N$ there exists a neighborhood V^P in W and functions $g_1^P, \ldots, g_k^P, h_1^P, \ldots, h_l^P: V^P \to \mathbb{R}$ (k fixed, l may depend on P) which are topologically independent at P and for which

$$N \cap V^P = \{x \in V^P \mid g_\mu^P(x) = 0, h_\nu^P(x) \geq 0 \text{ for all } \mu, \nu\}.$$

Then N is a ∂-manifold of dimension n.

5. Show that every compact n-manifold M is homeomorphic with a locally flat submanifold of some $\mathbb{R}^k$. (Hint: The proof of IV, 8.8 yields the required embedding $M \to \mathbb{R}^k$; remark that the graph of every map $M \to \mathbb{R}^l$ is locally flat in $M \times \mathbb{R}^l$). Extend the result to manifolds with countable base, using arguments as in Bos.

2. The Orientation Bundle of a Manifold

If M^n is a manifold we topologize the union $\bigcup_{P \in M} H_n(M, M-P)$ of its local homology groups; the resulting space $\tilde{M}$ is called the orientation bundle of M. It makes sense then to speak of continuous functions φ with $\varphi(P) \in H_n(M, M-P)$, $P \in M$ (*sections* of $\tilde{M}$; cf. 2.4), and this in turn allows to define the notion of an *orientation of* M (cf. 2.9). In VIII, 3 we shall use the group of all sections to provide a convenient description of the n-th homology of open sets in M.

2.1 Proposition and Definition. *The local homology groups $H_j(M^n, M^n-P; G)$ of an n-manifold M^n are zero for $j \neq n$, and $H_n(M, M-P; G) \cong G \cong H_n(M, M-P; \mathbb{Z}) \otimes G$. A generator o_P of $H_n(M, M-P; \mathbb{Z})$ is called orientation* (of M) *at P. There are exactly two orientations at every point P,* say o_P and $-o_P$.

For $G=\mathbb{Z}$ the proposition follows from IV, 2.2(c) and IV, 3.7 because M^n is locally homeomorphic with $\mathbb{R}^n$. For the general case one can apply the universal coefficient theorem or (simpler) use $S(M, M-P) \simeq (\mathbb{Z}, n)$. ∎

We now relate local homology classes in different (close-by) points.

2.2 Lemma. *Let $z, z' \in S(M; G)$ be cycles mod $M-P$, i.e. $\partial z, \partial z' \in S(M-P; G)$. Then there is a neighborhood V* of P *such that z, z' are cycles mod $M-Q$ for every $Q \in V$, i.e. $\partial z, \partial z' \in S(M-V; G)$. If the homology classes of z, z' agree at P, $[z]_P = [z']_P \in H(M, M-P; G)$, then they agree at all points Q of a neighborhood $V' \subset V$, i.e., $[z]_Q = [z']_Q$ for every $Q \in V'$.* (Remark. Using 5.18 this means $H(M, M-P) = \varinjlim H(M, M-V)$.)

Proof. The chains $\partial z, \partial z'$ are finite linear combinations (coefficients in G) of simplices σ with $\operatorname{im}(\sigma) \subset M-P$. Since $\operatorname{im}(\sigma)$ is compact there is a neighborhood V_σ of P such that $\operatorname{im}(\sigma) \subset M-V_\sigma$, and $V = \bigcap_\sigma V_\sigma$ is a neighborhood of P such that $\partial z, \partial z' \in S(M-V; G)$. If $[z]_P = [z']_P$ then a chain $c \in S(M; G)$ exists such that $z - z' - \partial c \in S(M-P; G)$, hence (as above) $z - z' - \partial c \in S(M-V'; G)$ for some neighborhood V' of P (which we may take within V). ∎

2.3 Definition and Proposition. We shall associate with every n-manifold M and every abelian group G a new manifold $\tilde{M} \otimes G$ and a covering map $\gamma_G: \tilde{M} \otimes G \to M$ such that $\gamma_G^{-1}(P) = H_n(M, M-P; G)$, for $P \in M$; in particular, $\tilde{M} \otimes G = \bigcup_{P \in M} H_n(M, M-P; G)$, as a set. As to the topology in $\tilde{M} \otimes G$, we consider pairs (V, z) where V is an open subset of M and $z \in Z_n(M, M-V; G)$ is a cycle mod $M-V$; we define

$$V_z = \{[z]_P \in H_n(M, M-P; G) | P \in V\},$$

and we assert that *the set of all such V_z is the base of a topology in $\tilde{M} \otimes G$. With respect to this topology the map γ_G is locally homeomorphic, and it is even a covering map* (cf. Massey, chap. V). *Furthermore, the maps $(u, v) \mapsto u \pm v$ of $D = \{(u, v) \in (\tilde{M} \otimes G) \times (\tilde{M} \otimes G) | \gamma_G u = \gamma_G v\}$ into $\tilde{M} \otimes G$ are continuous, i.e. addition and subtraction in $\tilde{M} \otimes G$ are continuous where defined.*

For $G=\mathbb{Z}$ we abbreviate $\gamma_{\mathbb{Z}}=\gamma$, $\tilde{M}\otimes\mathbb{Z}=\tilde{M}$. The map $\gamma\colon \tilde{M}\to M$ is called the *orientation-bundle* (or -sheaf) of M. *The map*

$$\beta\colon \tilde{M}\to\mathbb{Z},\quad \beta(u)=\|u\|=\text{absolute value of } u\in H_n(M,M-P)\cong\mathbb{Z},$$

is continuous, i.e. locally constant. *In particular, $\tilde{M}$ decomposes into a topological sum,*

$$\tilde{M}=\tilde{M}(0)\oplus\tilde{M}(1)\oplus\tilde{M}(2)\oplus\cdots,\quad \text{where } \tilde{M}(r)=\beta^{-1}(r).$$

The restricted maps $\gamma|\tilde{M}(r)\colon \tilde{M}(r)\to M$ are also covering maps.

Proof. Every $u\in\tilde{M}\otimes G$ lies in some V_z; indeed, if $z\in Z_n(M,M-P;G)$ represents u then, by 2.2, there is an open neighborhood V of P such that $z\in Z_n(M,M-V;G)$, hence $u\in V_z$. If $u\in(V'_{z'}\cap V''_{z''})$ then, again by 2.2, we can choose z and $V\subset(V'\cap V'')$ such that $[z]_Q=[z']_Q=[z'']_Q$ for all $Q\in V$, hence $u\in V_z\subset(V'_{z'}\cap V''_{z''})$. This proves that the set of all V_z is the base of a topology.

Next we show that γ_G is locally homeomorphic. Clearly γ_G maps V_z bijectively onto V, hence γ_G is open and locally bijective, and we have only to show continuity. Let then W be an open neighborhood of $P=\gamma_G(u)$. As we know already, u lies in some V_z, hence $(V\cap W)_z$ is a neighborhood of u which maps into W.

The map $(u,u')\mapsto u\pm u'$ takes $D\cap(V_z\times V_{z'})$ homeomorphically onto $V_{z\pm z'}$ (we just saw that both sets are homeomorphic with V under γ_G), and is therefore continuous.

It remains to show that β is locally constant, and γ_G is a covering. Given $P\in M$, choose a closed ball around P (in some coordinate neighborhood) and let V denote its interior. Then $M-V$ is a deformation retract of $M-Q$, for every $Q\in V$, hence $\iota_*^Q\colon H(M,M-V)\cong H(M,M-Q)$. If $z\in Z_n(M,M-V;\mathbb{Z})$ then $[z]_Q=\iota_*^Q[z]$, hence $\beta([z]_Q)=\|[z]_Q\|=\|[z]\|$ is independent of Q, hence β is constant in V_z, as asserted. Moreover, if we choose a generator $[z]$ of $H_n(M,M-V;\mathbb{Z})\cong\mathbb{Z}$ then $\gamma_G^{-1}(V)=\bigcup_{g\in G}V_{z\otimes g}$ is a decomposition into disjoint open sets $V_{z\otimes g}$, each of which maps homeomorphically onto V. Therefore γ_G (and each $\gamma|\tilde{M}(i)$) is a covering map. ∎

2.4 Definition. Let M^n be a manifold, $\gamma_G\colon\tilde{M}\otimes G\to M$ as in 2.3, and $A\subset M$. A map $s\colon A\to\tilde{M}\otimes G$ is called a *section* (of γ_G over A) if $\gamma_G\, s(P)=P$ for all $P\in A$. By 2.3, the sum or difference of two sections is again a section. The sections therefore form an abelian group which we denote by $\Gamma(A;G)$. Because γ_G is locally homeomorphic we have the following two properties.

(2.5) *Given $u\in\tilde{M}\otimes G$, there exists a neighborhood V of $\gamma_G(u)$ and a section $s\in\Gamma(V;G)$ such that $s\gamma_G(u)=u$.*

(2.6) *If $s, t \in \Gamma(A; G)$ agree at $P \in A$, then they agree in a whole neighborhood of P; in other words, $\{Q \in A | sQ = tQ\}$ is an open set in A.* ∎

If $[z] \in H_n(M, M-A; G)$ then $Q \mapsto [z]_Q$, $Q \in A$, is a section over A which we denote by $J_A[z]$ (continuity of $J_A[z]$ becomes obvious when restricting to sets V_z). In this way we get a homomorphism.

$$(2.7) \qquad J_A\colon H_n(M, M-A; G) \to \Gamma(A; G), \quad (J_A[z])(Q) = [z]_Q$$

which will play a fundamental role in VIII, 3. It is clearly *natural* with respect to inclusions, i.e.,

2.8 *If $A \subset A' \subset M$ then*

$$\begin{array}{ccc} H_n(M, M-A'; G) & \xrightarrow{J_{A'}} & \Gamma(A'; G) \\ \downarrow i_* & & \downarrow \rho \\ H_n(M, M-A; G) & \xrightarrow{J_A} & \Gamma(A; G) \end{array}$$

is a commutative diagram, where i = inclusion and ρ = restriction, $\rho(s) = s|A$. ∎

2.9 Definition. A section $O\colon A \to \tilde{M} = \tilde{M} \otimes \mathbb{Z}$ is called an *orientation of M along A* if $\beta O(P) = 1$ for all $P \in A$, i.e., if $O(P) \in H_n(M, M-P)$ is a generator (= orientation at P) for every $P \in A$. Or we can say, an orientation along A consists in selecting continuously an orientation at each $P \in A$. We say, *M is orientable along A* if such an O exists. In case $A = M$ we speak of *orientation* respectively *orientability* without further specification. If O is an orientation of M, and $V \subset M$ is an open set then $O|V$ is an orientation of V.

If $s \in \Gamma A = \Gamma(A; \mathbb{Z})$ is a section which is nowhere zero, $sP \neq 0$, then $P \mapsto sP/\|sP\|$ is an orientation. In particular, *M is orientable along A if a nowhere-zero section $s \in \Gamma A$ exists.*

If $O \in \Gamma A$ is an orientation along A then we get a homeomorphism

$$(2.10) \qquad A \times G \xrightarrow{\approx} \gamma_G^{-1} A, \quad \text{by } (P, g) \mapsto O(P) \otimes g.$$

In particular, $\tilde{M} \otimes G \approx M \times G$ if M is orientable. A section $s \in \Gamma(A; G)$ then takes the form of a locally constant function $s\colon A \to G$; the group *$\Gamma(A; G)$ becomes isomorphic with the group of locally constant functions $A \to G$.* If A is connected then locally constant functions are constant, hence $\Gamma(A;G)\cong G$. In particular, $\Gamma(M)\cong\mathbb{Z}$ if M is orientable and connected; *M has exactly two (opposite) orientations in this case.* For instance, if $M=\mathbb{S}^n$ and $\zeta\in H_n(\mathbb{S}^n;\mathbb{Z})$ is a generator then $J(\zeta)\in\Gamma\mathbb{S}^n$ is an orientation. This shows that our definition 2.4 of $\Gamma A=\Gamma(A;\mathbb{Z})$ for $A\subset\mathbb{S}^n$ is equivalent with the one we gave in IV, 6.2.

2.11 The Orientation-Covering. The manifold $\tilde{M}(1)$ is often called *orientation manifold* of M; its points are just the orientations at the various points of M. *The orientation manifold $\tilde{M}(1)$ is always orientable:* Its *canonical orientation* $\tilde{O}$ selects that orientation at $u\in\tilde{M}(1)$ which is mapped into u under γ_*: $H_n(\tilde{M}(1),\tilde{M}(1)-\{u\})\cong H_n(M,M-P)$ (this is well defined because $\gamma=\gamma_{\mathbb{Z}}$ is locally homeomorphic; cf. IV, 3.7). Of course, $\tilde{M}(1)$ is not distinguished from any other $\tilde{M}(i)$ with $i>0$. Indeed, $u\mapsto i\cdot u$ is a homeomorphism $\tilde{M}(1)\approx\tilde{M}(i)$ which commutes with γ.

The map $\gamma_1=\gamma|\tilde{M}(1)$: $\tilde{M}(1)\to M$ is a two-sheeted covering, and an orientation is a section of γ_1, hence M is orientable if and only if the covering γ_1 is trivial. Since two-sheeted covering are in one-one correspondence with subgroups of the fundamental group $\pi_1 M$ of index ≤ 2 (cf. Schubert III, 6.8) we get

2.12 Proposition. *If M is a connected manifold whose fundamental group $\pi_1 M$ has no subgroup of index 2 then M is orientable.* Because $H_1=\pi_1$ abelianized (cf. Schubert IV, 3.8) we can also replace $\pi_1 M$ by $H_1 M$ in this statement. ∎

If w: $[0,1]\to M$ is a path, and u an orientation at $w(0)$ then by Satz 1 in Schubert III, 6.3 a unique path $\tilde{w}$: $[0,1]\to\tilde{M}(1)$ exists such that $\gamma\tilde{w}=w$ and $\tilde{w}(0)=u$. We say $\tilde{w}(t)$ is the *continuation of $u=\tilde{w}(0)$ along w, and $\tilde{w}(1)$ is obtained from u by continuation along w.* This orientation $\tilde{w}(1)$ at $w(1)$ depends only on the homotopy class of w. The manifold M is orientable if and only if $\tilde{w}(1)$ is independent of w. In that case an orientation of M is obtained by choosing an orientation u at one point $P\in M$ and continuing u along all possible paths (assuming M connected). All of these assertions are proved in the theory of covering spaces (Godbillon VII–X, Massey V, Schubert III, 6). The proofs are simple, and even the reader who is not familiar with covering spaces, is invited to try for himself.

2.13 Orienting Products. Given manifolds M^m, N^n, we consider the following map between orientation bundles

$$\mu\colon\ \tilde{M}\times\tilde{N}\to\widetilde{M\times N},\quad \mu(u,v)=u\times v;$$

note that

$$u \times v \in H_{m+n}((M, M-P) \times (N, N-Q)) = H_{m+n}(M \times N, M \times N - (P, Q)).$$

Clearly $\gamma^{M \times N} \mu = \gamma^M \times \gamma^N$. If $V \subset M$, $W \subset N$ are open sets and

$$y \in Z_m(M, M-V), \quad z \in Z_n(N, N-W)$$

then $EZ(y \otimes z) \in Z_{m+n}(M \times N, M \times N - V \times W)$, and μ maps $V_y \times W_z$ homeomorphically onto $(V \times W)_{EZ(y \otimes z)}$ (recall that $[EZ(y \otimes z)] = [y] \times [z]$, where EZ is an Eilenberg-Zilber map). In particular, *μ is continuous.* If $A \subset M$, $B \subset N$, and $s \in \Gamma A$, $t \in \Gamma B$ then the composite $A \times B \xrightarrow{s \times t} \tilde{M} \times \tilde{N} \xrightarrow{\mu} \widetilde{M \times N}$ is also a section. There results a (bilinear) mapping

$$(2.14) \qquad (\Gamma A) \times (\Gamma B) \to \Gamma(A \times B), \quad (s, t) \mapsto \mu \circ (s \times t).$$

If $u \in H_m(M, M-P)$, $v \in H_n(N, N-Q)$ are generators then $u \times v$ is also a generator (cf. VII, 2.14). Therefore, if $s \in \Gamma A$, $t \in \Gamma B$ are orientations along A respectively B then $\mu \circ (s \times t)$ is an orientation along $A \times B$; it is called the *product orientation.* In particular, *the product of oriented manifolds is oriented* (by the product orientation). ∎

The square $M \times M$ of an orientable connected manifold M has a *canonical orientation*, namely $O \times O$ where O is any one of the two orientations of M. In particular, $\mathbb{C} = \mathbb{R} \times \mathbb{R}$ is canonically oriented, and therefore *$\mathbb{C}^n$ is canonically oriented* (by $O_{\mathbb{C}} \times O_{\mathbb{C}} \times \cdots \times O_{\mathbb{C}}$).

We now consider ∂-manifolds L^n. We want to relate the orientation bundles $\widetilde{iL}$ and $\widetilde{\partial L}$. Every open set $V \subset L$ is itself a ∂-manifold, and $iV = V \cap (iL)$, $\partial V = V \cap (\partial L)$. In analogy with 2.7 we define homomorphisms

$$(2.15) \qquad J_V^i \colon H_n(L, L-iV; G) \to \Gamma(iV; G), \quad (J_V^i[y])(P) = [y]_P,$$

where $y \in Z_n(L, L-iV; G)$, and $[y]_P$ is its class in

$$H(L, L-P; G) \overset{\text{exc}}{\cong} H_n(iL, iL-P; G), \quad P \in iV.$$

$$(2.16) \quad J_V^\partial \colon H_{n-1}(L-iV, L-V; G) \to \Gamma(\partial V; G), \quad (J_V^\partial[z])(Q) = [z]_Q,$$

where $z \in Z_{n-1}(L-iV, L-V; G)$, and $[z]_Q$ is its class in

$$H(L-iV, L-iV-Q; G) \overset{\text{exc}}{\cong} H(\partial L, \partial L - Q; G), \quad Q \in \partial V.$$

In diagram form, J_V^i, J_V^∂ are so defined that

$$
\begin{array}{ccc}
H_n(L, L-iV; G) & \xrightarrow{J_V^i} & \Gamma(iV; G) \\
\downarrow & & \downarrow e \\
H_n(L, L-P; G) & \xleftarrow{\cong} & H_n(iL, iL-P; G),
\end{array}
$$

(2.17)

$$
\begin{array}{ccc}
H_{n-1}(L-iV, L-V; G) & \xrightarrow{J_V^\partial} & \Gamma(\partial V; G) \\
\downarrow & & \downarrow e \\
H_{n-1}(L-iV, L-iV-Q; G) & \xleftarrow{\cong} & H_{n-1}(\partial L, \partial L-Q; G)
\end{array}
$$

are commutative for all $P\in iV$ resp. $Q\in\partial V$, where e assigns to each section s its value at P resp. Q.

We omit the easy continuity proofs (cf. parenthesis before 2.7), and also the verification of the following naturality properties.

2.18 Lemma. *If $V\subset V'\subset L$ are open sets then the following diagrams are commutative,*

$$
\begin{array}{ccc}
H_n(L, L-iV'; G) & \xrightarrow{J_{V'}^i} & \Gamma(iV'; G) \\
\downarrow & & \downarrow \rho \\
H_n(L, L-iV; G) & \xrightarrow{J_V^i} & \Gamma(iV; G)
\end{array}
\qquad
\begin{array}{ccc}
H_{n-1}(L-iV', L-V'; G) & \xrightarrow{J_{V'}^\partial} & \Gamma(\partial V'; G) \\
\downarrow & & \downarrow \rho \\
H_{n-1}(L-iV, L-V; G) & \xrightarrow{J_V^\partial} & \Gamma(\partial V; G),
\end{array}
$$

where $\rho(s')=s'|iV$ resp. $\rho(s')=s'|\partial V$. ∎

2.19 Proposition. *If L^n is a ∂-manifold then there exists a unique family of homomorphisms $\{\partial_V\colon \Gamma(iV; G)\to\Gamma(\partial V; G)\}_{V \text{ open in } L}$, which is natural with respect to inclusions $V\subset V'$* (i.e., $\rho\,\partial_{V'}=\partial_V\,\rho$) *and makes*

(2.20)

$$
\begin{array}{ccc}
H_n(L, L-iV; G) & \xrightarrow{J_V^i} & \Gamma(iV; G) \\
\downarrow \partial_* & & \downarrow \partial_V \\
H_{n-1}(L-iV, L-V; G) & \xrightarrow{J_V^\partial} & \Gamma(\partial V; G)
\end{array}
$$

commutative. Moreover, if $O\in\Gamma(iV)$ is an orientation of iV then $\partial_V O\in\Gamma(\partial V)$ is an orientation of ∂V, called induced orientation. *In particular, if iL is orientable then so is ∂L.* An orientation of iL is often called *orientation of L*, and L is called orientable if iL is so.

Proof. If $Q \in \partial L$ then there is an injection $\{x \in \mathbb{R}^n_+ \mid \|x\| \le 1\} \to L$ which takes $\{x \in \mathbb{R}^n_+ \mid \|x\| < 1\}$ homeomorphically onto an open neighborhood W of Q; in fact, these "half-balls" W form a neighborhood base at Q (just look at a coordinate neighborhood). If W is a half-ball then deformation retractions $L - iW \simeq L - P$ (for $P \in iW$), $L - W \simeq L - iW - Q$, $L - W \simeq L$ exist (deform radially from P resp. Q resp. $(0, 0, \ldots, -1) \in \mathbb{R}^n - \mathbb{R}^n_+$), hence $H(L, L - iW; G) \cong H(L, L - P; G)$,

$$H(L - iW, L - W; G) \cong H(L - iW, L - iW - Q; G),$$

and $\partial_*\colon\ H_n(L, L - iW; G) \cong H_{n-1}(L - iW, L - W; G)$, from the exact homology sequences of the appropriate triples. Also

$$e\colon\ \Gamma(iW; G) \cong H_n(iL, iL - P; G) \cong G,$$

$$e\colon\ \Gamma(\partial W; G) \cong H_{n-1}(\partial L, \partial L - Q; G) \cong G$$

because $\widetilde{iW \otimes G} \approx (iW) \times G$, $\widetilde{\partial W \otimes G} \approx (\partial W) \times G$ (recall that $iW \approx \mathbb{R}^n$, $\partial W \approx \mathbb{R}^{n-1}$ are orientable and use 2.10). It follows then from 2.17 that J^i_W, J^∂_W are isomorphic (all the other arrows are), and therefore 2.20 shows that $\partial_W = J^\partial_W \partial_* (J^i_W)^{-1}$ is unique and, being a composition of natural isomorphisms, is natural (with respect to inclusions of half-balls). Moreover, it is isomorphic and hence (for $G = \mathbb{Z}$) takes orientations (= generators) into orientations.

Let $\mathscr{W} = \{W\}$ denote the set of all half-balls. If $V \subset L$ is an arbitrary open set, and $s \in \Gamma(iV; G)$ then, by naturality, $\partial_V(s)$ must satisfy

(2.21) $\quad \partial_V(s) \mid \partial W = \partial_W(s \mid W), \quad$ for every half-ball $W \subset V$.

Since $\partial V = \bigcup_{W \subset V} \partial W$, $W \in \mathscr{W}$, this determines $\partial_V(s)$. We now want to show that, (i) the partial sections $\partial_W(s \mid W)$ match on the intersections $(\partial W) \cap (\partial W')$ (so that 2.21 defines ∂_V); further, (ii) $\partial_V(s \pm s') = (\partial_V s) \pm (\partial_V s')$, (iii) naturality $\rho\, \partial_{V'}(s') = \partial_V\, \rho(s')$ for $V \subset V'$, and (iv) commutativity of 2.20, i.e., $J^\partial_V \partial_*(\xi) = \partial_V J^i_V(\xi)$. Each of these formulas expresses an equality of sections in some ∂U where $U \subset L$ is open. In order to prove this equality at $Q \in \partial U$ one chooses a half-ball W such that $Q \in W \subset U$ and restricts to W where one already knows the assertion to be true. Similarly for the second part of 2.19: If $O \in \Gamma(iV)$ is an orientation then also $O \mid W$ for every half-ball $W \subset V$, hence $\partial_V(O) \mid \partial W = \partial_W(O \mid W)$ is an orientation, and hence also $\partial_V(O)$ because $\partial V = \bigcup_{W \subset V} \partial W$. ∎

2.22 Exercises. *1.* If X is any Hausdorff-space then the union

$$\tilde{X} = \bigcup_{P \in X} H_n(X, X - P)$$

of its local homology groups can be topologized and a projection $\gamma\colon \tilde{X} \to X$ can be defined as in 2.3. This map will still be locally homeomorphic but not necessarily a covering-map. If, for all $P \in X$,

$$H_j(X, X-P)=0 \quad \text{for } j \neq n, \qquad H_n(X, X-P) \cong \mathbb{Z},$$

and if γ is a covering map (equivalently: β is locally constant) then X is called a *generalized n-manifold.* How would you define a generalized ∂-manifold?

2. Let M^n, N^n be manifolds of the same dimension and $f: M \to N$ a map which is locally homeomorphic (e.g. a covering, or the inclusion of an open set). Show that f induces natural homomorphisms $f^*: \Gamma(B) \to \Gamma(f^{-1}B)$ for all $B \subset N$ (defined by $f_*^P[(f^*t)P]=t(fP)$; cf. IV, 3.4). If $t \in \Gamma B$ is an orientation of N along B then f^*t is an orientation of M along $f^{-1}B$. In particular, if N is orientable then so is M. If M, N are oriented by orientations O, O' then f is called *orientation preserving* (resp. reversing) if $f^*O'=O$ (resp. $f^*O'=-O$). In particular, this applies if $M=N$, $O=O'$.

3. Let $p: M \to N$ be a covering map. If N is orientable then so is M. If p is a regular covering (cf. Schubert III, 6.6) and M is orientable then N is orientable if and only if every covering transformation $M \to M$ is orientation-preserving.

4. Consider the map $\mu: \tilde{M} \times \tilde{N} \to \widetilde{M \times N}$ of 2.13 and verify $\gamma^{M\times N}(\mu(\tilde{P}, \tilde{Q}))=(\gamma^M(\tilde{P}), \gamma^N(\tilde{Q}))$, $\beta^{M\times N}(\mu(\tilde{P}, \tilde{Q}))=\beta^M(\tilde{P})\,\beta^N(\tilde{Q})$, for $\tilde{P} \in \tilde{M}$, $\tilde{Q} \in \tilde{N}$. Show that the restriction of μ defines a two-sheeted covering

$$\tilde{M}(1) \times \tilde{N}(1) \to \widetilde{(M \times N)}(1),$$

which is trivial if and only if one of M, N is orientable. Generalizing μ, define $(\tilde{M} \otimes G) \times (\tilde{N} \otimes G') \to \widetilde{(M \times N)} \otimes (G \otimes G')$ where G, G' are arbitrary abelian groups.

5. Show by an example (Möbius-strip) that the boundary ∂L of a ∂-manifold L can be orientable without L being orientable. Can it happen that $\partial_L(s) \in \Gamma(\partial L)$ is an orientation but $s \in \Gamma L$ is not? Show that the answer is no if every component of L has a non-empty boundary.

3. *Homology of Dimension $\geq n$ in n-Manifolds*

This § generalizes to arbitrary n-manifolds M^n what was proved for spheres $\mathbb{S}^n$ in IV, 6. Roughly speaking, we show that the homology of (pairs of) open sets in M^n vanishes above n and coincides with a suitable group of sections (VIII, 2) in dimension n. More generally, this holds for retracts of open sets.

3.1 Definition. Let $B \subset A \subset M$ where M is an n-manifold, $n>0$. Let $\Gamma(A,B;G)$ denote the group of sections of γ_G (2.4) over A which vanish on B, $\Gamma(A,B;G)=\{s\in\Gamma(A;G) \mid s|B=0\}$. There exists then a unique homomorphism J_{AB} which fills the following diagram,

$$\begin{array}{ccccc} H_n(M-B,M-A;G) & \longrightarrow & H_n(M,M-A;G) & \longrightarrow & H_n(M,M-B;G) \\ \Big\downarrow{\scriptstyle J_{AB}} & & \Big\downarrow{\scriptstyle J_A} & & \Big\downarrow{\scriptstyle J_B} \\ 0\to\Gamma(A,B;G) & \xrightarrow{\subset} & \Gamma(A;G) & \xrightarrow{\rho} & \Gamma(B;G), \end{array} \tag{3.2}$$

where $\rho(s)=s|B$, and J_A, J_B are defined in 2.7. This assertion is clear because the rows of 3.2 are exact and the right square is commutative. ∎

3.3 Proposition. *If $X\subset Y$ are subsets of M^n which are neighborhood retracts (e.g., if X, Y are open) then*

(a) $\qquad H_i(Y,X;G)=0 \quad$ for $i>n$,

(b) $\qquad J_{XY}\colon H_n(Y,X;G)\to\Gamma(M-X,M-Y;G)$

is a monomorphism whose image consists of all sections with bounded support, *i.e. of all sections $s\in\Gamma(M-X,M-Y;G)$ for which the set $\{P\in(M-X) \mid s(P)\neq 0\}$ is contained in a compact part of M.* If Γ_b denotes the group of these sections then we can also write

$$J_{XY}\colon H_n(Y,X;G)\cong\Gamma_b(M-X,M-Y;G).$$

Note that $\Gamma_b=\Gamma$ if $\overline{Y-X}$ is compact.

3.4. Corollary. *If M is an n-manifold and $C\subset M$ a closed connected subset then*

$$H_n(M,M-C;G)\cong\begin{cases} G & \textit{if } C \textit{ is compact and } M \textit{ orientable along } C,\\ G*\mathbb{Z}_2=\{g\in G \mid 2g=0\} & \\ & \textit{if } C \textit{ is compact and } M \textit{ not orientable along } C,\\ 0 & \textit{if } C \textit{ is not compact}.\end{cases}$$

In particular, *this applies to $C=M$ if M is connected.*

Proof. We have $H_n(M,M-C;G)\cong\Gamma_b(C;G)$. If C is not compact then $\Gamma_b(C;G)=0$. If C is compact then the sections of $\gamma_G\colon \tilde{M}\otimes G\to M$ over C can be identified with those components of $\gamma_G^{-1}C$ which are homeomorphic (via γ_G) with C; this follows because γ_G is a covering map. If M is orientable along C then $\gamma_G^{-1}C\approx C\times G$, hence $\Gamma_b(C;G)=$

$\Gamma(C;G)=G$. If M is not orientable along C then the orientation covering $\gamma_1\colon \tilde{M}(1)\to M$ is non-trivial over C, i.e., $\gamma_1^{-1}C$ is connected. The components of $\gamma_G^{-1}C$ are then of the form $\tilde{g}(\gamma_1^{-1}C)$, where

$$\tilde{g}\colon \tilde{M}(1)\to\tilde{M}\otimes G,\quad \tilde{g}(u)=u\otimes g,\quad\text{and}\quad g\in G$$

(use neighborhoods V_z as in 2.3 to prove continuity of $\tilde{g}$). The only points of $\tilde{g}(\gamma_1^{-1}C)$ which under γ_G map into $Q\in C$ are $u\otimes g$ and $(-u)\otimes g=u\otimes(-g)$, where $u\in H_n(M,M-Q;\mathbb{Z})$ is a generator. Thus $\gamma_G|\tilde{g}(\gamma_1^{-1}C)$ is homeomorphic if and only if $g=-g$, i.e., if and only if $2g=0$. ∎

3.5 Corollary. *If M is an n-manifold and $C\subset M$ a closed connected subset then the torsion subgroup of $H_{n-1}(M,M-C;\mathbb{Z})$ is of order two if C is compact and M non-orientable along C, and is zero otherwise.*

Proof. Let q be a non-zero integer. If C is compact and M orientable along C then

$$\mathbb{Z}_q\cong H_n(M,M-C;\mathbb{Z}_q)\cong H_n(M,M-C;\mathbb{Z})\otimes\mathbb{Z}_q\oplus H_{n-1}(M,M-C;\mathbb{Z})*\mathbb{Z}_q$$
$$\cong\mathbb{Z}_q\oplus H_{n-1}(M,M-C;\mathbb{Z})*\mathbb{Z}_q$$

(using 3.4 and the universal coefficient theorem), hence

$$H_{n-1}(M,M-C;\mathbb{Z})*\mathbb{Z}_q=0.$$

Similarly, $0=H_n(M,M-C;\mathbb{Z}_q)=H_{n-1}(M;\mathbb{Z})*\mathbb{Z}_q$ if C is not compact, or if M is not orientable along C and q is odd. Since

$$H_{n-1}(M,M-C;\mathbb{Z})*\mathbb{Z}_q=\{a\in H_{n-1}(M,M-C;\mathbb{Z})\,|\,qa=0\}$$

this shows that $H_{n-1}(M,M-C;\mathbb{Z})$ has no q-torsion in these cases. Finally, if C is compact and M non-orientable along C then (again by 3.4)

$$\mathbb{Z}_2\cong H_n(M,M-C;\mathbb{Z}_4)\cong H_{n-1}(M,M-C;\mathbb{Z})*\mathbb{Z}_4$$
$$=\{a\in H_{n-1}(M,M-C;\mathbb{Z})\,|\,4a=0\};$$

this easily implies that $H_{n-1}(M,M-C;\mathbb{Z})$ contains just one non-zero element of finite order. ∎

3.6 Corollary. *For any $A\subset M$ let $c_b(A)$ denote the number of bounded components of A* (i.e. components whose closure in M is compact). *If M is a connected n-manifold and $X\subset M$ is a neighborhood retract, $X\neq M$, then*

(3.7) $c_b(M-X)=c_b(M)+\dim(\ker[H_{n-1}(X;\mathbb{Z}_2)\to H_{n-1}(M;\mathbb{Z}_2)])$.

In particular, if $H_{n-1}(M;\mathbb{Z}_2)=0$ *then*

$$c_b(M-X)=c_b(M)+\dim H_{n-1}(X;\mathbb{Z}_2).$$

If M is orientable then we also have

$$(3.8)\quad c_b(M-X)=c_b(M)+\operatorname{rank}(\ker[H_{n-1}(X;\mathbb{Z})\to H_{n-1}(M;\mathbb{Z})]).$$

These results generalize the Jordan theorem IV, 7.2. Intuitively, they assert that every non-trivial cycle of X which bounds in M separates M. Note that $c_b(M)=1$ or 0 depending on wether M is compact or not.

Proof. As in IV, 7.1, one easily sees that $c_b(A)=\dim\Gamma_b(A;\mathbb{Z}_2)$; hence $c_b(M-X)=\dim H_n(M,X;\mathbb{Z}_2)$, and $c_b(M)=\dim H_n(M;\mathbb{Z}_2)$, by 3.3. Formula 3.7 now follows from the exact sequence

$$H_n(X;\mathbb{Z}_2)\to H_n(M;\mathbb{Z}_2)\to H_n(M,X;\mathbb{Z}_2)\to H_{n-1}(X;\mathbb{Z}_2)\to H_{n-1}(M;\mathbb{Z}_2)$$

because the first term vanishes: $H_n(X;\mathbb{Z}_2)\cong\Gamma_b(M,M-X;\mathbb{Z}_2)=0$, the latter because M is connected and $M-X\neq\emptyset$.

If M is orientable then we also have $c_b(A)=\operatorname{rank}\Gamma_b(A)$, and we can replace $\mathbb{Z}_2$ by $\mathbb{Z}$ in the preceding argument. ∎

3.9 Corollary. *Let* X^{n-1} *be a compact connected* $(n-1)$*-manifold,* M^n *an orientable connected n-manifold, and* $i\colon X\subset M$ *an embedding such that X is nulhomologous* mod 2 *in M* (this means $i_*\colon H_{n-1}(X;\mathbb{Z}_2)\to H_{n-1}(M;\mathbb{Z}_2)$ is zero). *Then X is orientable and is nulhomologous in M with integral coefficients.*

For instance, there is no embedding $P_{2k}\mathbb{R}\to\mathbb{R}^{2k+1}$, and every embedding $P_{2k}\mathbb{R}\to P_{2k+1}\mathbb{R}$ is isomorphic on $H_{2k}(-;\mathbb{Z}_2)$.

Proof. Comparing 3.7 and 3.8 shows

$$\begin{aligned}&\operatorname{rank}\ker[H_{n-1}(X;\mathbb{Z})\xrightarrow{i'_*}H_{n-1}(M;\mathbb{Z})]\\&\quad=\dim\ker[H_{n-1}(X;\mathbb{Z}_2)\xrightarrow{i_*}H_{n-1}(M;\mathbb{Z}_2)]=1,\end{aligned}$$

the latter by assumption $i_*=0$. It follows that $H_{n-1}(X;\mathbb{Z})=\mathbb{Z}$ (cf. 3.4) and $\operatorname{im}(i'_*)$ is finite; since $H_{n-1}(M;\mathbb{Z})$ is torsionfree (3.5) this implies $\operatorname{im}(i'_*)=0$. ∎

Proof of 3.3. Note first that the image of J is indeed contained in Γ_b because every homology class y has a representative chain in a compact part K of M (hence $(Jy)P\neq0\Rightarrow P\in K$).—We now proceed in several steps.

Case 1. X is open, $Y=M$ is an open subset of $\mathbb{S}^n$.

For integral coefficients, $G=\mathbb{Z}$, this is contained in IV, 6.4; one has to remark that $\Gamma_b(M-X)=\Gamma(\mathbb{S}^n-X, \mathbb{S}^n-Y)$. For arbitrary coefficients G the proof carries over word by word; in fact, the hardest step 5 of that proof can be omitted here because X, Y are assumed to be open.

Case 2. X is open, $Y=M=\bigcup_{i=1}^r V_i$ is a finite union of euclidean open subsets V_i (i.e. V_i is homeomorphic with an open subset of $\mathbb{R}^n$).

This case is reduced to case 1 by the Mayer-Vietoris principle IV, 6.7 which asserts.

(3.10) *If $(Y; X_1, X_2)$ is an excisive triad in M and if* (3.3a, b) *hold for (Y, X_1), (Y, X_2) and $(Y, X_1 \cup X_2)$ then also for $(Y, X_1 \cap X_2)$.*

In IV, 6.7 we assumed $G=\mathbb{Z}$ and $M=\mathbb{S}^n$ but the proof is exactly the same in the general case 3.10.

Getting back to our assumption, if $r=1$ we apply case 1. In general, we proceed by induction on r; if $r>1$ we have $M=V_1' \cup V_2'$, where V_k' is a union of less than r euclidean open sets. We can then find open sets W_k such that $\overline{W}_k \subset V_k'$ and $M=W_1 \cup W_2$. For instance, with some metric on M we can take

$$W_1=\{x \in M \mid 2 \cdot \text{distance}(x, M-V_1') > \text{distance}(x, M-V_2')\},$$

and similarly for W_2. Let $X_k=X \cup (M-\overline{W}_k)$. Then

$$H(M, X_k) \cong H(V_k', V_k' \cap X_k)$$

by excision. Therefore the inductive hypothesis (applied to the manifold V_k') gives $H_j(M, X_k) \cong H_j(V_k', V_k' \cap X_k)=0$ for $j>n$, and

$$H_n(M, X_k) \cong H_n(V_k', V_k' \cap X_k)=\Gamma_b(V_k'-X_k)=\Gamma_b(M-X_k),$$

the latter because $V_k'-X_k=M-X_k$. Thus 3.3a, b hold for (M, X_1), (M, X_2), and by the same argument also for $(M, X_1 \cup X_2)$. Therefore they hold for $(M, X_1 \cap X_2)$ by 3.10. But $X_1 \cap X_2=X$ because

$$(M-\overline{W}_1) \cap (M-\overline{W}_2)=M-(\overline{W}_1 \cup \overline{W}_2)=\emptyset.$$

Case 3. $Y=M$ as in case 2 (a finite union of euclidean open sets), X an arbitrary neighborhood retract.

As in IV, 6.4, this is the most difficult case. The proof is very similar to step 5 of IV, 6.4 but still the situation seems different enough to justify a repetition of the argument. Note that M and every neighborhood retract of M is an ENR (IV, 8.10).

We can assume M is connected (otherwise we argue for each component), and $X \neq M$. Then $\Gamma(M, M-X)=0$, hence $H_n X \to H_n M$ is the zero map because it factors: $H_n X \xrightarrow{J} \Gamma_b(M, M-X) \to \Gamma_b M \overset{J}{\cong} H_n M$. The homology sequence of (M, X) then shows $H_i(M, X) \cong H_{i-1} X$ for $i>n$. Now, if $U \neq M$ is an open set which retracts onto X then $H_{i-1} X \subset H_{i-1} U \cong H_i(M, U)=0$, the latter by case 2. This proves part (a) of 3.3.

For (b) we consider the diagram

$$
\text{(3.11)}\qquad
\begin{array}{ccccccc}
0 \to H_n M & \xrightarrow{\kappa_*} & H_n(M, X) & \longrightarrow & \hat{H}_{n-1} X & \to 0 \\
\downarrow{\scriptstyle\cong} & & \downarrow{\scriptstyle J} & & \downarrow{\scriptstyle \hat{J}} & \\
0 \to \Gamma_b M & \xrightarrow{\kappa'} & \Gamma_b(M-X) & \longrightarrow & \hat{\Gamma}(M-X) & \to 0,
\end{array}
$$

where $\hat{H}_{n-1} X = \ker(H_{n-1} X \to H_{n-1} M) \cong \operatorname{coker}(\kappa_*)$, $\hat{\Gamma}(M-X) = \operatorname{coker}(\kappa')$, and $\hat{J}$ is induced by J. It follows that $J = J_X$ is isomorphic if and only if $\hat{J} = \hat{J}_X$ is isomorphic. In particular, $\hat{J}_U$ is isomorphic for all open subsets U of M, by case 2.

Let $r\colon U \to X$ be a retraction of an open subset. If we choose U small enough then $i^X r \simeq i^U\colon U \to M$, where i^X, i^U are inclusions (cf. IV, 8.6), hence $i^X_* r_* = i^U_*$, hence r_* maps $\hat{H} U = \ker(i^U_*)$ into $\hat{H} X = \ker(i^X_*)$, and is a left inverse of $i_*\colon \hat{H} X \to \hat{H} U$. In particular, i_* is monomorphic. The diagram

$$
\text{(3.12)}\qquad
\begin{array}{ccc}
\hat{H}_{n-1} X & \xrightarrow{i_*} & \hat{H}_{n-1} U \\
\downarrow{\scriptstyle \hat{J}} & & \downarrow{\scriptstyle\cong} \\
\hat{\Gamma}(M-X) & \longrightarrow & \hat{\Gamma}(M-U)
\end{array}
$$

then shows that $\hat{J}$ is monomorphic.

To prove surjectivity we choose, for every $Q \in (M-X)$, an open set V_Q such that $X \subset V_Q \subset U-Q$ and $i_Q(r|V_Q) \simeq k_Q\colon V_Q \to U-Q$ where i_Q, k_Q are inclusions (this is possible by IV, 8.6); then $i_{Q*}(r|V_Q)_* = k_{Q*}$. We record the whole situation in the diagram

$$
\text{(3.13)}\qquad
\begin{array}{ccccccc}
\hat{H}_{n-1} X & \xrightarrow{j_{Q*}} & \hat{H}_{n-1} V_Q & \xrightarrow{k_{Q*}} & \hat{H}_{n-1}(U-Q) & \xrightarrow{l_{Q*}} & \hat{H}_{n-1} U \\
\downarrow{\scriptstyle J} & & {\scriptstyle\cong}\downarrow{\scriptstyle J} & & {\scriptstyle\cong}\downarrow{\scriptstyle J} & & {\scriptstyle\cong}\downarrow{\scriptstyle J} \\
\hat{\Gamma}(M-X) & \xrightarrow[j'_Q]{} & \hat{\Gamma}(M-V_Q) & \xrightarrow[k'_Q]{} & \hat{\Gamma}((M-U) \cup Q)) & \xrightarrow[l'_Q]{} & \hat{\Gamma}(M-U),
\end{array}
$$

$$
i_{Q*} = k_{Q*} j_{Q*}, \qquad i'_Q = k'_Q j'_Q, \qquad i' = l'_Q i'_Q.
$$

Let $\rho_Q = (r|V_Q)_* \hat{J}^{-1} j'_Q : \hat{\Gamma}(M-X) \to \hat{H}_{n-1} X$. Then

$$i'_Q \hat{J} \rho_Q = i'_Q \tag{3.14}$$

because

$$\begin{aligned} i'_Q \hat{J} \rho_Q &= i'_Q \hat{J} (r|V_Q)_* \hat{J}^{-1} j'_Q = \hat{J} i_{Q*} (r|V_Q)_* \hat{J}^{-1} j'_Q \\ &= \hat{J} k_{Q*} \hat{J}^{-1} j'_Q = k'_Q \hat{J} \hat{J}^{-1} j'_Q = k'_Q j'_Q = i'_Q. \end{aligned}$$

Composing (3.14) with l'_Q gives $(i'\hat{J})\rho_Q = i'$. The right side of this does not depend on Q, and $i' \hat{J} = \hat{J} i_*$ is monomorphic, hence $\rho = \rho_Q$ is independent of Q. We claim $\hat{J}\rho = \mathrm{id}$; in particular, $\hat{J}$ is epimorphic.

Let $s \in \Gamma_b(M-X)$, and $\hat{s} \in \hat{\Gamma}(M-X)$ its coset. Let $\sigma \in \Gamma_b(M-X)$ be a representative of $\hat{J}\rho(\hat{s})$. We have to show that s and σ differ only by a global section $t \in \Gamma_b M$. Now $i'_Q \hat{J} \rho \hat{s} = i'_Q \hat{s}$ by (3.14), hence $s|(M-U) \cup Q$ and $\sigma|(M-U) \cup Q$ differ only by a global section $t_Q \in \Gamma_b M$. All of these t_Q agree on $M-U$ (they equal $s-\sigma$ there) and therefore they all coincide (M is connected!), say $t_Q = t$. It follows that $sQ - \sigma Q = tQ$ for all $Q \in M-X$, as required.

Case 4. M as in case 2 and 3, X and Y arbitrary neighborhood retracts.

We can apply case 3 to both (M,X) and (M,Y). The exact homology sequence of the triple $X \subset Y \subset M$ then yields the result for (Y,X), as in step 6 of IV, 6.4.

Case 5. The general case.

Let $y \in H_k(Y,X)$, and let $z \in SY$ be a representative. Since z has a compact carrier there exists a finite union W of coordinate neighborhoods such that $z \in SW$. If $k > n$ then $[z] \in H_k(Y \cap W, X \cap W)$ is zero by case 4 (applied to the manifold W), hence $y = i_*[z] = 0$ (i = inclusion). If $k = n$ we consider the diagram

$$\begin{array}{ccc} H_n(Y \cap W, X \cap W) & \xrightarrow{i_*} & H_n(Y,X) \\ \cong \big\downarrow J & & \big\downarrow J \\ 0 \to \Gamma_b(W-X, W-Y) & \xrightarrow{e} & \Gamma_b(M-X, M-Y), \end{array} \tag{3.15}$$

where e is the map which extends every section (which vanishes outside of some compact set of W) by zero; clearly e is monomorphic. The left J is isomorphic by case 4 (applied to the manifold W) hence

$$Jy = 0 \Rightarrow eJ[z] = 0 \Rightarrow [z] = 0 \Rightarrow y = 0.$$

It remains to prove surjectivity of $J\colon H_n(Y,X)\to\Gamma_b(M-X,M-Y)$. If $s\in\Gamma_b(M-X,M-Y)$ then s vanishes outside of some compact set K of M, and K is contained in a finite union W of coordinate neighborhoods. Therefore s is in the image of e, hence also in the image of J, by 3.15. ∎

3.16 Exercises. *1.* If $X\subset Y$ are open subsets of a manifold M then $r\colon \Gamma_b(M-X,M-Y)\cong\Gamma_c(Y-X)$, where $\Gamma_c(Y-X)$ denotes the group of sections with *compact support*, i.e. of sections $s\in\Gamma(Y-X)$ such that $Y-X-s^{-1}(0)$ is compact. The isomorphism r is obtained by restricting, its inverse r^{-1} by trivially-extending sections.

The group $\Gamma_c(Y-X)$ is of interest because it depends only on $Y-X$ and the part of the orientation covering over $Y-X$, not on the ambient manifold M. For orientable Y, or with $\mathbb{Z}_2$-coefficients, it depends only on the space $Y-X$; it follows that, $H_n(Y,X)$ depends only on $Y-X$, in this case. We shall see in 7.14 that a similar result holds for all homology groups $H(Y,X)$.

2. If M is an n-manifold and G an abelian group then there is a unique family of homomorphisms $\alpha'\colon \Gamma(A;\mathbb{Z})\otimes G\to\Gamma(A;G)$, $A\subset M$, which is natural with respect to inclusions, and for one-point sets $A=P$ agrees with the isomorphism $H_n(M,M-P;\mathbb{Z})\otimes G\cong H_n(M,M-P;G)$ of 2.1.

3. If $X\subset Y$ are neighborhood retracts in an orientable n-manifold M then

$$\alpha'\colon \Gamma_b(M-X,M-Y;\mathbb{Z})\otimes G\cong\Gamma_b(M-X,M-Y;G),$$
$$\alpha\colon H_n(Y,X)\otimes G\cong H_n(Y,X;G),\qquad \text{for all } G.$$

The latter implies (by the argument of 3.5) that $H_{n-1}(Y,X;\mathbb{Z})$ is torsion-free.

*4**. If L is an n-dimensional ∂-manifold and $V\subset L$, define

$$J_V^i\colon H_n(L,L-iV;G)\to\Gamma_b(iV;G)$$

as in 2.15 (where $iV=V\cap iL$), and show that J_V^i is isomorphic if $L-iV$ is a neighborhood retract (hint: reduce to the absolute case 3.3 by attaching a collar to L). Find analogous conditions for V which imply isomorphisms $H_{n-1}(L-iV,L-V;G)\cong\Gamma_b(\partial V;G)$, where $\partial V=V\cap\partial L$.

4. Fundamental Class and Degree

For open subsets of $\mathbb{S}^n$ these notions have already played a rôle (IV, 5; VII, 2.14). We now briefly treat some generalizations to arbitrary manifolds.

4.1 Definition. Let M^n be an oriented n-manifold, $O \in \Gamma(M, \mathbb{Z})$ its orientation, and $K \subset M$ a compact set. There exists then a unique element $o_K \in H_n(M, M-K; \mathbb{Z})$ which under the isomorphism J of 3.3 corresponds to $(O|K) \in \Gamma(K; \mathbb{Z})$; thus o_K is characterized by the property that the inclusion homomorphism $i_*^P\colon H_n(M, M-K) \to H_n(M, M-P)$ takes o_K into $O(P)$, for every $P \in K$. The element o_K is called the *fundamental class around* K. In particular, if M is itself compact there is a fundamental class $o_M \in H_n(M; \mathbb{Z})$. If K is connected, non-empty, then $H_n(M, M-K; \mathbb{Z}) \cong \Gamma K \cong \mathbb{Z}$, and o_K is a generator of this group.

If M is not oriented (or not even orientable) we can take $\mathbb{Z}_2$-coefficients and still use the same definition. This remark applies to the whole §4: We speak of oriented manifolds, but the theory applies equally well to the non-oriented case after replacing $\mathbb{Z}$ by $\mathbb{Z}_2$.

4.2 Definition. Let $f\colon M'^n \to M^n$ be a continuous map between oriented n-manifolds, and let $K \subset M$ be a compact connected set ($\neq \emptyset$) such that $f^{-1}K$ is compact. Then $f_*\colon H_n(M', M'-f^{-1}K) \to H_n(M, M-K)$ takes the fundamental class $o_{f^{-1}K}$ into an integral multiple of o_K; this integer is called the *degree of f over K*, and is denoted by $\deg_K f$. In symbols, $f_*(o_{f^{-1}K}) = (\deg_K f)\, o_K$. If $K = \emptyset$ then $\deg_K f$ is not defined; we could agree that $\deg_\emptyset f = \mathbb{Z} =$ set of all integers.

For instance, if K is a point and M, M' are open subsets of $\mathbb{S}^n$ then the definition of $\deg_K f$ reduces to IV, 5. If $f^{-1}K = \emptyset$ (e.g. if K is a point $\notin \operatorname{im}(f)$) then $\deg_K f = 0$. If f is the inclusion map of an open subset M' of M (with $O' = O|M'$) then $\deg_K f = 1$ for every $K \subset M'$. More generally, if f is a homeomorphism of M' onto an open set of M then $\deg_K f = \pm 1$ for every $K \subset \operatorname{im}(f)$.—All of this is quite obvious (compare IV, 5.4).

It is sometimes convenient to replace $f^{-1}K$ by a larger compact set, as follows.

4.3 Proposition. *If $f\colon M' \to M$ and $K \subset M$ are as in 4.2, and if $K' \subset M'$ is any compact set containing $f^{-1}K$ then $f_*\colon H(M', M'-K') \to H(M, M-K)$ takes the fundamental class $o_{K'}$ into $(\deg_K f)\, o_K$.*

Proof. The inclusion homomorphism $H(M', M'-K') \to H(M', M'-f^{-1}K)$ takes $o_{K'}$ into $o_{f^{-1}K}$ (by definition of o). Therefore the composition $H(M', M'-K') \to H(M', M'-f^{-1}K) \xrightarrow{f_*} H(M, M-K)$ takes $o_{K'}$ into $f_*(o_{f^{-1}K}) = (\deg_K f)\, o_K$. ∎

4.4 Proposition. *If $f\colon M' \to M$ and $K \subset M$ are as in 4.2, and $I \subset K$ is also compact then $f_*\colon H(M', M'-f^{-1}I) \to H(M, M-I)$ takes $o_{f^{-1}I}$ into $(\deg_K f)\, o_I$. In particular, $\deg_I f = \deg_K f$ for every connected compact part I ($\neq \emptyset$) of K.*

For instance, this implies $\deg_K f = \deg_P f$ for every point $P \in K$. In particular, if $K \not\subset \operatorname{im}(f)$ then $P \notin \operatorname{im}(f)$ for some P, hence $\deg_K f = \deg_P f = 0$.

Proof. Consider the commutative diagram

$$\begin{array}{ccc} H(M', M' - f^{-1}K) & \xrightarrow{f_*} & H(M, M-K) \\ \downarrow{\scriptstyle i'_*} & & \downarrow{\scriptstyle i_*} \\ H(M', M' - f^{-1}I) & \xrightarrow{f_*} & H(M, M-I), \end{array}$$

where i, i' are inclusions. Chasing $o_{f^{-1}K}$ through the diagram gives $o_{f^{-1}K} \mapsto (\deg_K f)\, o_K \mapsto (\deg_K f)\, o_I$, respectively

$$o_{f^{-1}K} \mapsto o_{f^{-1}I} \mapsto f_*(o_{f^{-1}I}). \quad ∎$$

If $f^{-1}K$ is compact for every compact set $K \subset M$ then $f\colon M' \to M$ is called a *proper map.* E.g., every homeomorphism $M' \approx M$ is proper. If M' is itself compact then every continuous $f\colon M' \to M$ is proper.

4.5 Proposition and Definition. *If $f\colon M' \to M$ is a proper map of oriented manifolds and if M is connected then the number $\deg_K f$ is the same for all connected compact parts $K (\neq \emptyset)$ of M. It is called the degree of f, in symbols $\deg f$. The equality $f_*(o_{f^{-1}K}) = (\deg f)\, o_K$ holds for all compact sets $K \subset M$, whether they are connected or not.*

For instance, if M and M' are compact then $\deg f$ is characterized by the formula $f_*(o_{M'}) = (\deg f)\, o_M$. If M' is compact but M is not then $\deg f = 0$ because $\operatorname{im}(f) \neq M$. If M, M' are arbitrary again and $f\colon M' \approx M$ is homeomorphic then $\deg f = \pm 1$ (M being connected); according to these two cases f is called *orientation-preserving* or *-reversing*.

Proof. If $K^1, K^2 \subset M$ are arbitrary compact sets then we can find a connected compact set K in M which contains both K^1 and K^2 (cover $K^1 \cup K^2$ by a finite number of closed balls and connect these by paths). By 4.4, $f_*(o_{f^{-1}(K^\nu)}) = (\deg_K f)\, o_{K^\nu}$, for $\nu = 1, 2$; this implies the assertion. ∎

4.6 Corollary. *If $M'' \xrightarrow{g} M' \xrightarrow{f} M$ are maps of oriented n-manifolds, if g is proper, M' connected, and $K \subset M$ is a compact connected set $(\neq \emptyset)$ such that $f^{-1}K$ is compact then $\deg_K(fg) = (\deg g)(\deg_K f)$. In particular, if f is proper, too, and M connected then $\deg(fg) = (\deg g)(\deg f)$.*

Proof. $\deg(fg)\, o_K = (fg)_*\, o_{(fg)^{-1}K} = f_*[(\deg g)\, o_{f^{-1}K}] = (\deg g)(\deg_K f)\, o_K$, the second equality by the last part of 4.5. ∎

Generalizing IV, 5.8 we have the following

4.7 Proposition (Additivity). *If $f: M' \to M$ and $K \subset M$ are as in 4.2, and M' is a finite union of open sets M'_λ, $\lambda = 1, 2, \ldots, r$, such that the sets $K'_\lambda = (f^{-1}K) \cap M'_\lambda$ are mutually disjoint then $\deg_K f = \sum_{\lambda=1}^r \deg_K f^\lambda$, where $f^\lambda = f|M'_\lambda$* (n.b., *every K'_λ is compact because $f^{-1}K$ is the topological sum of the K'_λ*).

Proof. Consider the maps

$$\oplus_{\lambda=1}^r H(M'_\lambda, M'_\lambda - K'_\lambda) \xrightarrow{(i_*^\lambda)} H(M', M' - f^{-1}K) \xrightarrow{i_*^Q} H(M', M' - Q),$$

where the i^λ are inclusions and $Q \in f^{-1}K$. If we apply the composite to $\{o_{K'_\lambda}\}$ then all components $o_{K'_\lambda}$ go into zero except the component $o_{K'_\mu}$ for which $Q \in K'_\mu$; this one goes into o_Q. Therefore $i_*^Q[\{i_*^\lambda\}\{o_{K'_\lambda}\}] = o_Q$, for every $Q \in f^{-1}K$, hence $\{i_*^\lambda\}\{o_{K'_\lambda}\} = o_{f^{-1}K}$ (cf. 4.1). Now,

$$(\deg_K f) o_K = f_*(o_{f^{-1}K}) = f_*\{i_*^\lambda\}\{o_{K'_\lambda}\} = \{f_*^\lambda\}\{o_{K'_\lambda}\}$$
$$= \sum_{\lambda=1}^r f_*^\lambda(o_{K'_\lambda}) = \left(\sum_{\lambda=1}^r \deg_K f^\lambda\right) o_K. \quad \blacksquare$$

Proposition 4.7 can serve to interpret $\deg_P f$ as "number of points in $f^{-1}P$, counted with multiplicities". For this we refer the reader to the remarks after IV, 5.8; they carry over literally.

4.8 Example. If $X \subset \mathbb{R}^n$ is a compact connected $(n-1)$-manifold then $\mathbb{R}^n - X$ has two components, one bounded V, and one unbounded W (cf. 3.7). Let $B = V \cup X = \bar{V}$[8]. This is a neighborhood retract; indeed, if $\rho: U \to X$ is a neighborhood retraction for X then $r: U \cup V \to B$, $r|B = \mathrm{id}$, $r|U - V = \rho|U - V$, is a neighborhood retraction for B.—If X is locally flat (1.8) in $\mathbb{R}^n$ then B is easily seen to be a ∂-manifold with boundary $\partial B = X$.

A theorem of H. Hopf asserts that if $v(X)$ is a non-zero vector which depends continuously on $x \in X$ and points out of B then the degree of the map $X \to \mathbb{S}^{n-1}$, $x \mapsto \dfrac{v(x)}{\|v(x)\|}$, equals the Euler-characteristic of B. In order to make this precise, remark first that $H_n(B, X) \cong \Gamma_b(\mathbb{R}^n - X, W) \cong \Gamma V \cong \mathbb{Z}$. Let $o \in H_n(B, X)$ denote the generator which maps into

$$o_P \in H_n(B, B - P) = H_n(\mathbb{R}^n, \mathbb{R}^n - P), \quad \text{for } P \in V.$$

[8] **Proof.** Clearly $\bar{V} \subset (V \cup X)$. Assume $X \not\subset \bar{V}$; then X contains a small open $(n-1)$-ball D such that $D \cap \bar{V} = \emptyset$. It follows that $W \cup D$ is open in $\mathbb{R}^n$, and $\mathbb{R}^n - (X - D) = V \cup (W \cup D)$; in particular, V is a bounded component of $\mathbb{R}^n - (X - D)$. But $H_{n-1}(X - D; \mathbb{Z}_2) \cong \Gamma(X, D; \mathbb{Z}_2) = 0$, by 3.3; hence $\mathbb{R}^n - (X - D)$ has no bounded component, by 3.7.

Then $\partial_* o$ is a generator of $H_{n-1}X$ (because $H_n B \cong \Gamma_b(\mathbb{R}^n, W)=0$, $H_{n-1}B \cong H_n(\mathbb{R}^n, B) \cong \Gamma_b(\mathbb{R}^n - B)=0$, hence $\partial_*\colon H_n(B,X) \cong H_{n-1}X$). It is this generator of $H_{n-1}X$ and the analogous generator of $\mathbb{S}^{n-1}$ which we use to define the degree of $X \to \mathbb{S}^{n-1}$. The theorem of Hopf can now be formulated as follows.

4.9 Proposition. *If $\varphi\colon X \to B$ is a map which is homotopic to the inclusion map $\iota\colon X \to B$ and which has no fixed point ($\varphi \simeq \iota$, $\varphi(x) \neq x$) then the degree of $g\colon X \to \mathbb{S}^{n-1}$, $g(x) = \dfrac{x - \varphi x}{\|x - \varphi x\|}$, equals the Euler-characteristic of B,* $\deg(g) = \chi(B)$.

Proof. We can extend φ to a map $\Phi\colon B \to B$ such that $\Phi \simeq \mathrm{id}$. This is so because $X \subset B$ has the homotopy extension property (IV, 8.13, Exerc. 3); an ad hoc proof is as follows. Extend φ first to a neighborhood U of X in B, say $\psi\colon U \to B$, and take a deformation $d\colon U \times [0,1] \to B$ with $d(u,0) = \psi u$, $d(u,1) = u$; this is possible by 6.2. Choose a continuous function $\tau\colon B \to [0,2]$ such that $\tau | X = 0$, $\tau | B - U = 2$, and define

$$D\colon B \times [0,1] \to B$$

by

$$D(b,t) = \begin{cases} d(b, \mathrm{Min}(1, t + \tau u)), & \text{for } \tau u \le 1 \\ b, & \text{for } \tau u \ge 1. \end{cases}$$

Then $\Phi(b) = D(b,0)$ is the required map, and D is a homotopy $\Phi \simeq \mathrm{id}$.

Let F denote the fixed-point set of Φ. The fixed-point index I_Φ of Φ agrees with the fixed point index of $\Phi | V\colon V \to B \subset \mathbb{R}^n$ because $F \subset V$ (cf. VII, 5.11); the latter equals (cf. VII, 5.2) the local degree over 0 of $G\colon V \to \mathbb{R}^n$, where $Gv = v - \Phi v$. The fundamental class $o_F \in H_n(V, V-F) \cong H_n(B, B-F)$ is the image of $o \in H_n(B,X)$, as remarked above, therefore

$$G\colon (B,X) \to (\mathbb{R}^n, \mathbb{R}^n - \{0\}), \qquad Gb = b - \Phi b,$$

takes o into I_Φ-times the generator $o_0 \in H_n(\mathbb{R}^n, \mathbb{R}^n - \{0\})$.

It follows that $G|X\colon X \to \mathbb{R}^n - \{0\}$ takes $\partial_* o$ into I_Φ-times the generator of $H_{n-1}(\mathbb{R}^n - \{0\}) \cong H_{n-1}\mathbb{S}^{n-1}$. But $G|X$ is essentially g (up to homotopy), hence $\deg(g) = I_\Phi$. On the other hand, $I_\Phi = \Lambda(\Phi) = \Lambda(Id_B) = \chi(B)$, by VII, 6.6. ∎

4.10 Exercises. *1.* Let M be an orientable n-manifold and X a compact (but not necessarily connected) submanifold of dimension $n-1$ which bounds mod 2, i.e. whose mod 2 fundamental class $\bar{o}$ lies in the kernel of $i_*\colon H_{n-1}(X;\mathbb{Z}_2) \to H_{n-1}(M;\mathbb{Z}_2)$, i = inclusion. Then X is orientable, and it can be so oriented that its integral fundamental class

$$o \in \ker\left(i_*: H_{n-1}(X;\mathbb{Z}) \to H_{n-1}(M;\mathbb{Z})\right).$$

This generalizes 3.9. It can be proved by considering the diagram

$$\begin{array}{ccccccc}
\Gamma_b(M-X;\mathbb{Z}) \cong & H_n(M,X;\mathbb{Z}) & \longrightarrow & H_{n-1}(X;\mathbb{Z}) & \longrightarrow & H_{n-1}(M;\mathbb{Z}) \\
\downarrow & \downarrow & & \downarrow & & \downarrow \\
\Gamma_b(M-X;\mathbb{Z}_p) \cong & H_n(M,X;\mathbb{Z}_p) & \longrightarrow & H_{n-1}(X;\mathbb{Z}_p) & \longrightarrow & H_{n-1}(M;\mathbb{Z}_p),
\end{array}$$

first for $p=2$ and then also for other primes.

2. If M^m, N^n are oriented manifolds and $I \subset M$, $K \subset N$ are compact subsets then $o_{I\times K} = o_I \times o_K$ (compare VII, 2.15). If $f\colon M'^m \to M^m$, $g\colon N'^n \to N^n$ are maps of oriented manifolds then $\deg_{I\times K}(f\times g) = (\deg_I f)(\deg_K g)$, whenever these terms are defined.

3. If $\Theta\colon [0,1]\times M'^n \to M^n$ is a deformation (M, M' oriented), and $K\subset M$ is a compact connected set ($\neq\emptyset$) such that $\Theta^{-1}K$ is compact then $\deg_K(\Theta_0) = \deg_K(\Theta_1)$ (compare IV, 5.13, Exerc. 3).—Show that every complex-linear isomorphism $\varphi\colon \mathbb{C}^n \approx \mathbb{C}^n$ has degree $+1$ (hint: $\varphi \simeq \mathrm{id}$).

4. Every proper map $\mathbb{R}\to\mathbb{R}$ has degree 0 or ± 1. Determine the degree of $x \mapsto x^k$, $k=0,1,2,\ldots$ (for $x\in\mathbb{R}$, and also for $x\in\mathbb{C}$).

5*. Show: If $\tau\colon \mathbb{S}^n \to \mathbb{S}^n$ is an involution, $\tau\tau = \mathrm{id}$, $\tau \neq \mathrm{id}$, then $\tau P = -P$ for some $P\in\mathbb{S}^n$. Hints: For every $x\in\mathbb{S}^n$ let $f(x)$ be the center of the geodesic arc from x to τx; if $\tau x \neq -x$ this defines a map $f\colon \mathbb{S}^n\to\mathbb{S}^n$ such that $f(x) = f(\tau x)$ and $f\simeq\mathrm{id}$ (just deform along great arcs), hence $\deg(f)=1$. Let $M = \{x\in\mathbb{S}^n \mid \tau x \neq x\}$, and $M' = f^{-1}M$; then $\deg(f|M'\colon M'\to M) = \deg(f) = 1$. But $f|M'$ factors, $M' \xrightarrow{\pi} M'' \xrightarrow{f''} M$, where M'' is obtained from M' by identifying x with τx; the covering map π has degree 0 mod 2, hence $\deg(f|M') \equiv \deg(f'')\deg(\pi) \equiv 0 \bmod 2$.

More generally: If $\tau\colon \mathbb{S}^n\to\mathbb{S}^n$ is such that $\tau^k = \mathrm{id}$, $\tau\neq\mathrm{id}$, then $P + \tau P + \tau^2 P + \cdots + \tau^{k-1} P = 0$ for some $P\in\mathbb{S}^n$.

6*. For every manifold M let $\gamma_1\colon \tilde{M}_1 \to M$ denote its orientation covering (2.11), and $\tau\colon \tilde{M}_1\to\tilde{M}_1$, $\tau(u) = -u$, the *canonical involution.* A map $f\colon M'^n \to M^n$ is called *orientable* if $\tilde{f}_1\colon \tilde{M}'_1 \to \tilde{M}_1$ exists such that $\gamma_1\tilde{f}_1 = f\gamma'_1$ and $\tau\tilde{f}_1 = \tilde{f}_1\tau'$; any such $\tilde{f}_1$ is called an *orientation of f.* (For instance, if f is a homeomorphism of M' onto an open subset of M then f is orientable. If M, M' are oriented—by O, O'—then every f has a unique orientation $\tilde{f}_1$ such that $\tilde{f}_1 \circ O' = O \circ f$.) Show that the preceding theory of the degree (with integral coefficients) generalizes to orientable maps, replacing f by $\tilde{f}_1$ (n.b. $\tilde{M}_1$ is canonically oriented).

7. If $f: M' \to M$ is a map of oriented manifolds, and $K \subset M$ is compact connected, then it is sometimes possible to define $\deg_K f$ even though $f^{-1}K$ is not compact: Suppose $f^{-1}K$ is a disjoint union of sets K'_λ, $\lambda \in \Lambda$, which are both compact and open in $f^{-1}K$. Choose an open set $M'_\lambda \subset M'$ such that $K'_\lambda = (f^{-1}K) \cap M'_\lambda$, and put $\deg_K f = \sum_{\lambda \in \Lambda} \deg_K f^\lambda$ (where $f^\lambda = f|M'_\lambda$), whenever this makes sense ($\deg_K f$ may be an integer or $\pm\infty$). For instance, if f is a covering map of oriented (connected) manifolds then $\deg_P f =$ number of sheets (possibly ∞), for every $P \in M$.

5. *Limits*

Arguments about limits of spaces or groups have already been used in several instances (e.g., 3.3; 2.2; IV, 6.4) without explicitly saying so. This omission seemed justified because simple ad hoc proofs could be supplied, thus keeping the exposition short. Some of the deeper results on manifolds (VIII, 7), however, involve limits more essentially, and we now provide the necessary background.—More general limits (Kan-limits) are discussed in the appendix (cf. A.1).

5.1 Definition. A relation $\lambda \le \mu$ in a set Λ is called a *quasi-ordering* if

(i) $\lambda \le \lambda$ for every $\lambda \in \Lambda$ (*reflexivity*), and

(ii) $\lambda \le \mu \le \nu \Rightarrow \lambda \le \nu$ (*transitivity*).

A quasi-ordered set is *directed* if for every pair $\lambda, \lambda' \in \Lambda$ there exists $\mu \in \Lambda$ such that $\lambda \le \mu$ and $\lambda' \le \mu$.

5.2 Definition. If Λ is a quasi-ordered set and $\mathscr{K}$ a category then a *direct Λ-system* (in $\mathscr{K}$) is a function D which assigns to every $\lambda \in \Lambda$ an object $D_\lambda \in \mathscr{K}$, and to every pair $\lambda \le \mu$ in Λ a morphism $D^\mu_\lambda: D_\lambda \to D_\mu$ such that $\lambda \le \mu \le \nu \Rightarrow D^\nu_\mu D^\mu_\lambda = D^\nu_\lambda$, and $D^\lambda_\lambda = \mathrm{id}$. In other words, if we view Λ as a category, with objects $\{\lambda\}$ and with one morphism $\lambda \to \mu$ for each $\lambda \le \mu$ then *a direct system is the same as a functor $D: \Lambda \to \mathscr{K}$.*

A cofunctor $I: \Lambda \to \mathscr{K}$ is called an *inverse Λ-system* (in $\mathscr{K}$). It maps $\Lambda \ni \lambda \mapsto I^\lambda \in \mathscr{K}$, $(\lambda \le \mu) \mapsto (I^\mu_\lambda: I^\mu \to I^\lambda)$, and satisfies $I^\mu_\lambda I^\nu_\mu = I^\nu_\lambda$ for $\lambda \le \mu \le \nu$, $I^\lambda_\lambda = \mathrm{id}$.

A natural transformation $D \to D'$, respectively $I \to I'$, of functors is also called *transformation of direct* (*inverse*) *systems*. Direct (inverse) systems and their transformations form a category $\{\Lambda. \mathscr{K}\}$, resp. $\{\Lambda, \mathscr{K}^{\mathrm{op}}\}$. Replacing $\mathscr{K}$ by the dual category $\mathscr{K}^{\mathrm{op}}$ (or Λ by Λ^{op}) takes one into the other; in general, we can therefore restrict attention to direct systems.

5.3 Examples. If Λ is trivially ordered ($\lambda \le \mu \Leftrightarrow \lambda = \mu$) then Λ-systems are just Λ-families $\{D_\lambda\}_{\lambda\in\Lambda}$ of objects. If $\Lambda = \mathbb{N}$ = set of natural numbers with the usual order then Λ-systems are sequences $D_1 \to D_2 \to D_3 \to \cdots$, resp. $I^1 \leftarrow I^2 \leftarrow I^3 \leftarrow \cdots$ in $\mathscr{K}$. If Λ is the set of open subsets of a topological space Y, and $\le$ denotes inclusion then inverse Λ-systems are usually called *presheaves* (over Y, with values in $\mathscr{K}$). The last two examples Λ are directed, the first is not (if Λ has more than one element). Another example of a directed Λ is the set of (quasi-) compact subsets of Y (ordered by inclusion).

5.4 Definition. Let Λ be a quasi-ordered set, and $\mathscr{K}$ a category. Given an object $K \in \mathscr{K}$, consider the *constant functor* $\Lambda \to \mathscr{K}$ which takes every $\lambda \in \Lambda$ into K, and every relation $\lambda \le \mu$ into id_K; we use the same letter K to denote this functor. A transformation $\varphi: D \to K$ of direct systems then assigns to every $\lambda \in \Lambda$ a morphism $\varphi_\lambda: D_\lambda \to K$ such that $\varphi_\mu D^\mu_\lambda = \varphi_\lambda$ whenever $\lambda \le \mu$.

Let us fix a direct system D now. A transformation $u: D \to L$, where $L \in \mathscr{K}$, is called *universal* if

$$\mathscr{K}(L, K) \to \mathrm{Transf}(D, K), \qquad \psi \mapsto \psi \circ u, \qquad (\psi \circ u)_\lambda = \psi \circ u_\lambda,$$

is bijective for all $K \in \mathscr{K}$, i.e., if for every transformation $\varphi: D \to K$ there exists a unique morphism $\psi: L \to K$ such that $\varphi_\lambda = \psi\, u_\lambda$ for all $\lambda \in \Lambda$. For instance, if D itself is a constant functor, $D_\lambda = L$, $D^\mu_\lambda = \mathrm{id}_L$ then $u_\lambda = \mathrm{id}_L$ is universal.

If a universal transformation exists it is essentially unique; more precisely,

5.5 Proposition and Definition. *If $u: D \to L$, $u': D \to L'$ are two universal transformations then there is a unique morphism $\kappa: L \to L'$ such that $\kappa u = u'$, and this κ is an equivalence, $L \cong L'$.*

If a universal transformation $u: D \to L$ exists then L is called the (direct) *limit of* D; in symbols, $L = \varinjlim D$. We also write $L = \varinjlim \{D_\lambda | \lambda \in \Lambda\}$, or $L = \varinjlim \{D_\lambda\}$ if the morphisms D^μ_λ (and the index set Λ) are clear from the context. If $\varphi: D \to K$ is a transformation we also write $\{\varphi_\lambda\}: \varinjlim \{D_\lambda\} \to K$ for the corresponding morphism ψ.

Dually, for inverse systems I we have $u: L \to I$, $L = \varprojlim I = \varprojlim \{I^\lambda\}$, and $\{\varphi^\lambda\}: K \to \varprojlim \{I^\lambda\}$.

Proof of 5.5. By universality of u there is a unique $\kappa: L \to L'$ with $\kappa u = u'$; similarly for $\kappa': L' \to L$, $\kappa' u' = u$. It follows that $\kappa' \kappa u = u$, hence $\kappa' \kappa = \mathrm{id}_L$ because u is universal; similarly, $\kappa \kappa' = \mathrm{id}_{L'}$. ∎

5.6 (compare I, 1.15). If the ordering of Λ is trivial ($\lambda \leq \mu \Leftrightarrow \lambda = \mu$) then $\varinjlim D$ (if it exists) coincides with the *coproduct of the family* $\{D_\lambda\}_{\lambda \in \Lambda}$, in symbols $\bigsqcup_{\lambda \in \Lambda} D_\lambda$, and the morphisms $\iota_\mu \colon D_\mu \to \bigsqcup_{\lambda \in \Lambda} D_\lambda$ are the *injections of the cofactors.* In some concrete categories ($\mathscr{AG}$, or $\mathscr{Top}$, ...) $\bigsqcup D_\lambda$ is also denoted by $\oplus D_\lambda$, and is called the (direct, or topological) *sum,* with ι_μ the "inclusion of the summand D_μ".—Dually, $\varprojlim I$ coincides with the product $\prod_{\lambda \in \Lambda} I^\lambda$.

For every quasi-ordered set, $\varinjlim D$ can be thought of as a "quotient of $\bigsqcup D_\lambda$" and $\varprojlim I$ as a "subobject of $\prod I^\lambda$". We do not discuss these general notions (cf. Mitchell II, 2) but mention the following

5.7 Proposition. *Let* $D\colon \Lambda \to \mathscr{AG}$ *be a direct system of abelian groups (or modules, or complexes). Then* $\varinjlim D$ *is the quotient of* $\oplus_{\lambda \in \Lambda} D_\lambda$ *by the subgroup (-module, -complex) which is generated by all elements of the form* $(\iota_\lambda - \iota_\mu D_\lambda^\mu)(x_\lambda)$ *where* $x_\lambda \in D_\lambda$, $\lambda \leq \mu$, *and* ι_ν = *inclusion. A universal transformation* $v = \{v_\lambda\}$ *is given by composing*

$$D_\lambda \xrightarrow{\iota_\lambda} \oplus_{\nu \in \Lambda} D_\nu \xrightarrow{\text{proj}} \oplus_{\nu \in \Lambda} D_\nu / \{\iota_\lambda x_\lambda - \iota_\mu D_\lambda^\mu x_\lambda\}.$$

Dually,

$$\varprojlim I = \{x = \{x_\nu\} \in \textstyle\prod_{\nu \in \Lambda} I^\nu \mid x_\lambda = I_\lambda^\mu x_\mu \text{ for all } \lambda \leq \mu\}.$$

In particular, these limits always exist.

Proof. Put $L = \oplus_{\nu \in \Lambda} D_\nu / \{\iota_\lambda x_\lambda - \iota_\mu D_\lambda^\mu x_\lambda\}$. If $\{\varphi_\lambda \colon D_\lambda \to K\}$ is a transformation, $K \in \mathscr{AG}$, we have to construct $\psi\colon L \to K$ such that $\psi v_\lambda = \varphi_\lambda$ for $\lambda \in \Lambda$; since the images, $\operatorname{im}(v_\lambda)$, clearly generate L there is at most one such ψ. By the universal property I, 2.14 of direct sums we can find $\psi'\colon \oplus_{\nu \in \Lambda} D_\nu \to K$ such that $\psi' \iota_\lambda = \varphi_\lambda$. Further $\psi'(\iota_\lambda x_\lambda - \iota_\mu D_\lambda^\mu x_\lambda) = \varphi_\lambda x_\lambda - \varphi_\mu D_\lambda^\mu x_\lambda = \varphi_\lambda x_\lambda - \varphi_\lambda x_\lambda = 0$, hence ψ' vanishes on $\ker(\text{proj}\colon \oplus D_\nu \to L)$, hence $\psi\colon L \to K$ exists with $\psi\,\text{proj} = \psi'$, and $\psi v_\lambda = \psi\,\text{proj}\,\iota_\lambda = \psi' \iota_\lambda = \varphi_\lambda$.—Dually for $\varprojlim I$. ∎

5.8 Corollary. *If* t *is a covariant functor between abelian groups* (or modules, or complexes) *which is strongly additive and right exact* (cf. VI, 2.10) *then* $t(\varinjlim D) = \varinjlim(tD)$, *for every direct system* D.

Indeed, t commutes with sums and quotients (by assumption), hence with $\varinjlim$ by 5.7. ∎

We now discuss some functorial properties of $\varinjlim$.

5.9 Definition. Let $D\colon \Lambda \to \mathscr{K}$, $D'\colon \Lambda' \to \mathscr{K}$ be direct systems. If $\gamma\colon \Lambda \to \Lambda'$ is a map (not necessarily order-preserving) and $d_\lambda\colon D_\lambda \to D'_{\gamma\lambda}$, $\lambda \in \Lambda$, a family of morphisms then we say $d = \{d_\lambda\}_{\lambda\in\Lambda}$ *passes to the limit* if the composition of d with any transformation $\varphi'\colon D' \to K$, $K \in \mathscr{K}$, is a transformation $\varphi' d = \{\varphi'_{\gamma\lambda} d_\lambda\colon D_\lambda \to K\}$. Explicitly, for every transformation $\varphi'\colon D' \to K$ we must have $\varphi'_{\gamma\mu} d_\mu D^\mu_\lambda = \varphi'_{\gamma\lambda} d_\lambda$ whenever $\lambda \le \mu$; $\lambda, \mu \in \Lambda$.

If D' admits a universal transformation $u'\colon D' \to \varinjlim D'$ then d passes to the limit provided $u'd$ is a transformation (this is an easy exercise which will not be used later). *If also D has a limit, $u\colon D \to \varinjlim D$, and d passes to the limit then there is a unique morphism*

(5.10) $\quad \varinjlim d\colon \varinjlim D \to \varinjlim D'$ *such that* $(\varinjlim d)\, u_\lambda = u'_{\gamma\lambda} d_\lambda$,

for all $\lambda \in \Lambda$.

As a criterion we note

5.11 Lemma. *Let $d = \{d_\lambda\colon D_\lambda \to D'_{\gamma\lambda}\}_{\lambda\in\Lambda}$ be a family as in 5.9, and assume that for every relation $\lambda \le \mu$ in Λ there is an element $\rho \in \Lambda'$ such that $\gamma\lambda \le \rho$, $\gamma\mu \le \rho$ and $D'^{\rho}_{\gamma\mu} d_\mu D^\mu_\lambda = D'^{\rho}_{\gamma\lambda} d_\lambda$. Then d passes to the limit.*

Indeed, if $\varphi'\colon D' \to K$ is a transformation then

$$\varphi'_{\gamma\mu} d_\mu D^\mu_\lambda = \varphi'_\rho D'^{\rho}_{\gamma\mu} d_\mu D^\mu_\lambda = \varphi'_\rho D'^{\rho}_{\gamma\lambda} d_\lambda = \varphi'_{\gamma\lambda} d_\lambda. \quad \blacksquare$$

5.12 Proposition. *Let $D\colon \Lambda \to \mathscr{K}$, $D'\colon \Lambda' \to \mathscr{K}$, $D''\colon \Lambda'' \to \mathscr{K}$, and*

$$d = \{d_\lambda\colon D_\lambda \to D'_{\gamma\lambda}\}_{\lambda\in\Lambda}, \qquad d' = \{d'_\rho\colon D'_\rho \to D''_{\gamma'\rho}\}_{\rho\in\Lambda'}$$

as above. If both d and d' pass to the limit then so does $d'd = \{d'_{\gamma\lambda} d_\lambda\colon D_\lambda \to D''_{\gamma'\gamma\lambda}\}_{\lambda\in\Lambda}$. If, moreover, D, D', D'' have limits then $\varinjlim (d'd) = (\varinjlim d')(\varinjlim d)$.

Proof. If $\varphi''\colon D'' \to K$ is a transformation then so are $\varphi'' d'$ and $(\varphi'' d')d = \varphi''(d'd)$, because d' and d pass to the limit, hence $d'd$ passes to the limit. If the limits exist then

$$(\varinjlim d'd)\, u = u''(d'd) = (u''d')\, d = (\varinjlim d')\, u'd = (\varinjlim d')(\varinjlim d)\, u,$$

hence $\varinjlim (d'd) = (\varinjlim d')(\varinjlim d)$ by universality of u. $\blacksquare$

Proposition 5.12 allows to view $\varinjlim$ as a functor on the category of all d which pass to the limit. We leave the precise formulation to the reader, and consider two frequent special cases (5.13, 5.15).

5.13 Example 1 ($\gamma = \mathrm{id}$). If $d\colon D \to D'$ is a *transformation* of direct systems over the same $\Lambda = \Lambda'$ (cf. 5.2) then d always passes to the limit because compositions of transformations are transformations.

Suppose, for instance, $D\colon \Lambda \to \partial\mathcal{A}\mathcal{G}$ is a direct system of complexes. Then for every integer n the n-chains form a direct system of abelian groups $D^n\colon \Lambda \to \mathcal{A}\mathcal{G}$, and the n-th boundary $\{\partial^n_\lambda\colon D^n_\lambda \to D^{n+1}_\lambda\}_{\lambda\in\Lambda}$ is a transformation of direct systems (we use upper dimension indices n in order to distinguish them from the λ). We claim

(5.14) *If* $u=\{u_\lambda\colon D_\lambda \to L\}_{\lambda\in\Lambda}$ *is a universal transformation then* $u^n=\{u^n_\lambda\colon D^n_\lambda \to L^n\}_{\lambda\in\Lambda}$ *is also universal, hence* $(\varinjlim D)^n = \varinjlim (D^n)$. *Under this identification the boundary homomorphism* ∂^n of $\varinjlim D$ agrees with $\varinjlim\{\partial^n_\lambda\}$. The first assertion is quite obvious from 5.7; for a direct proof see Exerc. 4. Further, $(\partial^n)\, u^n_\lambda = u^{n-1}_\lambda\, \partial^n_\lambda = (\varinjlim\{\partial^n_\lambda\})\, u^n_\lambda$ because $u_\lambda\colon D_\lambda \to \varinjlim D$ is a chain map and because of 5.10; equality of the outside terms then shows $\partial^n \doteq \lim\{\partial^n_\lambda\}$ because u^n is universal.

5.15 Example 2 ($d_\lambda = \mathrm{id}$). Let Λ, Λ' be quasiordered sets and $\gamma'\colon \Lambda' \to \Lambda$ an order-preserving map ($\lambda' \le \mu' \Rightarrow \gamma'\lambda' \le \gamma'\mu'$). For every direct system $D\colon \Lambda \to \mathcal{K}$ the composite $D' = D\gamma'\colon \Lambda' \to \mathcal{K}$ is also a direct system ($D'_{\lambda'} = D_{\gamma'\lambda'}$, $D'^{\mu'}_{\lambda'} = D^{\gamma'\mu'}_{\gamma'\lambda'}$). Further, $d' = \{d'_{\lambda'} = \mathrm{id}\colon D'_{\lambda'} = D_{\gamma'\lambda'}\}$ passes to the limit because

$$\varphi_{\gamma'\mu'}\, d'_{\mu'}\, D'^{\mu'}_{\lambda'} = \varphi_{\gamma'\mu'}\, D^{\gamma'\mu'}_{\gamma'\lambda'} = \varphi_{\gamma'\lambda'} = \varphi_{\gamma'\lambda'}\, d'_{\lambda'},$$

for every transformation $\varphi\colon D \to K$. In this situation we write γ'_∞ instead of $\varinjlim d'$; thus $\gamma'_\infty\colon \varinjlim D' \to \varinjlim D$.

It is often important to know whether such a "change of parameters" $\gamma'\colon \Lambda' \to \Lambda$ leaves the limits unchanged, i.e. whether γ'_∞ is isomorphic. The following (5.16, 5.17) provides a useful criterion.

5.16 Definition. An order-preserving map $\gamma'\colon \Lambda' \to \Lambda$ between *directed* sets is called *cofinal* if for every $\lambda\in\Lambda$ there is a $\lambda'\in\Lambda$ such that $\lambda \le \gamma'\lambda'$. (For a generalization of this notion to non-directed sets or categories cf. A, 1.7 and A, 1.12, Exerc. 2).

If an inclusion map $\Lambda' \subset \Lambda$ is cofinal then we say Λ' *is cofinal in* Λ. For instance, every infinite subset of $\mathbb{N}$ is cofinal in $\mathbb{N}$. If Λ has an upper bound m ($\lambda \le m$ for all λ) then $\Lambda' = \{m\}$ is cofinal; in that case $\varinjlim D = D_m$ for every $D\colon \Lambda \to \mathcal{K}$.

5.17 Proposition. *If* $\gamma'\colon \Lambda' \to \Lambda$ *is cofinal, and* $D\colon \Lambda \to \mathcal{K}$ *is any direct system then composition with* γ' *defines a* 1-1-*correspondence* $\hat{\gamma}'$ *between transformations* $\varphi\colon D \to K$ *and transformations* $\varphi'\colon D' \to K$. *In formulas,*

$$\hat{\gamma}'\colon \mathrm{Transf}(D, K) \xrightarrow{\approx} \mathrm{Transf}(D', K), \qquad (\hat{\gamma}'(\varphi))_{\lambda'} = \varphi_{\gamma'\lambda'} = \varphi'_{\lambda'}.$$

Moreover, $u\colon D \to L$ *is universal if and only if* $u' = \hat{\gamma}'(u)\colon D' \to L$ *is universal, hence* $\gamma'_\infty\colon \varinjlim D' \cong \varinjlim D$, *provided one (and therefore both) of these limits exists.*

Proof. We construct an inverse to $\hat{\gamma}'$, say $\varepsilon\colon \mathrm{Transf}(D', K) \to \mathrm{Transf}(D, K)$. If $\varphi' \in \mathrm{Transf}(D', K)$ we define $(\varepsilon\,\varphi')_\lambda\colon D_\lambda \to K$, for $\lambda \in \Lambda$, by $(\varepsilon\,\varphi')_\lambda = \varphi'_{\lambda'} \cdot D_\lambda^{\gamma'\lambda'}$, where $\lambda' \in \Lambda'$ is so chosen that $\lambda \le \gamma'\lambda'$. This $(\varepsilon\,\varphi')_\lambda$ does not depend on the choice of λ', for if $\mu' \in \Lambda'$ is a second choice we can assume that $\mu' \ge \lambda'$ (because Λ' is directed), and then

$$\varphi'_{\mu'} \cdot D_\lambda^{\gamma'\mu'} = \varphi'_{\mu'} \cdot D_{\gamma'\lambda'}^{\gamma'\mu'} D_\lambda^{\gamma'\lambda'} = \varphi'_{\mu'} \cdot D_{\lambda'}^{\prime\mu'}\, D_\lambda^{\gamma'\lambda'} = \varphi'_{\lambda'} \cdot D_\lambda^{\gamma'\lambda'}.$$

We now claim that $\varepsilon\,\varphi' = \{(\varepsilon\,\varphi')_\lambda\}_{\lambda \in \Lambda}$ is a transformation $D \to K$. Indeed, if $\lambda \le \mu$ choose $\mu' \in \Lambda'$ such that $\mu \le \gamma'\mu'$ then $(\varepsilon\,\varphi')_\mu = \varphi'_{\mu'} \cdot D_\mu^{\gamma'\mu'}$, $(\varepsilon\,\varphi')_\lambda = \varphi'_{\mu'} \cdot D_\lambda^{\gamma'\mu'}$, hence $(\varepsilon\,\varphi')_\mu\, D_\lambda^\mu = (\varepsilon\,\varphi')_\lambda$, as required. Furthermore,

$$(\varepsilon\,\hat{\gamma}'(\varphi))_\lambda = (\hat{\gamma}'\,\varphi)_{\lambda'} \cdot D_\lambda^{\gamma'\lambda'} = \varphi_{\gamma'\lambda'} \cdot D_\lambda^{\gamma'\lambda'} = \varphi_\lambda,$$

and $(\hat{\gamma}'\,\varepsilon(\varphi'))_{\lambda'} = (\varepsilon\,\varphi')_{\gamma\lambda'} = \varphi'_{\lambda'} \cdot D_{\gamma\lambda'}^{\gamma\lambda'} = \varphi'_{\lambda'}$, hence $\hat{\gamma}'$, ε are reciprocal bijections.

In order to show that u, u' are simultaneously universal we consider the commutative diagram

$$\begin{array}{ccc} & \overset{\circ u}{\nearrow} & \mathrm{Transf}(D, K) \\ \mathscr{K}(L, K) & & \Big\| \wr\ \hat{\gamma}' \\ & \underset{\circ u'}{\searrow} & \mathrm{Transf}(D', K). \end{array}$$

Universality of u resp. u' means that the upper resp. lower arrow is bijective for all K. ∎

So far, the category $\mathscr{K}$ in which the limits were taken was essentially arbitrary (although some parts, like 5.7, were formulated for abelian groups). The following results, on the contrary, use special properties (of groups $\cdots$); they do not generalize to arbitrary $\mathscr{K}$. For instance, they do not dualize: the dual assertions for inverse limits of abelian groups are false (cf. Exerc. 5).

5.18 Proposition. *If Λ is a directed set, and $D\colon \Lambda \to \mathscr{AG}$ is a direct family of abelian groups (or modules, or complexes) then a transformation $v = \{v_\lambda\colon D_\lambda \to L\}_{\lambda \in \Lambda}$, is universal (hence $L = \varinjlim D$) if and only if the following two conditions hold.*

(i) $L = \bigcup_{\lambda \in \Lambda} \mathrm{im}(v_\lambda)$;

(ii) $\ker(v_\lambda) = \bigcup_{\lambda \le \mu} \ker(D_\lambda^\mu)$, for every $\lambda \in \Lambda$.

In words, *every $y \in L$ comes from some D_λ, and if $x \in D_\lambda$ is such that $v_\lambda(x) = 0$ then already $D_\lambda^\mu(x) = 0$ for some $\mu \ge \lambda$.*

Proof. Assume (i), (ii) hold. Given a transformation $\varphi\colon D\to K$, $K\in\mathscr{K}$, we have to construct $\psi\colon L\to K$ such that $\psi v_\lambda=\varphi_\lambda$, $\lambda\in\Lambda$; since the images, $\operatorname{im}(v_\lambda)$, cover all of L there is at most one such ψ. Note that $\varphi_\mu D^\mu_\lambda=\varphi_\lambda$ implies $\varphi_\lambda|\ker(D^\mu_\lambda)=0$; by condition (ii) there is therefore a unique homomorphism $\psi^\lambda\colon \operatorname{im}(v_\lambda)\to K$ such that $\psi^\lambda v_\lambda=\varphi_\lambda$. We claim that the family $\{\psi^\lambda\}_{\lambda\in\Lambda}$ matches to give $\psi\colon K\to L$. Suppose $\lambda\le\rho$; then

$$\operatorname{im}(v_\rho)\supset\operatorname{im}(v_\rho D^\rho_\lambda)=\operatorname{im}(v_\lambda),$$

and

$$\psi^\rho v_\lambda=\psi^\rho v_\rho D^\rho_\lambda=\varphi_\rho D^\rho_\lambda=\varphi_\lambda=\psi^\lambda v_\lambda,$$

hence $\psi^\rho|\operatorname{im}(v_\lambda)=\psi^\lambda$. If λ,λ' are any two elements of Λ then we can find $\rho\in\Lambda$ with $\rho\ge\lambda$, $\rho\ge\lambda'$, hence ψ^ρ is a common extension of $\psi^\lambda,\psi^{\lambda'}$. Therefore we can indeed define $\psi\colon L\to K$ by $\psi|\operatorname{im}(v_\lambda)=\psi^\lambda$, and get $\psi v_\lambda=\psi^\lambda v_\lambda=\varphi_\lambda$, as required.

Assume now $v\colon D\to L$ is universal; we have to prove (i), (ii). Remark first that $\bigcup_{\lambda\in\Lambda}\operatorname{im}(v_\lambda)$ is a subgroup (or -module, or -complex) of L because for any $\lambda,\lambda'\in\Lambda$ we have $[\operatorname{im}(v_\lambda)\cup\operatorname{im}(v_{\lambda'})]\subset\operatorname{im}(v_\rho)$, where $\rho\ge\lambda$, $\rho\ge\lambda'$. Consider then the projection $\pi\colon L\to L/\bigcup_{\lambda\in\Lambda}\operatorname{im}(v_\lambda)$; clearly $\pi v_\lambda=0$ for every $\lambda\in\Lambda$, hence $\pi=0$ by universality of v, hence $\bigcup_{\lambda\in\Lambda}\operatorname{im}(v_\lambda)=L$.

In order to prove (ii) we can assume that v is as in Proposition 5.7 (because any two universal transformations are equivalent). Now, if $v_\lambda(x_\lambda)=0$ then $\iota_\lambda x_\lambda\in\bigoplus_{\nu\in\Lambda}D_\nu$ is of the form $\iota_\lambda x_\lambda=\sum(\iota_\mu x_\mu-\iota_\rho D^\rho_\mu x_\mu)$. This sum being finite we can choose $m\in\Lambda$ such that $m\ge\lambda,\mu,\rho$ ($m\ge$ all indices which occur), so that the equality takes place in $\bigoplus_{\nu\le m}D_\nu$. Apply the homomorphism $\{D^m_\nu\}\colon\bigoplus_{\nu\le m}D_\nu\to D_m$ to the equality and get

$$D^m_\lambda x_\lambda=\sum(D^m_\mu x_\mu-D^m_\rho D^\rho_\mu x_\mu)=\sum(D^m_\mu x_\mu-D^m_\mu x_\mu)=0. \quad ∎$$

5.19 Corollary. *If A is an abelian group (module, complex) and $\{D_\lambda\}_{\lambda\in\Lambda}$ is a direct system of subgroups of A (Λ directed, $D^\mu_\lambda=$ inclusion) then $\{i_\lambda\colon D_\lambda\xrightarrow{\subset}\bigcup_{\nu\in\Lambda}D_\nu\}_{\lambda\in\Lambda}$ is universal, hence $\varinjlim\{D_\lambda\}=\bigcup_{\lambda\in\Lambda}D_\lambda$.* ∎

Another consequence of 5.18 is the following

5.20 Proposition. *Let Λ be a directed set, $C\colon\Lambda\to\partial\mathscr{AG}$ a direct system of complexes, and $u=\{u_\lambda\colon C_\lambda\to L\}_{\lambda\in\Lambda}$ a universal transformation ($L=\varinjlim C$). Then $Hu=\{Hu_\lambda\colon H(C_\lambda)\to HL\}_{\lambda\in\Lambda}$ is also universal, hence*

$$\varinjlim\{HC_\lambda\}=H(\varinjlim\{C_\lambda\}).$$

5.21 Corollary. *If Λ is a directed set, $D, D', D'': \Lambda \to \mathscr{A}\mathscr{G}$ are direct systems of abelian groups, and $D' \xrightarrow{\;\alpha\;} D \xrightarrow{\;\beta\;} D''$ are transformations such that $D'_\lambda \xrightarrow{\;\alpha_\lambda\;} D_\lambda \xrightarrow{\;\beta_\lambda\;} D''_\lambda$ is exact for every $\lambda \in \Lambda$ then*

$$\varinjlim D' \xrightarrow{\;\lim \alpha\;} \varinjlim D \xrightarrow{\;\lim \beta\;} \varinjlim D''$$

is also exact.

Indeed, just view $D'_\lambda \to D_\lambda \to D'_\lambda$ as a complex C_λ, with $C^0_\lambda = D_\lambda$; then $\varinjlim D' \to \varinjlim D \to \varinjlim D'$ agrees with $\varinjlim \{C_\lambda\}$ by 5.14. By assumption, $H^0 C_\lambda = 0$; hence by 5.20, $H^0(\varinjlim C) = 0$, as asserted.

Proof of 5.20. Let $[z] \in HL$, and $z \in ZL$ a representing cycle. By 5.18(i) we can find $\lambda \in \Lambda$ and $x \in C_\lambda$ such that $u_\lambda x = z$. Now $0 = \partial z = \partial u_\lambda x = u_\lambda(\partial x)$, hence by 5.18(ii) we can find $\nu \geq \lambda$ such $D^\nu_\lambda(\partial x) = 0$. Then $\partial(D^\nu_\lambda x) = D^\nu_\lambda \partial x = 0$, and $u_\nu(D^\nu_\lambda x) = u_\lambda x = z$, hence $z_\nu = D^\nu_\lambda x$ is a cycle whose homology class $[z_\nu]$ maps into $[z]$ under $H(u_\nu)$. Thus $\{H(u_\nu)\}_{\nu \in \Lambda}$ satisfies condition 5.18(i).

Assume now $[z_\lambda] \in HC_\lambda$ is such that $(H(u_\lambda))[z_\lambda] = 0$. Then $u_\lambda z_\lambda = \partial y$ for some $y \in L$. By 5.18(i) we can find $\nu \in \Lambda$ and $x \in C_\nu$ such that $u_\nu x = y$. Choose $\rho \in \Lambda$ such that $\rho \geq \lambda, \rho \geq \nu$. Then $u_\rho(D^\rho_\lambda z_\lambda - \partial D^\rho_\nu x) = u_\lambda z_\lambda - \partial u_\nu x = 0$. By 5.18(ii) we can find $\sigma \geq \rho$ such that $D^\sigma_\rho(D^\rho_\lambda z_\lambda - \partial D^\rho_\nu x) = 0$, hence $D^\sigma_\lambda z_\lambda = \partial(D^\sigma_\nu x)$, hence $H(D^\sigma_\lambda)[z_\lambda] = 0$. Thus $\{H(u_\lambda)\}_{\lambda \in \Lambda}$ satisfies condition 5.18(ii). ∎

5.22 Examples. Let Y be a topological space and let Λ be the set of all quasi-compact subspaces of Y. Inclusion defines an order relation, which turns Λ into a directed set (in fact, if $\lambda, \lambda' \in \Lambda$ then $(\lambda \cup \lambda') \in \Lambda$, and $\lambda, \lambda' \leq \lambda \cup \lambda'$). The function S which to every λ assigns its singular complex $S(\lambda)$ and to every pair $\lambda \leq \mu$ the inclusion map $S^\mu_\lambda: S(\lambda) \to S(\mu)$ is a direct system, and the inclusions $j_\lambda: S(\lambda) \to SY$ form a transformation. Clearly conditions 5.18(i), (ii) are satisfied (the image of every singular simplex is quasicompact), hence $SY = \varinjlim \{S(\lambda)\}$, hence $HY = \varinjlim \{H(\lambda)\}$ by 5.20. Similarly, $H(Y, B) = \varinjlim \{H(\lambda)\}$ if λ ranges over all *pairs* of quasi-compact sets in (Y, B).

For the same space Y let Λ' be a set of open subsets of Y which is directed under the inclusion ordering and whose union is the whole space, $\bigcup \lambda' = Y$. Every singular simplex lies in some λ', hence $SY = \bigcup S(\lambda') = \varinjlim \{S(\lambda')\}$, hence $HY = \varinjlim \{H(\lambda')\}$, as above for Λ. In fact, every compact set $\lambda \in \Lambda$ lies in some open set $\lambda' \in \Lambda'$. Choose a function $\gamma: \Lambda \to \Lambda'$ such that $\lambda \subset \gamma\lambda$ for every $\lambda \in \Lambda$, and let $d_\lambda: S(\lambda) \to S(\gamma\lambda)$ denote the inclusion map. One easily shows that $d = \{d_\lambda\}$ passes to the limit, and $\varinjlim d = \mathrm{id}_{SY}$; similarly for homology.

5.23 Exercises. *1.* Let Y be a topological space, and $\mathscr{X}=\{X\}$ a family of subspaces which is directed under inclusion and whose union is the whole space, $\bigcup X=Y$. Then the inclusions $\{j_X\colon X\to Y\}_{X\in\mathscr{X}}$ form a universal transformation ($Y=\varinjlim\{X\}$) if and only if Y has the finest topology for which all j_X are continuous. (Important example: $\mathscr{X}$ = set of all *compact* subspaces; if $Y=\varinjlim\mathscr{X}$ it is said to be a *k-space*. Cf. Kelley.) Show that $Y=\varinjlim\{X\}$ implies $Y\times L=\varinjlim\{X\times L\}$ for every locally compact L.

*2**. Let Y be a T_1-space (points are closed) and $\mathscr{V}=\{V\}$ a decomposition of Y into disjoint subsets ($\bigcup V=Y, V\cap V'\neq\emptyset\Rightarrow V=V'$). Let $\mathscr{X}$ be the set of all subspaces of Y which are *finite unions* of sets in $\mathscr{V}$. Then $\mathscr{X}$ is directed under inclusion. Show: if $Y=\varinjlim\mathscr{X}$ then every quasi-compact subset K of Y is contained in some $X\in\mathscr{X}$. For instance, $\mathscr{V}$ could be the set of all cells of a CW-decomposition; then this is V, 2.6. Or $\mathscr{V}$ could be the sequence of sets Y_n-Y_{n-1} where $\{Y_n\}_{n=1,2,\ldots}$ is an ascending sequence in Y whose union is Y; one finds that K must lie in some Y_n.— On the other hand, $\mathbb{R}=\varinjlim\mathscr{X}$, where $\mathscr{X}$ is the set of all countable subsets of $\mathbb{R}$, but $K=[0,1]$ does not lie in any $X\in\mathscr{X}$.

3. Consider the following sequence of homomorphisms

$$\mathbb{Z}_2\xrightarrow{2}\mathbb{Z}_4\xrightarrow{2}\mathbb{Z}_8\xrightarrow{2}\cdots\xrightarrow{2}\mathbb{Z}_{2^n}\xrightarrow{2}\cdots.$$

This can be viewed as a direct system ($\Lambda=\mathbb{N}$) in the category $\mathscr{FAG}$ of *finite* abelian groups, or in the category $\mathscr{AG}$ of all abelian groups. Show that in $\mathscr{FAG}$ the direct limit is zero whereas in $\mathscr{AG}$ it is not. Construct another sequence in $\mathscr{FAG}$ which does not possess a $\varinjlim$ (in $\mathscr{FAG}$).

4. If A is an abelian group let $\hat{A}$ be the following complex: $\hat{A}^n=0$ for $n\neq 0,-1$, $\hat{A}^0=\hat{A}^{-1}=A$, $\partial=\mathrm{id}\colon\hat{A}^{-1}\to\hat{A}^0$. Show that for any complex $C=\{\cdots\to C^{-1}\to C^0\to C^1\to\cdots\}$ the chain maps $C\to\hat{A}$ are in natural 1-1 correspondence with the homomorphisms $C^0\to A$. Use this remark to give a direct proof of 5.14, i.e. of $(\varinjlim\{C_\lambda\})^0=\varinjlim\{C_\lambda^0\}$.

5 (cf. Eilenberg-Steenrod, VIII, 5.5). Show that the inverse limit of the following inverse system I (over $\Lambda=\mathbb{N}$) is zero:

$$\mathbb{Z}\xleftarrow{3}\mathbb{Z}\xleftarrow{3}\mathbb{Z}\xleftarrow{3}\mathbb{Z}\xleftarrow{3}\cdots.$$

Let I'' denote the constant inverse system, $I''^{\lambda}=\mathbb{Z}_2$, $I''^{\lambda}_{\mu}=\mathrm{id}$. Then we have an exact sequence of transformations $0\to I\xrightarrow{2}I\to I''\to 0$ but the corresponding sequence of inverse limits $0\to 0\to 0\to\mathbb{Z}_2\to 0$ is not exact.

6. Čech Cohomology of Locally Compact Subsets of $\mathbb{R}^n$

The Čech cohomology $\check{H}X$ of a space X is usually defined (cf. Eilenberg-Steenrod IX) as being the direct limit of the system $\{H^*(N_\lambda)\}$, where λ ranges over the set of all open coverings of X (directed by refinement), and N_λ is the nerve of λ. If X is a locally compact subset of an ENR (=euclidean neighborhood retract) E then one can show by cofinality arguments (5.17) that $\check{H}X \cong \varinjlim \{H^* V\}$ where V ranges over the set of all (open) neighborhoods of X in E (directed by inverse inclusion). It is in this form that Čech groups naturally arise in the (co-)homology theory of manifolds. We shall therefore define $\check{H}X$ as $\varinjlim \{H^* V\}$, and we shall study the formal properties of this $\check{H}$ now.

6.1 Definition. Let $Y \subset X \subset E$, where E is an ENR (cf. IV, 8.5) and X, Y are locally compact. Let $\Lambda = \Lambda(X, Y)$ be the set of all pairs (V, W) of neighborhoods of X, Y such that $W \subset V$. Under reversed inclusion Λ is directed ($(V, W) \leq (\tilde{V}, \tilde{W}) \Leftrightarrow \tilde{V} \subset V$ and $\tilde{W} \subset W$), and $\{H^*(V, W)\}$ together with the restriction homomorphisms $\{H^*(V, W) \to H^*(\tilde{V}, \tilde{W})\}$ is a direct system of (graded) abelian groups; the coefficients for H^* are taken in some fixed abelian group G which, most of the time, we do not indicate. We define $\check{H}(X, Y) = \varinjlim \{H^*(V, W) | (V, W) \in \Lambda\}$ = *Čech cohomology of* X mod Y, and we denote by $u = u_{VW}$: $H^*(V, W) \to \check{H}(X, Y)$ the universal transformation. With coefficients and dimension indices $\check{H}^q(X, Y; G) = \varinjlim \{H^q(V, W; G)\}$ = *q-th Čech-cohomology group of* X mod Y *with coefficients in* G. As usual, we write $\check{H}(X, \emptyset) = \check{H}X$. We shall soon see (6.8) that $\check{H}(X, Y)$ only depends on (X, Y), *not on* E. Note that the set Λ' of *open* pairs $(V, W) \in \Lambda$ is cofinal in Λ so that we can replace Λ by Λ' whenever it is convenient (cf. 5.17).

The following lemma will serve us to turn $\check{H}$ into a functor.

6.2 Lemma. *Let E, E' be ENR's, and $X' \subset E'$ a locally compact subset.*

(a) *Every continuous map $f\colon X' \to E$ has an extension $F\colon U' \to E$ to some open neighborhood of X'; thus $F|X' = f$.*

(b) *If $F, G\colon E' \to E$ are two continuous maps, and $\vartheta_t\colon X' \to E$, $0 \leq t \leq 1$, is a homotopy between $F|X'$ and $G|X'$ then there is a homotopy $\Theta_t\colon U'' \to E$, defined on some open neighborhood U'' of X', such that $\Theta_0 = F|U''$, $\Theta_1 = G|U''$ and $\Theta_t|X' = \vartheta_t$.*

Remark. It is not hard to see that this means $\pi(X', E) \cong \varinjlim \{\pi(U', E)\}$ where $\pi(-, -)$ denotes homotopy classes of maps (compare 5.18).

Proof. We have maps $E \xrightarrow{i} O \xrightarrow{r} E$, where O is an open subset of some $\mathbb{R}^n$ and $ri = \mathrm{id}$. We can also assume that E' is contained in some euclidean space, and then we know from IV, 8.3 that X' has an open neighborhood E'' in E' such that X' is closed in E''.

(a) If $f\colon X' \to E$ is given then, by Tietze's extension lemma, $if\colon X' \to \mathbb{R}^n$ has an extension $\Phi\colon E'' \to \mathbb{R}^n$. Put $U' = \Phi^{-1}(O)$, and define $F\colon U' \to E$ by $F = r\Phi$.

(b) If $F, G\colon E' \to E$ and $\vartheta\colon F|X' \simeq G|X'$ are given we can use them to define a map d on the closed subspace $A = X' \times [0,1] \cup E'' \times \{0\} \cup E'' \times \{1\}$ of $E'' \times [0,1]$; in formulas, $d\colon A \to \mathbb{R}^n$, $d(x', t) = i\vartheta_t(x)$, $d(e'', 0) = iF(e'')$, $d(e'', 1) = iG(e'')$. By Tietze's lemma again, d admits an extension $D\colon E'' \times [0,1] \to \mathbb{R}^n$. Let $U'' = \{y \in E'' | D(y \times [0,1]) \subset O\}$. This is an open neighborhood of X', and $\Theta_t\colon U'' \to E$, $\Theta_t(y) = rD(y, t)$ is a deformation as required. ∎

6.3 Definition (of Induced Maps $\check{f}$). Let $Y \subset X \subset E$ and $Y' \subset X' \subset E'$ be as in 6.1 (locally compact subsets of ENR's), and let $f\colon (X', Y') \to (X, Y)$ be a map. By 6.2(a) there is an open neighborhood U' of X' and a map $F\colon U' \to E$ such that $F|X' = f$. If $W \subset V$ is a pair of (open) neighborhoods of $Y \subset X$ consider the composition

$$(6.4) \qquad F_{VW}\colon H^*(V, W) \xrightarrow{F^*} H^*(F^{-1}V, F^{-1}W) \xrightarrow{u'} \check{H}(X', Y'),$$

where u' is the universal transformation of the direct system which defines $\check{H}(X', Y')$. Since F^* commutes with inclusions $(\tilde{V}, \tilde{W}) \subset (V, W)$, $(F^{-1}\tilde{V}, F^{-1}\tilde{W}) \subset (F^{-1}V, F^{-1}W)$, we see that $\{F_{VW}\}$ is a transformation of the direct system $\{H^*(V, W)\}$ and therefore defines a homomorphism $\check{F}\colon \check{H}(X, Y) \to \check{H}(X', Y')$ of limits such that $\check{F}u = \{F_{VW}\}$.

Suppose $G\colon T' \to E$ is also an extension of f; we want to show $\check{G} = \check{F}$. In fact, let us consider the more general case where $G\colon T' \to E$ is a map of an open neighborhood T' of X' such that $G(X') \subset X$, $G(Y') \subset Y$ and $G|X' \simeq f\colon (X', Y') \to (X, Y)$ (i.e., $G|(X', Y')$ need not be equal to f, just homotopic). By 6.2(b) we can find an open neighborhood $U'' \subset (U' \cap T')$ of X' and a deformation $\Theta_t\colon U'' \to E$ between $G|U''$ and $F|U''$ such that $\Theta_t|X'\colon (X', Y') \to (X, Y)$. Now let $W \subset V$ be a pair of open neighborhoods of $Y \subset X$, as above, and define

$$V' = \{y \in U'' | \Theta_t y \in V \text{ for all } t\}, \qquad W' = \{y \in U'' | \Theta_t y \in W \text{ for all } t\}.$$

Then $W' \subset V'$ is a pair of open neighborhoods of $Y' \subset X'$, and $\Theta_t|V'$ is a homotopy $F|V' \simeq G|V'\colon (V', W') \to (V, W)$, hence $(F|V')^* = (G|V')^*$.

There results a commutative diagram

(6.5)

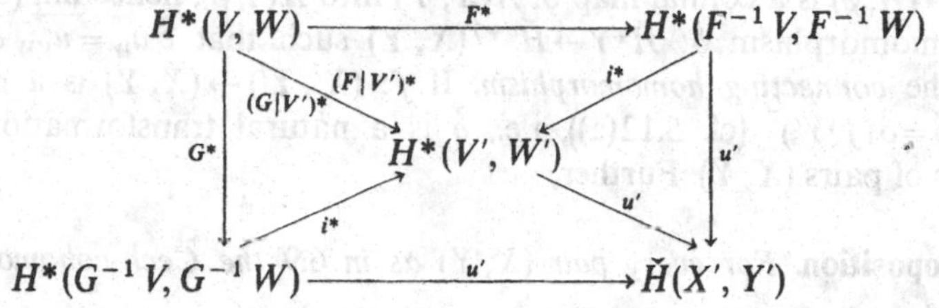

(i = inclusion). Equating the outer compositions gives $F_{VW} = G_{VW}$, hence $\check{F} = \check{G}$. *Thus* $\check{f} = \check{F}$: $\check{H}(X, Y) \to \check{H}(X', Y')$ *depends only on* f (not on the extension F).

We have now defined a graded group $\check{H}(X, Y)$ for every pair (X, Y) as in 6.1, and a homomorphism $\check{H}f = \check{f}$: $\check{H}(X, Y) \to \check{H}(X', Y')$ for every map $f: (X', Y') \to (X, Y)$, and we shall see in a moment, that $\check{H}$ is in fact a cofunctor. Moreover, the equation $\check{F} = \check{G}$ above used only $F|(X', Y') \simeq G|(X', Y')$, hence

6.6 Proposition. *If* $f, g: (X', Y') \to (X, Y)$ *are homotopic maps as in* 6.3 *then* $\check{f} = \check{g}$: $\check{H}(X, Y) \to \check{H}(X', Y')$. ∎

6.7 Proposition. *If* $(X'', Y'') \xrightarrow{f'} (X', Y') \xrightarrow{f} (X, Y)$ *are maps as in* 6.3 *then* $\check{H}(ff') = (\check{H}f')(\check{H}f)$; *if* $f = \mathrm{id}_{(X,Y)}$ *then* $\check{H}f$ *is the identity map of* $\check{H}(X, Y)$. *I.e.,* $\check{H}$ *is a cofunctor on the category whose objects are all pairs* (X, Y) *as in* 6.1 *and whose morphisms are (homotopy classes of) continuous maps (of pairs).*

This is rather obvious: If F is an extension of f, and F' an extension of f' then FF' is an extension of ff', and (in the notation of 6.3)

$$(FF')^\smile u = (FF')_{VW} = u''(FF')^* = u'' F'^* F^* = F'_{V'W'} F^* = \check{F}' u' F^* = \check{F}' \check{F} u,$$

hence $(FF')^\smile = \check{F}'\check{F}$. Further, Id_E extends $\mathrm{id}_{(X,Y)}$, hence $\check{H}(\mathrm{id}) = \mathrm{Id} = \mathrm{id}$. ∎

6.8 Corollary. *If* $Y \subset X \subset E$, $Y' \subset X' \subset E'$ *are as in* 6.3, *and* $f: (X', Y') \simeq (X, Y)$ *is a homotopy equivalence then* $\check{f}$: $\check{H}(X, Y) \cong \check{H}(X', Y')$; simply because a functor takes equivalences into equivalences. ∎

6.9 Definition (of the Connecting Homomorphism $\check{\delta}$). Let $Y \subset X$ be a locally compact pair in some ENR E (as above). Put $\check{H}Y = \check{H}(Y, \emptyset)$. We want to define $\check{\delta}$: $\check{H}^q Y \to \check{H}^{q+1}(X, Y)$. If we assign to each pair $W \subset V$ of (open) neighborhoods of $Y \subset X$ the homomorphism

$$\delta^* = \delta^*_{VW}: H^q W \to H^{q+1}(V, W),$$

we get a transformation of direct systems over $\Lambda(X, Y)$. Clearly, $(V, W) \mapsto (W, \emptyset)$ is a cofinal map of $\Lambda(X, Y)$ into $\Lambda(Y, \emptyset)$, hence $\underrightarrow{\lim}\{\delta^*_{VW}\}$ is a homomorphism $\check{\delta}$: $\check{H}^q Y \to \check{H}^{q+1}(X, Y)$ such that $\check{\delta} u_W = u_{VW} \delta^*_{VW}$, called the *connecting homomorphism.* If f: $(X', Y') \to (X, Y)$ is a map, then $\check{f}\check{\delta} = \check{\delta}(f|Y')\check{}$ (cf. 5.12(a)), i.e., $\check{\delta}$ is a natural transformation of functors of pairs (X, Y). Further,

6.10 Proposition. *For every pair* (X, Y) *as in* 6.9 *the Čech-cohomology sequence*

$$\cdots \xrightarrow{\check{j}} \check{H}^q X \xrightarrow{\check{i}} \check{H}^q Y \xrightarrow{\check{\delta}} \check{H}^{q+1}(X, Y) \xrightarrow{\check{j}} H^{q+1} X \xrightarrow{\check{i}} H^{q+1} Y \xrightarrow{\check{\delta}} \cdots$$

is exact.

Proof. Let $\Lambda = \Lambda(X, Y)$ be the directed set of all pairs $W \subset V$ of neighborhoods of $Y \subset X$. The maps

$$\Lambda(Y, \emptyset) \leftarrow \Lambda(X, Y) \rightarrow \Lambda(X, \emptyset), \qquad (W, \emptyset) \leftarrow\!\shortmid (V, W) \mapsto (V, \emptyset)$$

are cofinal, hence

$$\underrightarrow{\lim}\{H^* W | (V, W) \in \Lambda\} = \check{H} X \quad \text{and} \quad \underrightarrow{\lim}\{H^* V | (V, W) \in \Lambda\} = \check{H} Y,$$

by 5.17. Now,

$$\cdots \to H^q V \to H^q W \to H^{q+1}(V, W) \to H^{q+1} V \to H^{q+1} W \to \cdots$$

is exact for every $(V, W) \in \Lambda(X, Y)$, hence the corresponding Λ-$\underrightarrow{\lim}$-sequence is also exact, by 5.21. ∎

We now compare Čech-cohomology $\check{H}(X, Y)$ with ordinary cohomology $H^*(X, Y)$.

6.11 Definition. Let $Y \subset X$ be a locally compact pair in some ENR E. For every pair $W \subset V$ of neighborhoods of $Y \subset X$ the inclusion $(X, Y) \to (V, W)$ induces a homomorphism ρ_{VW}: $H^*(V, W) \to H^*(X, Y)$. The family $\{\rho_{VW}\}$ is a transformation of the direct system $\{H^*(V, W)\}$, hence defines a homomorphism ρ: $\check{H}(X, Y) = \underrightarrow{\lim}\{H^*(V, W)\} \to H^*(X, Y)$ such that $\rho\, u_{VW} = \rho_{VW}$, where u_{VW}: $H^*(V, W) \to \check{H}(X, Y)$ is the universal transformation.

If f: $(X', Y') \to (X, Y)$ is a map and F: $U' \to E$ an extension as in 6.3 then

$$f^* \rho u_{VW} = f^* \rho_{VW} = \rho_{F^{-1}V, F^{-1}W} F^* = \rho u'_{F^{-1}V, F^{-1}W} F^* = \rho F_{VW} = \rho \check{F} u_{VW}$$

(notation of 6.4), hence $f^* \rho = \rho \check{F} = \rho \check{f}$, i.e., ρ commutes with maps f, or more precisely: *ρ is a natural transformation of functors (of pairs (X, Y)).* Moreover, *ρ commutes with the connecting homomorphism, $\rho \check{\delta} = \delta^* \rho$.* Indeed, $\rho \check{\delta} u_W = \rho u_{VW} \delta^*_{VW} = \rho_{VW} \delta^*_{VW} = \delta^*_{XY} \rho_W = \delta^* \rho u_W$, for every pair $W \subset V$ of neighborhoods of $Y \subset X$.

In general, $\rho\colon \check{H}(X, Y) \to H^*(X, Y)$ is neither surjective nor injective (cf. Exerc. 3), however,

6.12 Proposition. *If $Y \subset X$ is a pair of ENR's then $\rho\colon \check{H}(X, Y) \cong H^*(X, Y)$, i.e., for ENR's Čech-cohomology coincides with ordinary cohomology.*

Proof. Since X is an ENR the Čech-cohomology $\check{H}X$ is the direct limit of $\{H^* V\}$, where V ranges over all neighborhoods of X in $E = X$; there is only one such V, namely $V = X$, hence $\rho\colon \check{H}X \cong H^*X$. Similarly, $\rho\colon \check{H}Y \cong H^*Y$. Moreover, because ρ is natural and commutes with connecting homomorphisms, it maps the Čech cohomology sequence 6.10 into the ordinary cohomology sequence. It is isomorphic on the absolute groups, hence also on the relative groups, by the five lemma. ∎

6.13 Mayer-Vietoris Sequence. Let $X_1, X_2 \subset X$ be topological spaces. We say *$X_1 \cap X_2$ separates X_1, X_2* provided $X_2 - X_1$ and $X_1 - X_2$ are both open (or both closed) in $X_1 \cup X_2 - X_1 \cap X_2$; in other words, if $X_1 \cup X_2 - X_1 \cap X_2$ decomposes as a topological sum $(X_2 - X_1) \oplus (X_1 - X_2)$. Still another way of putting this condition is, $\overline{X_2 - X_1} \cap (X_1 - X_2) = \emptyset = (X_2 - X_1) \cap \overline{X_1 - X_2}$.—For instance, *if X_1, X_2 are both open or both closed in $X_1 \cup X_2$ then $X_1 \cap X_2$ separates.* If X_1 is a closed hemisphere of $\mathbb{S}^n$ and X_2 the complementary open hemisphere then $X_1 \cap X_2 = \emptyset$ does *not* separate X_1, X_2.

We shall establish a Mayer-Vietoris sequence in Čech-cohomology for triads $(E; X_1, X_2)$ such that E is an ENR and X_1, X_2 are locally compact subspaces which are separated by $X_1 \cap X_2$. Note first that $X_1 \cap X_2$ and $X_1 \cup X_2$ are also locally compact (A_1, A_2 compact $\Rightarrow A_1 \cap A_2$, $A_1 \cup A_2$ compact). Let ΛX denote the directed system of all open neighborhoods of X in E. Consider the maps

$$\Lambda X_1 \times \Lambda X_2 \to \Lambda X_1, \Lambda X_2, \Lambda(X_1 \cup X_2), \Lambda(X_1 \cap X_2)$$

$$(V_1, V_2) \mapsto V_1, V_2, V_1 \cup V_2, V_1 \cap V_2.$$

All of them are cofinal and even surjective. For the first three this follows from $(V_1, E) \mapsto V_1$, $(E, V_2) \mapsto V_2$, $(V, V) \mapsto V$ for $V \in \Lambda(X_1 \cup X_2)$. For the last, $(V_1, V_2) \mapsto V_1 \cap V_2$, we choose open sets O_1, O_2 of E such that $(X_1 - X_2) \subset O_1$, $(X_2 - X_1) \subset O_2$, and $O_1 \cap O_2 = \emptyset$; this is possible because

$X_1 \cap X_2$ separates X_1, X_2; for instance, one can take for O_1 resp. O_2 the set of all points whose distance (in some metric) from $X_1 - X_2$ is smaller resp. larger than the distance from $X_2 - X_1$. Then for every $W \in \Lambda(X_1 \cap X_2)$ we have $(W \cup O_1, W \cup O_2) \mapsto (W \cup O_1) \cap (W \cup O_2) = W$.

For every pair $(V_1, V_2) \in \Lambda X_1 \times \Lambda X_2$ of open neighborhoods we have the exact $M-V$ sequence (cf. III, 8, resp. VI, 7.6)

$$\cdots \xrightarrow{d^*} H^q(V_1 \cup V_2) \xrightarrow{(i_1^*, i_2^*)} H^q V_1 \oplus H^q V_2 \xrightarrow{(j_1^*, -j_2^*)} H^q(V_1 \cap V_2)$$
$$\xrightarrow{d^*} H^{q+1}(V_1 \cup V_2) \to \cdots.$$

If $\tilde{V}_1 \subset V_1$, $\tilde{V}_2 \subset V_2$ then the $M-V$ sequence of (V_1, V_2) maps into the $M-V$ sequence of $(\tilde{V}_1, \tilde{V}_2)$, and we get a direct system of sequences, indexed by $\Lambda X_1 \times \Lambda X_2$. We can pass to the direct limit, and by cofinality (5.17) get the sequence

$$\cdots \xrightarrow{\check{d}} \check{H}^q(X_1 \cup X_2) \xrightarrow{(\check{i}_1, \check{i}_2)} \check{H}^q X_1 \oplus \check{H}^q X_2 \xrightarrow{(\check{j}_1, -\check{j}_2)} \check{H}^q(X_1 \cap X_2)$$
$$(6.14) \qquad \xrightarrow{\check{d}} \check{H}^{q+1}(X_1 \cup X_2) \to \cdots.$$

This is the absolute $M-V$ sequence in Čech-cohomology; by 5.21 *it is exact. It holds whenever $X_1 \cap X_2$ separates X_1, X_2* (and X_1, X_2 are locally compact subspaces of some ENR). ∎

6.15 Excision. With the same assumptions and notations as in 6.13 we have $H^*(V_1 \cup V_2, V_1) \cong H^*(V_2, V_1 \cap V_2)$ for every pair of open neighborhoods of X_1, X_2. Passing to direct limits this becomes

$$(6.16) \qquad \check{H}(X_1 \cup X_2, X_1) \cong \check{H}(X_2, X_1 \cap X_2);$$

in other words, *triads $(E; X_1, X_2)$ as in 6.13 are Čech-excisive.*

In order to compare this with the familiar excision theorem III, 7.4, let $B \subset A \subset X$ be subspaces of some ENR, and put $X_1 = A$, $X_2 = X - B$; then $X_1 \cup X_2 = X$, $X_1 \cap X_2 = A - B$, and 6.16 becomes

6.17 Proposition. *If $B \subset A \subset X$ are subspaces of some* ENR *E, if $B \subset \mathring{A}$, $\bar{B} \subset A$* (interior and closure with respect to X, not E), *and $A, X-B$ are locally compact, then $X, A-B$ are also locally compact and $\check{\imath}$: $\check{H}(X, A) \cong \check{H}(X-B, A-B)$, where i = inclusion.*

Indeed, the conditions $B \subset \mathring{A}$, $\bar{B} \subset A$ just mean that $X_1 \cap X_2 = A - B$ separates X_1, X_2. ∎

6.18 Continuity. One might think of constructing some kind of "super-Čech groups" by iterating the limit process 6.1. However, this leads to

Čech groups again, as we shall see now. Let F be a locally compact subspace of an ENR E, and $Y \subset X$ locally compact subspaces of F. Suppose Λ is a set of locally compact pairs in F such that

(i) *Λ is directed under inverse inclusion,*

(ii) $(M, N) \in \Lambda \Rightarrow Y \subset M$ and $X \subset N$,

(iii) *if (V, W) is a pair of open neighborhoods of (X, Y) then there is an $(M, N) \in \Lambda$ with $M \subset V$, $N \subset W$.*

Then $\{\check{H}(M, N)\}_{(M,N)\in\Lambda}$ together with the inclusion homomorphisms is a direct system (over Λ), and we can form $\varinjlim \{\check{H}(M, N)\}$. Furthermore, the inclusions (ii) induce homomorphisms $\sigma_{MN}: \check{H}(M, N) \to \check{H}(X, Y)$ which in the limit give $\sigma: \varinjlim \{\check{H}(M, N)\} \to \check{H}(X, Y)$. We claim

(6.19) $$\sigma: \varinjlim \{\check{H}(M, N)\} \cong \check{H}(X, Y).$$

This is an important special case of what is called *continuity of Čech-cohomology*. As an example for Λ one can take any cofinal system of pairs of locally compact neighborhoods of (X, Y). If Λ is a directed system of *compact* pairs then one can show (exercise!) that (ii) and (iii) are equivalent to $\bigcap_\Lambda (M, N) = (X, Y)$.

Proof of 6.19. We show that $\{\sigma_{MN}\}$ satisfies 5.18 (i), (ii). Let $y \in \check{H}(X, Y)$. There are open neighborhoods $W \subset V$ of $Y \subset X$ in E and $x \in H^*(V, W)$ such that $u_{VW}^{XY}(x) = y$, where $u_{VW}^{XY}: H^*(V, W) \to \check{H}(X, Y)$ is the universal transformation. Choose $(M, N) \subset (V, W)$. Then $u_{VW}^{MN}(x) \in \check{H}(M, N)$, and $\sigma_{MN}[u_{VW}^{MN}(x)] = u_{VW}^{XY}(x) = y$, hence σ satisfies 5.18 (i).

Assume now $x \in \check{H}(M, N)$ is such that $\sigma_{MN}(x) = 0$. Choose open neighborhoods (V, W) of (M, N) in E and $v \in H^*(V, W)$ such that $x = u_{VW}^{MN}(v)$. Then $u_{VW}^{XY}(v) = \sigma_{MN}\, u_{VW}^{MN}(v) = 0$, hence there are smaller neighborhoods (V', W') of (X, Y) such that $j^*(v) = 0$, where $j: (V', W') \subset (V, W)$. By (ii) and (i) we can find $(M', N') \in \Lambda$ such that $(M', N') \subset (V', W')$ and $(M', N') \subset (M, N)$. The second of these inclusions, k, takes x into $\check{k}(x) = \check{k}\, u_{VW}^{MN}(v) = u_{V'W'}^{M'N'}\, j^*(v) = 0$, hence σ satisfies 5.18 (ii).

As an application of 6.19 we prove

6.20 Proposition. *If $Y \subset X$ are locally compact subsets of some ENR E, and Y is compact then the space X/Y (which is obtained from X by identifying Y to a single point $\{Y\}$) is also locally compact in some ENR, and the projection map $\pi: (X, Y) \to (X/Y, \{Y\})$ induces isomorphisms*

$$\check{\pi}: \check{H}(X/Y, \{Y\}) \cong \check{H}(X, Y).$$

Proof. We first show that X/Y lies in some $\mathbb{R}^k$. By IV, 8.2 there is a homeomorphism φ of $X-Y$ onto a closed subset of some $\mathbb{R}^n$. Then $\Phi: X\to\mathbb{S}^n=\mathbb{R}^n\cup\{\infty\}$, $\Phi|X-Y=\varphi$, $\Phi(Y)=\infty$, is continuous, and if X is compact, it is an identification map, hence $X/Y\approx\operatorname{im}(\Phi)\subset\mathbb{S}^n\subset\mathbb{R}^{n+1}$. If X is not compact we choose an open neighborhood O of Y in X such that $\bar{O}$ is compact (closure in X), hence $\bar{O}/Y\subset\mathbb{R}^l$ as above; in particular, $O/Y\subset\mathbb{R}^l$. Since $X/Y-\{Y\}\approx X-Y$ is also contained in some $\mathbb{R}^n$, we know by IV, 8.8 that $X/Y=(X/Y-\{Y\})\cup(O/Y)$ lies in some $\mathbb{R}^k$.

Consider now the directed set Λ of all *compact* neighborhoods N of Y in X. It is cofinal in the set of all neighborhoods, hence $\check{H}(X,Y)=\varinjlim\{\check{H}(X,N)\}_{N\in\Lambda}$ by 6.19. Similarly, $\check{H}(X/Y,\{Y\})=\varinjlim\{\check{H}(X/Y,\tilde{N})\}$, where $\tilde{N}$ ranges over the set $\tilde{\Lambda}$ of all compact neighborhoods of $\{Y\}$ in X/Y. But $N\mapsto\pi(N)$, $\tilde{N}\mapsto\pi^{-1}(\tilde{N})$ are reciprocal bijections between Λ and $\tilde{\Lambda}$, and

$$\check{H}(X,N)\cong\check{H}(X-Y,N-Y)\overset{\pi}{\cong}\check{H}(X/Y-\{Y\},\tilde{N}-\{Y\})\cong\check{H}(X/Y,\tilde{N}),$$

where $\tilde{N}=\pi N$ (the outside isomorphisms by excision 6.17), hence

$$\varinjlim\{\check{H}(X,N)\}\cong\varinjlim\{\check{H}(X/Y,\tilde{N})\}.\ \blacksquare$$

6.21 $\smile$ **Products** can be introduced in Čech-cohomology simply by passing to limits with $\smile$-products in ordinary cohomology. Assume, for instance, $Y,Y'\subset X$ are locally compact subspaces of some ENR E. If $W,W'\subset V$ are corresponding open neighborhoods in E then

$$H^*(V,W)\times H^*(V,W')\xrightarrow{\smile}H^*(V,W\cup W')\xrightarrow{u}\check{H}(X,Y\cup Y')$$

is defined (for suitable coefficients) and passes to the limit as

$$\check{H}(X,Y)\times\check{H}(X,Y')\xrightarrow{\smile}\check{H}(X,Y\cup Y').$$

We leave all details to the reader, but point out that here, in contrast to VII, 8, *no excisiveness-conditions* have to be imposed on $(X;Y,Y')$.

$\frown$-products can be defined between Čech-cohomology classes x and either singular homology classes ζ or Čech-homology classes ζ' (which we did not discuss). The result $x\frown\zeta$ respectively $x\frown\zeta'$ is a singular resp. Čech-homology class. For details see 7.1, and also Exerc. 5.

6.22 Čech-Cohomology with Bounded (Compact) Supports. A subset B of a topological space E is called *bounded* (in E) if its closure $\bar{B}$ is compact. If $Y\subset X$ are locally compact (or locally closed; cf. IV, 8.3) subspaces of an ENR E we consider the set $\Omega=\Omega(X,Y)$ of all locally compact ω such that $Y\subset\omega\subset X$, and $X-\omega$ is bounded. Then Ω is directed under reversed inclusion ($\omega\leq\tilde{\omega}\Leftrightarrow\omega\supset\tilde{\omega}$). For every $\omega\in\Omega$ we have a graded group $\check{H}(X,\omega)$, and for every relation $\omega\leq\tilde{\omega}$ in Ω we have $\check{H}(X,\omega)\to\check{H}(X,\tilde{\omega})$. This constitutes a direct system whose direct limit is called the *Čech-*

cohomology of (X, Y) *with bounded supports*, in symbols $\check{H}_b(X, Y) = \varinjlim \{\check{H}(X, \omega) | \omega \in \Omega\}$. For instance, if $X - Y$ is itself bounded, i.e. if $Y \in \Omega$, then $\{Y\}$ is cofinal in Ω, hence $\check{H}_b(X, Y) = \check{H}(X, Y)$ in this case.

If X is *closed* in E then $\bar{B} \subset X$ for every $B \subset X$, hence Ω consists of all locally compact subspaces ω of X such that $Y \subset \omega$, and $X - \omega$ is bounded *in* X. Thus $\check{H}_b(X, Y)$, in this case, is independent of the embedding $X \subset E$; it is called the Čech-cohomology of (X, Y) with *compact supports*, and is denoted by $\check{H}_c(X, Y)$. As usual, we write $\check{H}_b(X, \emptyset) = \check{H}_b X$, $\check{H}_c(X, \emptyset) = \check{H}_c X$.

If we consider sets $\omega \in \Omega$ only which are *open in X* we obtain a directed subset $\Omega_0 = \Omega_0(X, Y) = \{\omega \in \Omega(X, Y) | \omega \text{ open in } X\}$. It need *not* be cofinal in Ω, but still

$$\check{H}_b(X, Y) \cong \varinjlim \{\check{H}(X, \omega) | \omega \in \Omega_0\}. \tag{6.23}$$

Proof. We show that the transformation $\{\check{H}(X, \omega_0) \to \check{H}_b(X, Y)\}_{\omega_0 \in \Omega_0}$ satisfies the criterion 5.18. If $y \in \check{H}_b(X, Y)$ then y comes from some $x \in \check{H}(X, \omega)$ with $\omega \in \Omega$; and x, by 6.19, comes from some $x_0 \in \check{H}(X, \omega_0)$ with $\omega_0 \in \Omega_0$, $\omega_0 \supset \omega$. Hence y comes from x_0, and we have verified 5.18 (i). Suppose now $x_0 \in \check{H}(X, \omega_0)$, $\omega_0 \in \Omega_0$, has zero-image in $\check{H}_b(X, Y)$; then it has zero-image in $\check{H}(X, \omega)$ for some $\omega \in \Omega$ with $\omega_0 \supset \omega$. By 6.19 again, it has zero-image in $\check{H}(X, \omega_0')$ for some $\omega_0' \in \Omega_0$ with $\omega_0 \supset \omega_0' \supset \omega$; this checks 5.18 (ii). ∎ (Note: The reader may analyse this proof and extract a general result about double limits.)

As a consequence of 6.23 we obtain

$$\check{H}_c(X, Y) \cong \check{H}_c(X - Y), \quad \textit{if } Y \textit{ is closed in } X. \tag{6.24}$$

Indeed, $\check{H}_c(X, Y)$ is the direct limit of $\{\check{H}(X, \omega_0)\}$, where ω_0 is an open neighborhood of Y with compact complement $X - \omega_0$, and $\check{H}(X, \omega_0) = \check{H}(X - Y, \omega_0 - Y)$ by excision 6.17. But $\varinjlim \{\check{H}(X - Y, \omega_0 - Y)\} = \check{H}_c(X - Y, \emptyset)$, by 6.23. ∎

6.25 Example. *If Y is closed in X, and $X - Y$ is a connected n-manifold then $\check{H}_c^n(X, Y; \mathbb{Z}) \cong \mathbb{Z}$ if $X - Y$ is orientable, and $\check{H}_c^n(X, Y; \mathbb{Z}) \cong \mathbb{Z}_2$ otherwise. In both cases, $\check{H}_c(X, Y; \mathbb{Z}_2) = \mathbb{Z}_2$.*

Proof. By 6.24 we can assume $Y = \emptyset$. Every compact set of X is contained in a connected compact set K (join by arcs), and if X is not-orientable we can even assume that it is not orientable along K (add orientation-reversing arcs). I.e., the family $\{K\}$ of these K is cofinal in the system of all compact sets, hence $\{X - K\}$ is cofinal in $\Omega_0(X, \emptyset)$, hence $\check{H}_c X =$

$\varinjlim\{\check{H}(X, X-K)\}$. Now

$$\check{H}^n(X, X-K;\mathbb{Z}) \cong H^n(X, X-K;\mathbb{Z})$$
$$\cong \operatorname{Hom}(H_n(X, X-K;\mathbb{Z}),\mathbb{Z}) \oplus \operatorname{Ext}(H_{n-1}(X, X-K;\mathbb{Z}),\mathbb{Z}).$$

If X is orientable then the 2nd term on the right is zero by 3.5, and the 1st term is $\mathbb{Z}$ by 3.4, hence $\check{H}^n(X, X-K;\mathbb{Z})\cong\mathbb{Z}$; this isomorphism is compatible with inclusions, hence $\check{H}^n_c X = \varinjlim\{\check{H}^n(X, X-K)\}\cong\mathbb{Z}$. If X is not orientable, hence not orientable along K, then the first term is zero, by 3.4, and the second term is $\mathbb{Z}_2$, by 3.5, hence $\check{H}^n(X, X-K;\mathbb{Z})\cong\mathbb{Z}_2$, hence $\check{H}^n_c X = \varinjlim\{\check{H}^n(X, X-K)\}\cong\mathbb{Z}_2$. Similarly for $\check{H}^n(X, Y;\mathbb{Z}_2)$. ∎

6.26 Induced Homomorphisms $\check{f}_b$: $\check{H}_b(X, Y)\to\check{H}_b(X', Y')$ can be defined for continuous maps f: $(X', Y')\to(X, Y)$ as in 6.3 provided $\omega\in\Omega_0(X, Y)$ implies $(f^{-1}\omega)\in\Omega_0(X', Y')$. Indeed,

$$\{\check{f}_\omega\colon \check{H}(X,\omega)\to\check{H}(X', f^{-1}\omega)\}_{\omega\in\Omega_0(X,Y)}$$

is then a family of maps which passes to the limit (5.9), and gives

$$\check{f}_b = \varinjlim\{\check{f}_\omega\}\colon \check{H}_b(X, Y)\to\check{H}_b(X', Y'). \tag{6.27}$$

This makes $\check{H}_b$ *a cofunctor on maps* f *as above.*

The condition $\omega\in\Omega_0(X, Y)\Rightarrow(f^{-1}\omega)\in\Omega_0(X', Y')$ means that subsets of $X-Y$ which are closed in X and bounded in E have counterimages $f^{-1}B$ which are bounded in E'. *It is always fulfilled, if the composite* $X'\xrightarrow{f}X\xrightarrow{\subset}E$ *is proper over* $E-Y$ (counterimages of compact subsets of $E-Y$ are compact). *In particular, it is fulfilled if* X *is closed in* E *and* f: $X'\to X$ *is proper over* $X-Y$.

6.28 Exercises. *1.* The definition of the functor $\check{H}(X, Y)$ and most of its properties do not really require Y to be locally compact (just X). Verify this assertion.

2. If $Z\subset Y\subset X$ are locally compact subspaces of an ENR, establish an exact sequence

$$\cdots\to\check{H}^q(X, Z)\to\check{H}^q(Y, Z)\to\check{H}^{q+1}(X, Y)\to\check{H}^{q+1}(X, Z)\to\cdots$$

in analogy to 6.10 and III, 3.4.

3. (a) Let $X_n\subset\mathbb{R}^2$ be the circle with radius $1/n$ and center $(0, 1/n)$ and let $X=\bigcup_{n=1}^{\infty}X_n$. Show that ρ: $\check{H}^1(X;\mathbb{Z}_2)\to H^1(X;\mathbb{Z}_2)$ is *not surjective* (hint: if $x\in\operatorname{im}(\rho)$ then $x|X_n=0$ for almost all n).

(b) Let $\Gamma=\{(x,y)\in\mathbb{R}^2 | y=\sin(1/x), x\neq 0\}$ = graph of $\sin(1/x)$, and let $X=\bar{\Gamma}$ its closure in $\mathbb{S}^2=\mathbb{R}^2\cup\{\infty\}$. Show that $H^1X=0$ but $\check{H}^1X\neq 0$ (hint: X has a cofinal sequence of neighborhoods each of which is homeomorphic with an annulus, hence $\check{H}^1(X;\mathbb{Z})\cong\mathbb{Z}$).

4. Prove. If $(X;X_1,X_2)$ is a triad such that $X_1, X_2, X_1\cap X_2, X_1\cup X_2$ are ENR's and $X_1\cap X_2$ separates X_1, X_2 (cf. 6.13) then the triad is *excisive* (III, 8.1). Hint: 6.16 and 6.12 show that

$$H^*(X_1\cup X_2, X_1)\cong H^*(X_2, X_1\cap X_2)$$

with arbitrary coefficients. This implies

$$H(X_2, X_1\cap X_2)\cong H(X_1\cup X_2, X_1)$$

by VI, 6.22, Exerc. 5 (see also VI, 7.22, Exerc. 5).

5*. Let $(X;X_1,X_2)$ be a triad such that X, X_1 are locally compact in some ENR E, and $\overline{X_1-X_2}\cap(X_2-X_1)=\emptyset$. If $W\subset V$ are open neighborhoods of $X_1\subset X$ in E, then $X_1'=(X_1\cup X_2)\cap W$ and $X_2'=(X_1\cup X_2)-\overline{X_1-X_2}$ are open in $X_1\cup X_2$, and $X_1\cup X_2=X_1'\cup X_2'$. We get

$$H^*(V,W)\times H(X,X_1\cup X_2)\to H^*(X,X_1')\times H(X,X_1'\cup X_2')$$
$$\xrightarrow{\frown} H(X,X_2')\to H(X,X_2),$$

and in the limit,

$$\check{H}(X,X_1)\times H(X,X_1\cup X_2)\xrightarrow{\frown} H(X,X_2')\to H(X,X_2).$$

Carry out the details.

6. If $Y\subset X$ are locally compact subsets of an orientable manifold show that $\Gamma_b(X,Y)$, as defined in 3.3, is isomorphic with $\check{H}^0_b(X,Y)$.

7*. Establish a natural exact sequence

$$\cdots\to\check{H}^q_b(X,Y)\to\check{H}^q_bX\to\check{H}^q_bY\to\check{H}^{q+1}_b(X,Y)\to\cdots,$$

for (X,Y) as in 6.22.

7. *Poincaré-Lefschetz Duality*

If M^n is a manifold, and $L\subset K$ are compact subsets of M we define (7.4) a natural bilinear pairing

$$\text{(7.1)}\qquad \frown:\ \check{H}^i(K,L)\times H_k(M,M-K)\to H_{k-i}(M-L,M-K),$$

simply by passing to limits with ordinary $\frown$-products; more generally, $L\subset K$ need only be *closed* in M provided K has a neighborhood in M which is an ENR. As coefficients we use an arbitrary (commutative) ring R with unit for $H(M, M-K)$, and an arbitrary R-module G for $\check{H}(K,L)$, $H(M-L, M-K)$. If M is oriented along K, and K is compact we denote by $o_K\in H_n(M, M-K)$ the fundamental class (with coefficients in R), i.e., the element which under J: $H_n(M, M-K, R)\cong\Gamma(K; R)$ corresponds to the orientation (in the notation of 4.1 this is $o_K\otimes 1$). We also use the notation o_K for non-oriented manifolds if the coefficient ring R has characteristic two (i.e., if $1=-1$ in R); as before, o_K corresponds to the canonical section $P\mapsto 1$ of $\tilde{M}\otimes R=M\times R$.

If we fix the second variable of the pairing 7.1 at $o=o_K$ we get a homomorphism $\frown o$: $\check{H}(K,L)\to H(M-L, M-K)$, and our main result asserts

7.2 Proposition (Duality theorem). *If $L\subset K$ are compact subsets of an n-manifold M then*

$$\frown o:\ \check{H}^i(K,L)\cong H_{n-i}(M-L, M-K).$$

The coefficients are arbitrary if M is oriented along K; otherwise, they are assumed to be of characteristic two. (N.B. $\check{H}(K,L)\cong H^*(K,L)$ if K, L are neighborhood retracts: cf. 6.12). The elements

$$x\in\check{H}^i(K,L) \quad\text{and}\quad \xi=x\frown o\in H_{n-i}(M-L, M-K)$$

are called (*Poincaré*) *dual* to each other.

This theorem has many interesting consequences and applications; some of them will be treated in the next §8. Also, several generalizations exist; some of them are indicated in 7.12, 7.16.

We now construct the $\frown$-product 7.1. Recall that $L\subset K$ are closed subsets of M, contained in some open ENR $E\subset M$. Consider the set $\Lambda=\Lambda(K,L)$ of all pairs $W\subset V$ of open neighborhoods of $L\subset K$. Then Λ is directed by reversed inclusion, and $\Lambda'=\{(V,W)\in\Lambda\,|\,V\subset E\}$ is cofinal in Λ. The latter implies

$$\varinjlim\{H^*(V,W)\}_\Lambda\cong\varinjlim\{H^*(V,W)\}_{\Lambda'}=\check{H}(K,L).$$

For $\xi\in H(M, M-K)$, consider the composite map

$$(7.3)\qquad \begin{aligned}\xi_{VW}:\ H^*(V,W)\to H^*(V-L, W-L)&\xrightarrow{\frown j_*^W\xi} H(V-L, V-K)\\ &\overset{\text{exc}}{\cong} H(M-L, M-K)\end{aligned}$$

where

$$j_*^W\colon H(M, M-K) \to H(M, (M-K)\cup W) \overset{\text{exc}}{\cong} H\big(V-L, (V-K)\cup(W-L)\big).$$

As (V, W) ranges over Λ the maps ξ_{VW}, with fixed ξ, constitute a transformation of the direct system $\{H^*(V, W)\}_{(V, W)\in\Lambda}$ into $H(M-L, M-K)$, hence a limit homomorphism

$$(7.4) \qquad \frown\xi\colon \check{H}(K, L) \to H(M-L, M-K), \qquad (\frown\xi)\circ u_{VW} = \xi_{VW},$$

where $u_{VW}\colon H^*(V, W) \to \check{H}(K, L)$ is the universal transformation. *The image of* $x\in\check{H}(K, L)$ *under* $\frown\xi$ *is denoted by* $x\frown\xi$, *and is called the (Čech) cap-product of* x *and* ξ. If $x\in\check{H}^i(K, L)$ and $\xi\in H_k(M, M-K)$ then $(x\frown\xi)\in H_{k-i}(M-L, M-K)$.

One has $(x+x')\frown\xi = x\frown\xi + x'\frown\xi$ by construction, and $x\frown(\xi+\xi') = x\frown\xi + x\frown\xi'$ because $(\xi+\xi')_{VW} = \xi_{VW} + \xi'_{VW}$; i.e., *the Čech cap-product* $\frown$ *is bilinear*.

If $i\colon (\tilde{K}, \tilde{L}) \subset (K, L)$ is an inclusion of closed pairs as above, then the diagram

$$(7.5)\qquad \begin{array}{ccccc} & & H^*(V-L, W-L) & \xrightarrow{\frown j_*^W \xi} & H(V-L, V-K) \cong H(M-L, M-K) \\ & \nearrow & \uparrow{\scriptstyle i'^*} & & \downarrow{\scriptstyle i'_*} \qquad\qquad \downarrow{\scriptstyle i'_*} \\ H^*(V, W) & & & & \\ & \searrow & & & \\ & & H^*(V-\tilde{L}, W-\tilde{L}) & \xrightarrow[\frown j_*^W i'_* \xi]{} & H(V-\tilde{L}, V-\tilde{K}) \cong H(M-\tilde{L}, M-\tilde{K}) \end{array}$$

(i' = inclusion) is commutative, the middle square by naturality VII, 12.6 of $\frown$-products. The top row of 7.5 is $\xi_{VW} = (\frown\xi)\circ u_{VW}$, the bottom row $(i'_*\xi)_{VW} = (\frown i'_*\xi)\circ\tilde{u}_{VW} = (\frown i'_*\xi)\circ\check{i}\circ u_{VW}$, the latter by definition of $\check{i}\colon \check{H}(K, L) \to \check{H}(\tilde{K}, \tilde{L})$. It follows that $i'_*(\frown\xi)\,u_{VW} = (\frown i'_*\xi)\,\check{i}\,u_{VW}$, hence $i'_*(\frown\xi) = (\frown i'_*\xi)\,\check{i}$ by universality of u, or

$$(7.6) \quad i'_*(x\frown\xi) = (\check{i}x)\frown(i'_*\xi), \qquad \text{for } \xi\in H(M, M-K),\ x\in\check{H}(K, L)$$

(naturality of $\frown$ *with respect to inclusions* i*).*

Proposition 7.2 asserts that $\frown\xi$ is isomorphic if $\xi = o_K$ = fundamental class along K. We establish the absolute case $L=\emptyset$ first; its proof, just as with 3.3, is based on a MV-principle, namely

7.7 Lemma. *If* $K_1, K_2 \subset M$ *are compact sets, and if 7.2 holds for* $(K_1, \emptyset)$, $(K_2, \emptyset)$, $(K_1\cap K_2, \emptyset)$ *then also for* $(K_1\cup K_2, \emptyset)$.

Proof. Let V_1, V_2 be open neighborhoods of K_1, K_2 and consider the diagram

$$\begin{CD}
H^* V_1 \oplus H^* V_2 @>>> H^*(V_1 \cap V_2) @>{d^*}>> H^*(V_1 \cup V_2) \\
@V{(\frown o_{K_1}) \oplus (\frown o_{K_2})}VV @V{\frown o_{K_1 \cap K_2}}VV @V{\frown o_{K_1 \cup K_2}}VV \\
H(M, M-K_1) \oplus H(M, M-K_2) @>>> H(M, M-K_1 \cap K_2) @>{d_*}>> H(M, M-K_1 \cup K_2)
\end{CD}$$

$$\tag{7.8}\begin{CD}
@>>> H^* V_1 \oplus H^* V_2 @>>> H^*(V_1 \cap V_2) \\
@. @V{(\frown o_{K_1}) \oplus (\frown o_{K_2})}VV @V{\frown o_{K_1 \cap K_2}}VV \\
@>>> H(M, M-K_1) \oplus H(M, M-K_2) @>>> H(M, M-K_1 \cap K_2).
\end{CD}$$

The rows are partial MV-sequences and therefore exact. The 1st, 3rd, and 4th square are commutative by naturality, VII, 12.6, of $\frown$-products (note that $H(M, M-K_1 \cap K_2) \cong H(V_1, V_1 - K_1 \cap K_2) \cong H(V_1 \cap V_2, V_1 \cap V_2 - K_1 \cap K_2)$). The second square is commutative by VII, 12.20 (one can assume $M = V_1 \cup V_2$, and one applies VII, 12.20 with $X_\mu = V_\mu$, $Y_\mu = M - K_\mu$, $\xi = o_{K_1 \cup K_2}$; the 2nd square of 7.8 is then just the outer part of the diagram VII, 12.21).

The set of all couples (V_1, V_2) is directed (by inverse inclusion), and the maps $(V_1, V_2) \mapsto V_1, V_2, V_1 \cap V_2, V_1 \cup V_2$ of this set into $\Lambda(K_1, \emptyset)$, $\Lambda(K_2, \emptyset)$, $\Lambda(K_1 \cap K_2, \emptyset)$, $\Lambda(K_1 \cup K_2, \emptyset)$ are cofinal. If we pass to the limit over $\{(V_1, V_2)\}$ then (by 5.17) the terms in the first row of 7.8 become Čech-groups, and the whole diagram becomes

$$\begin{CD}
\check{H} K_1 \oplus \check{H} K_2 @>>> \check{H}(K_1 \cap K_2) @>{d}>> \check{H}(K_1 \cup K_2) \\
@V{(\frown o_{K_1}) \oplus (\frown o_{K_2})}VV @V{\frown o_{K_1 \cap K_2}}VV @V{\frown o_{K_1 \cup K_2}}VV \\
H(M, M-K_1) \oplus H(M, M-K_2) @>>> H(M, M-K_1 \cap K_2) @>>> H(M, M-K_1 \cup K_2)
\end{CD}$$

$$\tag{7.9}\begin{CD}
@>>> \check{H} K_1 \oplus \check{H} K_2 @>>> \check{H}(K_1 \cap K_2) \\
@. @V{(\frown o_{K_1}) \oplus (\frown o_{K_2})}VV @V{\frown o_{K_1 \cap K_2}}VV \\
@>>> H(M, M-K_1) \oplus H(M, M-K_2) @>>> H(M, M-K_1 \cap K_2).
\end{CD}$$

The rows are still exact, by 5.21, (in fact, the first row is just 6.14), and the outer vertical arrows are isomorphic by assumption. Therefore the middle arrow $\frown o_{K_1 \cup K_2}$ is also isomorphic, by the five lemma. ∎

Proof of 7.2. We proceed in several steps, in analogy with IV, 6.4 and VIII, 3.3.

Case 1. $K=\emptyset$, or $K=P=$ a point.

If $K=\emptyset$ then all groups are zero. If $K=P$ then $\frown o_P$ takes the generator $1_P \in H^0 K=\check{H}^0 K$ into the generator $o_P \in H_n(M, M-P)$ by VII, 12.9, and all other groups are zero.

Case 2. $M=\mathbb{R}^n$, $K=\square=$ a cube, $L=\emptyset$.

Let $P\in\square$. Then $\square\simeq P$, $\mathbb{R}^n-\square\simeq\mathbb{R}^n-P$, hence $\check{H}K\cong H^*K\cong H^*P$, $H(\mathbb{R}^n, \mathbb{R}^n-\square)\cong H(\mathbb{R}^n, \mathbb{R}^n-P)$, and the result follows from Case 1 (using naturality 7.6 of $\frown$).

Case 3. $M=\mathbb{R}^n$, $K=$ union of finitely many (say r) cubes of a lattice (V, 3.4), $L=\emptyset$.

If $r=1$ then Case 2 applies. If $r>1$ then K is of the form $K=K_1\cup K_2$, where $K_1, K_2, K_1\cap K_2$ are unions of less than r cubes. We can apply an inductive hypothesis to $K_1, K_2, K_1\cap K_2$ and get the result for $K=K_1\cup K_2$ by the MV-principle 7.7.

Case 4. $M=\mathbb{R}^n$, K arbitrary compact, $L=\emptyset$.

Let $\{V\}$ be the directed set of all compact neighborhoods of K which are finite unions of cubes of a lattice. This set is cofinal in the set of all neighborhoods of K, hence $\check{H}K=\varinjlim\{\check{H}V\}$, by 6.19. Also $H(\mathbb{R}^n, \mathbb{R}^n-K)=\varinjlim\{H(\mathbb{R}^n, \mathbb{R}^n-V)\}$ because $\mathbb{R}^n-K=\bigcup_V(\mathbb{R}^n-V)$; cf. 2nd example 5.22. By Case 3 we have $\frown o_V\colon \check{H}V \xrightarrow{\cong} H(\mathbb{R}^n, \mathbb{R}^n-V)$, for every V; since $\frown$ is natural we can pass to the limit and get $\frown o_K\colon \check{H}K\cong H(\mathbb{R}^n, \mathbb{R}^n-K)$.

Case 5. M arbitrary, K arbitrary compact, $L=\emptyset$.

K is contained in a union of finitely many (say r) coordinate neighborhoods $\approx\mathbb{R}^n$. If $r=1$ then Case 4 applies because $H(M, M-K)\cong H(\mathbb{R}^n, \mathbb{R}^n-K)$. If $r>1$ then K is of the form $K=K_1\cup K_2$, where K_1, K_2 are compact sets which are covered by less than r coordinate neighborhoods. We can therefore apply an inductive hypothesis to K_1, K_2, $K_1\cap K_2$, and get the result for $K=K_1\cup K_2$ by the MV-principle 7.7.

Case 6. The general case.

Consider the diagram

$$\begin{array}{ccccccccc}
\check{H}K & \longrightarrow & \check{H}L & \xrightarrow{\delta} & \check{H}(K,L) & \longrightarrow & \check{H}K & \longrightarrow & \check{H}L \\
\big\downarrow{\scriptstyle\frown o_K} & & \big\downarrow{\scriptstyle\frown o_l} & & \big\downarrow{\scriptstyle\frown o_K} & & \big\downarrow{\scriptstyle\frown o_K} & & \big\downarrow{\scriptstyle\frown o_L} \\
H(M,M-K) & \rightarrow & H(M,M-L) & \xrightarrow{\partial_*} & H(M-L,M-K) & \rightarrow & H(M,M-K) & \rightarrow & H(M,M-L)
\end{array}$$

whose rows are the usual exact sequences (6.10 resp. III, 3.4). The 1st, 3rd and 4th square are commutative by naturality 7.6 of $\frown$-products. The 2nd square commutes because it is the direct limit of corresponding squares for open neighborhoods $W \subset V$ of $L \subset K$, and each of these commutes by VII, 12.22 (one can assume $V=M$, and one applies VII, 12.22 with $X=V$, $W'=M-L$, $V'=M-K$, $\xi=o_K$). The outer vertical arrows are isomorphic (by Case 5) hence also the middle arrow (by the five lemma). ∎

By a simple excision argument we can generalize 7.2 to *closed* subsets $L \subset K$ of M^n provided $\overline{K-L}$ *is compact*. Indeed, if C is any compact set such that $\overline{K-L} \subset C \subset M$ then

$$(7.10) \quad \begin{aligned} \check{H}(K,L) \cong \check{H}(K \cap C, L \cap C) &\overset{\frown o_{K \cap C}}{\cong} H(M-L \cap C, M-K \cap C) \\ &\cong H(M-L, M-K), \end{aligned}$$

the outside isomorphism by excision, the middle one by 7.2. (This seems to require that K lies in some ENR; however, if that is not the case we just define $\check{H}(K,L)$ by the first isomorphism 7.10.) We still denote the composite map 7.10 by $\frown o$ (although there is no $o_K \in H(M, M-K)$ if K is not compact). *Then, as before,* $\frown o$ *is natural with respect to inclusions* $i\colon (\tilde{K}, \tilde{L}) \xrightarrow{\subset} (K, L)$ *of pairs as above*, i.e.,

$$(7.11) \qquad i'_*(x \frown o) = (\check{i} x) \frown o,$$

for $x \in \check{H}(K,L)$, and $i'\colon (M-L, M-K) \xrightarrow{\subset} (M-\tilde{L}, M-\tilde{K})$. ∎

7.12 If K is any closed set in M^n let $\bar{\Omega}$ denote the set of all closed subsets A of K such that $\overline{K-A}$ is compact. Then $\{K-A\}_{A \in \bar{\Omega}}$ is a cofinal system of bounded subsets of K, i.e., $\bar{\Omega}$ is a cofinal subsystem of $\Omega(K, \emptyset)$ in the notation of 6.22, hence $\varinjlim \{\check{H}(K,A) | A \in \bar{\Omega}) = \check{H}_c K =$ *Čech cohomology of* K *with compact supports*. Also, $\varinjlim \{H(M-A, M-K) | A \in \bar{\Omega}\} = H(M, M-K)$ by 5.22, because $\bigcup_A (M-A) = M$. If M is oriented along K then $\frown o\colon \check{H}(K,A) \cong H(M-A, M-K)$ by 7.10, hence in the limit

$$(7.13) \qquad \frown o\colon \check{H}_c K \cong H(M, M-K).$$

In terms of representatives this isomorphism can be described as follows: Given $x \in \check{H}_c K$; it comes from some $x' \in \check{H}(K,A)$ where A is closed in K and $\overline{K-A}$ compact; the class x' in turn comes from some $y \in H^*(V, W)$, where $W \subset V$ are suitable open neighborhoods of $A \subset K$ in M. Then

$$(x \frown o) = (y \frown o_{K-W}) \in H(V, V-K) \cong H(M, M-K),$$

where $o_{K-W} \in H(V, (V-K) \cup W)$ is the fundamental class along $K-W$. Slightly more general than 7.13, and more symmetrical, we have

7.14 Proposition. *If $L \subset K \subset X$ are topological spaces such that L is closed in K, $K-L$ is closed in $X-L$, and $X-L$ is an n-manifold which is oriented along $K-L$ then*

$$\check{H}_c^i(K, L) \cong \check{H}_c^i(K-L) \overset{\frown o}{\cong} H(X-L, X-K),$$

the first isomorphism by 6.24, the second by 7.13 (with $M = X - L$). ∎

As an application, let V^{n-p} be a closed submanifold of M^n, let $L \subset K$ be closed subsets of V, and assume both M and V are oriented along $K-L$. (Take coefficients mod 2 in the non-oriented case). Applying 7.14 first in V then in M gives

$$(7.15) \qquad H_i(V-L, V-K) \cong \check{H}_c^{n-i-p}(K, L) \cong H_{i+p}(M-L, M-K).$$

The composite isomorphism is known as the *Thom isomorphism* (in homology). An important special case is $K=V$, $L=\emptyset$, then $H_i V \cong H_{i+p}(M, M-V)$. The reader can find more about the Thom isomorphism in § 11.

7.16 Exercises. *1.* The isomorphism $\check{H}(K, L) \cong H(M-L, M-K)$ of 7.10 does not really require K, L to be closed and $\overline{K-L}$ compact. Show that it holds if only L is closed *in* K, and $K \cap \overline{K-L}$ is compact (for $\check{H}(K, L)$ to make sense, $K-L$ should be locally compact in some ENR). Also it suffices that M is oriented along $K-L$.

2. If K is locally compact in some manifold M^n (but not necessarily closed) and M is oriented along K then $\check{H}_c^i K \cong H_{n-i}(M-\dot{K}, M-\overline{K})$, where $\dot{K} = \overline{K} - K$. Hint: write $K = A \cap O$, where A is closed and O is open in M. Then

$$\check{H}_c^i K \cong \check{H}_c^i(K \cup (M-O), M-O) \cong H_{n-i}(O, O-K) \cong H_{n-i}(M-\dot{K}, M-\overline{K}).$$

3.* If X is an ENR in some manifold M^n then one can find an open neighborhood U of X in M and a map $\rho\colon (M, U) \to (M, X)$ such that $\rho | X =$ inclusion, and the composition $(M, X) \xrightarrow{\subset} (M, U) \xrightarrow{\rho} (M, X)$ is homotopic to the identity map (use the technique of IV, 8.6, 8.7, and VIII, 6.2). Assume $M-X$ is locally compact and bounded (i.e. $\overline{M-X}$ compact), M oriented, and consider the composition

$$R\colon \check{H}(M-X) \to \check{H}(M-U) \overset{\frown o}{\cong} H(M, U) \xrightarrow{\rho_*} H(M, X).$$

I expect R to be isomorphic but I did not work out a complete proof (it seems to be rather delicate, probably along the lines of case 3 in the proof of 3.3). If $M-X$ is not bounded a similar result should be proved for cohomology with bounded supports $\check{H}_b(M-X)$; moreover, the result should extend to pairs (Y,X) of ENR's in M.

8. Examples, Applications

8.1 Poincaré Duality. If M^n is a *compact* manifold we can apply 7.2 with $K=M$, $L=\emptyset$. Then $\check{H}M=H^*M$ because M is the only neighborhood of K in M, and $o\in H_nM$, hence

(8.2) $$\frown o\colon\ H^i(M;G)\cong H_{n-i}(M;G);$$

the coefficients G are arbitrary if M is oriented, otherwise of characteristic two.

This special case of 7.2 is often referred to as *Poincaré duality*. By the universal coefficient theorem VI, 7.10, cohomology can be expressed in terms of homology; then 8.2 becomes

$$H_{n-i}(M;G)\cong \operatorname{Hom}_R(H_i(M;R);G)\oplus \operatorname{Ext}_R(H_{i-1}(M;R),G),$$

where R is a hereditary ring and G an R-module. For instance, if R is a field then we obtain

(8.3) $$H_{n-i}(M;R)\cong \operatorname{Hom}_R(H_i(M;R),R)=H_i(M;R)^*=\text{dual of } H_i(M;R);$$

this holds if either M is orientable or the coefficient field R has characteristic two.

8.4 Euler Characteristic of Manifolds. Let M^n be an arbitrary n-manifold again (not necessarily compact or oriented), and let $K\subset M$ be a compact ENR. Then the mod 2(co)-homology of K is finite (cf. V, 4.11), and $H^{n-i}K\cong\check{H}^{n-i}K\cong H_i(M,M-K)$; in particular

(8.5) $$\dim H_{n-i}(K;\mathbb{Z}_2)=\dim H_i(M,M-K,\mathbb{Z}_2).$$

This will imply

8.6 Proposition. *If $K\subset M^n$ is a compact ENR then $H(M;\mathbb{Z}_2)$ is finite if and only if $H(M-K;\mathbb{Z}_2)$ is finite, and in that case*

$$\chi_2 M=\chi_2(M-K)+(-1)^n\chi_2 K,$$

where χ_2 is the $\mathbb{Z}_2$-characteristic (VI, 7.19).

Recall (VI, 7.21) that on spaces with finitely generated homology χ_2 agrees with the Euler characteristic χ. In particular (cf. V, 4.11), $\chi_2 K = \chi K$.

Proof. The first assertion is clear from the mod 2 homology sequence of $(M, M-K)$ and 8.5. The second follows from $\chi_2 M = \chi_2(M-K) + \chi_2(M, M-K)$—cf. VI, 7.20—because

$$\begin{aligned}\chi_2(M, M-K) &= \sum(-1)^i \dim H_i(M, M-K; \mathbb{Z}_2)\\ &= (-1)^n \sum(-1)^{n-i} \dim H_{n-i}(K; \mathbb{Z}_2) = (-1)^n \chi_2 K. \quad \blacksquare\end{aligned}$$

8.7. Corollary. *If M is a compact manifold of odd dimensions then $\chi M = 0$. If $K \subset M$ is a compact ENR then $\chi K = \chi(M-K)$.*

Indeed, apply 8.6 with $K = M$ first, then with general K, and note that $\chi_2 = \chi$ here (VI, 7.21). $\blacksquare$

8.8 Corollary. *If L^{n+1} is a compact ∂ manifold then $\chi(\partial L) = (1 + (-1)^n)\chi L$. In particular, $\chi(\partial L)$ is always even.*

For instance, no even-dimensional projective space P_{2k} can be the boundary of a compact ∂-manifold (because χP_{2k} is odd).

Proof. Attach a collar to L (cf. 1.11) and get a manifold $M^{n+1} = L \cup (\partial L \times [0, 1))$. Then L is a compact deformation retract of M, and $M - L = \partial L \times (0, 1) \simeq \partial L$, hence

$$\chi L = \chi M = \chi(M-L) + (-1)^{n+1} \chi L = \chi(\partial L) - (-1)^n \chi L. \quad \blacksquare$$

8.9 Poincaré Duality in Cohomology. Dually to 7.1 one can define a bilinear pairing

(8.10) $\quad \smile : \check{H}^i(K, L) \times H^j(M-L, M-K) \to H^{i+j}(M, M-K)$

by passing to limits with ordinary $\smile$-products (assumptions as in 7.1). Indeed, fix $y \in H^j(M-L, M-K)$ and consider, for every pair $W \subset V$ of neighborhoods of $L \subset K$, the composition

$$\begin{aligned}y_{VW}\colon H^*(V, W) \to H^*(V-L, W-L) &\xrightarrow{\smile y} H^*(V-L, (W-L) \cup (V-K))\\ &\cong H^*(M, W \cup (M-K)) \to H^*(M, M-K).\end{aligned}$$

As (V, W) varies, this is a transformation of the direct system $\{H^*(V, W)\}$ into $H^*(M, M-K)$, hence a limit homomorphism

(8.11) $\quad \smile y\colon \check{H}(K, L) \to H^*(M, M-K).$

As in §7 we denote by $x\smile y\in H^{i+j}(M,M-K)$ the image of $x\in\check{H}^i(K,L)$ under $\smile y$. The reader can easily prove (we shall not use it) that $\smile$ is bilinear and is natural with respect to inclusions $(\tilde{K},\tilde{L})\subset(K,L)$. Further,

$$(8.12)\qquad \langle x\smile y,\xi\rangle=(-1)^{|x||y|}\langle y,x\frown\xi\rangle$$

for $x\in\check{H}(K,L)$, $y\in H^*(M-L,M-K)$, $\xi\in H(M,M-K)$, $\langle -,-\rangle$=scalar product.

Proof. There are neighborhoods $W\subset V$ of $L\subset K$, and $w\in H^*(V,W)$ such that $x=u\,w$ where $u\colon H^*(V,W)\to\check{H}(K,L)$ is the universal map (cf. 5.18 (i)). Then $x\smile y=y_{VW}(w)=w\smile y$, the last term omitting some inclusion maps. Similarly, $x\frown\xi=w\frown\xi$, hence

$$\langle x\smile y,\xi\rangle=\langle w\smile y,\xi\rangle=(-1)^{|w||y|}\langle y,w\frown\xi\rangle=(-1)^{|x||y|}\langle y,x\frown\xi\rangle. \quad\blacksquare$$

8.13 Proposition. *If M is an n-manifold, $L\subset K$ are compact ENR's in M, and homology is taken with coefficients in a field R then the composition*

$$H^i(K,L)\times H^{n-i}(M-L,M-K)\xrightarrow{\smile}H^n(M,M-K)\xrightarrow{\langle -,o_K\rangle}R$$

is a dual pairing provided M is oriented along K, or R is of characteristic two; the second arrow is the scalar product with o_K (n.b. $H^i(K,L)=\check{H}^i(K,L)$ because K, L are ENR's).

In particular, if M is compact, 8.13 applies with $K=M$, $L=\emptyset$; we get a dual pairing $\smile\colon H^iM\times H^{n-i}M\to H^nM\to R$. This is, of course, just another formulation of 8.2 (with field coefficients $G=R$); however, because *it involves cohomology only* it is sometimes more convenient to apply. As usual with dual pairings one has the notion of *dual bases*: If $B=\{b\}$ is a base of $H^*(M-L,M-K)$ then the dual base $\hat{B}=\{\hat{b}\}$ of $H^*(K,L)$ is defined by $\langle\hat{b}\smile a,o\rangle=\delta_{ab}$, $a,b\in B$; and vice versa. Clearly, $\hat{\hat{B}}=\{\pm b\}$.

Proof of 8.13. Our pairing is as follows

$$(x,y)\mapsto\langle x\smile y,o_K\rangle=\pm\langle y,x\frown o_K\rangle,$$

the latter by 8.12. But we know that the scalar product $\langle -,-\rangle$ is a dual pairing (VII, 1.7), and $x\mapsto x\frown o_K$ is isomorphic by 7.2. $\blacksquare$

8.14 Corollary. *If M is a compact orientable manifold of dimension $n\equiv 2 \bmod 4$ then the Euler characteristic χM is even.*

Proof. Consider the pairing $H^{n/2}M \times H^{n/2}M \to \mathbb{Q}$ with rational coefficients, $R=\mathbb{Q}$. This is a non-degenerate skew-symmetric ($n/2$ is odd!) bilinear form on the vector space $H^{n/2}M$. Such a form can only exist if $\dim(H^{n/2}M)$ is even. But

$$\begin{aligned}\chi M &= \textstyle\sum_{i=0}^{n}(-1)^i \dim H_i M = \sum_{i=0}^{n}(-1)^i \dim H^i M \\ &= -\dim H^{n/2}M + 2\textstyle\sum_{2i<n}(-1)^i \dim H^i M,\end{aligned}$$

the latter because $\dim(H^i M) = \dim(H^{n-i}M)$. ∎

8.15 Alexander Duality. If K is a compact subset of the sphere $\mathbb{S}^n$, and $P\in K$, $Q\in \mathbb{S}^n - K$ then

(8.16) $$\check{H}^{n-i}(K,P) \cong H_i(\mathbb{S}^n - P, \mathbb{S}^n - K) \overset{\partial_*}{\cong} H_{i-1}(\mathbb{S}^n - K, Q),$$

the latter because $H(\mathbb{S}^n - P, Q) = 0$. *If K is a neighborhood retract this becomes*

(8.17) $$\tilde{\check{H}}{}^{n-i}(K) \cong \tilde{H}_{i-1}(\mathbb{S}^n - K),$$

where $\tilde{H}$, as usual, denotes reduced homology. Formulas 8.16, 8.17 are known as *Alexander-duality*. They show, in particular, that $H(\mathbb{S}^n - K)$ depends only on K (in fact, on $\check{H}K$), and not on the way K is embedded in $\mathbb{S}^n$. For instance, if K is a compact connected $(n-1)$-manifold then $\mathbb{Z}_2 = \tilde{\check{H}}{}^{n-1}(K;\mathbb{Z}_2) \cong \tilde{H}_0(\mathbb{S}^n - K; \mathbb{Z}_2)$, hence $\mathbb{S}^n - K$ has two components (compare 3.6).

If X is a *closed* (proper) subset of $\mathbb{R}^n$, then $K = X \cup \{\infty\}$ is compact in $\mathbb{S}^n = \mathbb{R}^n \cup \{\infty\}$, hence (reading 8.16 backwards)

(8.18) $$\tilde{H}_{i-1}(\mathbb{R}^n - X) = \tilde{H}_{i-1}(\mathbb{S}^n - K) \cong \check{H}^{n-i}(K, \{\infty\}) \cong \check{H}_c^{n-i} X,$$

the latter by 6.24. Again it follows that $H(\mathbb{R}^n - X)$ depends only on X (in fact, on $\check{H}_c X$), and not on the way X is embedded as a *closed* subset of $\mathbb{R}^n$. For instance, if X is a connected $(n-1)$-manifold then $\check{H}_c^{n-1}(X;\mathbb{Z}_2) = \mathbb{Z}_2$ by 6.25, hence $\tilde{H}_0(\mathbb{R}^n - X; \mathbb{Z}_2) \cong \mathbb{Z}_2$, hence $\mathbb{R}^n - X$ has two components. Now use integral coefficients and find $\mathbb{Z} \cong \tilde{H}_0(\mathbb{R}^n - X; \mathbb{Z}) \cong \check{H}_c^{n-1} X$, hence X is orientable by 6.25. ∎

8.19 Künneth Relations for Čech Cohomology. If X, X' are locally compact subsets of ENR's then we can find oriented manifolds M, M' such that X, X' are closed subsets of M, M'; in fact, by IV, 8.2, we can find closed embeddings of X, X' in $\mathbb{R}^n, \mathbb{R}^{n'}$. Then $\check{H}_c X \cong H(M, M-X)$, $\check{H}_c X' \cong$

$H(M', M'-X')$, and

$$H_c(X\times X')\cong H(M\times M', M\times M'-X\times X')$$
$$=H[(M, M-X)\times(M', M'-X')].$$

To the last term we can apply the Künneth sequence VI, 12.12, and get a split-exact sequence

$$0\to(\check{H}_c X)\otimes(\check{H}_c X')\xrightarrow{\alpha}\check{H}_c(X\times X')\xrightarrow{\beta}(\check{H}_c X)*(\check{H}_c X')^+\to 0.$$

With indices it reads

$$\text{(8.20)}\quad \begin{aligned}&0\to\bigoplus_{j+k=r}(\check{H}_c^j X)\otimes(\check{H}_c^k X')\to\check{H}_c^r(X\times X')\\&\to\bigoplus_{j+k=r+1}(H_c^j X)*(H_c^k X')\to 0.\end{aligned}$$

This is the exact Künneth sequence for Čech cohomology with compact supports. It is natural (up to sign) with respect to proper maps; a proof of naturality is indicated in Exercise 5. *Just as the ordinary Künneth sequence,* VI,12.12, *it splits (un-naturally). The coefficients G, G' for $\check{H}_c X$, $\check{H}_c X'$ can be arbitrary (modules over a hereditary ring) provided $G*G'=0$; the coefficients for $\check{H}_c(X\times X')$ are $G\otimes G'$. In particular, $\check{H}_c(X\times X')\cong(\check{H}_c X)\otimes(\check{H}_c X')$ if field coefficients are used throughout.* ∎

One can also prove Künneth relations for *relative* groups $\check{H}_c(X, Y)$, where X is as above and Y is closed in X. However, *this reduces to the absolute case* because $H_c(X, Y)\cong H_c(X-Y)$ by 6.24.

8.21 Exercises. *1.* Construct a compact connected orientable 4-manifold with prescribed Euler-characteristic.

Hint: If M, N are 4-manifolds remove a small open ball in each and paste the remaining ∂-manifolds along their boundaries; the result is a 4-manifold $M+N$ with $\chi(M+N)=\chi M+\chi N-2$. Now start with $P_2\mathbb{C}$, $\mathbb{S}^1\times\mathbb{S}^3$, and form iterated sums.

Orientable manifolds of dimension $4k$ (respectively $4k+2$), $k\geq 1$, with prescribed (even) characteristic can be constructed as multiple products of 4-manifolds (with $\mathbb{S}^2$).

2. If M is a compact oriented n-manifold, $d\colon M\to M\times M$ the diagonal map then $d_*(o)\in H_n(M\times M; R)$ is called the *diagonal class*, and its dual $\mu\in H^n(M\times M; R)$ the *dual diagonal class*, $\mu\frown(o\times o)=d_*(o)$. Assume R is a field, $B=\{b\}$ a base of $H^*(M; R)$, and $\{\hat{b}\}$ the dual base, defined by $\langle\hat{b}\smile a, o\rangle=\delta_{ab}$, $a, b\in B$. Show that $\mu=\Sigma_{b\in B}(-1)^{|b|}\hat{b}\times b$. *Hint:* Put

$\mu=\sum \lambda_{ab}(\hat{a}\times b)$ and compute the coefficients $\lambda_{ab}\in R$ from

$$\langle (a\times\hat{b})\smile\mu, o\times o\rangle=\langle a\times\hat{b}, \mu\frown(o\times o)\rangle=\langle a\times\hat{b}, d_* o\rangle$$
$$=\langle d^*(a\times\hat{b}), o\rangle=\langle a\smile\hat{b}, o\rangle=\pm\langle\hat{b}\smile a, o\rangle=\pm\delta_{ab}.$$

3. If $f\colon M\to M'$ is a map between compact oriented manifolds of dimensions n, n', let γ_f (=*class of the graph*) denote the image of $o\in H_n M$ under $HM \xrightarrow{d_*} H(M\times M) \xrightarrow{(f\times \mathrm{id})_*} H(M'\times M)$, and denote by $\gamma^f\in H^{n'}(M'\times M)$ the dual class, $\gamma^f\frown(o'\times o)=\gamma_f$; coefficients in a field R. Let $B=\{b\}$, $B'=\{b'\}$ be bases of $H^* M$, $H^* M'$ and $\{\hat{b}\}$, $\{\hat{b}'\}$ the dual bases. Show: If $f^*(b')=\sum_{b\in B}\lambda^b_{b'}\, b$, $\lambda^b_{b'}\in R$, then

$$\gamma^f=\sum_{b\in B,\, b'\in B'}(-1)^{|b|}\lambda^b_{b'}(\hat{b}'\times b).$$

In other words, the components of γ^f with respect to the base $\{\hat{b}'\times b\}$ agree (up to sign) with the matrix coefficients of f^*; in particular, f^* is determined by γ^f. Compare $\gamma^f\in(H^* M')\otimes(H^* M)$ with $\Theta^{-1}(f^*)\in(H^* M')\otimes(H^* M)$ as defined in VII, 6.1 (where $(H^i M')^* = H^{n'-i} M'$ by 8.13).

4*. Construct a chain map ψ (of degree $n-1$) which induces Alexander duality 8.17, and show $\psi\colon \check{S}^* K\simeq S(\mathbb{S}^n-K)$, K being a compact neighborhood retract.

5. *Naturality of the Künneth Sequence* 8.20. Show first that 8.20 does not depend on the ambient manifold M, M'; this reduces to considering $X\subset M_1\subset M_2$. Next prove naturality for inclusions $X\xrightarrow{\subset} Y$. The general proper map $X\to Y$ can be factored $X\xrightarrow{i}\mathbb{S}^k\times Y\xrightarrow{p} Y$, where i is a proper inclusion, p = projection, and k is so large that $\check{p}$ is isomorphic in the dimensions which matter; then $\check{p}=(\check{j})^{-1}$, where $j\colon Y\to\mathbb{S}^k\times Y$ is an inclusion.

9. *Duality in ∂-Manifolds*

For simplicity we treat *compact* ∂-manifolds L^n only (a generalization is indicated in Exerc. 3). We denote by M^n the manifold which is obtained from L by attaching a collar along ∂L (cf. 1.11), i.e. $M=L\cup_j(\partial L\times[0,1))$. We remark that $(M, M-iL)\simeq(L,\partial L)$, simply by shrinking each segment $[0,1)$ to 0. We also remark that $M-iL$ is a neighborhood retract in M (proof: ∂L is covered by finitely many coordinate neighborhoods U_k. Therefore $M-iL$ is covered by $\{U_k\times[0,1)\}$. Each of these being a euclidean half-space, it follows from IV, 8.10 that $M-iL$ is an ENR). Let R be a ring (of characteristic 2 if L is not orientable), and pick an orientation

$$O\in\Gamma(iL; R)\overset{3.3}{\cong} H_n(M, M-iL; R)\cong H_n(L,\partial L; R).$$

Then $\partial_*\colon H_n(L,\partial L;R)\to H_{n-1}(\partial L;R)$ maps O into a fundamental cycle $o=\partial_* O$ of ∂L (cf. 2.19).

9.1 Proposition. *The following diagram is commutative, and all vertical arrows are isomorphic.*

$$\begin{array}{ccccccccccc}
\cdots\to & H^{n-i-1}L & \longrightarrow & H^{n-i-1}(\partial L) & \xrightarrow{\delta^*} & H^{n-i}(L,\partial L) & \longrightarrow & H^{n-i}L & \longrightarrow & H^{n-i}(\partial L) & \to\cdots \\
& \Big\downarrow{\scriptstyle (-1)^{n-i-1}\frown O} & & \Big\downarrow{\scriptstyle \frown o} & & \Big\downarrow{\scriptstyle (-1)^{n-i}\frown O} & & \Big\downarrow{\scriptstyle (-1)^{n-i}\frown O} & & \Big\downarrow{\scriptstyle \frown o} & \\
\cdots\to & H_{i+1}(L,\partial L) & \xrightarrow{\partial_*} & H_i(\partial L) & \longrightarrow & H_i L & \longrightarrow & H_i(L,\partial L) & \longrightarrow & H_{i-1}(\partial L) & \to\cdots
\end{array}$$

The rows are, of course, exact. All coefficients are taken in an arbitrary R-module G.

Proof. The 1st, 2nd, and 4th square commute by ∂-compatibility VII, 12.13–14, the 3rd square by naturality VII, 12.6 of $\frown$-products. The maps $\frown o$ are isomorphic by 8.2. It suffices therefore, by the five lemma, to prove $\frown O\colon H^*L\cong H(L,\partial L)$. This follows from the diagram

$$\begin{array}{ccccc}
H^*L & \cong & H^*M & \cong & H^*L \\
\Big\downarrow{\scriptstyle \frown O} & & \Big\downarrow{\scriptstyle \frown O} & & {\scriptstyle \cong}\Big\downarrow{\scriptstyle \frown O'} \\
H(L,\partial L) & \cong & H(M,M-i\,L) & \cong & H(M,M-L).
\end{array}$$

The horizontal maps are induced by inclusions, all of them homotopy equivalences. The class $O'\in H_n(M,M-L)$ is, by definition, the image of O under the lower right isomorphism. The corresponding section $J(O')\in\Gamma L$ takes the value 1 at every point of $i\,L$ (it agrees with JO there), and therefore, by continuity, takes the value 1 at every point of $\overline{i\,L}=L$. Therefore, JO' is an orientation along L, and O' the fundamental class along L, hence $\frown O'$ is isomorphic by 7.2. ∎

9.2 As an application of 9.1 we prove Thom's index theorem. We have to recall first some elementary facts about real quadratic forms. If V is an r-dimensional vector-space over $\mathbb{R}$, and $Q\colon V\to\mathbb{R}$ is a quadratic form then there is a base in V such that $Q(x)=x_1^2+x_2^2+\cdots+x_p^2-x_{p+1}^2-x_{p+2}^2-\cdots-x_{p+q}^2$, where $\{x_i\}$ are the coordinates of x. The number $\sigma(Q)=p-q$ is called the *signature of* Q; it does not depend on the choice of the base. If a is the maximal dimension of a linear subspace on which Q vanishes then

$$|\sigma(Q)|=2r-2a-(p+q). \tag{9.3}$$

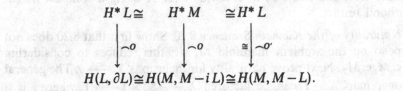

(If $p \geq q$ then *one* such subspace is given by the equations $x_i = x_{i+p}$ for $1 \leq i \leq q$, $x_i = 0$ for $q < i \leq p$; it is not hard to see that it is of maximal dimension.) In particular, *if Q is non-degenerate;* i.e. if $p+q=r$, then

(9.4) $$|\sigma(Q)| = r - 2a.$$

9.5 Definition. Let M^n be a compact oriented manifold, $o \in H_n(M; \mathbb{Z})$ its fundamental class. If $n = 4k$ then the quadratic form

$$Q_M\colon H^{2k}(M; \mathbb{R}) \xrightarrow{\smile\text{-square}} H^{4k}(M; \mathbb{R}) \xrightarrow{\langle -, o\rangle} \mathbb{R}, \quad Q_M(x) = \langle x \smile x, o\rangle,$$

is non-degenerate (8.13 with $K = M$, $L = \emptyset$, $i = 2k$). Its signature is called *signature of* M, in symbols $\sigma M = \sigma(Q_M)$. If $n \not\equiv 0 \bmod 4$ then $\sigma M = 0$, by convention.

The signature is an important tool in the theory of manifolds. One of its basic properties is the following.

9.6 Proposition. *If L^{4k+1} is a compact oriented ∂-manifold then $\sigma(\partial L) = 0$.*

Proof. Let $A = \operatorname{im}(i^*\colon H^*(L; \mathbb{R}) \to H^*(\partial L; \mathbb{R}))$, and consider the following portion of the diagram 9.1.

$$\begin{array}{ccccc} H^*L & \xrightarrow{i^*} & H^*(\partial L) & \xrightarrow{\delta^*} & H^*(L, \partial L) \\ & & \big\downarrow{\frown o} & & \big\downarrow{\cong} \\ & & H(\partial L) & \xrightarrow{i_*} & HL. \end{array}$$

We have

$$\begin{aligned} x \in A \;&\Leftrightarrow\; \delta^* x = 0 \Leftrightarrow i_*(x \frown o) = 0 \\ &\overset{\text{VII, 1.7}}{\Leftrightarrow}\; \langle H^*L, i_*(x \frown o)\rangle = \{0\} \overset{\text{VII, 1.8}}{\Leftrightarrow} \langle A, x \frown o\rangle = \{0\} \\ &\overset{\text{VII, 12.8}}{\Leftrightarrow}\; \langle A \smile x, o\rangle = \{0\}, \end{aligned}$$

i.e., with respect to the dual pairing $(x, y) \mapsto \langle x \smile y, o\rangle$ the vector space A is its own annihilator; in particular, $\dim A^{4k-i} = \dim H^i - \dim A^i$, hence $2 \cdot \dim A^{2k} = \dim H^{2k}$. The quadratic form $Q(x) = \langle x \smile x, o\rangle$ vanishes on A^{2k}, hence $|\sigma M| = |\sigma Q| \leq \dim H^{2k} - 2 \cdot \dim A^{2k} = 0$, the inequality by 9.4. ∎

Suppose, for instance, M^{4k} is a compact oriented manifold such that $r = \dim H^{2k}(M; \mathbb{R})$ *is odd* (e.g. $M = P_{2k}\mathbb{C}$, $P_{2l}\mathbb{H}$). Then σM is odd by 9.4, hence M cannot be the boundary of any oriented ∂-manifold L^{4k+1}. Of course, this follows more simply from 8.8 because χM is odd, but we can refine the result here: If $l \cdot M$ denotes the l-fold topological sum $M \oplus M \oplus \cdots \oplus M$, each summand with the same orientation, then $\sigma(l \cdot M) = l(\sigma M)$ is also not zero ($l > 0$), hence $l \cdot M$ does not bound either. In general, $\sigma(M^{4k} \oplus N^{4k}) = \sigma M + \sigma N$, because $H^{2k}(M \oplus N) = H^{2k} M \oplus H^{2k} N$ is a direct sum decomposition which splits the quadratic form, $Q_{M \oplus N} = Q_M \oplus Q_N$. If we reverse the orientation of M (notation: $-M$) then $Q_{-M} = -Q_M$, hence $\sigma(-M) = -\sigma(M)$. The formula $\sigma(l \cdot M) = l(\sigma M)$ makes sense then and is true for every integer l.

In cobordism theory (cf. Milnor 1962) one introduces an equivalence relation ("cobordism") between (compact oriented) n-manifolds by

$$M \sim N \Leftrightarrow M \oplus (-N) \approx \partial L \quad \text{for some compact oriented } L.$$

The set of equivalence classes is denoted by Ω^n. Taking topological sums turns Ω^n into an abelian group, and 9.6 together with the preceding remarks shows that σ defines a non-trivial homomorphism $\Omega^{4k} \to \mathbb{Z}$.

9.7 Proposition. *If M, N are compact oriented manifolds, and $M \times N$ is taken with the product orientation then $\sigma(M \times N) = (\sigma M)(\sigma N)$.* (For a generalization to fibered manifolds cf. Chern-Hirzebruch-Serre.)

Proof. Let $m = \dim M$, $n = \dim N$. If $m + n \not\equiv 0 \bmod 4$ then $\sigma(M \times N) = 0 = (\sigma M)(\sigma N)$. Assume then $m + n = 4p$. We can decompose the quadratic form $Q = Q_{M \times N}$ according to

$$H^{2p}(M \times N) = (H^{m/2} M) \otimes (H^{n/2} N)$$
$$\oplus_{2i<m} [(H^i M) \otimes (H^{2p-i} N) \oplus (H^{m-i} M) \otimes (H^{n-2p+i} N)],$$

where the first summand is zero if m or n is odd; coefficients are taken in $\mathbb{R}$. The decomposition follows from VII, 8.18; products of factors in different summands never contribute to the top dimension $4p = m + n$. Therefore,

$$\sigma(M \times N) = \sigma(Q | H^{m/2} M \otimes H^{n/2} N) + \textstyle\sum_{2i<m} \sigma(Q | [H^i M \otimes H^{2p-i} N \oplus H^{m-i} M \otimes H^{n-2p+i}]). \tag{9.8}$$

Fix $i < m/2$, choose bases A of $H^i M$, B of $H^{2p-i} N$, and let $\hat{A}, \hat{B}$ the dual bases of $H^{m-i} M$, $H^{n-2p+i} N$. Consider then the base $\{a \otimes b + \hat{a} \otimes \hat{b}\} \cup \{a \otimes b - \hat{a} \otimes \hat{b}\}$ of $H^i M \otimes H^{2p-i} N \oplus H^{m-i} M \otimes H^{n-2p+i} N$, where $a \in A$, $b \in B$, and $\{\hat{a}\}$, $\{\hat{b}\}$ are the dual bases.

From VII, 8.16 it follows that the product of any two different base-elements is zero, and $(a \otimes b + \hat{a} \otimes \hat{b})^2 = 2(a \otimes b) \smile (\hat{a} \otimes \hat{b}) = -(a \otimes b - \hat{a} \otimes \hat{b})^2$. Thus, the number of positive squares equals the number of negative squares, hence $\sigma(Q | [H^i M \otimes H^{2p-i} N \oplus H^{m-i} M \otimes H^{n-2p+i} N]) = 0$, hence

$\sigma(M\times N)=\sigma(Q|H^{m/2}M\otimes H^{n/2}N)$, by 9.8. If m or n is odd this is zero. If $m/2$ (hence $n/2$) is odd then $Q|H^{m/2}M\otimes H^{n/2}N=0$, by skew-symmetry VII, 8.7 of $\smile$-products. Thus, we are left with the case $m=4r$, $n=4s$, and the familiar assertion that the signature is multiplicative with respect to the tensor product of quadratic forms. Its proof is simple: If A is a base for $H^{2r}M$ such that Q_M has normal form (sum of positive minus sum of negative squares), and B is an analogous base for $H^{2s}N$ then $A\times B=\{a\otimes b\}_{a\in A,\,b\in B}$ does the same for $H^{2r}M\otimes H^{2s}N$. But then

$$\sigma M=\textstyle\sum_{a\in A}\langle a\smile a, o_M\rangle,$$
$$\sigma N=\textstyle\sum_{b\in B}\langle b\smile b, o_N\rangle,$$
$$\begin{aligned}\sigma(M\times N)=\sigma(Q|H^{2r}M\otimes H^{2s}N)&=\textstyle\sum_{a,b}\langle a\otimes b\smile a\otimes b, o_M\times o_N\rangle\\ &=\textstyle\sum_{a,b}\langle (a\smile a)\otimes(b\smile b), o_M\times o_N\rangle\\ &=\textstyle\sum_{a,b}\langle a\smile a, o_M\rangle\langle b\smile b, o_N\rangle\\ &=(\sigma M)(\sigma N).\quad\blacksquare\end{aligned}$$

Proposition 9.7 shows, for instance, that every product

$$P_{2n_1}\mathbb{C}\times P_{2n_2}\mathbb{C}\times\cdots\times P_{2n_k}\mathbb{C}$$

has signature $\sigma=1$. One can easily show that the product operation $\times$ is compatible with cobordism and turns $\Omega=\oplus_{i=0}^{\infty}\Omega_i$ into a ring; then 9.7 asserts that $\sigma\colon\Omega\to\mathbb{Z}$ is a *ring homomorphism.* If one considers differentiable manifolds only then Thom showed that the products of complex projective spaces as above generate a free abelian subgroup of finite index in $\Omega_{4n}^{\text{diff}}$, where $n=2(n_1+n_2+\cdots+n_k)$, and that Ω_i^{diff} is finite if $i\not\equiv 0$ (4). The complete structure of Ω^{diff} is also known (cf. Milnor 1962).

9.9 Exercises. *1.* If L^n is an oriented compact ∂-manifold with fundamental class $O\in H_n(L,\partial L;R)$, R a field, then

$$H^i(L,\partial L;R)\times H^{n-i}(L;R)\to H^n(L,\partial L;R)\xrightarrow{\langle -,O\rangle}R$$

is a dual pairing (compare 9.1 and 8.13).

2. Let $L=L^n$ be a compact oriented ∂-manifold whose boundary ∂L is the disjoint union of two $(n-1)$-manifold, $\partial L=\partial_1L\otimes\partial_2L$. Consider the diagram

$$\text{(9.10)}\quad\begin{array}{ccccccccc}H^{n-i-1}L&\longrightarrow&H^{n-i-1}(\partial_1L)&\longrightarrow&H^{n-i}(L,\partial_1L)&\longrightarrow&H^{n-i}L&\longrightarrow&H^{n-i}(\partial_1L)\\ \downarrow{\scriptstyle(-1)^{n-i-1}\frown O}&&\downarrow{\scriptstyle\frown o_1}&&\downarrow{\scriptstyle(-1)^{n-i}\frown O}&&\downarrow{\scriptstyle(-1)^{n-i}\frown O}&&\downarrow{\scriptstyle\frown o_1}\\ H_{i+1}(L,\partial L)&\longrightarrow&H_i(\partial_1L)&\longrightarrow&H_i(L,\partial_2L)&\longrightarrow&H_i(L,\partial L)&\longrightarrow&H_{i-1}(\partial_1L)\end{array}$$

whose first row is the cohomology sequence of $(L, \partial_1 L)$, whose second row is the homology sequence of $(L, \partial L, \partial_2 L)$, and where $O \in H_n(L, \partial L)$, $o_1 \in H_{n-1}(\partial_1 L)$ are fundamental classes. Show that 9.10 commutes. (It agrees with 9.1 if $\partial_2 L = \emptyset$.) The outside vertical arrows are isomorphic (9.1), hence $H^{n-i}(L, \partial_1 L) \cong H_i(L, \partial_2 L)$.

*3**. Let L^n be an oriented ∂-manifold (possibly non-compact) and $K \subset L$ a compact set. Attach a collar along ∂L and get a manifold M^n. Then $\check{H}K \cong H(M, M-K)$, $\check{H}(K, \partial K) \cong H(M-\partial K, M-K)$, $\check{H}(\partial K) \cong H(M, M-\partial K)$ by duality. Show that if the situation at $K \cap \partial L$ is sufficiently regular then $H(M, M-K) \cong H(L, L-iK)$, $H(M-\partial K, M-K) \cong H(L, L-K)$, $H_r(M, M-\partial K) \cong H_{r-1}(\partial L, \partial L - \partial K)$. For instance, if every point of $\partial K = K \cap \partial L$ has a neighborhood U in L such that $(U, U \cap K)$ is homeomorphic with $(\partial U, \partial K) \times [0, 1)$ then the situation *is* sufficiently regular.

4. Show that for every compact oriented manifold M the signature σM and the Euler-characteristic χM are congruent mod 2.

5. If L^{2n+1} is an orientable ∂-manifold then

$$\dim H_n(\partial L; R) = 2 \dim \ker [i_*: H_n(\partial L; R) \to H_n(L; R)]$$

for every field R, i.e., *every second generator of* $H_n \partial L$ *is killed by* i_* (proof as for 9.6). If M^{2n+1} is obtained by doubling L then $\dim H_n(\partial L; R) \leq 2 \dim H_{n+1}(M; R)$.

*6**. If M is a compact oriented n-manifold which admits an injective map $i: M \to V$ into some $(n+1)$-manifold V such that $i_*(o_M) = 0$ then $\sigma M = 0$.

10. Transfer

If $f: M' \to M$ is a map between oriented manifolds then we can transform the induced (co-)homology homomorphisms f_* resp. $\check{f}$ by Poincaré-duality. The resulting maps $f^! = D^{-1} f_* D'$ resp. $f_! = D' \check{f} D^{-1}$ are called *transfer* homomorphisms (also *Umkehr*-homomorphisms). If f is a covering map then $f^!, f_!$ agree with what is called transfer in the homology theory of groups; this may justify the name.

In this §10 we use transfers to deduce geometric properties of maps f which satisfy $f_*^{-1}(o_K) \neq 0$ (where K is compact in M, and o_K its fundamental class). In §11 they will be studied for inclusion maps. Their multiplicative properties are formulated in Exercise 4.

We begin with a *naturality property of* $\frown$*-products*. Let $f: M' \to M$ be a map of manifolds, let $K \subset M$ be a closed set such that both K and

$f^{-1}K$ lie in some ENR (e.g. if they are compact, or if the manifolds are ENR). For every closed set $L \subset K$ we have homomorphisms

$$\check{f}\colon \check{H}(K, L) \to \check{H}(f^{-1}K, f^{-1}L),$$

$$f_*\colon H(M' - f^{-1}L, M' - f^{-1}K) \to H(M - L, M - K),$$

and we assert

(10.1) $f_*((\check{f}x) \frown \eta) = x \frown (f_*\eta)$, for $x \in \check{H}(K, L)$, $\eta \in H(M', M' - f^{-1}K)$.

This follows from naturality VII, 12.6 of ordinary $\frown$-products by passing to limits. Indeed, $x \in \check{H}(K, L)$ is represented (in the sense of 5.18(i)) by some $v \in H^*(V, W) \cong H^*(V - L, W - L)$, where $W \subset V$ are neighborhoods of $L \subset K$, and $\check{f}x$ is represented by $(f^*v) \in H^*(f^{-1}V, f^{-1}W)$. Further, $x \frown (f_*\eta) = v \frown (j_* f_*\eta) = v \frown (f_* j_*\eta)$, and $(\check{f}x) \frown \eta = (f^*v) \frown (j_*\eta)$, by Definition 7.3, 7.4. Therefore 10.1 coincides with $f_*((f^*v) \frown (j_*\eta)) = v \frown (f_* j_*\eta)$, which holds by VII, 12.6. ∎

10.2 Proposition. *If $f\colon M' \to M$ is a map of oriented manifolds of dimension m' resp. m, if $K \subset M$ is a compact set (whose counterimage $f^{-1}K$ lies in some ENR) such that r-times the fundamental cycle $o_K \in H_m(M, M - K)$ is the f_*-image of some $\eta \in H_m(M', M' - f^{-1}K)$ (i.e. $f_*^{-1}(r\, o_K) \neq \emptyset$) then for every compact set $L \subset K$ there is a sequence of homomorphisms*

(10.3) $\check{H}^i(K, L) \xrightarrow{\check{f}} \check{H}^i(f^{-1}K, f^{-1}L) \to \check{H}_c^{i+(m'-m)}(f^{-1}K, f^{-1}L) \xrightarrow{f^!} \check{H}^i(K, L)$

whose composite equals r-times the identity map ($\check{H}_c$ = Cech-cohomology with compact supports; the coefficients of η, o_K should be taken in a ring R, the coefficients of 10.3 in any R-module). The result holds for non-oriented M', M if $1 + 1 = 0$ in R.

Proof. The composition

$$\check{H}^i(K, L) \xrightarrow{\check{f}} \check{H}^i(f^{-1}K, f^{-1}L)$$
$$\xrightarrow{\frown \eta} H_{m-i}(M' - f^{-1}L, M' - f^{-1}K) \xrightarrow{f_*} H_{m-i}(M - L, M - K)$$

takes $x \in \check{H}^i(K, L)$ into $f_*((\check{f}x) \frown \eta) = x \frown (f_*\eta) = x \frown (r\, o_K) = r(x \frown o_K)$. Composing further with $(\frown o_K)^{-1}\colon H_{m-i}(M - L, M - K) \cong \check{H}^i(K, L)$ takes x into rx. If we now replace $H_{m-i}(M' - f^{-1}L, M' - f^{-1}K)$ by the isomorphic group $\check{H}_c^{m'-m+i}(f^{-1}K, f^{-1}L)$, cf. 7.14, we get the required sequence 10.3. ∎

10.4 Remarks. Proposition 10.2 has interesting geometric consequences. One can define the *dimension* of K to be the largest i such that $\check{H}^i(K, L; G) \neq 0$ for some $L \subset K$ (cf. Nagami, §§ 35–39, for more precision). Then 10.2 implies that *the dimension of $f^{-1}K$ exceeds that of K*

by at least $(m'-m)$, *provided* $o_K \in \mathrm{im}(f_*)$. Also, it should be noted that $r\,o_K \in \mathrm{im}(f_*)$ implies $r\,o_I \in \mathrm{im}(f_*)$ for every compact subset $I \subset K$ (e.g., I = a point). If M is itself compact and $r\,o_M \in \mathrm{im}(f_*)$ then these remarks apply to every compact part K of M. In particular, $r H^i M \neq 0$ implies $H^i M' \neq 0$ and $H^{i+m'-m} M' \neq 0$. For instance, if $M' = \mathbb{S}^m$ is a sphere $(m'=m)$ and $f\colon M' \to M$ is a map of degree r then $r H^i M = 0$ for $0 < i < m$.

One can use 10.2 to study the problem of local sections: A *local section* of $f\colon M' \to M$ at $P \in M$ is a map $\sigma\colon U \to M'$ of a neighborhood U of P such that $f\sigma = \mathrm{id}$. If σ exists then $f_* \sigma_*(o_P) = o_P$, hence $o_P \in \mathrm{im}(f_*)$, hence $\check{H}_c^{m'-m}(f^{-1}P; \mathbb{Z})$ contains a direct summand $\cong \check{H}^0(P;\mathbb{Z}) = \mathbb{Z}$; in particular, $\dim(f^{-1}P) \geq m'-m$. *Thus one can sometimes tell, just by looking at* $f^{-1}P$, *that* f *admits no local section at* P (e.g. if $m' > m$ and $f^{-1}P$ is finite). For the sake of non-topologists we formulate the following special case (where M', M are open subsets of euclidean spaces): *Let* $f_j(x_1, x_2, \ldots, x_n) = b_j$, $j = 1, \ldots, m$, *be* m *continuous equations in* $n \geq m$ *unknowns. Suppose they can be solved continuously in a neighborhood* U *of* $P \in \mathbb{R}^m$, *i.e. there are continuous functions* $\sigma_k(y_1, \ldots, y_m)$, $k = 1, \ldots, n$, *defined* for $y \in U$, with values σy in an open set V of $\mathbb{R}^n$ such that $f_j(\sigma_1 y, \sigma_2 y, \ldots, \sigma_n y) = y_j$. Then for every $b = (b_1, \ldots, b_m) \in U$ the solutions $\{x \in V\}$ of $f_j(x_1, \ldots, x_n) = b_j$ form a set of dimension at least $n-m$;

$$\dim\{x \in V \mid f_j(x_1, \ldots, x_n) = b_j \text{ for all } j\} \geq n-m.$$

10.5 Definition. The homomorphism

$$f^!\colon \check{H}_c^i(f^{-1}K, f^{-1}L) \to \check{H}_c^{j-(m'-m)}(K, L)$$

which appears in 10.3 depends only on f, not on η. It is called the (*cohomology*) *transfer*. As the proof of 10.2 shows it is obtained by composing

$$\begin{aligned}(10.6)\quad &\check{H}_c^j(f^{-1}K, f^{-1}L) \xrightarrow{\frown o'} H_{m'-j}(M'-f^{-1}L, M'-f^{-1}K)\\ &\xrightarrow{f_*} H_{m'-j}(M-L, M-K) \xrightarrow{(\frown o)^{-1}} \check{H}_c^{m-m'+j}(K, L),\end{aligned}$$

i.e., *it is the transform of* f_* *under Poincaré duality. It is defined for every map* $f\colon M' \to M$ *of oriented manifolds* and every closed pair (K, L) in M (n.b., in 10.3 we assumed K to be compact; then $\check{H}(K, L) = \check{H}_c(K, L)$).

Dually, we can define the *homology transfer*

$$f_!\colon H_j(V, U) \to H_{j+(m'-m)}(f^{-1}V, f^{-1}U)$$

by composing

$$\begin{aligned}(10.7)\quad &H_j(V, U) \xrightarrow{(\frown o)^{-1}} \check{H}_c^{m-j}(M-U, M-V) \xrightarrow{f_c} \check{H}_c^{m-j}(M'-f^{-1}U, M'-f^{-1}V)\\ &\xrightarrow{\frown o'} H_{m'-m+j}(f^{-1}V, f^{-1}U).\end{aligned}$$

This is defined for *open pairs* (V, U) in M such that f is *proper over* $(M-U)-(M-V)=V-U$ (cf. 6.26). Heuristically, $f_!$ should be thought of as "taking the counterimage" (cf. 10.10, 10.12).

Both transfers compose functorially and commute with inclusion maps; more precisely,

10.8 Proposition. *Let $M'' \xrightarrow{f'} M' \xrightarrow{f} M$ be maps of oriented manifolds.*

(a) *If (K, L) is a closed pair in M then the composite*

$$\begin{aligned}\check{H}_c^j(f'^{-1}f^{-1}K, f'^{-1}f^{-1}L) &\xrightarrow{f'^!} \check{H}_c^{j-m''+m'}(f^{-1}K, f^{-1}L)\\ &\xrightarrow{f^!} \check{H}_c^{j-m''+m}(K, L)\end{aligned}$$

agrees with $(ff')^!$, i.e. $(ff')^! = f^! f'^!$.

(b) *If (V, U) is an open pair in M such that f is proper over $V-U$ and f' is proper over $f^{-1}V - f^{-1}U$ then ff' is proper over $V-U$ and*

$$(ff')_! = f'_! f_!\colon H_j(V, U) \to H_{j+m''-m}(f'^{-1}f^{-1}V, f'^{-1}f^{-1}U).$$

If $f\colon M \to M$ is the identity map, then $f^! = \mathrm{id}$, $f_! = \mathrm{id}$.

This follows immediately from the definitions 10.6, 10.7 because f_* and $\check{f}_c$ compose functorially. ∎

10.9 Proposition. *Let $f\colon M' \to M$ be a map of oriented manifolds.*

(a) *If $i\colon (\tilde{K}, \tilde{L}) \xrightarrow{\subset} (K, L)$ is an inclusion of closed pairs in M then the diagram*

$$\begin{array}{ccc} \check{H}_c(f^{-1}K, f^{-1}L) & \xrightarrow{f^!} & \check{H}_c(K, L) \\ \downarrow{\scriptstyle \check{i}'_c} & & \downarrow{\scriptstyle \check{i}_c} \\ \check{H}_c(f^{-1}\tilde{K}, f^{-1}\tilde{L}) & \xrightarrow{f^!} & \check{H}_c(\tilde{K}, \tilde{L}) \end{array}$$

commutes (i' = inclusion).

(b) *If $i\colon (\tilde{V}, \tilde{U}) \xrightarrow{\subset} (V, U)$ is an inclusion of open pairs in M, and if f is proper over $V-U$ and $\tilde{V}-\tilde{U}$ then the diagram*

$$\begin{array}{ccc} H(\tilde{V}, \tilde{U}) & \xrightarrow{f_!} & H(f^{-1}\tilde{V}, f^{-1}\tilde{U}) \\ \downarrow{\scriptstyle i_*} & & \downarrow{\scriptstyle i'_*} \\ H(V, U) & \xrightarrow{f_!} & H(f^{-1}V, f^{-1}U) \end{array}$$

commutes (i' = inclusion).

Since $f_*, \check{f}_c$ are functorial the proof reduces to showing that $\frown o$ resp. $\frown o'$ commute with inclusions. This is essentially 7.6 plus a passage to limits ($\check{H}_c$ being a limit of groups $\check{H}$). We leave the details to the reader. ∎

As an interesting application of $f_!$ and 10.9(b) we mention the following.

10.10 Proposition. *If $f\colon M' \to M$ is a map between oriented manifolds of the same dimension, and if (V, U) is an open pair in M such that f is proper over $V-U$ and of degree r over $V-U$ (i.e. of the same degree r over every $P \in V-U$; cf. 4.2) then the composition*

$$\text{(10.11)} \qquad H_j(V, U) \xrightarrow{f_!} H_j(f^{-1}V, f^{-1}U) \xrightarrow{f_*} H_j(V, U)$$

is r-times the identity map.

For instance, if $f\colon M' \to M$ is a covering map of (connected) oriented manifolds and if the number of sheets is $r < \infty$ then f is proper (over M) of degree $\pm r$. If ρ' resp. ρ is the fundamental group of M' resp. M then HM' resp. HM can be interpreted as homology of ρ' resp. ρ (with chain-complex coefficients). Further, f imbeds ρ' in ρ with index r, and $f_!$ can be identified with the usual transfer $H(\rho) \to H(\rho')$; cf. Cartan-Eilenberg, XII, 8(2); proposition 10.10 becomes XII, 8(6) l.c.

If, for another example, $f\colon \mathbb{R}^n \to M^n$ is a proper map of degree r then, by 10.10, $r(HM) \subset f_*(H\mathbb{R}^n)$, hence $r(\tilde{H}M) = 0$. In particular, only acyclic manifolds M can receive proper maps of degree ± 1 from $\mathbb{R}^n$ (in fact, M must be contractible; cf. Exerc. 3).

Proof of 10.10. If $M-U$ is compact then the proof is as for 10.2: Any $h \in H(V, U)$ can be written as $h = y \frown o$ with $y \in \check{H}(M-U, M-V)$, and $f_!(h) = (\check{f} y) \frown o'$, hence $f_* f_!(h) = f_*((\check{f} y) \frown o') = y \frown (f_* o') = r(y \frown o) = rh$, the 3rd equation by 4.5. Assume next that $V-U$ is compact. Let $B = M - \overline{V-U}$; then $[M-(U \cup B)] \subset (M-B)$ is compact, hence 10.10 holds for $(V \cup B, U \cup B)$, hence $i_*(f_* f_!) \overset{10.9(b)}{=} (f_* f_!)\, i_* = r\, i_*$. But $i_*\colon H(V, U) \cong H(V \cup B, U \cup B)$ by excision, hence $f_* f_! = r\,\mathrm{id}$ holds for (V, U).

Consider then the general case. Given $h \in H(V, U)$ there is an open set $W \subset V$ with compact closure $\overline{W}$ such that h is in the image of $i_*\colon H(W \cup U, U) \to H(V, U)$, say $h = i_*(k)$. Then

$$(f_* f_!)\, h = (f_* f_!)(i_* k) = i_*(f_* f_! k) = i_*(r k) = r(i_* k) = r h,$$

the 2nd equation by 10.9(b), the 3rd because $\overline{(W \cup U) - U}$ is compact. ∎

10.12 Corollary. *Let M be an oriented manifold, and let $i\colon W \to M$ denote the inclusion of an open subset. If (V, U) is an open pair in M such that $(V-U) \subset W$ then i is proper over $V-U$ and of degree 1, hence*

$$H(V, U) \xrightarrow{i_!} H(V \cap W, U \cap W) \xrightarrow{i_*} H(V, U)$$

is the identity map by 10.10. But $i_*\colon H(V \cap W, U \cap W) \cong H(V, U)$ by excision, hence $i_! = i_*^{-1}$ in this case. ∎

10.13 Corollary. *Let $f\colon M' \to M$ be a map of oriented manifolds and $i\colon W \to M$ the inclusion of an open subset. Put $W' = f^{-1}W$, $i'\colon W' \xrightarrow{\subset} M'$, and $f^W = f|W'\colon W' \to W$. If (V, U) is an open pair in W such that f is proper over $V-U$ then we have two transfers, $f_!^M$ and $f_!^W$. We claim, they are equal, $f_!^M = f_!^W\colon H(V, U) \to H(f^{-1}V, f^{-1}U)$. In particular, in order to compute $f_!\colon H(V, U) \to H(f^{-1}V, f^{-1}U)$ we can always replace $f\colon M' \to M$ by $f^V\colon f^{-1}V \to V$.*

Proof. We have $f^M i' = i f^W$, hence $i'_! f_!^M = (f^M i')_! = (i f^W)_! = f_!^W i_!$ by 10.8(b). But $i_!\colon H(V, U) \to H(V \cap W, U \cap W) \cong H(V, U)$ and

$$i'_!\colon H(f^{-1}V, f^{-1}U) \to H(f^{-1}V, f^{-1}U)$$

are identity maps by 10.12. Hence $f_!^M = f_!^W$, as asserted. ∎

10.14 Exercises. *1.* Let $\Gamma = \{(x, y) \in \mathbb{R}^2 \mid y = \sin(1/x), x \neq 0\}$ = graph of $\sin(1/x)$, and let $X = \bar{\Gamma}$ its closure in $\mathbb{S}^2 = \mathbb{R}^2 \cup \{\infty\}$. Construct a map $f\colon \mathbb{S}^2 \to \mathbb{S}^2$ of degree 1 such that $X = f^{-1}(\mathbb{S}^1)$ = counterimage of a circle. This shows that the *singular* cohomology of $f^{-1}\mathbb{S}^1$ can be zero (whereas $\check{H}^1(f^{-1}\mathbb{S}^1) \neq 0$ whenever degree$(f) \neq 0$). In the same spirit, construct a map $g\colon \mathbb{S}^2 \to \mathbb{S}^1$ such that $X = g^{-1}(P)$ for some $P \in \mathbb{S}^1$, and g admits a local section at P.

2. Dualize proposition 10.2.

3. Let $f\colon M' \to M$ be a mapping of connected manifolds, and let $p\colon \tilde{M} \to M$ be the covering which corresponds to the subgroup $f_*(\pi_1 M')$ of the fundamental group $\pi_1 M$, so that the index $\iota = [\pi_1 M\colon f_*(\pi_1 M')]$ equals the number of sheets of p. The map f lifts to $\tilde{f}\colon M' \to \tilde{M}$, $p\tilde{f} = f$ (cf. Schubert III.6 for the theory of coverings). If $\dim M' = \dim M$, and f is proper of degree r then $\tilde{f}$ is proper and $r = \iota \cdot \deg(\tilde{f})$; in particular, ι divides r. For instance, if $M' = \mathbb{R}^n$ then $\pi_1 M' = 1$, hence $\iota = [\pi_1 M : 1]$ = order of $\pi_1 M$, hence $[\pi_1 M : 1]$ divides r. In particular, if $f\colon \mathbb{R}^n \to M^n$ has degree ± 1, then $\pi_1 M = \{1\}$. Since also $\tilde{H}M = 0$ by 10.10, it follows (cf. Hu 1965, VII, 8.5) that M *must be contractible.*

4*. *The multiplicative properties of transfers* are expressed by the following formulas

(10.15) $$f_!(x\frown\xi)=(\check{f}x)\frown(f_!\xi),$$

(10.16) $$f^!(\check{f}x\smile y)=x\smile f^! y,$$

(10.17) $$f_*(y\frown f_!\xi)=(-1)^{(m-|\xi|)(m-m')}(f^! y)\frown\xi.$$

Formula 10.15 holds for $x\in\check{H}_c(K,L)$, $\xi\in H(M,M-K)$, if $L\subset K$ are closed in M^m and $f\colon M'^{m'}\to M^m$ is proper over $K-L$. This requires defining $\frown$-products for Čech-cohomology classes with compact support which can be done by composing $\check{H}_c\to\check{H}\xrightarrow{\frown\xi}H$. Slightly more general, the formula applies to $x\in\check{H}_c(K,L_1)$, $\xi\in H(M-L_2,M-K)$, if $L_1,L_2\subset K$ are closed in M, and $(x\frown\xi)\in H(M-L_1\cup L_2,M-K)$ is defined as before, replacing M by $M-L_2$. Similarly for 10.17. Formula 10.16 holds for $x\in\check{H}(K,L_1)$, $y\in\check{H}(f^{-1}K,f^{-1}L_2)$, where $L_1,L_2\subset K$ are as above (but f need not be proper). It requires defining $\smile$-products for Čech classes as indicated in 6.21.

According to our sign rule VI, 9.8 for commuting graded objects we should expect a sign $(-1)^{|f_!||x|}=(-1)^{(m'-m)|x|}=(-1)^{|f^!||x|}$ in 10.15, 10.16, and $(-1)^{|y|(m'-m)}$ in 10.17 (since $f_!,f^!$ are maps of degree $\pm(m-m')$). In fact, in a more systematic treatment we should redefine transfers $f^!$ resp. $f_!$ by multiplying the composition 10.6 resp. 10.7 with $(-1)^{j(m'-m)}$ resp. $(-1)^{(m-j)(m'-m)}$ (for inclusions we shall do just that in § 11); this would produce the expected signs in formulas as above.

A way of remembering 10.15 is to say that $f_!$ is a homomorphism of $\check{H}K$-modules, where $\check{H}K$ operates on $H(M,M-K)$ via $\frown$, and on $H(M',M'-f^{-1}K)$ via $\check{f}$ and $\frown$. Similarly for 10.16, whereas 10.17 expresses a duality.

5. Show that the middle arrow $\check{H}^i(f^{-1}K,f^{-1}L)\to\check{H}_c^{i+(m'-m)}(f^{-1}K,f^{-1}L)$ of 10.3 is the $\smile$-product (6.21) with a fixed element z of $\check{H}_c^{m'-m}(f^{-1}K)$, namely the Poincaré-dual of η $(z\frown o'=\eta)$.

11. Thom Class, Thom Isomorphism

Let M^{n+k} be an oriented manifold, N^n an oriented submanifold with inclusion map $e\colon N\to M$, and assume $\bar{N}=N$ (N is closed in M). Then for every closed pair (X,A) in N the transfer $e_!$ is the composite

(11.1) $$H_q(M-A,M-X)\xrightarrow{(\frown o_M)^{-1}}\check{H}_c^{n+k-q}(X,A)\xrightarrow{\frown o_N}H_{q-k}(N-A,N-X).$$

For reasons which will appear later in this § we modify the definition of $e_!$ by a sign $(-1)^{k(n+k-q)}$, i.e. *from now on* $e_!$: $H_q(M-A, M-X) \to H_{q-k}(N-A, N-X)$ *will denote* $(-1)^{k(n+k-q)}$*-times the composition* 11.1. *It is isomorphic*, by 7.14 (arbitrary coefficients, mod 2 if M or N are not oriented; we don't have to assume $\bar{N}=N$ provided we take subsets $A \subset X$ of N which are closed in M).

For small dimensions 11.1 implies

(11.2) $H_q(M-A, M-X)=0 \quad$ for $q<k=\dim M-\dim N$,

(11.3) $H_k(M-A, M-X; \mathbb{Z}) \cong H_0(N-A, N-X; \mathbb{Z}) \cong$ *free abelian group generated by the components of* $N-A$ *which lie in* X.

In particular, $H_k(M, M-N; \mathbb{Z}) \cong H_0(N; \mathbb{Z})$ is freely generated by elements v_λ which correspond to the components N_λ of N. We call v_λ the *transverse class of* N_λ (in M). In the decomposition

$$H_k(M, M-N) \cong \oplus_\lambda H_k(M, M-N_\lambda)$$

it is a generator of $H_k(M, N-N_\lambda; \mathbb{Z}) \cong \mathbb{Z}$. If N is connected, we also write v_N or v_N^M for its transverse class.

The isomorphisms $e_!$ *commute with inclusions. In more detail, if* $(\tilde{X}, \tilde{A}) \subset (X, A)$ *are closed pairs in* N *then*

(11.4)
$$\begin{array}{ccc} H_q(M-A, M-X) & \overset{e_!}{\cong} & H_{q-k}(N-A, N-X) \\ \downarrow j_* & & \downarrow j_* \\ H_q(M-\tilde{A}, M-\tilde{X}) & \overset{e_!}{\cong} & H_{q-k}(N-\tilde{A}, N-\tilde{X}) \end{array}$$

(j = *inclusion*) *commutes, by* 10.9(b).

If V *is an open set in* M *then*

(11.5)
$$\begin{array}{ccc} H_q(V-A, V-X) & \overset{(e|V\cap N)_!}{\cong} & H_{q-k}((V\cap N)-A, (V\cap N)-X) \\ \downarrow j_* & & \downarrow j_* \\ H_q(M-A, M-X) & \overset{e_!}{\cong} & H_{q-k}(N-A, N-X) \end{array}$$

commutes for every closed pair (X, A) *in* N; in fact, this is just 10.9(b), using $(e|V\cap N)_! = e_!$ (10.13). ∎ In particular, for $q=k$, $A=\emptyset$, $X=N$, we see that j_*: $H_k(V, V-N) \to H_k(M, M-N)$ *takes transverse classes into transverse classes*,

(11.6) $$j_*(v_\lambda) = v_\lambda,$$

where $\tilde{\lambda} \subset \lambda$ are components of $(V \cap N)$ respectively N. From 11.6 it follows that *the transverse classes* $v_\lambda \in H_k(M, M-N)$ *have arbitrarily small representatives.* In fact, if $P \in N_\lambda$, and V is any open neighborhood of P in M then v_λ is the image of $v_{\tilde{\lambda}} \in H_k(V, V - V \cap N)$, where $\tilde{\lambda}$ is the component of P in $V \cap N$.

11.7 As an illustration, consider the case $N = \mathbb{R}^n \times \{0\} \subset \mathbb{R}^n \times \mathbb{R}^k = M$. Then $(M, M-N) = \mathbb{R}^n \times (\mathbb{R}^k, \mathbb{R}^k - \{0\})$, and $H_k(M, M-N; \mathbb{Z}) \cong \mathbb{Z}$, $H_i(M, M-N) = 0$ for $i \neq k$. If $\sigma\colon \Delta_k \to \mathbb{R}^{n+k} = \mathbb{R}^n \times \mathbb{R}^k$ is any non-degenerate affine simplex which meets N in exactly one interior point (i.e., σ is "transverse" to $\mathbb{R}^n \times \{0\}$) then σ is a relative cycle of M mod $M-N$ whose homology class $[\sigma]$ generates $H_k(M, M-N; \mathbb{Z})$, hence $[\sigma] = \pm v_N$.

If, in the general case again, the embedding $e\colon N \to M$ is *flat* at $P \in N$ (cf. 1.8) then, by definition, P has an open neighborhood V in M such that $(V, V \cap N) \approx (\mathbb{R}^n \times \mathbb{R}^k, \mathbb{R}^n \times \{0\})$. The transverse class v_N (assuming N connected) is then the image of $v_{V \cap N}$ which, in turn, is given as above. This provides an intuitive idea of v_N for a fairly general class of embeddings.

If we apply the universal coefficient formula (VI, 7.10) to 11.2, 11.3 we find

(11.8) $\quad H^q(M-A, M-X) = 0, \qquad$ for $q < k = \dim M - \dim N$;

(11.9) $\quad H^k(M-A, M-X; G) \cong \mathrm{Hom}(H_0(N-A, N-X; \mathbb{Z}), G)$

$\cong$ direct product of as many factors G as there are components of $N - A$ in X.

11.10 Proposition and Definition. Using 11.3 the elements of

$$H^k(M, M-N; G) = \mathrm{Hom}(H_0(N; \mathbb{Z}), G)$$

can be described as follows: *For every component N_λ of N choose an element $g_\lambda \in G$; then there is a unique class $y \in H^k(M, M-N; G)$ such that $\langle y, v_\lambda \rangle = g_\lambda$ for every λ (v_λ = transverse class).* ∎

In particular, there is a unique class $\tau = \tau_N^M \in H^k(M, M-N; \mathbb{Z})$ such that $\langle \tau, v_\lambda \rangle = 1$ for every λ; it is called the *Thom class* of N (in M). The image of τ under $e^*\colon H^k(M, M-N; \mathbb{Z}) \to H^k(N; \mathbb{Z})$ is called *Euler class* of e, or *normal Euler class* of N in M; it is denoted by $\chi = \chi_N^M$. We also write τ respectively χ for the image (under $\otimes$) of the Thom- resp. Euler class in $H^k(M, M-N; R)$ resp. $H^k(M; R)$, where R is any ring. The name originates from a special case: If $e\colon N \to N \times N$ is the diagonal embedding and N is compact then one can show that $\langle \chi_N^{N \times N}, o_N \rangle$ equals the Euler characteristic of N (cf. Exerc. 3).

Both, Thom and Euler class, are natural with respect to inclusions $j: V \to M$ of open subsets, i.e.

$$(11.11) \qquad j^*(\tau_N^M) = \tau_{V\cap N}^V, \qquad (j|V\cap N)^*(\chi_N^M) = \chi_{V\cap N}^V.$$

The first formula follows from 11.6 and the definition of τ, the second from the first. ∎

11.12 Proposition. *If $e: N \to M$ can be deformed into a mapping $f: N \to M$ whose image lies in $M-N$ then $\chi_N^M = 0$. In other words, χ_N^M can be viewed as an* obstruction for deforming N into $M-N$ (cf. also 11.25). *Also, if $H^k(M;\mathbb{Z})=0$ then $\chi_N^M=0$. For instance, this applies* if $M=\mathbb{R}^{n+k}$.

Proof. In both cases $e^*: H^k(M, M-N;\mathbb{Z}) \to H^k(N;\mathbb{Z})$ vanishes: in the first case because $e^* = f^*$ factors through $H^k(M-N, M-N)=0$, in the second case because e^* factors through $H^k(M;\mathbb{Z})=0$. ∎

The following proposition relates intersections to $\smile$-products; such relations will be studied in more detail in § 13.

11.13 Proposition. *Let N_1, N_2 be oriented submanifolds ($N_p = N_p$) of an oriented manifold M such that $N = N_1 \cap N_2$ is a connected manifold. Suppose N_1, N_2 intersect transversally at some point $P \in N$, meaning that P has an open neighborhood V in M such that*

$$(V; V\cap N_1, V\cap N_2) \approx (\mathbb{R}^{k_2}\times\mathbb{R}^n\times\mathbb{R}^{k_1}, \{0\}\times\mathbb{R}^n\times\mathbb{R}^{k_1}, \mathbb{R}^{k_2}\times\mathbb{R}^n\times\{0\});$$

in particular, the dimensions of N, N_1, N_2, M, are $n, n+k_1, n+k_2, n+k_1+k_2$. Then N is orientable, and $\tau_N^M = \pm(\tau_{N_1}^M \smile \tau_{N_2}^M)$. Also $\pm\chi_N^M = (e_1^*\chi_{N_1}^M)\smile(e_2^*\chi_{N_2}^M)$, where $e_p: N \to N_p$ denotes inclusion. (As to the signs, the 1st part of the proof will show how to compute them in terms of given orientations of M, N_1, N_2.)

Proof. Assume first $V=M$, hence

$$(M; N_1, N_2) = (\mathbb{R}^{k_2}\times\mathbb{R}^n\times\mathbb{R}^{k_1}; \{0\}\times\mathbb{R}^n\times\mathbb{R}^{k_1}, \mathbb{R}^{k_2}\times\mathbb{R}^n\times\{0\}),$$

and $N = N_1\cap N_2 = \{0\}\times\mathbb{R}^n\times\{0\}$. Let $o_l \in H_l(\mathbb{R}^l, \mathbb{R}^l - 0;\mathbb{Z}) \approx \mathbb{Z}$, and $\mu_l \in H^l(\mathbb{R}^l, \mathbb{R}^l - 0;\mathbb{Z})$ be generators, hence $\langle\mu_l, o_l\rangle = \pm 1$. Let 1 also denote the generators of the various groups $H_0(-,\mathbb{Z})$ and $H^0(-,\mathbb{Z})$. Then $o_{k_2}\times 1\times 1$, $1\times 1\times o_{k_1}$, $o_{k_2}\times 1\times o_{k_1}$ are generators of $H_{k_2}(M, M-N_1;\mathbb{Z}), H_{k_1}(M, M-N_2;\mathbb{Z}), H_{k_1+k_2}(M, M-N;\mathbb{Z})$(cf. VII, 2.14), so they agree, up to sign, with the transverse classes of N_1, N_2, N. It follows that $\pm\tau_{N_1}^M = \mu_{k_2}\times 1\times 1$, $\pm\tau_{N_2}^M = 1\times 1\times\mu_{k_1}$, $\pm\tau_N^M = \mu_{k_2}\times 1\times\mu_{k_1}$, hence $\tau_N^M = \pm(\tau_{N_1}^M)\smile(\tau_{N_2}^M)$ by VII, 8.15.

In the general case let $\nu^V_{V\cap N}$ the transverse class of $V\cap N$, and

$$j\colon (V; V-N_1, V-N_2, V-N) \to (M; M-N_1, M-N_2, M-N)$$

the inclusion. Then

$$\begin{aligned}\langle \tau^M_{N_1} \smile \tau^M_{N_2}, j_*(\nu^V_{N\cap V})\rangle &= \langle (j^* \tau^M_{N_1}) \smile (j^* \tau^M_{N_2}), \nu^V_{N\cap V}\rangle \\ &= \langle \tau^V_{N_1\cap V} \smile \tau^V_{N_2\cap V}, \nu^V_{N\cap V}\rangle = \pm\langle \tau^V_{N\cap V}, \nu^V_{N\cap V}\rangle = \pm 1.\end{aligned}$$

In particular, $j_*(\nu^V_{N\cap V})$ must be of infinite order, hence

$$\check{H}^n_c N \cong H_{k_1+k_2}(M, M-N)$$

is an infinite group, hence N is orientable, by 6.25 or 11.29. But then $j_*(\nu^V_{N\cap V}) = \nu^M_N$ by 11.6, hence $\langle \tau^M_{N_1} \smile \tau^M_{N_2}, \nu^M_N\rangle = \pm 1$, hence $\tau^M_{N_1} \smile \tau^M_{N_2} = \pm \tau^M_N$ by definition of the latter.

As to the Euler class,

$$\chi^M_N = e^*(\tau^M_N) = \pm e^*(\tau^M_{N_1} \smile \tau^M_{N_2}) = \pm (e_1^* \tau^M_{N_1}) \smile (e_2^* \tau^M_{N_2}) = \pm\, \chi^M_{N_1} \smile \chi^M_{N_2}. \quad \blacksquare$$

We now show that the transfer $e_!$ can be approximated by $\tau\frown$, the cap-product with the Thom class. More precisely,

11.14 Proposition. *Let $e\colon N^n \to M^{n+k}$ as above (closed inclusion of oriented manifolds), let $X \subset N$ be closed and $W \subset M$ open such that $(N-X) \subset W \subset (M-X)$. Then the composition*

$$H_q(M, M-X) \xrightarrow{e_!} H_{q-k}(N, N-X) \xrightarrow{i_*} H_{q-k}(M, W)$$

agrees with $\tau\frown$, i.e. for every $h \in H(M, M-X)$ we have $i_* e_!(h) = \tau \frown h$, where i = inclusion (note that $M-X = (M-N) \cup W$, so that $\tau \frown h \in H(M, W)$).

The proof requires some preliminaries. We show first

11.15 Lemma. *If X is compact* (in the situation of 11.14) *then*

$$e_!(w \frown o^M_X) = (-1)^{k|w|}(i^* w) \frown o^N_X \quad \textit{for } w \in H^*(M, W),$$

where o^M_X respectively o^N_X denotes the fundamental class of X in M respectively N. (Compare this with 10.15.)

Proof. If z is the image of w under the composition

$$H^*(M, W) \xrightarrow{i^*} H^*(N, N-X) = \check{H}(N, N-X) \to \check{H}_c N$$

then $w \frown o^M_X = z \frown o^M$ and $(i^* w) \frown o^N_X = z \frown o^N$, by definition of the right sides (cf. 7.13 and the explanation thereafter), hence

$$e_!(w \frown o^M_X) = e_!(z \frown o^M) = (-1)^{k|w|}\, z \frown o^N = (-1)^{k|w|}(i^* w) \frown o^N_X. \quad \blacksquare$$

11.16 Corollary. *If $X = P$ is a point (in the situation 11.14) and $w \in H^n(M, W)$ then $w \frown o_P^M = (-1)^{kn} \langle i^* w, o_P^N \rangle v_P$, where v_P denotes the transverse class of the component of P.*

Indeed, $(-1)^{kn} e_!(w \frown o_P^M) = (i^* w) \frown o_P^N$ is $\langle i^* w, o_P^N \rangle$-times the homology class of the point P, by VII, 12.8; hence the assertion by definition of v_P. ∎

11.17 Lemma. *If X is compact (in 11.14) and $r: (M, W) \to (N, N-X)$ is a retraction then $r_*(\tau_N \frown o_X^M) = o_X^N$.*

Proof. Assume first $X = P$ is a point. Let $\mu \in H^n(N, N-P; \mathbb{Z})$ be the generator with $\langle \mu, o_P^N \rangle = 1$. Then $(r^* \mu) \frown o_P^M = (-1)^{kn} \langle \mu, o_P^N \rangle v_P = (-1)^{kn} v_P$ by 11.16, hence

$$\begin{aligned}\langle \mu, r_*(\tau \frown o_P^M) \rangle &= \langle r^* \mu, \tau \frown o_P^M \rangle = \langle (r^* \mu) \smile \tau, o_P^M \rangle \\ &= (-1)^{kn} \langle \tau, (r^* \mu) \frown o_P^M \rangle = \langle \tau, v_P \rangle = 1 .\end{aligned}$$

Since $r_*(\tau \frown o_P^M)$ is a multiple of o_P^N this proves $r_*(\tau \frown o_P^M) = o_P^N$.

In the general case consider the commutative diagram

$$\begin{array}{ccc} H(M, r^{-1}(N-P)) & \xleftarrow{j_*} & H(M, W) \\ \downarrow{\scriptstyle r_*} & & \downarrow{\scriptstyle r_*} \\ H(N, N-P) & \xleftarrow{j_*^P} & H(N, N-X), \end{array}$$

where $P \in X$, and j^P, j are inclusions. Comparing the images of $\tau \frown o_X^M$ gives $j_*^P(r_*(\tau \frown o_X^M)) = r_*(\tau \frown o_P^M) = o_P^N$, using naturality of $\frown$ and the first part of the proof. This equation holds for all $P \in X$, therefore $r_*(\tau \frown o_X^M) = o_X^N$ by Definition 4.1 of the latter. ∎

We now prove a special case of 11.14, namely

11.18 Proposition. *If X is compact (in 11.14) then $i_*(o_X^N) = \tau_N^M \frown o_X^M$.*

In particular, if $W = M - X$ then (by Definition 7.4) the right side agrees with $\chi_X^M \frown o_X^M$, where χ_X^M is the image of τ_N^M under $H^k(M, M-N) \to H^k M \to \check{H}^k X$; the proposition then asserts *that* χ_X^M $(= \chi_N^M | X)$ *is the Poincaré-dual of* $i_*(o_X^N) \in H_n(M, M-X)$. If M is itself compact and $X = N$, $W = \emptyset$ then $\tau_N^M \frown o_N^M = \tau_N^M \frown (j_* o_M^M) = (j^* \tau_N^M) \frown o_M^M$, where $j =$ inclusion, hence $(j^* \tau_N^M) \in H^k M$ is the Poincaré-dual of $(i_* o_N^N) \in H_n M$.

Proof. Assume first M and N are ENR's, and let $r: M' \to N$ be a neighborhood retraction; put $W' = r^{-1}(N-X) \cap W$, so that $r: (M', W') \to$

$(N, N-X)$. If we choose (M', W') small enough then the composite $(M', W') \xrightarrow{r} (N, N-X) \xrightarrow{i} (M, W)$ is homotopic to the inclusion mapping j (cf. IV, 8.7), hence $j_* = i_* r_*$. Apply i_* to 11.17 (with M replaced by M') and get

$$i_*(o_X^N) = i_* r_*(\tau_N^{M'} \frown o_X^{M'}) = j_*(j^* \tau_N^M \frown o_X^{M'}) = \tau_N^M \frown j_* o_X^{M'} = \tau_N^M \frown o_X^M.$$

In the general case we can find an open subset $M' \subset M$, such that M' and $N' = N \cap M'$ are ENR's, and $X \subset N'$ (because M, N are locally ENR and X is compact; cf. IV, 8.10); put $W' = W \cap M'$. Then

$$\text{(11.19)} \qquad \begin{array}{ccc} (N', N'-X) & \xrightarrow{i'} & (M', W') \\ \downarrow{\scriptstyle j'} & & \downarrow{\scriptstyle j} \\ (N, N-X) & \xrightarrow{i} & (M, W) \end{array}$$

is a commutative diagram of inclusion maps, hence

$$\begin{aligned} i_*(o_X^N) &= i_* j'_*(o_X^{N'}) = j_* i'_*(o_X^{N'}) = j_*(\tau_{N'}^{M'} \frown o_X^{M'}) = j_*(j^* \tau_N^M \frown o_X^{M'}) \\ &= \tau_N^M \frown j_* o_X^{M'} = \tau_N^M \frown o_X^M \quad , \end{aligned}$$

the third equation by the first part of the proof. ∎

Proof of 11.14. If $h \in H(M, M-X)$ then h is dual in M to some class $x \in \check{H}_c X$, and $(-1)^{k|x|} e_! h$ is the dual of x in N. For some closed $A \subset X$ such that $X - A$ is compact the class x has a representative x' in $\check{H}(X, A)$, because $\check{H}_c X = \varinjlim \check{H}(X, A)$. This class x', in turn, has a representative y in $H^*(M', M'')$, where (M', M'') are suitable open neighborhoods of (X, A) in M (because $\check{H}(X, A) = \varinjlim H^*(M', M'')$). By the remarks following 7.13, the Poincaré-dual h of x has the form $h = j_*(y \frown o_{X-M''}^{M'})$, where $j: (M', M'-X) \to (M, M-X)$ denotes inclusion. Similarly, with notations as in 11.19, we have that $i'^* y$ represents x in N, and the dual $(-1)^{k|x|} e_! h$ of x in N is given by

$$e_! h = (-1)^{k|y|} j'_*(i'^* y \frown o_{X-N''}^{N'}).$$

Hence

$$\begin{aligned} i_* e_! h &= (-1)^{k|y|} i_* j'_*(i'^* y \frown o_{X-N''}^{N'}) = (-1)^{k|y|} j_* i'_*(i'^* y \frown o_{X-N''}^{N'}) \\ &= (-1)^{k|y|} j_*(y \frown i'_* o_{X-N''}^{N'}) \overset{11.18}{=} (-1)^{k|y|} j_*(y \frown \tau_{N'}^{M'} \frown o_{X-M''}^{M'}) \\ &= j_*(\tau_{N'}^{M'} \frown y \frown o_{X-M''}^{M'}) = \tau_N^M \frown j_*(y \frown o_{X-M''}^{M'}) = \tau_N^M \frown h \end{aligned}$$

(note that $\tau_{N'}^{M'} = j^* \tau_N^M$, by 11.11). ∎

11.20 Corollary of 11.14. *Let $e\colon N^n \overset{\subset}{\longrightarrow} M^{n+k}$ be as in 11.14, and assume N is a neighborhood retract in M, say $r\colon M' \to N$, $r'e = \mathrm{id}$, M' open in M. Then for every closed $X \subset N$ the composition*

$$(11.21)\qquad \begin{aligned} H_q(M, M-X) \overset{\text{exc}}{\cong} H_q(M', M'-X) &\xrightarrow{\tau_N^{M'}\frown} H_{q-k}(M', M'-r^{-1}X) \\ &\xrightarrow{r_*} H_{q-k}(N, N-X) \end{aligned}$$

agrees with $e_!$; in particular, it is isomorphic. Dually, the composition

$$(11.22)\qquad \begin{aligned} e^!\colon H^{q-k}(N, N-X) \xrightarrow{r^*} H^{q-k}(M', M'-r^{-1}X) \\ \xrightarrow{\tau_N^{M'}\smile} H^q(M', M'-X) \overset{\text{exc}}{\cong} H^q(M, M-X) \end{aligned}$$

is isomorphic. Both isomorphisms are named after *Thom*.

If, moreover, the composite $M' \xrightarrow{r} N \xrightarrow{e} M$ is homotopic to the inclusion map $j\colon M' \to M$ then the Thom maps are isomorphisms of H^*M-modules, i.e.,

$$(11.23)\qquad e_!(y\frown h) = (-1)^{k|y|}(e^*y)\frown(e_!h), \qquad e^!(x\smile e^*y) = (e^!x)\smile y$$

for $y \in H^*M$, $h \in H(M, M-X)$, $x \in H^*(N, N-X)$.

Note that the assumptions of 11.20 are always satisfied if M, N are ENR's (IV, 8). If the retraction r, or the homotopy $er \simeq j$, exist in a neighborhood of X only, then one can still (by excision) prove the conclusions of 11.20 for this particular X. Also, the assumption $er \simeq j$ is not really essential for the first equation 11.23 (one can use a $\varinjlim$-argument for homology) but I don't know about the second.

Proof. If we compose $\tau_N^{M'}\frown = i'_* e'_!$ (cf. 11.14) with r_* we get $r_*(\tau_N^{M'}\frown) = e'_!$ which proves the first part (by naturality 11.5 of $e_!$). Now choose a representative cocycle t of $\tau_N^{M'}$; then the following are chain maps

$$\begin{aligned} S(M', M'-X) \simeq SM'/S\{M'-N, M'-r^{-1}X\} &\xrightarrow{t\frown} S(M', M'-r^{-1}X) \\ &\xrightarrow{r} S(N, N-X) \end{aligned}$$

(strictly speaking, we must shift dimension indices and introduce signs to make this correct). Their composite induces isomorphisms on homology—namely $r_*(\tau\frown) = e'_!$—and is therefore a homotopy equivalence (II, 4.3). Hence the dual composite

$$\begin{aligned} S^*(N, N-X) &\xrightarrow{r^*} S^*(M', M'-r^{-1}X) \\ &\xrightarrow{t\smile} (SM'/S\{M'-N, M'-r^{-1}X\})^* \cong S^*(M', M'-X) \end{aligned}$$

is also a homotopy equivalence. Since it induces $(\tau\smile)\,r^*$ on (co-)homology this must be isomorphic, as asserted in the second part of the proposition.

Finally,

$$\begin{aligned} e_!(y\frown h) &= r_*(\tau_N^{M'}\frown j_*^{-1}(y\frown h)) = r_*(\tau_N^{M'}\frown j^* y\frown j_*^{-1}h) \\ &= r_*(\tau_N^{M'}\frown r^* e^* y\frown j_*^{-1}h) = (-1)^{k|y|}\, r_*(r^* e^* y\frown \tau_N^{M'}\frown j_*^{-1}h) \\ &= (-1)^{k|y|}\, e^* y\frown r_*(\tau_N^{M'}\frown j_*^{-1}h) = (-1)^{k|y|}\, e^* y\frown e_! h, \end{aligned}$$

and dually for $e^!$. ∎

11.24 Corollary. *Put $X=N$ in 11.20. Then $e^!(1)=\tau_N^M$, and $e^!(\chi_N^M)=\tau_N^M\smile\tau_N^M$. In particular, $2\chi_N^M=0$ if k is odd* (since $\tau\smile\tau=(-1)^k\tau\smile\tau$).

Indeed, $e^!(1)=\tau\smile r^*(1)=\tau\smile 1=\tau$, hence

$$e^!(\chi_N^M)=e^!(1\smile e^*\tau)=e^!(1)\smile\tau=\tau\smile\tau. \;∎$$

I don't know, whether the last conclusion ($2\chi=0$) of 11.24 holds without assuming r, or $e\,r\simeq j$.

There are many other interesting consequences of 11.14. We discuss a few of them now, others in § 12.

11.25 Proposition. *If, as before, $e\colon N^n\to M^{n+k}$ is a closed inclusion of oriented manifolds and $X\subset N$ is closed then the composite map*

$$H_q(N,N-X)\xrightarrow{e_*}H_q(M,M-X)\xrightarrow{e_!}H_{q-k}(N,N-X)$$

agrees with $\chi_N^M\frown$, i.e. we have $e_!\,e_*\,\xi=\chi\frown\xi$ for $\xi\in H_q(N,N-X)$. In particular, $e_*\,\xi=0\Leftrightarrow\chi\frown\xi=0$.

For instance, if X is a compact ENR in N then $e_*(o_X^N)=0\Leftrightarrow\chi\frown o_X^N=0\Leftrightarrow\chi|X=0$, the latter by Poincaré-duality. In other words, the fundamental cycle in N around X is homologous in M to something in $M-X$ ("can be pushed into $M-X$") if and only if $\chi|X=0$. In particular, this applies to $X=N$ if N is compact.

Proof. As in 11.14, we consider open sets $W\subset M$ such that

$$N-X\subset W\subset M-X;$$

let $i\colon(N,N-X)\to(M,W)$ the inclusion map. Then $i_*\,e_!\,e_*\,\xi=\tau\frown e_*\,\xi=i_*(e^*\tau\frown\xi)=i_*(\chi\frown\xi)$, the first equation by 11.14, the second by naturality of $\frown$. This proves the assertion if i_* is monomorphic. For any $\xi\in H(N,N-X)$ we can find an open pair (M',W') in $(M,M-X)$ such that $(N',N'-X)=(N\cap M',N\cap M'-X)$ is a retract of (M',W') and $\xi\in\operatorname{im}(I_*\colon H(N',N'-X)\to H(N,N-X))$, say $\xi=I_*\,\xi'$ (because ξ is represented by a chain with compact carrier, and M,N are locally ENR;

cf. also proof of 11.18). Then $i'_*: H(N', N'-X) \to H(M', W')$ is monomorphic, hence $e'_! e'_* \xi' = \chi_N^{M'} \frown \xi'$ by the first part of the proof. Now apply I_* to this equation and get the required result by naturality of $e_!$ and χ. ∎

11.26 Proposition. *Let $N \xrightarrow{e} M \xrightarrow{d} L$ be closed inclusion maps of oriented manifolds. Assume the Thom class $\tau_N^M \in H^*(M, M-N)$ has an extension $\tilde{\tau}_N^M \in H^*(\tilde{L}, \tilde{W})$ to some open neighborhood $(\tilde{L}, \tilde{W})$ of $(M, M-N)$ in $(L, L-N)$. Then $\tau_N^L = \tilde{\tau}_N^M \smile \tau_M^L$, $d^*(\tau_N^L) = \tau_N^M \smile \chi_M^L$, and $\chi_N^L = \chi_N^M \smile e^*(\chi_M^L)$. The extension $\tilde{\tau}$ always exists if M is a neighborhood retract in L: just take $\tilde{\tau}_N^M = r^* \tau_N^M$, where $r\colon \tilde{L} \to M$ is a neighborhood retraction, and $\tilde{W} = r^{-1}(M-N)$.* I don't know whether $\tilde{\tau}$ always exists, nor whether the last equation between Euler classes always holds.

Proof. Let N_λ be any component of N and v_λ its transverse class; put $k = \dim M - \dim N$, $h = \dim L - \dim N$. We have

$$\langle \tilde{\tau}_N^M \smile \tau_M^L, v_\lambda^L \rangle = \langle \tilde{\tau}_N^M, \tau_M^L \frown v_\lambda^L \rangle \overset{11.14}{=} \langle \tilde{\tau}_N^M, d_*(d_! v_\lambda^L) \rangle = \langle \tilde{\tau}_N^M, d_* v_\lambda^M \rangle$$

$$= \langle d^* \tilde{\tau}_N^M, v_\lambda^M \rangle = \langle \tau_N^M, v_\lambda^M \rangle = 1,$$

the 3rd equation because $e_!(v_\lambda^M)$ and $e_!(d_! v_\lambda^L) = (d\,e)_!(v_\lambda^L)$ coincide (they both equal the class of a point in N_λ). This shows $\tilde{\tau}_N^M \smile \tau_M^L = \tau_N^L$, by definition of the latter. Apply d^* to this equation and get $d^*(\tau_N^L) = d^*(\tilde{\tau}_N^M) \smile d^*(\tau_M^L) = \tau_N^M \smile \chi_M^L = \tau_N^M \smile \chi_M^L$. Now apply e^* and get

$$\chi_N^L = \chi_N^L = (d\,e)^*(\tau_N^L) = e^*\, d^*(\tau_N^L) = e^*(\tau_N^M) \smile e^*(\chi_M^L) = \chi_N^M \smile e^*(\chi_M^L). \;∎$$

11.27 So far in this § we have only considered oriented manifolds. As usual, analogous results hold (with the same proofs) in the non-oriented case if coefficients mod 2 are used. In order to get a finer theory one has to use *local coefficients* (cf. Steenrod 1951, § 31). We shall not go into this; we shall, however, deduce some of the easier integral results directly, essentially by reduction to the oriented case. Assume then $N^n \subset M^{n+k}$ are manifolds (not necessarily orientable), $\bar{N} = N$. For every component N_λ of N, choose an open set $M' \subset M$ such that M' and $N' = N \cap M'$ are orientable, and $N'_\lambda = N_\lambda \cap M' \neq \emptyset$; let $v'_\lambda \in H_n(M', M'-N')$ be a transverse class of N'_λ in M', and let $v_\lambda \in H_n(M, M-N)$ denote its image. We know (11.6) that $\pm v_\lambda$ is in fact the transverse class of N_λ in M if these are orientable; so we continue to call it so, even if they are not orientable.

11.28 Proposition. *Let $N^n \subset M^{n+k}$ as above, and let X be a closed subset of N. Then $H_j(M, M-X; \mathbb{Z}) = 0$ for $j < k$, and $H_k(M, M-X; \mathbb{Z})$ is gener-*

ated by the transverse classes ν_λ of components N_λ which lie in X. In contrast to 11.3, however, these classes are no longer *free* generators; some of them may be of order 2 (cf. 11.29).

Proof. The proposition is true if both M and N are orientable. Suppose M_1, M_2 are open subsets in M such that the proposition holds for (M_1, N_1), (M_2, N_2), $(M_1 \cap M_2, N_1 \cap N_2)$, where $N_q = M_q \cap N$. We then consider the Mayer-Vietoris sequence (III, 8.22)

$$H_j(M_1, M_1 - X_1) \oplus H_j(M_2, M_2 - X_2) \to H_j(M_1 \cup M_2, M_1 \cup M_2 - X_1 \cup X_2)$$
$$\to H_{j-1}(M_1 \cap M_2, M_1 \cap M_2 - X_1 \cap X_2),$$

where $X_q = M_q \cap X$. We can apply the proposition to the outside terms and thus, by exactness, prove it for the middle term, i.e., we can conclude that the proposition holds for $(M_1 \cup M_2, N_1 \cup N_2)$. By induction, the proposition then holds for every finite union $(M', N') = \bigcup_{q=1}^{r} (M_q, N_q)$ of orientable pairs. Since $H(M, M-X) = \varinjlim H(M', M'-X)$, this proves the proposition by passage to the limit. ∎

11.29 Proposition. *Let M^{n+k} be an orientable manifold and $N^n \subset M^{n+k}$ a non-orientable connected submanifold, $\bar{N} = N$. We know* (11.28) *that the transverse class ν_N generates $H_k(M, M-N; \mathbb{Z})$. We assert, that ν_N is of order two, hence $H_k(M, M-N; \mathbb{Z}) \cong \mathbb{Z}_2$.*

Proof. Pick $P \in N$ and an orientation $o_P \in H_n(N, N-P)$ at this point. Consider all oriented connected open subsets of N which contain P, with the given orientation at P. Their union is N, and since N is not orientable there must be two of them, say N' and N'', whose orientations disagree at some other point $Q \in N' \cap N''$. Orient M and pick open subsets $M', M'' \subset M$ such that $N' = N \cap M'$, $N'' = N \cap M''$. Their transverse classes $\nu' \in H(M', M'-N')$, $\nu'' \in H(M'', M''-N'')$ are defined then (not only up to sign) and they are both images of the transverse class

$$\nu_P \in H(M' \cap M'', M' \cap M'' - N' \cap N''),$$

hence they map to the same generator $\nu_N \in H_k(M, M-N)$. The transverse class $\nu_Q \in H(M' \cap M'', M' \cap M'' - N' \cap N'')$, on the other hand, maps with opposite signs into $\pm\nu'$, $\mp\nu''$ because the orientations of N', N'' disagree at Q. The images of ν', ν'' in $H_k(M, M-N)$ must therefore be of opposite sign, hence $\nu_N = -\nu_N$, or $2\,\nu_N = 0$.

It remains to show that $\nu_N \neq 0$, or $H_k(M, M-N; \mathbb{Z}) \neq 0$. But

$$H_k(M, M-N; \mathbb{Z}) \otimes \mathbb{Z}_2 \cong H_k(M, M-N; \mathbb{Z}_2) \cong \mathbb{Z}_2,$$

the first isomorphism by 11.28 and the universal coefficient theorem, the second because mod 2 the isomorphisms 11.1, 11.3 hold for non-oriented manifolds. ∎

11.30 Exercises. *1**. Let $N_1, N_2 \subset M$ be oriented manifolds, $\bar{N}_p = N_p$, and assume $N_1 \cap N_2$ is a connected manifold with $\dim(N_1 \cap N_2) = \dim N_1 + \dim N_2 - \dim M$. Prove: If $\tau_{N_1} \smile \tau_{N_2} \neq 0$ then $N_1 \cap N_2$ is orientable, $\tau_{N_1} \smile \tau_{N_2} = \mu \tau_{N_1 \cap N_2}$ for some integer μ, and $e_2^*(\tau_{N_1}^M) = \mu \tau_{N_1 \cap N_2}^{N_2}$, $e_{21}^*(\chi_{N_1}^M) = \mu \chi_{N_1 \cap N_2}^{N_2}$, where e_2, e_{21} are inclusions. The integer μ is called the *intersection multiplicity* ($=0$ if $\tau_{N_1} \smile \tau_{N_2} = 0$). Show that μ can be determined locally, i.e. in any open set W of M which meets $N_1 \cap N_2$. If N_1, N_2 intersect transversally at some $P \in N_1 \cap N_2$ then μ was shown to be ± 1 in 11.13; compute the sign in terms of the orientations of $N_1 \cap N_2$, N_1, N_2, M.

2. Let $N \subset M$, $N' \subset M'$ be oriented manifolds ($\bar{N} = N$, $\bar{N}' = N'$) of dimension n, m, n', m'. Consider $N \times N' \subset M \times M'$ with the product orientations and prove $\tau_{N \times N'}^{M \times M'} = (-1)^{m(m'-n')} \tau_N^M \times \tau_{N'}^{M'}$.

3. Let N be a compact oriented manifold, orient $N \times N$ with the product orientation, and consider the diagonal embedding $e\colon N \to N \times N$. Prove that $\langle \chi_N^{N \times N}, o_N \rangle = \chi(N) =$ Euler-characteristic of N. Hint: Use 11.18 and 8.21, Exerc. 2.

4. Let $f\colon N \to M$ be a map of compact oriented manifolds, and define $\bar{\tau} \in H^*M$ by $\bar{\tau} \frown o_M = f_*(o_N)$. Prove that $f_* f_!(h) = \bar{\tau} \frown h$ for every $h \in HM$. Compare this with 11.14.

5*. Recall (4.10, Exerc. 6) that a map $f\colon N \to M$ of manifolds is called *orientable* if it lifts to a map $\tilde{f}\colon \tilde{N} \to \tilde{M}$ of orientation-covers which commutes with the canonical involutions of $\tilde{N}, \tilde{M}$. Show that a closed inclusion map $e\colon N^n \to M^{n+k}$ is orientable if and only if every transverse class v_λ of N in M has infinite order. This, in turn, holds if and only if a class $\tau \in H^k(M, M - N; \mathbb{Z})$ exists such that $\langle \tau, v_\lambda \rangle = 1$ for every λ.

12. The Gysin Sequence. Examples

The Gysin sequence is a consequence of 11.14. It relates the (co-)homology of N and $M - N$ provided $e_*\colon HN \cong HM$. This assumption may appear very restrictive, however, even if it is not satisfied one can usually find an open neighborhood M' of N in M such that $HN \cong HM'$, and replace M by M'; in fact, I know of no instance where M' doesn't exist. We discuss some examples to show how the Gysin sequence can be used to determine $H(M - N)$ from HN (Stiefel manifolds), or HN from $H(M - N)$ (Grassmann manifolds).

12.1 Proposition. *Let $e\colon N^n \subset M^{n+k}$ be oriented manifolds as in § 11 ($\overline{N}=N$), let $M' \subset M$ be an open set and $N' = N \cap M'$. If $i_*\colon H(N,N') \cong H(M,M')$ then*

$$\tau_N^M \frown\colon H_r(M, M' \cup (M-N)) \cong H_{r-k}(M, M'), \tag{12.2}$$

$$\tau_N^M \smile\colon H^{r-k}(M, M') \cong H^r(M, M' \cup (M-N)) \tag{12.3}$$

(arbitrary coefficients; mod 2 if M or N is not oriented). *Moreover, there are* (dual) *exact sequences*

$$\begin{aligned} \cdots \to H_{r-k+1}(N,N') \xrightarrow{\sigma_*} H_r(M-N, M'-N') \xrightarrow{\rho_* = i_*^{-1} j_*} H_r(N,N') \\ \xrightarrow{\chi_N^M \frown} H_{r-k}(N,N') \xrightarrow{\sigma_*} H_{r-1}(M-N, M'-N') \to \cdots, \end{aligned} \tag{12.4}$$

$$\begin{aligned} \cdots \leftarrow H^{r-k+1}(N,N') \xleftarrow{\sigma^*} H^r(M-N, M'-N') \xleftarrow{\rho^* = j^*(i^*)^{-1}} H^r(N,N') \\ \xleftarrow{\chi_N^M \smile} H^{r-k}(N,N') \xleftarrow{\sigma^*} H^{r-1}(M-N, M'-N') \leftarrow \cdots. \end{aligned} \tag{12.5}$$

The maps ρ are induced by inclusions, as indicated; σ will be defined during the proof. Both sequences are named after *Gysin*. The case $M' = \emptyset = N'$ is of particular importance; the Gysin sequences then relate the (co-)homology of N and $M-N$.

Proof. We know from 11.14 that, up to sign, the map $\tau \frown$ agrees with the composition $H(M, M' \cup (M-N)) \xrightarrow{e_!} H(N,N') \xrightarrow{i_*} H(M,M')$, and by assumption this is isomorphic. As in the proof of 11.20 we now choose a representative cocycle t of τ and we conclude that

$$t \frown\colon S(M, M' \cup (M-N)) \to S(M, M')$$

is a chain homotopy equivalence. The dual chain homotopy equivalence $t \smile$ induces $\tau \smile$ on cohomology, which is therefore isomorphic. This proves 12.2, 12.3.

For the Gysin sequence we consider the diagram

$$\begin{array}{ccccccccc} \to H(M' \cup (M-N), M') & \xrightarrow{j_*^2} & H(M,M') & \xrightarrow{\bar{j}_*} & H(M, M' \cup (M-N)) & \xrightarrow{\partial_*} & H(M' \cup (M-N), M') \to \\ \uparrow {\scriptstyle \cong\, j_*^1} & & \uparrow {\scriptstyle \cong\, i_*} & & \uparrow {\scriptstyle \cong\, (\tau\frown)^{-1} i_*} & & \uparrow {\scriptstyle \cong\, j_*^1} \\ \to H(M-N, M'-N') & \xrightarrow{\rho_*} & H(N,N') & \xrightarrow{\chi\frown} & H(N,N') & \xrightarrow{\sigma_*} & H(M-N, M'-N') \to, \end{array} \tag{12.6}$$

where the first row is the homology sequence of the triple

$$(M, M' \cup (M-N), M').$$

The vertical arrows are isomorphic (j_*^1 by excision), and ρ_* resp. σ_* are so defined as to make the first resp. third square commutative. The second row is the required Gysin-sequence 12.4. In order to prove that it is exact it suffices to verify that the middle square is commutative (the first row being exact). This follows from

$$i_*(\chi \frown h) = i_*((i^* \tau) \frown h) = \tau \frown (i_* h) = \tau \frown (\tilde{j}_* i_* h).$$

The cohomology Gysin sequence 12.5 is obtained dually: just replace H by H^* and $\frown$ by $\smile$ in 12.6, and reverse all arrows. ∎

The graded (co-)homology groups which appear in the Gysin sequences 12.4, 12.5 can be viewed as modules over $H^* M$ (via inclusion, and $\frown$- or $\smile$-products); then we have the following

12.7 Complement to 12.1. *All the maps which appear in the Gysin sequence are (graded) homomorphisms of graded H^*M-modules* (n.b. for graded maps φ this involves a sign: $\varphi(y \cdot h) = (-1)^{|\varphi||y|} y \cdot \varphi(h)$).

This can be refined: The operation of H^*M factors through $H^*M \to \check{H}(M-M')$ (by excision and passage to the limit), and the maps of the Gysin sequence are in fact homomorphisms of $\check{H}(M-M')$-modules. We leave these details as an exercise to the reader.

Proof of 12.7. The first row of 12.6 consists of H^*M-homomorphisms (cf. VII, 12.19). It suffices therefore to show that the vertical arrows are H^*M-homomorphisms. This is clear for j_*^1 and i_*, we must only prove it for $\tau\frown$. But

$$\tau \frown (y \frown h) = (\tau \smile y) \frown h = (-1)^{k|y|}(y \smile \tau) \frown h = (-1)^{k|y|} y \frown (\tau \frown h),$$

as asserted. ∎

We now formulate some consequences and special cases of the Gysin sequence which will be used in our examples. For simplicity, we consider the absolute case $M' = \emptyset = N'$ and *co*homology only. All coefficients are taken in a fixed commutative ring R (of characteristic two if M or N is not oriented); we write χ both for the (integral) Euler class and its image in $H^k(N; R)$.

12.8 Proposition. *Let e: $N^n \subset M^{n+k}$ be oriented manifolds, $\bar{N} = N$, such that e_*: $HN \cong HM$. Then*

(i) *ρ^*: $H^r N \to H^r(M-N)$ is monomorphic for $r<k$, and epimorphic for $r<k-1$. The kernel of ρ^*: $H^k(N;\mathbb{Z}) \to H^k(M-N;\mathbb{Z})$ is a cyclic group with generator χ; if N is connected, then*

$$\operatorname{coker}[\rho^*: H^{k-1}(N;\mathbb{Z}) \to H^{k-1}(M-N;\mathbb{Z})]$$

is zero resp. $\cong \mathbb{Z}$ depending on whether the order of χ is infinite or finite.

(ii) *If A is a subset of $H^* N$ such that $\{\rho^* a\}_{a\in A}$ generates $H^*(M-N)$ as a ring then $A\cup\{\chi\}$ generates $H^* N$ as a ring.*

(iii) *If $\chi_N^M=0$ (hence σ^* epimorphic) and $u\in H^{k-1}(M-N)$ is any element such that $\sigma^*(u)=1\in H^0 N$ then $H^*(M-N)\cong(H^* N)\cdot 1\oplus(H^* N)\cdot u$, as $H^* N$-modules (cf. 12.7). With dimension indices,*

$$H^r(M-N)\cong(H^r N)\oplus(H^{r-k-1} N).$$

Note that $\chi_N^M=0$ whenever k is odd and $H^k(N;\mathbb{Z})$ has no 2-torsion (cf. 11.24).

(iv) *Suppose there are homomorphisms $\gamma^r\colon H^r(M-N)\to H^r N$ such that $\rho^*\gamma^r=\mathrm{id}$ for $r\leq s$, and $\gamma(y\smile z)=(\gamma y)\smile(\gamma z)$ for $|y|+|z|\leq s$. This means, $\gamma=\{\gamma^r\}$ is a multiplicative right inverse of ρ^*, up to dimension s; put $\gamma^r=0$ for $r>s$. Let $H^*(M-N)[x]$ denote the (graded) polynomial ring over $H^*(M-N)$ in one indeterminate x such that $|x|=k$. Then*

$$\Gamma\colon H^*(M-N)[x]\to H^* N,\quad \Gamma(\sum a_j x^j)=\sum\gamma(a_j)\chi^j,$$

is a ring-isomorphism up to dimension s. In dimension $s+1$ the kernel of Γ consists of all constant polynomials, $\ker(\Gamma^{s+1})=H^{s+1}(M-N)$.

Proof. Part (i) follows immediately from the Gysin sequence 12.5 since $H^j N=0$ for $j<0$, and $H^0(N;\mathbb{Z})\cong\mathbb{Z}$ if N is connected. Part (ii) follows by induction on dimension: If $y\in H^* N$, we can find a polynomial $p(a)$ in elements $a\in A$, such that $\rho^*(p)=\rho^*(y)$, hence $(y-p)\in\ker(\rho^*)$, hence $y-p=\chi\smile q$ by 12.5, where $|q|=|p|-k<|p|$. By induction, q is a polynomial in χ and elements a, hence $y=p(a)+\chi\smile q(a,\chi)$, as asserted.

If $\chi=0$ then 12.5 reduces to a short exact sequence of $H^* N$-modules $0\to H^* N\xrightarrow{\rho^*}H^*(M-N)\xrightarrow{\sigma^*}H^* N\to 0$, and the map $y\mapsto(\rho^* y)\smile u$, $y\in H^* N$, is a right inverse of σ^* (using 12.7); this proves part (iii). Part (iv) is similar to (ii); in fact, the argument for (ii) shows that Γ is epimorphic (up to dimension s). Suppose now $\sum_{j\geq 0} a_j x^j$ is an element of $\ker(\Gamma)$ of dimension $r\leq s$. Then $\sum_{j\geq 0}\gamma(a_j)\chi^j=0$; apply ρ^*, use $\rho^*(\chi)=0$, and get $a_0=\rho^*\gamma(a_0)=0$. Therefore $\Gamma(\sum_{j>0} a_j x^{j-1})\smile\chi=\sum_{j\geq 0}\gamma(a_j)\chi^j=0$, and $\dim(\sum_{j>0} a_j x^{j-1})=r-k$. The partial Gysin sequence

$$H^{r-1} N\xrightarrow{\rho^*}H^{r-1}(M-N)\xrightarrow{\sigma^*}H^{r-k} N\xrightarrow{\chi\smile}H^r N$$

shows that $\chi\smile$ is monomorphic (because ρ^* is epimorphic), hence $\Gamma(\sum_{j>0} a_j x^{j-1})=0$, hence $\sum_{j>0} a_j x^{j-1}=0$ by inductive hypothesis, hence $\sum_{j\geq 0} a_j x^j=(\sum_{j>0} a_j x^{j-1})\,x=0$. The same argument applies if $\dim(\sum a_j x^j)=s+1$ provided $a_0=0$; hence $\ker(\Gamma^{s+1})=H^{s+1}(M-N)$, as asserted. ∎

12.9 Example. Stiefel Manifolds. If F is one of the (skew) fields $\mathbb{R}, \mathbb{C}, \mathbb{H}$ let $V_{pq}F$ denote the set of all linearly independent p-tuples of vectors in F^{p+q},

$$V_{pq}F = \{(v_1, v_2, \ldots, v_p) \in F^{p+q} \times F^{p+q} \times \cdots \times F^{p+q} \mid \{v_i\} \text{ linearly independent over } F\}.$$

Clearly, $V_{pq}F$ is an open subset of $(F^{p+q})^p \approx F^{p(p+q)}$, hence is a manifold of dimension $dp(p+q)$, where $d=1,2,4$ as $F=\mathbb{R}, \mathbb{C}, \mathbb{H}$. It is known as (real, complex, quaternionic) *Stiefel manifold*; its elements are also called *p-frames* in F^{p+q}. Note that $V_{p0}F$ is the set of all bases of F^p; it can be identified with the linear group $Gl(F, p)$.—As an application of 12.8 (iii) we prove

12.10 Proposition. *The complex Stiefel manifold $V_{pq}\mathbb{C}$ and the product of spheres $\mathbb{S}^{2q+1} \times \mathbb{S}^{2q+3} \times \cdots \times \mathbb{S}^{2q+2p-1}$ have isomorphic integral cohomology rings, i.e. $H^*(V_{pq}\mathbb{C}; \mathbb{Z}) = E(\sigma^{2q+1}, \ldots, \sigma^{2q+2p-1})$ is an exterior algebra (cf. VII, 10.15) over $\mathbb{Z}$ with generators $\{\sigma^j\}$ of dimension $j=2q+1$, $2q+3, \ldots, 2q+2p-1$.*

For $V_{pq}\mathbb{H}$ there is a similar result and proof (just replace 2 by 4), whereas for $V_{pq}\mathbb{R}$ the situation is more complicated (cf. 12.11).

Proof, by induction on p, starting at $V_{0,p+q}$ = a point (or $V_{1,p+q-1}\mathbb{C} \simeq \mathbb{S}^{2p+2q-1}$). Let

$$M_{pq} = V_{pq}\mathbb{C} \times \mathbb{C}^{p+q} = \{(v_1, \ldots, v_p, v_{p+1}) \mid v_i \in \mathbb{C}^{p+q}, (v_1, \ldots, v_p) \text{ independent}\},$$

and

$$N_{pq} = \{(v_1, \ldots, v_p, v_{p+1}) \in M_{pq} \mid v_{p+1} \text{ dependent on } (v_1 \ldots v_p)\}.$$

Clearly N_{pq} is a closed subset of M_{pq}, and $M_{pq} - N_{pq} = V_{p+1,q-1}\mathbb{C}$. Further,

$$V_{pq}\mathbb{C} \times \mathbb{C}^p \to N_{pq}, \quad (v_1, \ldots, v_p, \lambda_1, \ldots, \lambda_p) \mapsto (v_1, \ldots, v_p, v_{p+1} = \textstyle\sum_{i=1}^p \lambda_i v_i),$$

is a homeomorphism, hence N_{pq} is a manifold of dimension

$$n = 2p(p+q) + 2p,$$

and is homotopy equivalent with $V_{pq}\mathbb{C}$. Also, M_{pq} is a manifold of dimension $2p(p+q)+2(p+q) = n+2q$, and is homotopy equivalent with $V_{pq}\mathbb{C}$, hence $H_* M_{pq} \cong H_* V_{pq}\mathbb{C} \cong H_* N_{pq}$. We can therefore apply the Gysin sequence to the inclusion $N_{pq} \subset M_{pq}$. The Euler class χ_N^M lies in

$$H^{2q} N_{pq} \cong H^{2q}(V_{pq}\mathbb{C}) = 0,$$

the latter by the inductive hypothesis $H^*(V_{pq}\mathbb{C})=E(\sigma^{2q+1},\ldots)$. It follows from 12.8(iii) that

$$\begin{aligned}H^*(V_{p+1,q-1}\mathbb{C})&=H^*(M_{pq}-N_{pq})\\ &\cong E(\sigma^{2q+1},\ldots,\sigma^{2q+2p-1})\cdot 1\oplus E(\sigma^{2q+1},\ldots,\sigma^{2q+2p-1})\cdot\sigma^{2q-1}\end{aligned}$$

as E-modules. To finish the proof we must only show that $\sigma^{2q-1}\smile\sigma^{2q-1}=0$. But $2(\sigma^{2q-1}\smile\sigma^{2q-1})=0$ because $2q-1$ is odd, and $H^*(V_{p+1,q-1}\mathbb{C})$ is torsionfree by the formula above. ∎

12.11 For real Stiefel manifolds $V_{pq}\mathbb{R}$ one might expect an analogous result (replacing 2 by 1). However, this is false in general; the above proof breaks down because the Euler class $\chi_N^M\in H^{q+1}(V_{pq}\mathbb{R})$ is not always zero. Taking coefficients $\mathbb{Z}_2$ avoids this difficulty (i.e. $\chi_N^M=0$ then; cf. Exerc. 6) but then one can no longer prove $\sigma\smile\sigma=0$ as above. Still, 12.8(iii) will show that *elements* $\sigma^j\in H^j(V_{pq}\mathbb{R};\mathbb{Z}_2)$ *exist for* $j=q+1,q+2,\ldots,q+p-1$ *such that the monomials* $\{\sigma^{j_1}\smile\sigma^{j_2}\smile\cdots\smile\sigma^{j_r}\}$, $q+1\le j_1<j_2<\cdots<j_r\le q+p-1$, *form a* $\mathbb{Z}_2$*-base of* $H^*(V_{pq}\mathbb{R};\mathbb{Z}_2)$. As to the values of $\sigma^j\smile\sigma^j$, we refer the reader to Epstein-Steenrod IV, 4.

Avoiding all questions about χ_N^M and $\sigma\smile\sigma$ one can apply 12.8(i) to the embeddings $N_{pq}\mathbb{R}\subset M_{pq}\mathbb{R}$ and get inductively

12.12 Proposition. $H^r(V_{pq}\mathbb{R};\mathbb{Z}_2)=0$ *for* $0<r\le q$. *More generally, using the homology Gysin sequence,* $H_r(V_{pq}\mathbb{R};\mathbb{Z})=0$ *for* $0<r\le q$. ∎

12.13 Example. Grassmann Manifolds. If F is one of the (skew-)fields $\mathbb{R},\mathbb{C},\mathbb{H}$ let $G_{pq}=G_{pq}F$ denote the set of all p-dimensional linear subspaces of F^{p+q}. For instance, $G_{0q}F$ consists of just one element, and $G_{1q}F=$ set of 1-dimensional subspaces of $F^{q+1}=P_qF$ (cf. V, 3.5). We shall see that there is a natural topology which turns $G_{pq}F$ into a compact manifold of dimension dpq, generalizing P_qF.

Let $\alpha\colon F^{p+q}\to F^p$ be any linear epimorphism, and let

$$G^\alpha_{pq}=\{g\in G_{pq}|\alpha(g)=F^p\}=\{g\in G_{pq}|g\cap\ker(\alpha)=\{0\}\}.$$

If $g\in G^\alpha_{pq}$ then there is a unique linear map $\zeta\colon F^p\to F^{p+q}$ such that $\zeta(F^p)=g$ and $\alpha\zeta=\mathrm{id}$, i.e. the correspondence $\zeta\mapsto\zeta(F^p)$ is a bijection $\varphi_\alpha\colon\{\zeta\in\mathscr{L}(F^p,F^{p+q})|\alpha\zeta=\mathrm{id}\}\approx G^\alpha_{pq}$. We use this to topologize G^α_{pq}, so that φ_α becomes a homeomorphism (where $\mathscr{L}(F^p,F^{p+q})$, the space of linear maps $F^p\to F^{p+q}$, has the usual topology $\approx F^{p(p+q)}$). If we fix *one* $\zeta_0\colon F^p\to F^{p+q}$ with $\alpha\zeta_0=\mathrm{id}$, then adding ζ_0 defines a homeomorphism

$$\zeta_0+\colon\ \mathscr{L}(F^p,\ker(\alpha))\approx\{\zeta\in\mathscr{L}(F^p,F^{p+q})|\alpha\zeta=\mathrm{id}\},\qquad \xi\mapsto\zeta_0+\xi,$$

hence $G^\alpha_{pq} \approx \mathscr{L}(F^p, \ker(\alpha)) \approx F^{pq} \approx \mathbb{R}^{dpq}$, where $d=1,2,4$ as $F=\mathbb{R}, \mathbb{C}, \mathbb{H}$. Clearly every $g \in G_{pq}$ lies in some G^α_{pq}; in fact, any *two* $g, g' \in G_{pq}$ lie in a common G^α_{pq} (an easy exercise in linear algebra). If $\alpha, \beta\colon F^{p+q} \to F^p$ are two linear epimorphisms then

$$\varphi_\alpha^{-1}(G^\beta_{pq}) = \{\zeta \mid \alpha\zeta = \mathrm{id}\} \cap \{\zeta \mid \ker(\beta\zeta) = \{0\}\};$$

clearly $\{\zeta \mid \ker(\beta\zeta) = \{0\}\}$ is an open set, hence $G^\alpha_{pq} \cap G^\beta_{pq}$ is open in G^α_{pq} (or G^β_{pq}). We now choose the finest topology in G_{pq} for which all inclusions $G^\alpha_{pq} \to G_{pq}$ are continuous, i.e. we say $U \subset G_{pq}$ is open if and only if every $U \cap G^\alpha_{pq}$ is open in G^α_{pq}. As every $G^\alpha_{pq} \cap G^\beta_{pq}$ is open it follows that the inclusions $G^\beta_{pq} \to G_{pq}$ are open maps, i.e. G^β_{pq} is open in G_{pq}, and its topology (from φ_β) agrees with the subspace topology. Any two $g, g' \in G_{pq}$ lie in a common $G^\alpha_{pq} \approx F^{pq}$; they have disjoint neighborhoods there and hence in G_{pq}. Therefore, $G_{pq}F$ is a manifold of dimension dpq, known as (real, complex, quaternionic) *Grassmann manifold*.

Another description of $G_{pq}F$ is as follows. Let $Gl(p+q, F)$ denote the group of all linear isomorphisms $F^{p+q} \to F^{p+q}$; this is an open subset of $\mathscr{L}(F^{p+q}, F^{p+q})$, and it is a topological group under composition. If $g \in G_{pq}$ and $\psi \in Gl(p+q)$ then $\psi(g) \in G_{pq}$, hence a map (an *operation*) $Gl(p+q) \times G_{pq} \to G_{pq}$, $(\psi, g) \mapsto \psi(g)$, which is easily seen to be continuous. The operation is *transitive*, i.e. if we fix $g_0 \in G_{pq}$ then every $g \in G_{pq}$ is of the form $\psi(g_0)$. Therefore, the mapping $\pi\colon Gl(p+q) \to G_{pq}$, $\pi(\psi) = \psi(g_0)$, induces a continuous bijection $\bar{\pi}\colon Gl(p+q)/Gl(p, p+q) \to G_{pq}$, where $Gl(p, p+q)$ is the subgroup of all ψ such that $\psi(g_0) = g_0$ (the *isotropy group* of g_0), and the coset space on the left is taken with the quotient topology. If $F = \mathbb{C}$ or $\mathbb{H}$ then $Gl(p+q)$ is connected, hence G_{pq} *is connected*. If $F = \mathbb{R}$ we get the same result using the group $Gl^+(p+q)$ of orientation preserving linear isomorphisms instead of $Gl(p+q)$. If $U(p+q, F) \subset Gl(p+q, F)$ denotes the subgroup of all isometries of F^{p+q} (with respect to the metric $\sum x_i \bar{x}_i$) then already $U(p+q)$ operates transitively, i.e. the composite $\pi'\colon U(p+q) \subset Gl(p+q) \xrightarrow{\pi} G_{pq}$ is surjective. But $U(p+q)$ is compact, hence $G_{pq} = \mathrm{im}(\pi')$ *is compact*, and π' is an identification map. The latter implies that π *is also an identification map*, hence $\bar{\pi}$ is a homeomorphism $Gl(p+q)/Gl(p, p+q) \approx G_{pq}$.

If $F = \mathbb{C}$ *or* $\mathbb{H}$ *then* $G_{pq}F$ *is orientable*. Indeed, fix an orientation o of G_{pq} at g_0, and define a mapping $\tilde{\pi}\colon Gl(p+q) \to \tilde{G}_{pq}$ (= orientation covering of G_{pq}; cf. 2.11) by $\tilde{\pi}(\psi) = \psi_*(o)$, where ψ is viewed as a homeomorphism $G_{pq} \to G_{pq}$. The definition of $\tilde{G}_{pq}$ (2.3, 2.11) easily shows that $\tilde{\pi}$ is continuous. The restriction $\tilde{\pi} \mid Gl(p, p+q)$ to the isotropy group can only assume the two values $o, -o$; since it is continuous and $Gl(p, p+q)$ is *connected* (an easy exercise, compare Pontrjagin, § 65, Beispiel 108;

here $F \neq \mathbb{R}$ is essential) this restriction must be constant. Similarly, $\tilde{\pi}$ is constant on every coset of $Gl(p, p+q)$, hence $\tilde{\pi}$ induces a map

$$G_{pq} \approx Gl(p+q)/Gl(p, p+q) \to \tilde{G}_{pq},$$

i.e., an orientation of G_{pq}. ∎

12.14 We now study the *cohomology of Grassmann manifolds.* For simplicity, we assume $F=\mathbb{C}$; the case $F=\mathbb{H}$ is similar (but less important), whereas $F=\mathbb{R}$ presents more difficulties (even with coefficients $\mathbb{Z}_2$; cf. Exerc. 5). The following auxiliary spaces will be used.

(i) $L_{pq}=G_{pq}\times\mathbb{C}^{p+q}$. This is an oriented manifold of dimension $2(pq+p+q)$; we pick the canonical orientation on $\mathbb{C}^r$ (cf. 2.13), and the resulting canonical orientation on G_{pq} (using $G_{pq}^{\alpha}\approx\mathbb{C}^{pq}$).

(ii) $M_{pq}=\{(g,v)\in G_{pq}\times\mathbb{C}^{p+q}=L_{pq}\,|\,v\in g\}$. Clearly, this is a closed subset of L_{pq}. If we let $M_{pq}^{\alpha}=\{(g,v)\in M_{pq}\,|\,g\in G_{pq}^{\alpha}\}$ then

$$\{\zeta\in\mathscr{L}(\mathbb{C}^p,\mathbb{C}^{p+q})\,|\,\alpha\zeta=\mathrm{id}\}\times\mathbb{C}^p\to M_{pq}^{\alpha},\qquad(\zeta,z)\mapsto(\zeta(\mathbb{C}^p),\zeta(z)),$$

is a homeomorphism, hence $M_{pq}^{\alpha}\approx G_{pq}^{\alpha}\times\mathbb{C}^p\approx\mathbb{C}^{pq}\times\mathbb{C}^p$, hence M_{pq} is a manifold of dimension $2(pq+p)$. It can be oriented just as G_{pq} above, or one can verify that the local product orientations in $M_{pq}^{\alpha}\approx G_{pq}^{\alpha}\times\mathbb{C}^p$ match.

(iii) $N_{pq}=\{(g,v)\in G_{pq}\times\mathbb{C}^{pq}\,|\,v=0\}$. Clearly $N_{pq}\approx G_{pq}$ is a closed submanifold of M_{pq}, and it is a deformation retract of L_{pq} as well as of M_{pq} (by the deformation (g,tv), $0\le t\le 1$), hence $H^*L\cong H^*M\cong H^*N$. Further,

$$(12.15)\qquad (L_{pq}-M_{pq})\simeq(M_{p+1,q-1}-N_{p+1,q-1}).$$

Proof. If $(g,v)\in(L_{pq}-M_{pq})$, let $[g,v]$ the $(p+1)$-dimensional subspace of C^{p+q} spanned by g and v, and define

$$r\colon (L_{pq}-M_{pq})\to(M_{p+1,q-1}-N_{p+1,q-1})$$

by $r(g,v)=([g,v],v)$. Define j in the other direction by $j(g,v)=(v\perp g,v)$, where $v\perp g$ denotes the p-subspace of g which is orthogonal to v. Then $rj(g,v)=([v\perp g,v],v)=(g,v)$, hence $rj=\mathrm{id}$; and $jr(g,v)=(v\perp[g,v],v)$. Now, $v\perp[g,v]$ and g are both transversal (=independent) to v. We can deform one into the other by moving every point along the segment parallel to v, hence $jr\simeq\mathrm{id}$. ∎

12.16 Definition. By induction on p we define elements

$$c_i=c_i^p\in H^{2i}(G_{pq}\mathbb{C};\mathbb{Z})\qquad\text{for } 0\le i\le p,$$

called *Chern classes*, as follows. Put $c_0^0=1$. If $p>0$, and c_i^{p-1} is already defined, consider the Gysin sequence of $N_{pq}\subset M_{pq}$; by 12.8(i) we have $\rho^*: H^r N_{pq}\cong H^r(M_{pq}-N_{pq})$ for $r\leq 2(p-1)$. Now define $c_p^p\in H^{2p}(N_{pq};\mathbb{Z})=H^{2p}(G_{pq};\mathbb{Z})$ as being the Euler class of $N_{pq}\subset M_{pq}$, and c_i^p for $0\leq i<p$ as being the image of c_i^{p-1} under the composite

$$H^{2i}G_{p-1,q+1}\cong H^{2i}L_{p-1,q+1}\to H^{2i}(L_{p-1,q+1}-M_{p-1,q+1})$$
$$\overset{12.15}{\cong} H^{2i}(M_{pq}-N_{pq})\overset{\rho^*}{\cong} H^{2i}N_{pq}=H^{2i}G_{pq}.$$

Clearly $c_0^p=1$ for all p. One often puts $c_i^p=0$ for $i>p$, and one writes $c=c^p=\sum_{i=0}^p c_i^p=\sum_{i=0}^\infty c_i^p=\sum_{i=0}^\infty c_i$; this element of $\oplus_i H^{2i}G_{pq}$ is called the *total Chern class*.

If one associates with every p-space $g\in G_{pq}$ the orthogonal q-space $g^\perp\in G_{qp}$ one gets the *duality homeomorphism* $D: G_{pq}\approx G_{qp}$. The elements $\bar{c}_i^p=D^*c_i^q$ are called *dual Chern classes*. One can show that $c\smile\bar{c}=1$, i.e. $\sum_i c_i\smile\bar{c}_{n-i}=0$ for $n>0$. One can also show that the classes $\{c_i^p\}$ generate the ring H^*G_{pq}; if one uses both $\{c_i^p\}$ and $\{\bar{c}_j^p\}$ to generate H^*G_{pq} then $c\smile\bar{c}=1$ is a system of defining relations. For these facts we refer the reader to Borel 1953, 31.1 (see also Exerc. 3); here we shall only prove the following

12.17 Proposition. *Let $\mathbb{Z}[x_1,x_2,\ldots,x_p]$ denote the graded polynomial ring in generators x_i of dimension $2i$. The ring homomorphism*

$$C: \mathbb{Z}[x_1,x_2,\ldots,x_p]\to H^*(G_{pq}\mathbb{C};\mathbb{Z}),\quad C(x_i)=c_i^p,\quad 1\leq i\leq p,$$

is isomorphic up to (including) dimension $2q$, i.e. up to $2q$ the c_i^p are algebraically independent generators.

Proof by induction on p. The case $p=0$ is clear. Assume then $p>0$, and $\{c_i^{p-1}\}$, for $1\leq i\leq p-1$, are algebraically independent generators of $H^*(G_{p-1,q+1})$. The map $H^jL_{p-1,q+1}\to H^j(L_{p-1,q+1}-M_{p-1,q+1})$ is isomorphic for $j\leq 2q$ because

$$H^j(L_{p-1,q+1},L_{p-1,q+1}-M_{p-1,q+1})\overset{11.22}{\cong} H^{j-(2q+2)}M_{p-1,q+1}=0$$

for $j\leq 2q+1$ (cf. also 12.8(i), applied to $M_{p-1,q+1}\subset L_{p-1,q+1}$), hence the composite

$$\varphi: H^*G_{p-1,q+1}\cong H^*L_{p-1,q+1}$$
$$\to H^*(L_{p-1,q+1}-M_{p-1,q+1})\cong H^*(M_{pq}-N_{pq})$$

is isomorphic up to $2q$. We can therefore define a ring homomorphism $\gamma: H^*(M_{pq}-N_{pq}) \to H^* G_{pq}$ in dimensions $\leq 2q$ by $\gamma\varphi(c_i^{p-1})=c_i^p, 0\leq i<p$, and by definition of c_i^p we have $(\rho^*\gamma)(\varphi c_i^{p-1})=\varphi c_i^{p-1}$, hence $\rho^*\gamma=\mathrm{id}$ in dimensions $\leq 2q$. Now we have only to apply 12.8 (iv), and get

$$H^* G_{pq} \cong H^*(M_{pq}-N_{pq})[x_p] \cong \mathbb{Z}[x_1,\ldots,x_{p-1}][x_p] = \mathbb{Z}[x_1,\ldots,x_p]$$

in dimensions $\leq 2q$, as asserted. ∎

12.18 Exercises. *1.* Put $H^{**}X=\oplus_{i=0}^{\infty} H^i(X;\mathbb{Z})$. In the situation of 12.8, deduce from the Gysin-sequence that

$$\operatorname{rank}(H^{**}N)<\infty \Leftrightarrow \operatorname{rank}(H^{**}(M-N))<\infty$$

(similarly for dimensions if coefficients are taken in a field). In this case, kernel and cokernel of $\chi\smile: H^{**}N \to H^{**}N$ have equal rank, and the Gysin-sequence shows $\ker(\chi\smile)\cong\operatorname{coker}(\rho^*)$, $\operatorname{coker}(\chi\smile)\cong\operatorname{im}(\rho^*)$, hence $\operatorname{rank}\operatorname{im}(\rho^*)=\operatorname{rank}\operatorname{coker}(\rho^*)=\frac{1}{2}\operatorname{rank} H^{**}(M-N)$.—Under the same assumption, the Euler-characteristic $\chi(M-N)$ equals $(1+(-1)^k)\chi(N)$.

2. In the situation 12.8, if $\ker(\chi\smile: H^j(N;\mathbb{Z})\to H^{j+k}(N;\mathbb{Z}))$ is torsionfree for $j>r$ then $H^j(N;\mathbb{Z})$ is torsionfree for $j>r$ (proof: If z is a torsion element of maximal dimension then $\chi\smile z=0$, hence $|z|\leq r$). If also $\operatorname{coker}(\chi\smile: H^j(N;\mathbb{Z})\to H^{j+k}(N;\mathbb{Z}))$ is torsionfree, for $j\geq r$, then $H^i(M-N;\mathbb{Z})$ is torsionfree for $i\geq r+k$ (this uses the Gysin sequence).

3.* Show: *The cohomology ring $H^*(G_{pq}\mathbb{C};\mathbb{Z})$ is generated by the Chern classes $\{c_i^p\}$, and $H^*(G_{pq};\mathbb{Z})$ is a free (abelian) group. Proof* by induction on p: Consider the Gysin sequence of $M_{p-1,q+1}\subset L_{p-1,q+1}$, and in it the ring homomorphism $\bar\rho^*: H^{**}M\to H^{**}(L-M)$; since $\{c_i^{p-1}\}_{i<p}$ generate $H^{**}G_{p-1,q+1}$, their images $\{\bar\rho^* c_i^{p-1}\}$ generate $\operatorname{im}(\bar\rho^*)$; since $\operatorname{coker}(\bar\rho^*)\cong\ker(\chi_M^L\smile)$ is free, $H^{**}(L-M)\cong\operatorname{im}(\bar\rho^*)\oplus\operatorname{coker}(\bar\rho^*)$. In the Gysin sequence of $N_{pq}\subset M_{pq}$, the map

$$\rho^*: H^{**}N_{pq}\to H^{**}(M_{pq}-N_{pq})\cong H^{**}(L_{p-1,q+1}-M_{p-1,q+1})$$

takes c_i^p into $\bar\rho^*(c_i^{p-1})$, by definition of c_i^p, hence $\operatorname{im}(\rho^*)\supset\operatorname{im}(\bar\rho^*)$. But $\operatorname{rank}\operatorname{im}(\bar\rho^*)=\operatorname{rank}\operatorname{im}(\rho^*)$ by Exerc. 1, hence $\operatorname{im}(\bar\rho^*)=\operatorname{im}(\rho^*)$ because $\operatorname{coker}(\bar\rho^*)$ is free, hence $\{\rho^*(c_i^{p-1})\}_{i<p}$ are ring generators of $\operatorname{im}(\rho^*)$ and $\ker(\chi_N^M\smile)\cong\operatorname{coker}(\rho^*)\cong\operatorname{coker}(\bar\rho^*)$ is free, hence $\{c_i^p\}_{i\leq p}$ are ring generators of $H^*N_{pq}=H^*G_{pq}$ (12.8 (ii)), and H^*N_{pq} is free (Exerc. 2).

4. Let $g_0\in G_{pq}F$, and $\zeta: F^p\cong g_0$ a linear isomorphism. Use 11.6 to show that the map $\zeta_0: (F^p, F^p-0)\to(M_{pq}, M_{pq}-N_{pq})$, $\zeta_0(v)=(g_0,\zeta v)$ (cf. 12.14 for M, N) takes a generator of $H_{dp}(F^p, F^p-0)$ into $\pm$ the transverse class v_N^M (coefficients $\mathbb{Z}$ for $F=\mathbb{C}$ or $\mathbb{H}$, and $\mathbb{Z}_2$ for $F=\mathbb{R}$).

5*. Let $\bar{\rho}\colon M_{pq}-N_{pq}\to G_{p-1,q+1}$ be the map which to every $(g,v)\in M-N$ assigns $v\perp g$, the orthogonal complement of v in g (cf. proof of 12.15). For $F=\mathbb{C}$ the recursive definition of Chern classes can be summarized by $\rho^*(c_i^p)=\bar{\rho}^*(c_i^{p-1})$, $i<p$. For $F=\mathbb{R}$ one can define analogous classes $w_i^p\in H^i(G_{pq}\mathbb{R};\mathbb{Z}_2)$, the *Stiefel-Whitney* classes, by $\rho^*(w_i^p)=\bar{\rho}^*(w_i^{p-1})$, $i<p$, and $w_p^p=$Euler class of $N_{pq}\subset M_{pq}$. However, for $i=p-1$ there is a difficulty: ρ^* might not be surjective in dimension $p-1$ (it is injective!); one has to prove therefore that $\bar{\rho}^*\,w_{p-1}^{p-1}\in\operatorname{im}(\rho^*)$. The Gysin sequence of $N_{pq}\subset M_{pq}$ shows that this is equivalent with the vanishing of $\delta^*\,\bar{\rho}^*\,w_{p-1}^{p-1}\in H^p(M_{pq},M_{pq}-N_{pq})$. This group is generated by the Thom class τ_N^M, hence $\delta^*\,\bar{\rho}^*\,w_{p-1}^{p-1}=\lambda\tau$, and $\lambda=\langle\delta^*\,\bar{\rho}^*\,w_{p-1}^{p-1},v_N^M\rangle=\langle w_{p-1}^{p-1},\bar{\rho}_*\,\partial_*\,v_N^M\rangle$. By Exerc. 4, $\partial_*\,v$ is represented by the unit sphere S^{p-1} of $g_0\in G_{pq}$; on this sphere $\bar{\rho}$ satisfies $\bar{\rho}(x)=\bar{\rho}(-x)$, hence $\bar{\rho}|S^{p-1}$ factors through the identification map $S^{p-1}\to P_{p-1}\mathbb{R}$, hence $\bar{\rho}_*\,\partial_*\,v=0$, hence $\lambda=0$.

After this extra argument the theory of Stiefel-Whitney classes is parallel to that for Chern classes. In particular, the same proof as for 12.17 and Exerc. 3 shows that *the ring $H^*(G_{pq}\mathbb{R};\mathbb{Z}_2)$ is generated by $\{w_i^p\}_{i\le p}$, and that these classes are algebraically independent (over $\mathbb{Z}_2$) in dimensions $\le q$.*

6*. Let $\bar{V}_{pq}\mathbb{R}$ be obtained from the Stiefel manifold $V_{pq}\mathbb{R}$ by identifying p-frames if their vectors differ only by sign, $(v_1,\ldots,v_p)\sim(\pm v_1,\ldots,\pm v_p)$. Let $\bar{M}_{pq}=\bar{V}_{pq}\times\mathbb{R}^{p+q}$, $\bar{N}_{pq}\subset\bar{M}_{pq}$ be obtained similarly from the manifolds $M_{pq}=V_{pq}\times\mathbb{R}^{p+q}$, $N_{pq}\subset M_{pq}$ which occur in the proof of 12.10 (with $\mathbb{C}$ replaced by $\mathbb{R}$). Show as in Exerc. 4 that the identification map π takes the transverse class v_N^M into $v_{\bar{N}}^{\bar{M}}$ (coefficients $\mathbb{Z}_2$), and therefore takes $\chi_{\bar{N}}^{\bar{M}}$ into χ_N^M. On the other, show that $\pi^*\colon H^q(\bar{N};\mathbb{Z}_2)\to H^q(N;\mathbb{Z}_2)$ is zero (similar to $\delta^*\,\bar{\rho}^*\,v=0$ in Exerc. 5), hence $\chi_N^M=0$. This allows to prove the remarks in 12.11.

7. Show that $G_{pq}\mathbb{R}$ is orientable if and only if $p+q$ is even. Hint: The special orthogonal group $SO(p+q)$ (=isometries of determinant $+1$) acts transitively on $G_{pq}\mathbb{R}$; the isotropy group of any point $g\in G_{pq}$ has two components. Study the isotropy group in a neighborhood G_{pq}^α of g and argue as for $G_{pq}\mathbb{C}$ (at the end of 12.13).

13. Intersection of Homology Classes

If X, Y are subsets of an (oriented) manifold M^n we might hope to express properties of the geometric situation near $X\cap Y$ by an intersection pairing $H_i X\times H_j Y\to H_{i+j-n}(X\cap Y)$ which generalizes the intersection numbers of VII, 4. Examples (Exerc. 3) show that this does not quite work. We can, however, assign to every $\xi\in H_i X$, $\eta\in H_j Y$ a coherent system of intersection classes $(\xi\bullet\eta)_U\in H_{i+j-n}\,U$, where U ranges over all neighborhoods of $X \cap Y$, i.e. we can define an intersection pairing $H_k X \times H_j Y \to \varprojlim \{H_{k+j-n} U\}$. If $X \cap Y$ is an ENR we can retract the $(\xi \bullet \eta)_U$ to $H(X \cap Y)$, and hence get an intersection class $\xi \bullet \eta$ in $H_{k+j-n}(X \cap Y)$, as desired.

We begin with open sets, and we shall later proceed to the general case by taking suitable limits. All homology groups will have coefficients in a fixed commutative ring R (of characteristic two if M^n is not oriented).

13.1 Definition. Let $M = M^n$ denote an oriented manifold, and $d\colon M \to M \times M$ the diagonal embedding. For arbitrary open pairs (V, S), (W, T) in M we consider the maps

$$H_i(V,S) \otimes H_j(W,T) \xrightarrow{\times} H_{i+j}(V \times W, S \times W \cup V \times T)$$

(13.2) $$\longrightarrow H_{i+j}(V \times W \cup (M \times M - dM), S \times W \cup V \times T \cup (M \times M - dM))$$

$$\xrightarrow{d_!} H_{i+j-n}(V \cap W, (S \cap W) \cup (V \cap T)),$$

where the transfer $d_!$ is defined as in 10.5. The composite map 13.2 (or the corresponding bilinear map), multiplied by $(-1)^{n(n-j)}$, is called the *intersection pairing*, and is denoted by a heavy dot $\bullet$.

With elements

(13.3) $$\xi \bullet \eta = (-1)^{n(n-|\eta|)} d_!(\xi \times \eta), \quad \xi \in H(V,S), \quad \eta \in H(W,T).$$

On the right side the unmarked arrow of 13.2 (which is induced by inclusion) does not appear; similar abbreviations (omission of inclusion maps) will also be used in other places of this §.

Naturality of $\times$-products (VII, 2.7) and transfer (VIII, 10.9(b)) imply

13.4 Proposition (naturality of $\bullet$ with respect to inclusions).

(a) *If* $i_1\colon (V,S) \to (\tilde V, \tilde S)$, $i_2\colon (W,T) \to (\tilde W, \tilde T)$,

$$i\colon (V \cap W, (S \cap W) \cup (V \cap T)) \to (\tilde V \cap \tilde W, (\tilde S \cap \tilde W) \cup (\tilde V \cap \tilde T)),$$

are inclusion maps of open pairs in M then

$$(i_{1*}\xi) \bullet (i_{2*}\eta) = i_*(\xi \bullet \eta).$$

(b) *If* (V,S), (W,T), ξ, η *are as above, and* $L \subset M$ *is an open set containing* $V \cup W$ *then the two intersection-products* $\xi \underset{M}{\bullet} \eta$ *and* $\xi \underset{L}{\bullet} \eta$ *agree.* ∎

For instance, in (b) we can always take $L = V \cup W$. In 13.4(a) we can take $(\tilde V, \tilde S) = (V, S \cup (V - \overline{V \cap W}))$, $(\tilde W, \tilde T) = (W, T \cup (W - \overline{W \cap V}))$, and we see that the intersection pairing factors

$$H(V,S)\times H(W,T)\to H(V,(V-\overline{V\cap W})\cup S)\times H(W,(W-\overline{V\cap W})\cup T)$$
$$\to H(V\cap W,(S\cap W)\cup(V\cap T)).$$

If U is any neighborhood of $\overline{V\cap W}$ then the middle term is isomorphic, by excision, with

$$H(U\cap V,U\cap[(V-\overline{V\cap W})\cup S])\times H(U\cap W,U\cap[(W-\overline{V\cap W})\cup T]);$$

thus, we may say that *the intersection-product $\xi\bullet\eta$ depends only on the "parts" of ξ,η in U, where U is any neighborhood of $\overline{V\cap W}$* (compare remark after VII, 4.5).

If we substitute the Definition 10.5 of $d_!$ in 13.3 we find that $\xi\bullet\eta=(-1)^{n(n-|\eta|)}(d^*z)\frown o$, where o is a suitable fundamental class, and z, the dual of $\xi\times\eta$, is defined by $z\frown(o\times o)=\xi\times\eta$. If x,y are the duals of ξ,η then $x\frown o=\xi$, $y\frown o=\eta$, hence $(x\times y)\frown(o\times o)=(-1)^{n|y|}(x\frown o)\times(y\frown o)=(-1)^{n|y|}\xi\times\eta$, by VII, 12.17; hence $z=(-1)^{n|y|}x\times y$, and $d^*z=(-1)^{n|y|}x\smile y$ by VII, 8.14. Altogether, *if x,y are Poincaré-duals of $\xi\in H(V,S)$, $\eta\in H(W,T)$ then*

$$\xi\bullet\eta=(x\smile y)\frown o=x\frown(y\frown o)=x\frown\eta. \tag{13.5}$$

These formulas justify the choice of signs in 13.3.—Unfortunately, our argument for 13.5 contains imprecisions: We have omitted several inclusion maps, we never really defined cup-products $x\smile y$ of Čech-classes, and some of the formulas which we applied to products of Čech-classes were only proved for singular cohomology classes. However, by definition of Čech-cohomology, every element in $\check{H}$ is represented by singular cohomology classes, and the elements x,y,z above should be understood as representatives in this sense. The formulas then make sense (some inclusions are still missing, though), and are correct (cf. also Exerc. 1 for a complete formulation of the equation $\xi\bullet\eta=x\frown\eta$).

The formulas 13.5 and the properties of $\smile$- or $\frown$-products imply the following

13.6 Naturality. *If $f\colon M'\to M$ is a map between oriented manifolds, and $(V,S),(W,T)$ are open pairs in M over which f is proper then the transfer maps $f_!$ of* VIII, 10.5 *satisfy*

$$f_!(\xi\bullet\eta)=(f_!\xi)\bullet(f_!\eta),\quad\text{for } \xi\in H(V,S),\ \eta\in H(W,T).$$

13.7 Corollary. *If M',M are oriented manifolds of the same dimension and $f\colon M'\to M$ is a map of degree r* (cf. 4.5) *then $rf_*(\xi'\bullet\eta')=(f_*\xi')\bullet(f_*\eta')$, for* *elements*

$$\xi' \in \operatorname{im}(f_!\colon H(V,S) \to H(f^{-1}V, f^{-1}S)),$$
$$\eta' \in \operatorname{im}(f_!\colon H(W,T) \to H(f^{-1}W, f^{-1}T)).$$

In particular, if $f\colon M' \approx M$ is an orientation preserving (resp. reversing) homeomorphism then $f_*(\xi' \bullet \eta') = (f_* \xi') \bullet (f_* \eta')$ resp. $= -(f_* \xi') \bullet (f_* \eta')$, for arbitrary elements $\xi' \in H(V', S')$, $\eta' \in H(W', T')$ and arbitrary open pairs (V', S'), (W', T') in M'.

Indeed, if $\xi' = f_! \xi$, $\eta' = f_! \eta$ then

$$\begin{aligned} r f_*(\xi' \bullet \eta') &= r f_*((f_! \xi) \bullet (f_! \eta)) = r f_* f_! (\xi \bullet \eta) = r^2 (\xi \bullet \eta) = (r\xi) \bullet (r\eta) \\ &= (f_* f_! \xi) \bullet (f_* f_! \eta) = (f_* \xi') \bullet (f_* \eta'), \end{aligned}$$

the 3rd and 5th equation by 10.10. If $f\colon M' \approx M$, then $r = \pm 1$, hence $f_! = \pm f_*^{-1}$ is isomorphic (10.10), hence the assertion.

13.8 Commutativity. $\xi \bullet \eta = (-1)^{(n-|\xi|)(n-|\eta|)} \eta \bullet \xi$.

13.9 Associativity. $(\xi \bullet \eta) \bullet \zeta = \xi \bullet (\eta \bullet \zeta)$.

13.10 Units. $o \bullet \xi = \xi = \xi \bullet o$, if $\xi \in H(V, S)$, and $o \in H(M, M-K)$ is the fundamental class around some compact set K such that $(V-S) \subset K \subset M$.

13.11 Stability. The following diagram is commutative

$$\begin{array}{ccc}
H(V,S) \otimes H(W,T) & \xrightarrow{\ \bullet\ } & H(V \cap W, (S \cap W) \cup (V \cap T)) \\
 & & \downarrow \partial_* \\
\Big\downarrow {\scriptstyle (\partial_* \otimes \mathrm{id},\, \pm \mathrm{id} \otimes \partial_*)} & & H((S \cap W) \cup (V \cap T), S \cap T) \\
 & & \uparrow {\scriptstyle (i_{1*},\, i_{2*})} \\
HS \otimes H(W,T) \oplus H(V,S) \otimes HT & \xrightarrow{\ \bullet \oplus \bullet\ } & H(S \cap W, S \cap T) \oplus H(V \cap T, S \cap T),
\end{array}$$

where i_1, i_2 are inclusions. More precisely,

$$(13.12) \qquad \partial_*(\xi \bullet \eta) = i_{1*}[(\partial_* \xi) \bullet \eta] + (-1)^{(n-|\xi|)} i_{2*}[\xi \bullet (\partial_* \eta)].$$

13.13 Multiplicativity. *If M^n, $M'^{n'}$ are oriented manifolds, and ξ, η* resp. ξ', η' are homology classes of open pairs in M resp. M' then

$$(\xi \times \xi') \bullet (\eta \times \eta') = (-1)^{(n'-|\xi'|)(n-|\eta|)} (\xi \bullet \eta) \times (\xi' \bullet \eta').$$

In particular, if M, M' are compact with fundamental classes o, o', and $p, p'\colon M \times M' \to M, M'$ are the projection maps then $p_! \xi = \xi \times o'$, $p'_! \eta' = (-1)^{n(n'-|\eta'|)} o \times \eta'$, hence

$$\xi \times \eta' = (-1)^{n(n'-|\eta'|)} (p_! \xi) \bullet (p'_! \eta').$$

This expresses the homology $\times$-products in terms of transfers (VIII, 10) and intersections.

As remarked before, 13.6–13.13 follow from 13.5 and the properties of $\smile$- and $\frown$-products (cf. VII, 8 and 12). We leave the details to the reader but we point out that for the proofs one can assume that the open sets $V, W, \ldots$ in question are *bounded*, all of them contained in one compact set K say (because every homology class $\xi, \eta, \ldots$ is represented by a chain with compact carrier), and then $o = o_K \in H(M, M-K)$. ∎

We now extend the intersection-pairing from open to arbitrary subsets $X, Y, \ldots$ of M. In general, $\xi \bullet \eta$ will no longer be a homology class in $X \cap Y$ but rather a coherent family of homology classes $\{(\xi \bullet \eta)_U\}$, where U ranges over all (open) neighborhoods of $X \cap Y$, i.e., $\xi \bullet \eta$ will be an element of an inverse limit of homology groups. For neighborhood retracts (in particular, for open sets) this limit will turn out to be isomorphic with ordinary homology.

13.14 Definition. If M is a manifold as above, and (X, A) is an arbitrary pair of subsets we define

$$\hat{H}(X, A) = \varprojlim \{H(U, R) \mid (U, R) \text{ neighborhoods of } (X, A) \text{ in } M\},$$

with coefficients in an arbitrary abelian group. An element $\kappa \in \hat{H}_k(X, A)$ is then, by definition, a family $\{\kappa_{UR} \in H_k(U, R)\}$ such that $i_* \kappa_{UR} = \kappa_{\tilde{U}\tilde{R}}$ whenever $(U, R) \subset (\tilde{U}, \tilde{R})$; i = inclusion. In particular, if $\xi \in H(X, A)$, and $\iota^{UR}\colon (X, A) \to (U, R)$ denotes inclusion then $\{\iota_*^{UR} \xi\}$ is such a family, hence a homomorphism

(13.15) $$\iota\colon H(X, A) \to \hat{H}(X, A), \quad (\iota \xi)_{UR} = \iota_*^{UR} \xi.$$

Clearly, every inclusion map $j\colon (X, A) \to (\tilde{X}, \tilde{A})$ induces a homomorphism $j = \hat{H}(j)\colon \hat{H}(X, A) \to \hat{H}(\tilde{X}, \tilde{A})$, and $\hat{H}$ thereby becomes a functor on the category of inclusion maps j. Moreover, $\iota\colon H \to \hat{H}$ is a natural transformation.

13.16 Remarks. For locally compact sets (X, A) one can, as in 6.3, turn $\hat{H}$ into a functor on arbitrary continuous maps (not just inclusions), at least if X lies in some ENR. In particular, $\hat{H}(X, A)$ does not depend then on the ambient manifold M, but only on (X, A); it agrees with *Čech-homology* (proof as for A, 3.11; or by A, 3.16, Exerc. 3). We shall not go into this, mainly because of the following result which reduces $\hat{H}$ to H in many interesting cases.

13.17 Proposition (compare 6.12). *If X and some neighborhood of X in M are ENR's, and if A is a (relatively) open subset of X then*

$$\iota\colon H(X, A) \cong \hat{H}(X, A).$$

More generally, the same conclusion holds if both X and A are ENR's but the proof is more complicated (cf. Exerc. 4).

Proof. We can assume that M itself is an ENR and that X is a retract of M, say $r\colon M \to X$, $ri = \mathrm{id}$ (otherwise we replace M by an open subset). Put $N = r^{-1}A$; then $r\colon (M, N) \to (X, A)$, $i\colon (X, A) \to (M, N)$, $ri = \mathrm{id}$. For every $\zeta \in \hat{H}(X, A) = \varprojlim \{H(U, R)\}$, define $\rho(\zeta) = r_*(\zeta_{MN})$; then $\rho\iota(\xi) = \rho\iota_*^{MN}(\xi) = r_* i_*(\xi) = \xi$, hence $\rho\colon \hat{H}(X, A) \to H(X, A)$ is a left inverse of ι. In order to show that $\iota\rho = \mathrm{id}$ we recall (IV, 8.7) that every open neighborhood U of X contains an open neighborhood V such that $\iota^U(r|V) \simeq i_V^U\colon V \to U$; the homotopy can be chosen to be constant on X (cf. IV, 8.6). If $S \subset V$ denotes the set of points whose deformation path lies within R then S is open, $rS \subset A \subset S \subset R$, and we have the same homotopy of pairs $\iota^{UR}(r|V) \simeq i_{VS}^{UR}\colon (V, S) \to (U, R)$, hence $\iota_*^{UR}(r|V)_* = i_{VS*}^{UR}$. Therefore,

$$\begin{aligned}(\iota\rho(\zeta))_{UR} &= \iota_*^{UR} r_* \zeta_{MN} = \iota_*^{UR} r_* i_{VS*}^{MN} \zeta_{VS} = \iota_*^{UR}(r|V)_* \zeta_{VS} \\ &= i_{VS*}^{UR} \zeta_{VS} = \zeta_{UR}, \qquad \text{or } \iota\rho = \mathrm{id}. \quad \blacksquare\end{aligned}$$

13.18 Definition. Let M be an oriented manifold and $(X, A), (Y, B)$ arbitrary pairs of subsets of M. Consider compact pairs $(X', A') \subset (X, A)$, $(Y', B') \subset (Y, B)$ and a pair (U, R) of open neighborhoods of

$$(X \cap Y, (A \cap Y) \cup (X \cap B)).$$

Choose open pairs (V, S), (W, T) such that

$$(X', A') \subset (V, S); \quad (Y', B') \subset (W, T);$$
$$(V \cap W, (S \cap W) \cup (V \cap T)) \subset (U, R);$$

and consider the composition

$$\begin{aligned}(13.19)\qquad & H(X', A') \times H(Y', B') \to H(V, S) \times H(W, T) \\ & \quad \xrightarrow{\cdot} H(V \cap W, (S \cap W) \cup (V \cap T)) \to H(U, R).\end{aligned}$$

By naturality 13.4, this composite does not depend on the choice of S, T, V, W; we write $(\xi'\bullet\eta')_{UR}$ for the image of $(\xi',\eta')\in H(X',A')\times H(Y',B')$. Also by 13.4, the composite is compatible with inclusion maps of the outer terms, i.e. if $(U,R)\subset(\tilde{U},\tilde{R})$ then $i_*(\xi'\bullet\eta')_{UR}=(\xi'\bullet\eta')_{\tilde{U}\tilde{R}}$, and if $(X',A')\subset(X'',A'')(Y',B')\subset(Y'',B'')$ then $(\xi'\bullet\eta')_{UR}=(\xi''\bullet\eta'')_{UR}$, where ξ'',η'' are the images of ξ',η'. The former means that the family $\{(\xi'\bullet\eta')_{UR}\}$ is an element of $\varprojlim\{H(U,R)\}=\check{H}(X\cap Y,(A\cap Y)\cup(X\cap B))$; the latter means that $(\xi',\eta')\mapsto\{(\xi'\bullet\eta')_{UR}\}$ is a map of the direct system $\{H(X',A')\times H(Y',B')\}$, which is indexed by all compact pairs (X',A'), (Y',B') in (X,A), (Y,B). We can therefore pass to the direct limit; since $\varinjlim\{H(X',A')\}=H(X,A)$, $\varinjlim\{H(Y',B')\}=H(Y,B)$ by 5.22, we obtain a map

$$H(X,A)\times H(Y,B)\xrightarrow{\ \bullet\ }\check{H}(X\cap Y,(A\cap Y)\cup(X\cap B)) \tag{13.20}$$

which we still call the *intersection pairing*, and denote by $\bullet$, i.e. the image of $(\xi,\eta)\in H_i(X,A)\times H_j(Y,B)$ is $\xi\bullet\eta\in\check{H}_{i+j-n}(X\cap Y,(A\cap Y)\cup(X\cap B))$. Note that this is compatible with 13.1 since $\check{H}=H$ on open sets (even neighborhood retracts).

We repeat the definition of $\xi\bullet\eta$ for arbitrary pairs: *Given* $\xi\in H(X,A)$, $\eta\in H(Y,B)$, *choose compact pairs* $(X',A')\subset(X,A)$, $(Y',B')\subset(Y,B)$ *and elements* $\xi'\in H(X',A')$, $\eta'\in H(Y',B')$ *such that* $\xi'\mapsto\xi$, $\eta'\mapsto\eta$; *then* $(\xi\bullet\eta)_{UR}=(\xi'\bullet\eta')_{UR}$ *is the image of* (ξ',η') *under the composition* 13.19.

The general (13.20) and the special (13.2) intersection pairing thus differ only by some homomorphisms which are induced by inclusions. The properties 13.4–13.13 of special intersections therefore generalize, although some of them become more complicated to formulate. Associativity $(\xi\bullet\eta)\bullet\zeta=\xi\bullet(\eta\bullet\zeta)$, for instance, does not even make sense unless $\check{H}\cong H$ on some of the intersections (but one can modify the formulation; cf. Exerc. 6). For stability one has to define connecting homomorphisms $\partial:\check{H}(X,A)\to\check{H}\hat{A}$. We omit this and mention only the following naturality properties.

13.21 Proposition. (a) *If* $i_1:(X,A)\to(\tilde{X},\tilde{A})$, $i_2:(Y,B)\to(\tilde{Y},\tilde{B})$, *are inclusions then*

$$(i_{1*}\xi)\bullet(i_{2*}\eta)=\underset{\frown}{i}(\xi\bullet\eta).$$

(b) *If* $L\subset M$ *is an open set containing* $X\cup Y$ *then*

$$\xi\bullet_M\eta=\xi\bullet_L\eta.$$

(c) *If* M',M *are oriented manifolds of the same dimension, and* $f: M'\approx M$ *is an orientation-preserving resp.-reversing homeomorphism then*

$$f_*(\xi'\bullet\eta')=(f_*\xi')\bullet(f_*\eta'),\quad\text{resp.}\quad=-(f_*\xi')\bullet(f_*\eta').$$

These follow from 13.4 and 13.7. ▮ As explained after 13.4, parts (a) and (b) imply that $\xi \bullet \eta$ *can be computed in any neighborhood of* $\overline{X \cap Y}$.

13.22 Example. Let N_1, N_2 be oriented submanifolds of dimensions n_1, n_2 of an oriented m-manifold M^m, and assume $N = N_1 \cap N_2$ is *a compact* connected orientable manifold of dimension $n = n_1 + n_2 - m$. The intersection $o_N^{N_1} \bullet o_N^{N_2}$ of the fundamental classes

$$o_N^{N_\nu} \in H_{n_\nu}(N_\nu, N_\nu - N; \mathbb{Z}), \quad \nu = 1, 2,$$

is an element of $H_n(N; \mathbb{Z})$—because N is an ENR—hence $o_N^{N_1} \bullet o_N^{N_2} = \mu o_N^N$ with $\mu \in \mathbb{Z}$. Using 11.18 the integer μ can easily be identified with the intersection multiplicity of 11.30 Exerc. 1. Independently of § 11 but in the same spirit as 11.13 we prove

13.23 Proposition. *If* N_1, N_2 *intersect transversally at some point* $P \in N = N_1 \cap N_2$ *then* $o_N^{N_1} \bullet o_N^{N_2} = \pm o_N^N$. *More generally,* $o_K^{N_1} \bullet o_K^{N_2} = \pm o_K^N$ *for every compact set* $K \subset N$. *The sign* $\pm$ *is the same for all* K; *the proof will show how to express it in terms of given orientations.*

Proof. Assume first $K = P$. The intersection $o_P^{N_1} \bullet o_P^{N_2}$ can be computed in any neighborhood U of P (by naturality 13.21 of $\bullet$, using excisions $H(X, X-P) \cong H(U \cap X, U \cap X - P)$). By assumption, there is some neighborhood U of P in which N_1, N_2 look like coordinate subspaces of $\mathbb{R}^m$; more precisely,

$$(U; U \cap N_1, U \cap N_2; P) \approx (\mathbb{R}^{k_2} \times \mathbb{R}^n \times \mathbb{R}^{k_1}; \{0\} \times \mathbb{R}^n \times \mathbb{R}^{k_1}, \mathbb{R}^{k_2} \times \mathbb{R}^n \times \{0\}; \{(0,0,0)\}),$$

with $n_\nu = n + k_\nu$. Therefore, if $o_l \in H_l(\mathbb{R}^l, \mathbb{R}^l - \{0\})$ and $[0] \in H_0\{0\}$ are generators we have to show that

$$(13.24) \qquad ([0] \times o_n \times o_{k_1}) \bullet (o_{k_2} \times o_n \times [0]) = \pm [0] \times o_n \times [0].$$

By naturality 13.21 (a), it suffices to prove this equality for the intersection pairing

$$\begin{aligned} &H_{n+k_1}(\mathring{\mathbb{B}}^{k_2} \times (\mathbb{R}^n, \mathbb{R}^n - \{0\}) \times (\mathbb{R}^{k_1}, \mathbb{R}^{k_1} - \mathbb{B}^{k_1})) \\ &\times H_{n+k_2}((\mathbb{R}^{k_2}, \mathbb{R}^{k_2} - \mathbb{B}^{k_2}) \times (\mathbb{R}^n, \mathbb{R}^n - \{0\}) \times \mathring{\mathbb{B}}^{k_1}) \\ &\xrightarrow{\ \bullet\ } H_n(\mathring{\mathbb{B}}^{k_1} \times (\mathbb{R}^n, \mathbb{R}^n - \{0\}) \times \mathring{\mathbb{B}}^{k_2}). \end{aligned}$$

But here we are dealing with *open* sets in $\mathbb{R}^m$, and we can apply 13.5. The Poincaré-dual of $[0] \times o_n \times o_{k_1}$ resp. $o_{k_2} \times o_n \times [0]$ is represented by a generator $\mu_1 \in H^{k_2}((\mathbb{R}^{k_2}, \mathbb{R}^{k_2} - \{0\}) \times \mathbb{R}^n \times \mathbb{R}^{k_1})$ resp.

$$\mu_2 \in H^{k_1}(\mathbb{R}^{k_2} \times \mathbb{R}^n \times (\mathbb{R}^{k_1}, \mathbb{R}^{k_1} - \{0\})),$$

and $\mu_1 \smile \mu_2$ is a generator of

$$H^{k_1+k_2}((\mathbb{R}^{k_2}, \mathbb{R}^{k_2}-\{0\}) \times \mathbb{R}^n \times (\mathbb{R}^{k_1}, \mathbb{R}^{k_1}-\{0\}))$$

by VII, 9.2. By 13.5, this proves 13.24.

For the general case consider the diagram

$$(13.25)\qquad \begin{array}{ccc}
H_{n_1}(N_1, N_1-P) \times H_{n_2}(N_2, N_2-P) & \xrightarrow{\ \bullet\ } & H_n(N, N-P) \\
\uparrow & & \uparrow\cong \\
H_{n_1}(N_1, N_1-N) \times H_{n_2}(N_2, N_2-N) & \xrightarrow{\ \bullet\ } & H_n(N) \\
\downarrow & & \downarrow \\
H_{n_1}(N_1; N_1-K) \times H_{n_2}(N_2, N_2-K) & \xrightarrow{\ \bullet\ } & H_n(N, N-K),
\end{array}$$

where all vertical arrows are induced by inclusion; they take fundamental classes into fundamental classes. The diagram commutes by naturality of intersections. The upper right vertical arrow is isomorphic because N is connected. The upper square now shows $o_N^{N_1} \bullet o_N^{N_2} = \pm o_N^N$, the lower square thereafter $o_K^{N_1} \bullet o_K^{N_2} = \pm o_K^N$. ∎

13.26 Applications. Since intersection-products are (essentially) Poincaré-dual to $\smile$- or $\frown$-products (13.5) they will not produce more results than the latter (less, in fact, because they are only defined in manifolds). However, they are closer to geometric intuition and therefore possess considerable heuristic value; they often indicate how to turn an intuitive geometric result on intersections into a rigorous one. For instance, if we slightly deform two non-parallel planes in $\mathbb{R}^3$ then the deformed figures will still intersect in a continuum—why?

Intersection-products can also serve to *compute* $\smile$-products in manifolds M. Suppose, for instance, $x, y \in H^* M$ are dual to $\xi, \eta \in HM$, hence $x \smile y$ dual to $\xi \bullet \eta$. If ξ, η have simple representative cycles, or representative cycles in simple subsets X, Y then it may be easy to compute $\xi \bullet \eta$ (using the properties of $\bullet$, or comparing with other manifolds), or at least we can say that $\xi \bullet \eta$ has a representative in (close-by) $X \cap Y$. In particular, ξ, η, might be represented by submanifolds N_1, N_2 which intersect transversally; then $\pm \xi \bullet \eta$ is represented by $N_1 \cap N_2$ (cf. 13.23). For instance, in projective space P_n one easily shows (by CW-decomposition V, 3.5) that $H(P_n)$ is freely generated (mod 2 in the real case) by the homology classes of projective subspaces P_k, $k \le n$. Any two projective subspaces P, P' of the same dimension represent the same homology class, $[P]=[P']$, because one can transform P into P' by a projective transformation φ with $\varphi \simeq \mathrm{id}$. In computing $[P_k]\bullet[P_j]$ one can therefore assume that P_k, P_j are in general position (and therefore intersect transversally), hence $[P_k]\bullet[P_j]=\pm[P_{k+j-n}]$. This determines $\bullet$ in HP_n, and therefore $\smile$ in H^*P_n.

For another application consider an oriented manifold M and a compact orientable submanifold N. We want to know whether N is nulhomologous in M ($i_* o_N = 0$?), and we have the following criterion: *If another oriented submanifold N' of M exists such that N' and N intersect transversally and $N \cap N'$ is not nulhomologous in N' then N is not nulhomologous in M.*

Proof. Suppose $i_* o_N = 0$. Then we can find a bounded open set $\tilde{M} \subset M$ such that $\tilde{\imath}\colon N \subset \tilde{M}$ and $\tilde{\imath}_* o_N = 0$. Take a compact set $K \subset N'$ which contains $N' \cap \tilde{M}$ and consider the diagram

$$\begin{array}{ccc} (HN)\times H(N',N'-K) & \xrightarrow{\tilde{\imath}_*\times \mathrm{id}} & (H\tilde{M})\times H(N',N'-K) \\ \downarrow{\scriptstyle\bullet} & & \downarrow{\scriptstyle\bullet} \\ H(N\cap N') & \xrightarrow{\tilde{\jmath}_*} & H(\tilde{M}\cap N'). \end{array}$$

It commutes by 13.21(a), hence $\tilde{\jmath}_*(o_N \bullet o_K^{N'}) = (\tilde{\imath}_* o_N)\bullet o_K^{N'} = 0$. But $o_N \bullet o_K^{N'} = o_N \bullet o_{N'\cap N}^{N'} = \pm o_{N\cap N'}$, by 13.23; hence $N \cap N'$ bounds in $\tilde{M} \cap N'$ and therefore in N', a contradiction. ∎

13.27 Remark. In VII, 4 we defined intersection numbers $\xi \circ \eta$ for homology classed $\xi \in H(X,A), \eta \in (Y,B)$, such that $|\xi|+|\eta| = n$ and $A \cap Y = \emptyset = X \cap B$. In the present context we can define *intersection numbers* $I(\xi,\eta) \in A$ for such classes by

$$I(\xi,\eta) = \gamma(\xi\bullet\eta)_M = \langle 1, (\xi\bullet\eta)_M\rangle, \tag{13.28}$$

where γ = augmentation. We shall see in a moment (13.29) that this is compatible with VII, 4. We can also define *local* intersection numbers if $X \cap Y$ decomposes; more precisely, if $\{V_l\}$, $l=1,2,\ldots$, are mutually disjoint open sets in M^n such that $(X\cap Y) \subset \bigcup_l V_l$ then $(\xi\bullet\eta)_V \in HV \cong \bigoplus_l HV_l$ has components $(\xi\bullet\eta)_V^l \in HV_l$ whose augmentation values are called the local intersection numbers, $I_l(\xi,\eta) = \gamma(\xi\bullet\eta)_V^l$. Clearly, the global intersection number $I(\xi,\eta)$ is the sum of the local intersection numbers $I_l(\xi,\eta)$.

13.29 Proposition. *Let* $(X,A),(Y,B)$ *be pairs of sets in* $\mathbb{R}^n$ *such that* $A\cap Y=\emptyset=X\cap B$, *and let* $\xi\in H_{n-k}(X,A)$, $\eta\in H_k(Y,B)$. *Recall* (VII, 4) *that* $\xi\circ\eta\in H_n(\mathbb{R}^n,\mathbb{R}^n-\{0\})$, *whereas* $(\xi\bullet\eta)_U\in H_0U$ *for every neighborhood* U *of* $X\cap Y$. *We claim:*

$$\mu\frown(\xi\circ\eta)=(\eta\bullet\xi)_{\mathbb{R}^n},$$

where $\mu=\mu_n\in H^n(\mathbb{R}^n,\mathbb{R}^n-\{0\})$ *is the generator such that* $\langle\mu,o_n\rangle=1$. *In other words,* $\xi\circ\eta=\lambda o_n$, *where* $\lambda\in R$ *is the value of the augmentation on* $\eta\bullet\xi$.

Proof. By definition, $(\xi\bullet\eta)_{\mathbb{R}^n}=(-1)^{n|\eta|}d_!j_*(\xi\times\eta)$ [9], where $d\colon\mathbb{R}^n\to\mathbb{R}^n\times\mathbb{R}^n$ is the diagonal and

$$j\colon(X\times Y,A\times Y\cup X\times B)\to(\mathbb{R}^n\times\mathbb{R}^n,\mathbb{R}^n\times\mathbb{R}^n-d\mathbb{R}^n)$$

is inclusion; therefore

$$d_*(\xi\bullet\eta)_{\mathbb{R}^n}=(-1)^{n|\eta|}d_*d_!j_*(\xi\times\eta)=(-1)^{n|\eta|}\tau\frown j_*(\xi\times\eta),$$

where τ is the Thom-class of d (cf. 11.14). On the other hand, $\xi\circ\eta=(-1)^{|\eta|}i_*^{-1}j_*(\xi\times\eta)$ by VII, 4.14, where

$$i\colon(\mathbb{R}^n,\mathbb{R}^n-\{0\})\to(\mathbb{R}^n\times\mathbb{R}^n,\mathbb{R}^n\times\mathbb{R}^n-d\mathbb{R}^n),\quad ix=(x,0);$$

hence $i_*(\xi\circ\eta)=(-1)^{|\eta|}j_*(\xi\times\eta)$. Now,

$$\begin{aligned}d_*(\eta\bullet\xi)_{\mathbb{R}^n}&=(-1)^{|\xi||\eta|}d_*(\xi\bullet\eta)_{\mathbb{R}^n}=(-1)^{|\eta|}\tau\frown j_*(\xi\times\eta)\\&=\tau\frown i_*(\xi\circ\eta)=i_*(i^*\tau\frown(\xi\circ\eta)).\end{aligned}$$

Since obviously $i_*=d_*\colon H_0\mathbb{R}^n\cong H_0(\mathbb{R}^n\times\mathbb{R}^n)$, this proves $(\eta\bullet\xi)_{\mathbb{R}^n}=i^*\tau\frown(\xi\circ\eta)$. Now let $\xi=o_n$ denote the generator of $H_n(\mathbb{R}^n,\mathbb{R}^n-\mathbb{B}^n)\cong H_n(\mathbb{R}^n,\mathbb{R}^n-\{0\})$, and $\eta=[0]$ the generator of $H_0\mathring{\mathbb{B}}^n\cong H_0\mathbb{R}^n$. Then $(\eta\bullet\xi)_{\mathbb{R}^n}=[0]$ by 13.10, and $\xi\circ\eta=o_n$ by VII, 4.11. It follows that $[0]=i^*\tau\frown o_n$, hence $i^*\tau=\mu$. ∎

13.30 Exercises. *1**. Show that the intersection pairing 13.2 for open pairs in M agrees with the following composition

$$\begin{aligned}&H(V,S)\times H(W,T)\xrightarrow{(\frown o)^{-1}\times\mathrm{id}}\check{H}_c(M-S,M-V)\times H(W,T)\\&\longrightarrow\check{H}(W-(W\cap S)\cup T,W-(W\cap V)\cup T)\times H(W,(W\cap S)\cup T)\\&\xrightarrow{\frown(7.1)}H((W\cap V)\cup T,(W\cap S)\cup T)\overset{\mathrm{exc}}{\cong}H(W\cap V,(W\cap S)\cup(T\cap V)).\end{aligned}$$

[9] We use $d_!$ from §11 here, not 10.5, because we'll apply 11.14.

This is a rigorous formulation of the equation $\xi \bullet \eta = x \frown \eta$ of 13.5. We could have used this composite to define intersection-products but we found it to be more cumbersome than 13.2; in particular, it is not symmetric. On the other hand, it is the unsymmetry which indicates a refinement: we did *not really use that* W, T *are open*: W *can be arbitrary*, T *relatively open in* W (verify this assertion). It follows that in the limit construction 13.18 we can always take $W = Y'$ (and T relatively open in W), and for (U, R) we can take neighborhoods of

$$(X \cap Y, (A \cap Y) \cup (X \cap B)) \quad \textit{in } Y$$

(instead of M), i.e. we arrive at an intersection pairing $H(X, A) \times H(Y, B) \to \varprojlim \{H(U, R)\}$, where (U, R) ranges over all neighborhoods of

$$(X \cap Y, (A \cap Y) \cup (X \cap B)) \quad \textit{in } Y.$$

2. If M^n is an oriented manifold and A is an open subset such that $M - A$ is compact then $H(M, A)$, suitably indexed, is a commutative graded ring (with $\bullet$ as multiplication), having o_{M-A} as unit. If Y is any subset of M, if $B \subset Y$ is relatively open, and $(Y \cap A) \subset B$, then $H(Y, B)$ is an $H(M, A)$-module, with respect to $\bullet$ (as refined in Exerc. 1). Compare this with VII, 8.17.

3. Write $S^1 = \mathbb{R} \cup \{\infty\}$. Let $\Gamma = \{(x, y) \in S^1 \times \mathbb{R} \mid y = \sin(1/x), x \neq 0\} =$ graph of $\sin(1/x)$, and let $Z = \bar{\Gamma}$ its closure in $S^1 \times \mathbb{R}$. Construct a function $f\colon S^1 \times \mathbb{R} \to \mathbb{R}$ such that $f^{-1}(t) = S^1 \times \{t\}$ for $|t| \geq 2$ and $f^{-1}(0) = Z$ (compare 10.14 Exerc. 1). In the manifold $M^3 = S^1 \times \mathbb{R} \times \mathbb{R}$, consider the subspaces $X = S^1 \times \mathbb{R} \times \{0\}$, $Y = \text{graph}(f)$; then $X \cap Y \approx Z$, hence $H_1(X \cap Y) = 0$. On the other hand, if $A = \{(x, y, z) \in X \mid |y| \geq 2\}$, $B = \{(x, y, z) \in Y \mid |y| \geq 2\}$ then $H_2(X, A) \cong \mathbb{Z} \cong H_2(Y, B)$, and the intersection class $(\xi \bullet \eta)_M$ of any two generators is a generator of $H_1 M \cong \mathbb{Z}$. It follows that $(\xi \bullet \eta)_M$ is not in the image of $H_1(X \cap Y) \to H_1 M$.—In order to get an analogous example with $A = \emptyset = B$ one can replace $\mathbb{R}$ by $\mathbb{R}/\{t \mid |t| \geq 2\}$ in the above, i.e., identify the subset $|t| \geq 2$ of $\mathbb{R}$ to a point.

4. Show that the map $\iota\colon H(X, A) \to \hat{H}(X, A)$ of 13.15 is isomorphic if X and A are ENR's. Hint: As in 7.16 Exerc. 3 show that there are open neighborhoods (V, S) of (X, A) and a map $\sigma\colon (V, S) \to (X, A)$ such that the composite $(X, A) \xrightarrow{\subset} (V, S) \xrightarrow{\sigma} (X, A)$ is homotopic to the identity map. In other words, up to homotopy (X, A) is a retract of an open pair. Now use 13.16, 13.17.

5. Let X, Y be subsets of an oriented manifold M which are separated by $X \cap Y$ (cf. 6.13), i.e. such that $X - Y$, $Y - X$ are both open (or both closed) in $(X \cup Y) - (X \cap Y)$. For every open neighborhood U of $X \cap Y$ we can find open neighborhoods V, W of X, Y such that $(V \cap W) \subset U$, hence an intersection pairing $(HV) \times (HW) \to HU$, and by passage to limits

$\diamondsuit\colon (\hat{H}X)\times(\hat{H}Y)\to \hat{H}(X\cap Y)$. Show that the pairing • of 13.20 factors as follows:

$$(HX)\times(HY)\xrightarrow{\iota\times\iota}(\hat{H}X)\times(\hat{H}Y)\xrightarrow{\diamondsuit}\hat{H}(X\cap Y).$$

Generalize to relative homology.

6. Show that the intersection pairing $\diamondsuit$ of Exerc. 5 is associative. For the pairing • of 13.20 associativity does not make sense; instead we have the following (more cumbersome) relation

$$((\xi\bullet\eta)_U\bullet\zeta)_W=(\xi\bullet(\eta\bullet\zeta)_V)_W,$$

which holds for $\xi\in HX, \eta\in HY, \zeta\in HZ$, and open neighborhoods U, V, W of $X\cap Y$, $Y\cap Z$, $X\cap Y\cap Z$ such that $(X\cap V)\subset W$, $(U\cap Z)\subset W$. Verify this assertion and its generalization to relative homology.

Show that $x\frown(\xi\bullet\eta)=(x\frown\xi)\bullet\eta$ for $x\in H^*(X,A_1)$, $\xi\in H(X,A_1\cup A_2)$, $\eta\in H(Y,B)$, and *open* sets X, A_ν, Y, B. Generalize to arbitrary sets.

Appendix: Kan- and Čech-Extensions of Functors

A.1 Limits of Functors

In VIII, 5 we treated limits of directed systems D. We found it convenient then to think of D as a functor. Here we take up that point of view, and discuss the general theory (due to D. Kan, Chap. II) of limits of functors. In more comprehensive treatments, our limits are called colimits (left limits, direct limits) but since we don't use any others we simply speak of "limits".

1.1 Definition. Let $\Lambda, \mathscr{K}$ be categories, and $D: \Lambda \to \mathscr{K}$ a functor. Any object $K \in \mathscr{K}$ defines a constant functor $\Lambda \to \mathscr{K}$ which we denote by the same letter K. A (natural) transformation $\varphi: D \to K$ is then a family $\{\varphi_\lambda: D\lambda \to K\}_{\lambda \in \Lambda}$ of morphism in $\mathscr{K}$ such that $\varphi_\lambda = \varphi_\mu(D\alpha)$, for all morphisms $\alpha: \lambda \to \mu$ in Λ. Such a transformation $u: D \to L$, where $L \in \mathscr{K}$, is called *universal* if for every transformation $\varphi: D \to K$ there is a unique morphism $\psi: L \to K$ such that $\varphi = \psi u$. In formulas,

$$(1.2) \qquad \mathscr{K}(L, K) \approx \mathrm{Transf}(D, K), \qquad \psi \mapsto \psi u.$$

If $u: D \to L$, $u': D \to L'$ are two universal transformations then there is a unique equivalence $\kappa: L \approx L'$ such that $u' = \kappa u$, i.e. universal transformations are essentially unique (proof as for VIII, 5.5). They may not exist; if they do then L is called the *limit of* D; in symbols, $L = \mathit{lim}(D)$. If $\varphi: D \to K$ is a transformation we also write $\varphi: \lim(D) \to K$ for the corresponding morphism.

We can consider all functors D with range $\mathscr{K}$ whose limit exists. Then "lim" should be a $\mathscr{K}$-valued functor on a category with objects $\{D\}$ and as many morphisms as possible. We define these morphisms now; there are too many of them, however, to form a category in the usual sense.

1.3 Definition. Let $D: \Lambda \to \mathscr{K}$, $D': \Lambda' \to \mathscr{K}$ be functors. Consider pairs (γ, d), where γ is a function which assigns to every object $\lambda \in \Lambda$ an object

$\gamma(\lambda)\in\Lambda'$, and $d=\{d_\lambda\colon D(\lambda)\to D'(\gamma\lambda)\}_{\lambda\in\Lambda}$ is a family of morphisms. We say (γ, d) *passes to the limit* if the composition of d with any transformation $\varphi'\colon D'\to K$, $K\in\mathscr{K}$, is a transformation $\varphi' d\colon D\to K$, i.e. if the compositions $\{D\lambda\xrightarrow{d_\lambda} D'(\gamma\lambda)\xrightarrow{\varphi'_{\gamma\lambda}} K\}_{\lambda\in\Lambda}$ form a transformation, for every $\varphi'\in\mathrm{Transf}(D', K)$, $K\in\mathscr{K}$. If D has a limit, say $u\colon D\to\lim(D)$, then $\varphi' d$ defines a unique morphism $\psi\colon \lim(D)\to K$ such that $\psi u=\varphi' d$ (by Definition 1.1); in particular, if also D' has a limit, $u'\colon D'\to\lim(D')$, then a unique morphism $\lim(d)\colon \lim(D)\to\lim(D')$ exists such that $\lim(d)\, u=u' d$.

1.4 Lemma. *Let (γ, d) be a pair as in 1.3. Assume that for every morphism $\alpha\colon\lambda\to\mu$ in Λ there exist morphisms $\gamma\lambda\xrightarrow{a}\rho\xleftarrow{b}\gamma\mu$ in Λ' such that*

(1.5)

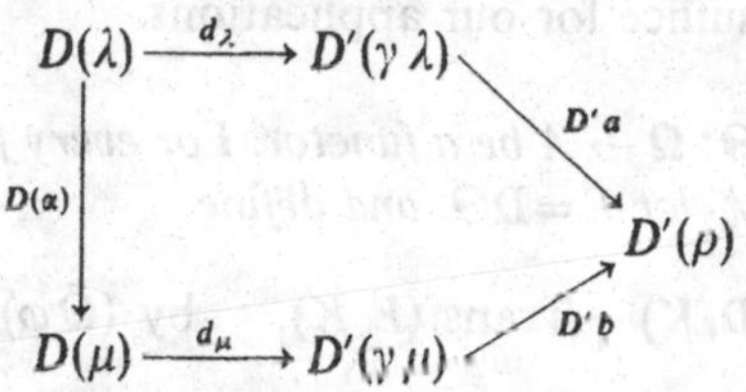

commutes. Then (γ, d) passes to the limit.

Indeed, if $\varphi'\colon D'\to K$ is a transformation then $\varphi=\varphi' d$ satisfies

$$\varphi_\lambda=\varphi'_{\gamma\lambda}\, d_\lambda=\varphi'_\rho(D' a)\, d_\lambda=\varphi'_\rho(D' b)\, d_\mu(D\alpha)=\varphi'_{\gamma\mu}\, d_\mu(D\alpha)=\varphi_\mu(D\alpha),$$

i.e. φ is a transformation. ∎

1.6 Proposition. *Let $D\colon\Lambda\to\mathscr{K}$, $D'\colon\Lambda'\to\mathscr{K}$, $D''\colon\Lambda''\to\mathscr{K}$ be functors, and (γ, d), (γ', d') pairs as in 1.3 ($\gamma\colon\Lambda\to\Lambda'$, $\gamma'\colon\Lambda'\to\Lambda''$, $d_\lambda\colon D\lambda\to D'(\gamma\lambda)$, $d'_{\lambda'}\colon D'(\lambda')\to D''(\gamma'\lambda')$). If both (γ, d) and (γ', d') pass to the limit then so does $(\gamma', d')(\gamma, d)=(\gamma'\gamma, d' d)$, where $(d' d)_\lambda=d'_{\gamma\lambda}\, d_\lambda$. If, moreover, D, D', D'' have limits then $\lim(d' d)=(\lim d')(\lim d)$.*

The proof is as for VIII, 5.12. ∎

For instance, every natural transformation d between functors $D, D'\colon\Lambda\to\mathscr{K}$ passes to the limit ($\gamma=\mathrm{id}$; compare VIII, 5.13). If $\Theta\colon\Omega\to\Lambda$ is a functor then Θ composes with every $D\colon\Lambda\to\mathscr{K}$ to give $E=D\Theta\colon\Omega\to\mathscr{K}$, and the identity morphisms $d_\omega\colon E\omega\to D(\Theta\omega)$, $\omega\in\Omega$, pass to the limit ($\gamma=\Theta$, $d_\omega=\mathrm{id}$; compare VIII, 5.15). As in VIII, 5.15, we write $\Theta_\infty\colon\lim E\to\lim D$ instead of $\lim(d)$ in this situation. Generalizing VIII, 5.16 we define

1.7 Definition. A functor $\Theta\colon\Omega\to\Lambda$ is called *weakly cofinal* if every $\lambda\in\Lambda$ admits a morphism $\lambda\to\Theta(\omega)$, for some $\omega\in\Omega$. It is called *strongly cofinal* if, moreover, every pair of morphisms $\Theta\omega_1\leftarrow\lambda\to\Theta\omega_2$ (where

$\omega_i \in \Omega$) can be completed to a commutative diagram

(1.8) 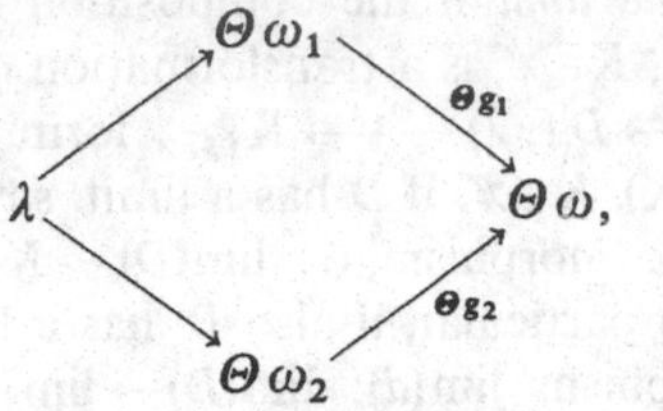

where $g_i\colon \omega_i \to \omega$ are morphisms in Ω. A subcategory Ω of Λ is called (weakly, strongly) cofinal if the inclusion functor is so.

For finer forms of cofinality cf. Exerc. 2. We have chosen the crudest versions here which suffice for our applications.

1.9 Proposition. *Let $\Theta\colon \Omega \to \Lambda$ be a functor. For every functor $D\colon \Lambda \to \mathcal{K}$ and every object $K \in \mathcal{K}$ let $E = D\Theta$, and define*

(1.10) $\hat{\Theta}\colon \mathrm{Transf}(D, K) \to \mathrm{Transf}(E, K)$, by $(\hat{\Theta}\varphi)_\omega = \varphi_{\Theta\omega}$,

i.e. the map $\hat{\Theta}$ assigns to every transformation $\varphi\colon D \to K$ the transformation $\psi = \hat{\Theta}(\varphi)\colon E \to K$ such that $\psi_\omega = \varphi_{\Theta\omega}$.

(i) *If Θ is weakly cofinal then $\hat{\Theta}$ is injective.*

(ii) *If Θ is strongly cofinal then $\hat{\Theta}$ is bijective. Moreover, $u\colon D \to L$ is universal if and only if $v = \hat{\Theta}(u)\colon E \to L$ is universal, hence $\Theta_\infty\colon \lim(E) \cong \lim(D)$, if one of these limits exists.*

Proof. Assume Θ is weakly cofinal, and let $\varphi \in \mathrm{Transf}(D, K)$ a transformation. Every $\lambda \in \Lambda$ admits a morphism $f\colon \lambda \to \Theta\omega$, hence $\varphi_\lambda = \varphi_{\Theta\omega}(Df) = (\hat{\Theta}\varphi)_\omega(Df)$. This expresses φ in terms of $\hat{\Theta}\varphi$. It shows that $\hat{\Theta}$ is injective, and it also indicates how to construct an inverse I of $\hat{\Theta}$, namely as follows. For every transformation $\psi \in \mathrm{Transf}(E, K)$ we should define $I\psi \in \mathrm{Transf}(D, K)$ by $(I\psi)_\lambda = \psi_\omega(Df)$. In general, this will depend on the choice of (ω, f), but it doesn't if Θ is strongly cofinal. Because then, if $f_1\colon \lambda \to \Theta\omega_1, f_2\colon \lambda \to \Theta\omega_2$ are two choices we can find $g_1\colon \omega_1 \to \omega \leftarrow \omega_2\colon g_2$ such that $(\Theta g_1) \circ f_1 = (\Theta g_2) \circ f_2$, hence

$$\psi_{\omega_i} \circ (Df_i) = \psi_\omega \circ (Eg_i) \circ (Df_i) = \psi_\omega \circ (D\Theta g_i) \circ (Df_i) = \psi_\omega \circ D((\Theta g_i) \circ f_i),$$

and this does not depend on i. We have to show that $I\psi = \{(I\psi)_\lambda\}_{\lambda \in \Lambda}$ is a transformation, and that I is inverse to $\hat{\Theta}$.

Let $e\colon \mu \to \lambda$ a morphism in Λ, and $f\colon \lambda \to \Theta\omega$ as above. Then

$$(I\psi)_\mu = \psi_\omega \circ D(f \circ e) = \psi_\omega \circ (Df) \circ (De) = (I\psi)_\lambda \circ (De),$$

showing that $I\psi$ is a transformation. Furthermore,

$$(I\hat{\Theta}\varphi)_\lambda=(\hat{\Theta}\varphi)_\omega\circ(Df)=\varphi_{\Theta\omega}\circ(Df)=\varphi_\lambda,$$

and

$$(\hat{\Theta}I\psi)_\omega=(I\psi)_{\Theta\omega}=\psi_\omega\circ D(\mathrm{id})=\psi_\omega,$$

hence $I\hat{\Theta}=\mathrm{id}$, $\hat{\Theta}I=\mathrm{id}$. The last assertion of 1.9 (ii) (universality) follows as in VIII, 5.17. ∎

When do limits of functors $\Lambda\to\mathscr{K}$ exist? If the category Λ is *small* (i.e. the class of objects is a set) then one has similar criterions as in the case of quasi-ordered sets (compare VIII, 5.7). If it is not small one might still be able to use the following generalization of VIII, 5.7.

1.11 Proposition. *If Λ is an arbitrary category, Ω a small category, and $\Theta\colon\Omega\to\Lambda$ a weakly cofinal functor then every functor $D\colon\Lambda\to\mathscr{AG}$ (=category of abelian groups, or modules, or complexes ...) admits a limit, namely the quotient of $\oplus_{\nu\in\Omega}D(\Theta\nu)$ by the subgroup (-module, -complex ...) which is generated by all elements of the form $(\iota_{\omega_1}(Df_1)-\iota_{\omega_2}(Df_2))(x_\lambda)$, where $\iota_\omega=$ inclusion, $x_\lambda\in D(\lambda)$, $\lambda\in\Lambda$, $\omega_1,\omega_2\in\Omega$, and $\Theta\omega_1\xleftarrow{f_1}\lambda\xrightarrow{f_2}\Theta\omega_2$ are morphisms in Λ.*

Proof. Let $L=\oplus_{\nu\in\Omega}D(\Theta\nu)/\{\iota_{\omega_1}(Df_1)\,x_\lambda-\iota_{\omega_2}(Df_2)\,x_\lambda\}$. Every $\lambda\in\Lambda$ admits a morphism $f\colon\lambda\to\Theta\omega$, for some $\omega\in\Omega$. Let $u_\lambda\colon D\lambda\to L$ be the composition

$$D\lambda\xrightarrow{Df}D(\Theta\omega)\xrightarrow{\iota_\omega}\oplus_{\nu\in\Omega}D(\Theta\nu)\xrightarrow{p}L,$$

where $p=$projection. This does not depend on (ω,f); for if $\Theta\omega_1\xleftarrow{f_1}\lambda\xrightarrow{f_2}\Theta\omega_2$ are two choices then $\iota_{\omega_1}(Df_1)-\iota_{\omega_2}(Df_2)$ maps $D\lambda$ into the kernel of p. Furthermore, $u=\{u_\lambda\}$ is a transformation, for if $g\colon\mu\to\lambda$ is a morphism in Λ then $u_\lambda(Dg)=p\,\iota_\omega(Df)(Dg)=p\,\iota_\omega D(fg)=u_\mu$. We assert that u is universal. Indeed, if $\varphi_\lambda\colon D\lambda\to K$ is a transformation, then clearly $\psi'=\{\varphi_{\Theta\nu}\}_{\nu\in\Omega}\colon\oplus_{\nu\in\Omega}D(\Theta\nu)\to K$ is a morphism such that $\psi'\iota_\omega(Df)=\varphi_{\Theta\omega}(Df)=\varphi_\lambda$, and $\psi'|\ker(p)=0$; hence ψ' passes to the quotient L and induces $\psi\colon L\to K$ such that $\psi u_\lambda=\psi'\iota_\omega(Df)=\varphi_\lambda$. This ψ is unique because the images of u_λ generate L (the images of $u_{\Theta\nu}$ generate already). ∎

1.12 Exercises. *1.* Let π be a group, and think of π as a category with a single object e whose endomorphisms are the elements of π. Show that a functor $D\colon\pi\to\mathscr{AG}$ is the same as a left operation of π on an abelian group $G=De$, and that $\lim D=G_\pi=$quotient of G by the subgroup generated by $\{g-(Dx)\,g\}$, $g\in G$, $x\in\pi$.

2. Our notion 1.7 of strong cofinality is rather crude: even the identity functor $\Theta = \mathrm{id}\colon \Lambda \to \Lambda$ may not be strongly cofinal, although Θ_∞ is (trivially) isomorphic. Clearly, $\Theta\colon \Omega \to \Lambda$ should be called *cofinal* if $\Theta_\infty\colon \lim(D\Theta) \cong \lim(D)$ for every functor $D\colon \Lambda \to \mathscr{K}$, and arbitrary $\mathscr{K}$. Show that Θ is cofinal if (i) every λ admits $\lambda \to \Theta\omega$, and (ii) every pair of morphisms $\Theta\omega \xleftarrow{\beta} \lambda \xrightarrow{\beta'} \Theta\omega'$ can be connected by a commutative diagram

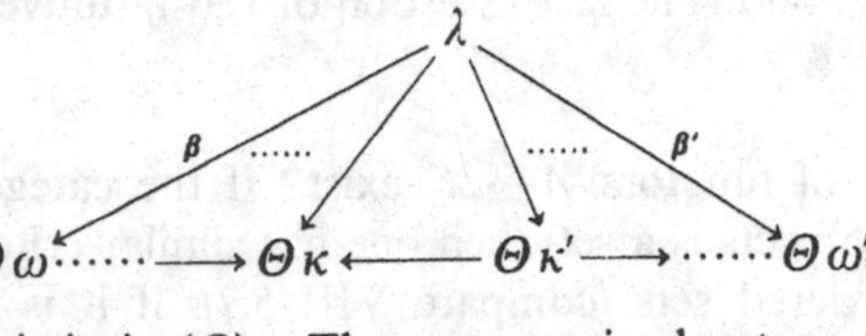

in Λ whose row is in $\mathrm{im}(\Theta)$.—The converse is also true (cf. MacLane 1972, IX.3).

3. A category Λ is said to be *directed* if (i) every pair of objects λ_1, λ_2 admits morphisms $\lambda_1 \to \mu \leftarrow \lambda_2$, and (ii) every pair of morphisms $\alpha_1, \alpha_2\colon \lambda \to \mu$ is equalized by some $\beta\colon \mu \to \nu$, i.e. $\beta\alpha_1 = \beta\alpha_2$ (compare Verdier, 2.7).
This generalizes the notion of a directed set in such a way that the exactness properties VIII, 5.18–5.20 extend to limits of functors $\Lambda \to \mathscr{AG}$.

A.2 Polyhedrons under a Space, and Partitions of Unity

We shall be concerned with extending functors from polyhedrons to more general spaces. Following D. Kan, we can obtain extensions by taking limits of functors from the category of polyhedrons under a space; we deduce the relevant properties of such categories. Following E. Čech, we can obtain extensions by taking limits of direct systems which are indexed by sets of coverings (cf. Eilenberg-Steenrod, IX); the connection between the two methods is provided by partitions of unity (numerations).

2.1 Definitions. For every topological space A we define the category $\mathscr{P}^A$ of polyhedrons under A as follows. An object of $\mathscr{P}^A$ (a *polyhedron under A*) is a homotopy class of maps $\xi\colon A \to R_\xi$ whose range R_ξ is a polyhedron. A morphism from ξ to η is a homotopy class of maps $\alpha\colon R_\xi \to R_\eta$ such that $\alpha\xi \simeq \eta$. Composition is given by composing maps α.

This construction $\mathscr{P}^A$ *is functorial in A*, i.e. every map $f\colon B \to A$ induces a functor

(2.2) $$\mathscr{P}^f\colon \mathscr{P}^A \to \mathscr{P}^B, \quad [\xi] \mapsto [\xi f], \quad [\alpha] \mapsto [\alpha],$$

where brackets as usual denote homotopy classes. Clearly $\mathscr{P}^{fg}=(\mathscr{P}^g)(\mathscr{P}^f)$, and $\mathscr{P}^{\mathrm{id}}=\mathrm{Id}$, *so* $A\mapsto\mathscr{P}^A$ *is a cofunctor from spaces to categories.*

Rather than $\mathscr{P}^A$ we shall use the dual category which we denote by $\Lambda=\Lambda_A$, so $\Lambda_A^{\mathrm{op}}=\mathscr{P}^A$.

2.3 Proposition. *If A is a polyhedron then Λ_A has a strongly cofinal subcategory Ω consisting of a single object, namely* $\mathrm{id}\colon A\to A$, *and a single morphism,* id_A.

If A is dominated by a polyhedron, say $A\xrightarrow{i}P\xrightarrow{r}A$, with $ri\simeq\mathrm{id}$, then Λ_A has a strongly cofinal subcategory Ω consisting of a single object, namely $A\xrightarrow{i}P$, and two morphisms, id_P *and* ir.

Proof. In the first case, every object $\xi\colon A\to R_\xi$ of Λ admits a unique morphism into $\mathrm{id}\colon A\to A$, namely ξ (remember to reverse arrows since $\mathscr{P}^A=\Lambda_A^{\mathrm{op}}$!). In the second case, $\xi\colon A\to R_\xi$ admits the morphism ξr into $i\colon A\to P$, and any two morphisms of $\xi\colon A\to R_\xi$ into $i\colon A\to P$ become equal after composition with ir. ∎

2.4 Proposition. *Let A be a locally closed subset of a polyhedron P, and let Ω denote the set of all open neighborhoods of A in P, directed by reversed inclusion. As a category, Ω is dual to the category of inclusion maps of open neighborhoods of A in P. We have a functor*

(2.5) $$\Theta\colon\Omega\to\Lambda_A,\qquad \Theta V=[A\xrightarrow{\mathrm{incl}}V],\qquad \Theta\iota=[\iota],$$

where V denotes open neighborhoods of A, and ι inclusion maps between such neighborhoods. This functor Θ is strongly cofinal.

Proof. Recall first that open subsets of polyhedrons are polyhedrons[10] so that ΘV is indeed a polyhedron under A, and Θ is a functor from Ω to Λ. Since A is locally closed there is an open neighborhood Q such that A is closed in Q. Now if $[\xi]$ is an object in Λ we can, by Tietze's extension lemma, extend the map $\xi\colon A\to R_\xi$ to an open neighborhood V of A in Q, i.e. we can factor ξ as $A\xrightarrow{\subset}V\to R_\xi$; this proves weak cofinality. Suppose then we have two such factorizations, i.e. two maps $\eta_1\colon V_1\to R_\xi\leftarrow V_2\colon\eta_2$ of open neighborhoods V_i which on A agree (up to $\simeq$) with ξ. Again from Tietze's extension lemma we get (cf. proof of VIII, 6.2(b)) an open neighborhood $V\subset(V_1\cap V_2)$ on which η_1,η_2 are homotopic; this proves strong cofinality. ∎

[10] For polyhedrons in $\mathbb{R}^n$ see Alexandroff-Hopf III, 3.2. A general proof follows Spanier, p. 149, Exerc. 3, using Whitehead's (1939) subdivision theorem 35.

We are going to compare Λ_A to the usual Čech category of open coverings of A but we need some preliminaries first. Recall that a family of continuous functions, $\pi=\{\pi_j\colon A\to[0,1]\}_{j\in J}$, is called *locally finite* if every point $a\in A$ has a neighborhood V such that $\pi_j|V=0$ for all but finitely many j. It is *point-finite* if for every $a\in A$ the set $\{j\,|\,\pi_j(a)\neq 0\}$ is finite. It is a *partition of unity* if $\sum_j \pi_j(a)=1$, for every $a\in A$; in particular, the set $\{j\,|\,\pi_j(a)\neq 0\}$ must then be countable, for every $a\in A$.

2.6 Lemma. *If $\pi=\{\pi_j\}_{j\in J}$ is a partition of unity (not necessarily point-finite) and $\varepsilon>0$ then every point $a\in A$ has a neighborhood in which only finitely many π_j have values $\geq\varepsilon$.*

Proof. Given $a\in A$, we can choose a finite subset $F\subset J$ such that $\sum_{j\in F}\pi_j(a)>1-\varepsilon$. Let $V=\{x\in A\,|\,\sum_{j\in F}\pi_j(x)>1-\varepsilon\}$. This is a neighborhood of a in which π_j can assume values $\geq\varepsilon$ only if $j\in F$ (because $\sum_{j\in J}\pi_j=1$). ∎

2.7 Corollary. *If $\pi=\{\pi_j\}_{j\in J}$ is a partition of unity then*

$$\mu(x)=\mathrm{Sup}_{j\in J}\{\pi_j x\}=\mathrm{Max}_{j\in J}\{\pi_j x\}$$

is a continuous function, and $\mu(x)>0$.

Proof. Since $\sum_j \pi_j(x)=1$, we must have $\mu(x)>0$. By 2.6, $\mathrm{Sup}_{j\in J}\{\pi_j\}$ agrees, locally, with the maximum of finitely many among the π_j, and is therefore continuous. ∎

2.8 Proposition (Mather). *If $\pi=\{\pi_j\}_{j\in J}$ is a partition of unity (not necessarily point-finite) then there exists a locally finite partition of unity $\rho=\{\rho_j\}_{j\in J}$ such that $\rho_j^{-1}(0,1]\subset\pi_j^{-1}(0,1]$*, for all $j\in J$. (Any such ρ will be called an *improvement* of π.)

Proof. Let $\sigma_j(x)=\mathrm{Max}(0,2\pi_j(x)-\mu(x))$, where $\mu(x)=\mathrm{Max}_{j\in J}\{\pi_j(x)\}$, as in 2.7. Then σ_j is continuous, and $\sigma_j^{-1}(0,1]\subset\pi_j^{-1}(0,1]$. Let $a\in A$ and $\varepsilon=\frac{1}{4}\mu(a)$. By 2.7 and 2.8 we can find a neighborhood V of a and a finite set $F\subset J$ such that $\mu(x)>2\varepsilon$ and $\pi_j(x)<\varepsilon$ for $x\in V$, $j\notin F$; hence $\sigma_j(x)=0$ for $x\in V$, $j\notin F$; hence $\{\sigma_j\}_{j\in J}$ is locally finite. On the other hand, $\pi_k(a)=\mu(a)$ for some $k\in J$, hence $\sigma_k(a)=\pi_k(a)=\mu(a)>0$, hence $\sum_{j\in J}\sigma_j(a)>0$ for all $a\in A$. Therefore $\rho_j(x)=\sigma_j(x)/\sum_{i\in J}\sigma_i(x)$ is a partition, as required. ∎

2.9 Corollary [Dowker, §16, Thm. 1]. *If A is a simplicial space and if A' is obtained from A by taking the strong topology* (V, 7.14) *then the identity map $\iota\colon A\to A'$ is a homotopy equivalence. In fact, a homotopy* *inverse* κ: $A' \to A$ *and deformations,* $\kappa\iota \simeq \mathrm{id}$, $\iota\kappa \simeq \mathrm{id}'$ *can be so chosen that no point leaves the closure of its carrier simplex.*

Proof. The barycentric coordinates π_j: $A' \to [0,1]$, $j \in J =$ set of vertices, constitute a (point finite) partition of unity π (cf. V, 7.13). By 2.8, π admits a locally finite improvement ρ. Consider the map κ: $A' \to A$ which is defined by $\pi_j \iota \kappa = \rho_j$. It is continuous, because $\iota\kappa$ is continuous (due to the strong topology), and every point has a neighborhood whose image lies in a finite (i.e. compact) simplicial subspace (on these the topologies of A, A' agree). The deformation $x \mapsto x_t$, defined by

$$\pi_j(x_t) = t\,\rho_j(x) + (1-t)\,\pi_j(x),$$

is continuous as a map D': $A' \times [0,1] \to A'$ because the $\pi_j D'$ are continuous, and it is continuous as a map D: $A \times [0,1] \to A$ because $D|X \times [0,1]$ is continuous for every finite simplicial subspace $X \subset A$. Hence D: $\kappa\iota \simeq \mathrm{id}$, D': $\iota\kappa \simeq \mathrm{id}$. ∎

2.10 Definitions. Let A be a topological space. A covering $\mathscr{U}$ of A is called *numerable* if a partition of unity $\pi = \{\pi_U\}_{U \in \mathscr{U}}$ exists such that $\pi_U^{-1}(0,1] \subset U$ for every $U \in \mathscr{U}$. We say, π is a *numeration* of $\mathscr{U}$, provided it is also point-finite. By 2.8, *if* $\mathscr{U}$ *is numerable then it even admits a locally finite numeration.*

The set Ω_A of all numerable coverings of A is directed by *refinement* (recall that $\mathscr{U} \geq \mathscr{V}$ if every $U \in \mathscr{U}$ is contained in some $V \in \mathscr{V}$). As usual, we think of Ω_A as a *category*.—For instance, if A is paracompact then every open covering is numerable, if A is normal then every locally finite open covering is numerable.

The *nerve* $\nu\mathscr{U}$ of a numerable covering $\mathscr{U}$ is a simplicial space (cf. V, 7) whose n-simplices correspond to $(n+1)$-tuples $(U_0, U_1, \ldots, U_n)$ such that $U_j \in \mathscr{U}$ and $\bigcap_j U_j \neq \emptyset$; in particular, the vertices of $\nu\mathscr{U}$ are just the non-empty sets $U \in \mathscr{U}$ (actually, this is the description of $\nu\mathscr{U}$ by its vertex schema; cf. V, 7.15). If we take $\nu\mathscr{U}$ with the strong topology then *a numeration* π *of* $\mathscr{U}$ *is the same as a map* π: $A \to \nu\mathscr{U}$ *such that* $\pi(U)$ *is contained in the open star of the vertex* U (cf. V, 7, Exerc. 4), namely, π maps $a \in A$ into the point whose barycentric coordinates are $\{\pi_U(a)\}$. If we take $\nu\mathscr{U}$ with the weak topology, as we normally do, then π: $A \to \nu\mathscr{U}$ need not be continuous. It is continuous, however, if $\{\pi_U\}$ is locally finite because then every $a \in A$ has a neighborhood whose image under π lies in a finite simplicial subspace of $\nu\mathscr{U}$. The homotopy class of this map does not depend on the choice of π; if π' is a second choice then $(1-t)\,\pi + t\,\pi'$, $0 \leq t \leq 1$, is a deformation of π into π' (compare 2.22).

Suppose now $\mathscr{U} \geq \mathscr{V}$ are numerable coverings. We can choose a map $\mu: \mathscr{U} \to \mathscr{V}$ such that $U \subset (\mu U)$ for every $U \in \mathscr{U}$. There is a unique simplicial map $v_{\mathscr{V}}^{\mathscr{U}}: v\mathscr{U} \to v\mathscr{V}$ which on vertices agrees with μ (cf. V, 7.11); its homotopy class does not depend on the choice of μ, as follows by "linear deformation" as above (cf. also 2.22). In barycentric coordinates we have $(v_{\mathscr{V}}^{\mathscr{U}} x)_V = \sum_{\mu U = V} x_U$, for $x \in v\mathscr{U}$. This implies $(v_{\mathscr{V}}^{\mathscr{U}})\pi = \rho$, where $\pi: A \to v\mathscr{U}$ is any numeration of $\mathscr{U}$, and $\rho: A \to v\mathscr{V}$ is the numeration such that $\rho_V = \sum_{\mu U = V} \pi_U$. Since the homotopy class of ρ is unique we always have $(v_{\mathscr{V}}^{\mathscr{U}})\pi \simeq \rho$, for any choice of μ, π, ρ. We can therefore define a functor

$$(2.11)\quad \Theta: \Omega_A \to \Lambda_A, \qquad \Theta\mathscr{U} = [\pi: A \to v\mathscr{U}], \qquad \Theta(\mathscr{U} \geq \mathscr{V}) = [v_{\mathscr{V}}^{\mathscr{U}}],$$

which we call the *numeration-functor* ($\Lambda^{\text{op}} = \mathscr{P}^A$, $[\,]$ = homotopy class).

2.12 Proposition. *The numeration functor is weakly cofinal.*— In fact, it is also strongly cofinal but that is more difficult to prove (cf. Exerc. 5), and we don't really need it.

Proof. Let $[\xi: A \to R_\xi]$ an object of Λ_A. We have to construct a numerable covering $\mathscr{U}$ such that, up to homotopy, ξ factors through $\pi: A \to v\mathscr{U}$. Take a triangulation of R_ξ, let J denote the set of vertices, $\beta_j: R_\xi \to [0,1]$, $j \in J$, the corresponding barycentric coordinates, and let $\mathscr{U}$ denote the set of all sets $(\beta_j\,\xi)^{-1}(0,1]$. By definition, we can assign to every $U \in \mathscr{U}$ an index $j(U) \in J$ such that $U = (\beta_{j(U)}\,\xi)^{-1}(0,1]$. Then $\pi = \{\pi_U = \beta_{j(U)}\,\xi\}_{U \in \mathscr{U}}$ is a numeration of $\mathscr{U}$, and $U \mapsto j(U)$ defines a simplicial map $f: v\mathscr{U} \to R_\xi$ such that $f\pi \simeq \xi$. ∎

2.13 Corollary. *For every topological space A the category $\Lambda_A = (\mathscr{P}^A)^{\text{op}}$ admits a weakly cofinal functor $\Omega_A \to \Lambda_A$ whose domain Ω_A is small.* ∎

2.14 Definition. Let $\alpha: B \to A$ a continuous map. If $\mathscr{U}$ is a numerable covering of A then $\alpha^{-1}\mathscr{U} = \{\alpha^{-1} U\}_{U \in \mathscr{U}}$ is a numerable covering of B; indeed, if π is a numeration of $\mathscr{U}$ then $\pi\alpha$ is a numeration of $\alpha^{-1}\mathscr{U}$. If $\mathscr{U} \geq \mathscr{V}$ then clearly $\alpha^{-1}\mathscr{U} \geq \alpha^{-1}\mathscr{V}$. We can therefore define a functor

$$(2.15)\qquad \Omega_\alpha: \Omega_A \to \Omega_B, \qquad \Omega_\alpha\,\mathscr{U} = \alpha^{-1}\mathscr{U}.$$

Clearly $\Omega_{\alpha\beta} = \Omega_\beta\,\Omega_\alpha$, and $\Omega_{\text{id}} = \text{id}$, thus Ω is a cofunctor from spaces to categories (in fact, to directed sets).

If $\mathscr{U}$ is as above then to every (non-empty) $W \in \alpha^{-1}\mathscr{U} = \Omega_\alpha\mathscr{U}$ we can assign a set $\mu W \in \mathscr{U}$ such that $\alpha^{-1}(\mu W) = W$, and μ defines a simplicial map of nerves $v_\alpha^{\mathscr{U}}: v\Omega_\alpha\mathscr{U} \to v\mathscr{U}$ which on vertices agrees with μ. Its homotopy class does not depend on the choice of μ, and the diagram

$$\begin{array}{ccc} B & \xrightarrow{\alpha} & A \\ \downarrow{\scriptstyle \pi\alpha} & & \downarrow{\scriptstyle \pi} \\ \nu\,\Omega_\alpha\,\mathscr{U} & \xrightarrow{\nu_\alpha^{\mathscr{U}}} & \nu\,\mathscr{U} \end{array} \tag{2.16}$$

homotopy-commutes, both assertions by contiguity (cf. 2.22). Similarly,

$$\nu_{\mathscr{V}}^{\mathscr{U}}\,\nu_\alpha^{\mathscr{U}} \simeq \nu_\alpha^{\mathscr{V}}\,\nu_{\alpha^{-1}\mathscr{V}}^{\alpha^{-1}\mathscr{U}} \text{ if } \mathscr{U} \geq \mathscr{V}, \quad \text{and} \quad \nu_{\alpha\beta}^{\mathscr{U}} \simeq \nu_\alpha^{\mathscr{U}}\,\nu_\beta^{\alpha^{-1}\mathscr{U}} \text{ if } C \xrightarrow{\beta} B \xrightarrow{\alpha} A.$$

2.17 Numerable Coverings of Products (compare Eilenberg-Steenrod IX, 5). Let A, B topological spaces and $\mathscr{U}$ a numerable covering of A. A function $\mathscr{S}$ which assigns to every $U \in \mathscr{U}$ a numerable covering $\mathscr{S}U$ of B is called a *stacking function* (on $\mathscr{U}$). The set of all sets $U \times V$, where $U \in \mathscr{U}$ and $V \in \mathscr{S}U$ is then a covering of $A \times B$, which we denote by $\mathscr{U} \times \mathscr{S}$. It is numerable, in fact, if $\pi^{\mathscr{U}} = \{\pi_U^{\mathscr{U}}\}$ is a numeration of $\mathscr{U}$, and $\pi^{\mathscr{S}U} = \{\pi_V^{\mathscr{S}U}\}$ is a numeration of $\mathscr{S}U$ then

$$\pi_{U\times V}^{\mathscr{U}\times\mathscr{S}}\colon A \times B \to [0, 1], \qquad \pi_{U\times V}^{\mathscr{U}\times\mathscr{S}}(a, b) = \pi_U^{\mathscr{U}}(a) \cdot \pi_V^{\mathscr{S}U}(b),$$

is a numeration $\pi^{\mathscr{U}\times\mathscr{S}}$ of $\mathscr{U} \times \mathscr{S}$.—Coverings of the form $\mathscr{U} \times \mathscr{S}$ will be called *stacked coverings* (of $A \times B$, over $\mathscr{U}$).

Given $\mathscr{U}$, a stacking function $\mathscr{S}$, and for every $U \in \mathscr{U}$ a numeration $\pi^{\mathscr{S}U}$ of $\mathscr{S}U$, we define a continuous map

$$\pi^{\mathscr{S}}\colon (\nu\mathscr{U}) \times B \to \nu(\mathscr{U} \times \mathscr{S}), \quad \pi^{\mathscr{S}}(\{x_U\}, b) = \{y_{U\times V} = x_U \cdot \pi_V^{\mathscr{S}U}(b)\}, \tag{2.18}$$

where $\nu\mathscr{U}$, $\nu(\mathscr{U} \times \mathscr{S})$ are the nerves of $\mathscr{U}$, $\mathscr{U} \times \mathscr{S}$, and points $x \in \nu\mathscr{U}$, $y \in \nu(\mathscr{U} \times \mathscr{S})$ are described by their barycentric coordinates, $x = \{x_U\}$, $y = \{y_{U\times V}\}$. Any such map $\pi^{\mathscr{S}}$ will be called a *stack-numeration*.

One can easily show that the homotopy class of $\pi^{\mathscr{S}}$ depends only on $\mathscr{S}$ (not on the $\pi^{\mathscr{S}U}$), and that the composition

$$A \times B \xrightarrow{\pi^{\mathscr{U}} \times \mathrm{id}} (\nu\mathscr{U}) \times B \xrightarrow{\pi^{\mathscr{S}}} \nu(\mathscr{U} \times \mathscr{S})$$

is homotopic to $\pi^{\mathscr{U}\times\mathscr{S}}$, where $\pi^{\mathscr{U}}, \pi^{\mathscr{U}\times\mathscr{S}}$ are numerations of $\mathscr{U}, \mathscr{U} \times \mathscr{S}$. We shall not use this, and we therefore leave the proof to the reader, as an exercise. One can also show (cf. Exerc. 6) that, *for compact B, every numerable covering of $A \times B$ admits a stacked refinement*. Of this we only need the special case $B = [0, 1]$, and there we have the following more explicit result.

2.19 Proposition. *If A is a space, and $\mathscr{W}$ is a numerable covering of $A\times[0,1]$ then there exists a numerable covering $\mathscr{U}$ of A, and a function $r\colon \mathscr{U}\to\mathbb{Z}$ with values $rU>1$, such that every set*

$$U\times\left[\frac{i-1}{rU},\frac{i+1}{rU}\right],\quad \textit{where } U\in\mathscr{U}, i\in\mathbb{Z},\ 0<i<rU,$$

is contained in some $W\in\mathscr{W}$. In particular, there is a stacked covering which refines $\mathscr{W}$, namely $\left\{U\times\left[\frac{i-1}{rU},\frac{i+1}{rU}\right]\right\}$, with $U\in\mathscr{U}$ and $i=1,2,\ldots,rU-1$.

Proof. For every $(r-1)$-tupel $(W_1,\ldots,W_{r-1})$ in $\mathscr{W}$ we define

$$U(W_1,\ldots,W_{r-1}) = \left\{a\in A\,\middle|\, a\times\left[\frac{i-1}{r},\frac{i+1}{r}\right]\subset W_i \text{ for } i=1,2,\ldots,r-1\right\}. \tag{2.20}$$

Clearly, $U(W_1,\ldots,W_{r-1})\times\left[\frac{i-1}{r},\frac{i+1}{r}\right]\subset W_i$ for all i, hence it suffices to show that $\mathscr{U}=\{U(W_1,\ldots,W_{r-1})\}$ is a numerable covering of A. Take a locally finite numeration $\pi=\{\pi_W\}$ of $\mathscr{W}$, and define

$$\begin{aligned}&\rho_U\colon A\to[0,1],\\ &\rho_U(a)=\mathrm{Min}_{i=1,\ldots,r-1}\,\mathrm{Min}\left\{\pi_{W_i}(a,t)\,\middle|\, t\in\left[\frac{i-1}{r},\frac{i+1}{r}\right]\right\},\end{aligned} \tag{2.21}$$

for $U=U(W_1,\ldots,W_{r-1})$. If $\rho_U(a)>0$ then $\pi_{W_i}(a,t)>0$, hence $(a,t)\in W_i$ for all $t\in\left[\frac{i-1}{r},\frac{i+1}{r}\right]$ and all i, hence $a\in U(W_1,\ldots,W_{r-1})$; this proves $\rho_U^{-1}(0,1]\subset U$. Every $(a,t)\in A\times[0,1]$ has a neighborhood which is contained in some $\pi_W^{-1}(0,1]$ and which meets only a finite number of these sets. Since $[0,1]$ is compact, we can find, for every $a\in A$,

(i) sets $W_1,\ldots,W_{r-1}\in\mathscr{W}$ such that $a\times\left[\frac{i-1}{r},\frac{i+1}{r}\right]\subset\pi_{W_i}^{-1}(0,1]$, $i=1,\ldots,r-1$;

(ii) a neighborhood V of a such that $V\times[0,1]$ meets only a finite number of sets $\pi_W^{-1}(0,1]$.

Property (i) shows that, for every $a\in A$, we have at least one

$$U=U(W_1,\ldots,W_{r-1})\quad\text{with } \rho_U(a)>0.$$

Property (ii) implies that, for *fixed* r, the family $\{\rho_{U(W_1,\dots,W_r)}\}$ is locally finite. Let $\rho_r(a)=\mathrm{Max}\{\rho_{U(W_1,\dots,W_s)}(a)|s<r\}$, and define

$$\pi'_{U(W_1,\dots,W_r)}(a)=\mathrm{Max}\{0,\rho_{U(W_1,\dots,W_r)}(a)-r\,\rho_r(a)\}.$$

Clearly, $\pi'^{-1}_U(0,1]\subset\rho_U^{-1}(0,1]\subset U$. Given $a\in A$, let k be minimal such that $\rho_U(a)>0$ for some $U=U(W_1,\dots,W_k)$; then $\pi'_U(a)=\rho_U(a)>0$. Moreover, if we choose $N>k$ such that $N\rho_U(a)>1$ then $N\rho_U(x)>1$ for all x in a neighborhood V' of a; in this neighborhood we have $r\rho_r>1$ for all $r\geq N$, hence all $\pi'_{U(W_1,\dots,W_r)}$ with $r\geq N$ vanish in V'. This shows that the family of all π'_U is locally finite. To make it a numeration of $\mathscr{U}$, simply divide each π'_U by the sum of all of these functions. ∎

2.22 Remark. If A is a simplical space, and X an arbitrary topological space then two maps $f_0, f_1\colon X\to A$ are said to be *contiguous* if, for all $x\in X$, the pair $f_0(x)$, $f_1(x)$ is contained in a single simplex of A. *Contiguous maps are homotopic.* This is familiar for simplicial maps (Eilenberg-Steenrod VI.3), but we also used it in another case (after 2.16). Define

$$f_t\colon X\to A,\quad 0\leq t\leq 1,\quad \text{by } \pi f_t(x)=(1-t)\,\pi f_0(x)+t\,\pi f_1(x),$$

where $\pi\colon A\to[0,1]$ is any barycentric coordinate. This "linear deformation" $\{f_t\}$ may not be continuous in the weak topology of A, however, it obviously is continuous in the strong topology, hence $f_0\simeq f_1$ by Dowker's theorem 2.9.

2.23 Exercises. *1.* If $\{\pi_j\colon A\to[0,1]\}_{j\in J}$ is a partition of unity (not necessarily point-finite), and I is any subset of J then

$$\pi_I\colon A\to[0,1],\quad \pi_I(a)=\textstyle\sum_{j\in I}\pi_j(a),$$

is continuous.

2. If $\mathscr{V}$ is a covering of A which admits a numerable refinement then $\mathscr{V}$ is numerable. If A is a polyhedron, and $\mathscr{V}$ is any open covering of A then A can be so triangulated (cf. J.H.C. Whitehead 1939, Thm. 35) that every open star is contained in some $V\in\mathscr{V}$. The set of open stars is then numerable (by barycentric coordinates) and refines $\mathscr{V}$, hence $\mathscr{V}$ is numerable. This (together with 2.8) shows that A is paracompact.

3 (compare Eilenberg-Steenrod II.8). If $\mathscr{U},\mathscr{V}$ are numerable coverings of A, B then $\mathscr{U}\times\mathscr{V}=\{U\times V\}$, $U\in\mathscr{U}$, $V\in\mathscr{V}$, is a numerable covering of $A\times B$. The projections $\mathscr{U}\times\mathscr{V}\to\mathscr{U},\mathscr{V}$ define simplicial maps of nerves $\nu(\mathscr{U}\times\mathscr{V})\to\nu\mathscr{U},\nu\mathscr{V}$, hence a map $r\colon\nu(\mathscr{U}\times\mathscr{V})\to(\nu\mathscr{U})\times(\nu\mathscr{V})$. We also have a map $i\colon(\nu\mathscr{U})\times(\nu\mathscr{V})\to\nu(\mathscr{U}\times\mathscr{V})$, namely $(i(x,y))_{U\times V}=x_U\cdot y_V$, where $\{x_U\}$ denotes the family of barycentric coordinates of

$x \in v\mathscr{U}$, etc. Show that $ri = \mathrm{id}$, $ir \simeq \mathrm{id}$, and that the diagram

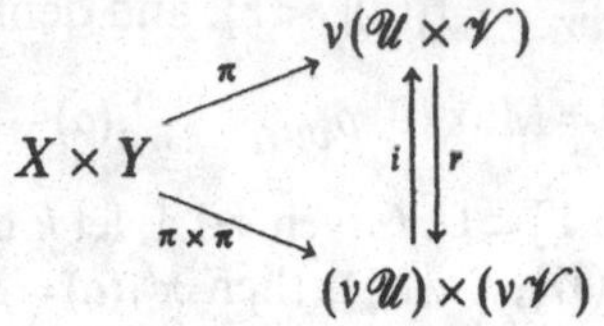

homotopy-commutes. *Corollary: The product of two polyhedrons has the homotopy type of a polyhedron.*

4. Let $\pi = \{\pi_j\}_{j \in J}$ a partition of unity on A. A subset $S \subset J$ is called a *simplex of* π if a point $a \in A$ exists such that $\pi_j a \neq 0$ for all $j \in S$. Every simplex of π is countable; further, π is point-finite iff every simplex is finite. The partition is said to be *barycentric* if

(i) for every simplex $S \subset J$ and every family $\{a_s\}_{s \in S}$ such that $0 \leq a_s \leq 1$, $\sum_s a_s = 1$, there is a unique point $a \in A$ with $a_s = \pi_s(a)$, for all $s \in S$.

(ii) A has the topology induced by π, i.e. the coarsest topology for which every π_j is continuous.

If, moreover, every simplex is finite then we say π is *finitely-barycentric.* A finitely-barycentric partition π is just about the same as a triangulation of A; in fact, it is a homeomorphism $\pi: A \approx v\mathscr{U}$, where $\mathscr{U} = \{\pi_j^{-1}(0, 1]\}$, and $v\mathscr{U}$ is taken with the strong topology. A general barycentric partition π might be thought of as a "triangulation" in which simplices of *countably-infinite dimension* are admitted. Let $A_f \subset A$ be the subspace which consists of all points a such that $\{j \in J \mid \pi_j a \neq 0\}$ is finite. Then $\pi \mid A_f$ is finitely-barycentric; use 2.8 to show that the inclusion map $A_f \to A$ is a homotopy-equivalence.

5*. *For every topological space A the numeration* functor $\Omega_A \to \Lambda_A$ *is strongly cofinal.* We know already (2.12) that it is weakly cofinal. What remains to be shown is that the following diagram can always be completed (dotted arrows),

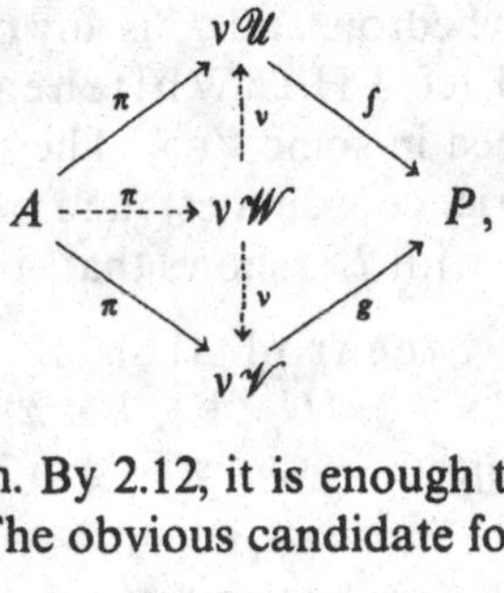

where P is a polyhedron. By 2.12, it is enough to fill in a polyhedron Q instead of a nerve $v\mathscr{W}$. The obvious candidate for Q is as follows,

$$Q = \{(x, y, \omega) \in (v\mathscr{U}) \times (v\mathscr{V}) \times P^{[0,1]} \mid \omega(0) = f(x),\ \omega(1) = g(y)\};$$

it is clear how to define the dotted arrows. The only trouble is to show that Q is (homotopy equivalent to) a polyhedron. For this one may consult Milnor 1959.—The same reference is needed to show that Λ_A is a directed category (1.12, Exerc. 3).

6*. *If A is an arbitrary space, B a compact space, and $\mathscr{W}$ a numerable covering of $A \times B$ then there exists a stacked covering of $A \times B$ which refines $\mathscr{W}$.* This easily implies 2.19. The proof is along the same lines as that for 2.19, although more complicated. We give some indications. Consider the set J of all functions $j\colon \mathscr{K}_j \to \mathscr{W}$, where $\mathscr{K}_j$ is a finite numerable covering of B by compact sets. For every $j \in J$, let $U_j = \{a \in A \mid a \times K \subset j(K)$ for all $K \in \mathscr{K}_j\}$. One can show that $\mathscr{U} = \{U_j\}_{j \in J}$ is a numerable covering of A; clearly, $U_j \times K \subset j(K) \in \mathscr{W}$, hence $\{U_j \times K\}$ is a stacked covering (with stacking function $U_j \mapsto \mathscr{K}_j$) which refines $\mathscr{W}$.

In order to prove that $\mathscr{U}$ is numerable one can (as for 2.19) use the functions

$$\rho_j\colon A \to [0,1], \qquad \rho_j(a) = \operatorname{Min}_{K \in \mathscr{K}_j} \operatorname{Min}\{\pi_{jK}(a,t) \mid t \in K\},$$

where $\pi = \{\pi_W\}_{W \in \mathscr{W}}$ is a locally finite numeration of $\mathscr{W}$; one well-orders J, defines $\pi_j'(u) = \operatorname{Max}\{0, \rho_j(a) - \operatorname{Sup}_{i<j} \rho_i(a)\}$, and divides each π_j' by the sum of all of these functions. Another way to prove numerability of $\mathscr{U}$ is to assume first that A is paracompact. If $\mathscr{W}$ is open then $\mathscr{U}$ is easily seen to be an open covering, hence numerable. For instance, this applies if $A = P^B$, where P is a polyhedron in the strong topology, because then P^B is metric. The general case then follows because $\mathscr{W}$ is refined by the counterimage of a numerable covering on $(\nu\mathscr{W})^B \times B$ under a continuous map $\tilde{\pi} \times \mathrm{id}\colon A \times B \to (\nu\mathscr{W})^B \times B$.

A.3 Extending Functors from Polyhedrons to More General Spaces

As before, we denote by $\mathscr{T}\!op$ the category of topological spaces and continuous maps, and we let $\mathscr{P}\!ol \subset \mathscr{T}\!op$ be the full subcategory of polyhedrons (=triangulable spaces). We consider homotopy-invariant[11] cofunctors $F\colon \mathscr{P}\!ol \to \mathscr{K}$, and we shall be concerned with the problem of extending F to $\mathscr{T}\!op$. The range-category $\mathscr{K}$ is assumed to possess limits of arbitrary direct systems (in the sense of VIII, 5), and is otherwise arbitrary. For instance, $\mathscr{K}$ may stand for the category of abelian groups, or the category of sets.

3.1 Definition. If $F\colon \mathscr{P}\!ol \to \mathscr{K}$, $G\colon \mathscr{T}\!op \to \mathscr{K}$, *are cofunctors then* G is called an *extension* of F if $G|\mathscr{P}\!ol$ is equivalent with F.

[11] This means: $\alpha \simeq \beta \Rightarrow F\alpha = F\beta$, for continuous maps α, β.

3.2 Lemma. *If G: $\mathcal{T}op \to \mathcal{K}$ is an extension of F: $\mathcal{P}ol \to \mathcal{K}$ then G is equivalent to some cofunctor G': $\mathcal{T}op \to \mathcal{K}$ such that $G'|\mathcal{P}ol = F$.* In view of this we shall usually assume $G|\mathcal{P}ol = F$ when speaking of extensions.

Proof. Let Φ: $G|\mathcal{P}ol \cong F$ be an equivalence. Define G' on objects A resp. morphisms α of $\mathcal{T}op$ as follows.

$$G'A = \begin{cases} FA & \text{if } A \in \mathcal{P}ol, \\ GA & \text{if } A \notin \mathcal{P}ol, \end{cases}$$

$$G'\alpha = \begin{cases} F\alpha, & \text{if range}(\alpha) \in \mathcal{P}ol \text{ and domain}(\alpha) \in \mathcal{P}ol, \\ \Phi(G\alpha), & \text{if range}(\alpha) \in \mathcal{P}ol \text{ and domain}(\alpha) \notin \mathcal{P}ol, \\ (G\alpha)\Phi^{-1}, & \text{if range}(\alpha) \notin \mathcal{P}ol \text{ and domain}(\alpha) \in \mathcal{P}ol, \\ G\alpha, & \text{if range}(\alpha) \notin \mathcal{P}ol \text{ and domain}(\alpha) \notin \mathcal{P}ol. \end{cases}$$

Obviously, G' is a cofunctor such that $G' \cong G$, and $G|\mathcal{P}ol = F$. ∎

3.3 Definition. Let F: $\mathcal{P}ol \to \mathcal{K}$ be a homotopy-invariant cofunctor. For every topological space A we consider the category $\mathcal{P}^A$ of polyhedrons under A, its dual Λ_A, and the functor $F \circ R$: $\Lambda_A \to \mathcal{K}$, where R = range. Thus $F \circ R$ assigns to the object $[\xi: A \to R_\xi]$ of Λ_A the object $F(R_\xi)$ of $\mathcal{K}$. We shall see below (3.8) that the limit of $F \circ R$ (in the sense of 1.1) always exists; we denote it by $F^K A = \lim(F \circ R)$.

Every continuous map α: $B \to A$ induces a functor Λ_α: $\Lambda_A \to \Lambda_B$ (cf. 2.2), hence a limit morphism (cf. remark after 1.6) which we denote by $F^K \alpha = (\Lambda_\alpha)_\infty$: $F^K A \to F^K B$. Since $\Lambda_{\alpha\beta} = \Lambda_\beta \Lambda_\alpha$, and $\Lambda_{\mathrm{id}} = \mathrm{id}$, we see that F^K is a cofunctor, F^K: $\mathcal{T}op \to \mathcal{K}$. In fact, we shall see (3.8, 3.7) that F^K is an extension of F; it is called the *Kan-extension*.

3.4 Definition (compare 2.10). Let F: $\mathcal{P}ol \to \mathcal{K}$ be a homotopy invariant cofunctor. For every topological space A let Ω_A the directed set of numerable coverings of A. If we assign to every numerable covering $\mathcal{U} \in \Omega_A$ the value of F on the nerve $\nu\mathcal{U}$, and to every refinement $\mathcal{U} \geq \mathcal{V}$ the induced morphism $F(\nu_{\mathcal{V}}^{\mathcal{U}})$: $F(\nu\mathcal{V}) \to F(\nu\mathcal{U})$, we obtain a direct system whose limit we denote by $\check{F}A = \lim(F \circ \nu) = \lim\{F\nu\mathcal{U}\}_{\mathcal{U} \in \Omega_A}$.

If α: $B \to A$ is a continuous map then (cf. 2.15, 2.16) Ω_α: $\Omega_A \to \Omega_B$ is order-preserving, and the family of morphisms $F(\nu_\alpha^{\mathcal{U}})$: $F(\nu\mathcal{U}) \to F(\nu\Omega_\alpha\mathcal{U})$, $\mathcal{U} \in \Omega_A$, passes to the limit (e.g. by VIII, 5.11). The induced limit-morphism $\check{F}\alpha$ is given by

$$\text{(3.5)} \qquad \check{F}\alpha: \check{F}A \to \check{F}B, \quad (\check{F}\alpha)\, u_A^{\mathcal{U}} = u_B^{\Omega_\alpha \mathcal{U}} (F\nu_\alpha^{\mathcal{U}}),$$

where u_A resp. u_B is the universal transformation for $\check{F}A$ resp. $\check{F}B$. One easily verifies that $\check{F}(\alpha\beta) = (\check{F}\beta)(\check{F}\alpha)$, and $\check{F}(\mathrm{id}) = \mathrm{id}$, i.e. that $\check{F}$ is a cofunctor, $\check{F}: \mathcal{Top} \to \mathcal{K}$. In fact, we shall see (cf. 3.7) that $\check{F}$ is an extension of F; it is called the *Čech-extension.* Moreover, we shall prove (3.8) that $\check{F} \sim F^K$.

3.6 Proposition. *If* $F: \mathcal{Pol} \to \mathcal{K}$ *is homotopy-invariant then so is* $\check{F}: \mathcal{Top} \to \mathcal{K}$.

Proof. It is enough to show that $\check{F} j_0 = \check{F} j_1$, where $j_t: A \to A \times [0,1]$, $j_t(a) = (a, t)$. Consider stacked coverings $\mathcal{U} \times \mathcal{S} = \left\{ U \times \left[\frac{i-1}{r}, \frac{i+1}{r}\right] \right\}$ of $A \times [0,1]$ as in 2.19, $U \in \mathcal{U}$, $i = 1, 2, \ldots, r(U)-1$. Then $j_t^{-1}(\mathcal{U} \times \mathcal{S}) = \mathcal{U}$, for every t. Moreover, we assert that the maps $v_{j_0}^{\mathcal{U}\times\mathcal{S}}, v_{j_1}^{\mathcal{U}\times\mathcal{S}}: v\mathcal{U} \to v(\mathcal{U} \times \mathcal{S})$ of 2.16 are homotopic, in fact, that the stack numeration $\pi^{\mathcal{S}}: (v\mathcal{U}) \times [0,1] \to v(\mathcal{U} \times \mathcal{S})$ of 2.18 is a deformation of $v_{j_0}^{\mathcal{U}\times\mathcal{S}}$ into $v_{j_1}^{\mathcal{U}\times\mathcal{S}}$. For this, one observes that for fixed $U \in \mathcal{U}$ the endpoints 0 resp. 1 of $[0,1]$ lie in only one set of the covering $\mathcal{S}U$, namely in $\left[0, \frac{2}{r}\right]$ resp. $\left[\frac{r-2}{r}, 1\right]$; hence the only function $\pi_i^{\mathcal{S}U}$ which is not zero on 0 resp. 1 is $\pi_1^{\mathcal{S}U}$, $\pi_1^{\mathcal{S}U}(0) = 1$, resp. $\pi_{r-1}^{\mathcal{S}U}$, $\pi_{r-1}^{\mathcal{S}U}(1) = 1$. The definition 2.18 therefore shows that the maps $\pi^{\mathcal{S}} | (v\mathcal{U}) \times \{0\}$, $\pi^{\mathcal{S}} | v\mathcal{U} \times \{1\}$, coincide with the simplicial maps

$$U \mapsto U \times \left[0, \frac{2}{r}\right], \qquad U \mapsto U \times \left[\frac{r-2}{r}, 1\right]$$

of $v\mathcal{U}$ into $v(\mathcal{U} \times \mathcal{S})$, and these maps agree with $v_{j_0}^{\mathcal{U}\times\mathcal{S}}, v_{j_1}^{\mathcal{U}\times\mathcal{S}}$, by definition of the latter.

Since F is homotopy-invariant we get $F(v_{j_0}^{\mathcal{U}\times\mathcal{S}}) = F(v_{j_1}^{\mathcal{U}\times\mathcal{S}})$, hence (cf. 3.5) $(\check{F} j_0)\, u^{\mathcal{U}\times\mathcal{S}} = u^{\mathcal{U}} F(v_{j_0}^{\mathcal{U}\times\mathcal{S}}) = u^{\mathcal{U}} F(v_{j_1}^{\mathcal{U}\times\mathcal{S}}) = (\check{F} j_1)\, u^{\mathcal{U}\times\mathcal{S}}$, where u denotes universal transformations. Every numerable covering $\mathcal{W}$ of $A \times [0,1]$ admits a refinement of the form $\mathcal{U} \times \mathcal{S}$, by 2.19. It follows that

$$(\check{F} j_0)\, u^{\mathcal{W}} = (\check{F} j_0)\, u^{\mathcal{U}\times\mathcal{S}} (F\, v_{\mathcal{W}}^{\mathcal{U}\times\mathcal{S}}) = (\check{F} j_1)\, u^{\mathcal{U}\times\mathcal{S}} (F\, v_{\mathcal{W}}^{\mathcal{U}\times\mathcal{S}}) = (\check{F} j_1)\, u^{\mathcal{W}},$$

hence $\check{F} j_0 = \check{F} j_1$. ∎

Recall (2.10) that every $\mathcal{U} \in \Omega_A$ admits a unique (up to $\simeq$) numeration $\pi^{\mathcal{U}}: A \to v\mathcal{U}$. Let $F: \mathcal{Pol} \to \mathcal{K}$ be a homotopy-invariant cofunctor.

3.7 Proposition. *If P is a polyhedron then* $\{F\pi^{\mathcal{U}}: F v\mathcal{U} \to FP\}_{\mathcal{U}\in\Omega_P}$ *is a universal transformation (for* $F \circ v$*), hence* $\check{F}P = FP$. *This isomorphism is natural, i.e.* $\check{F}$ *is an extension of* F. *For every topological space A the transformation* $\{\check{F}\pi^{\mathcal{U}}: F v\mathcal{U} \to \check{F}A\}_{\mathcal{U}\in\Omega_A}$ *is universal.*

Proof. If $\mathcal{T}$ is a triangulation of P then for every vertex v of $\mathcal{T}$ we have the barycentric coordinate $\hat{v}: P \to [0,1]$, and the sets $\hat{v}^{-1}(0,1]$ constitute a numerable covering $\mathscr{V}\mathscr{T}$ of P. In fact, the barycentric coordinates $\{\hat{v}\}$ form a numeration $\pi^{\mathscr{V}\mathscr{T}}: P \to v\mathscr{V}\mathscr{T}$ which is also a simplicial homeomorphism; in particular, $F(\pi^{\mathscr{V}\mathscr{T}})$: $F(v\mathscr{V}\mathscr{T}) \cong FP$.

If $\mathscr{S}, \mathscr{T}$ are two triangulation of P such that $\mathscr{V}\mathscr{S} \geq \mathscr{V}\mathscr{T}$ then we say $\mathscr{S}$ is a *refinement* of $\mathscr{T}$, and we write $\mathscr{S} \geq \mathscr{T}$. For instance, every subdivision (Spanier 3.3) of $\mathscr{T}$ is a refinement of $\mathscr{T}$. If $\mathscr{S} \geq \mathscr{T}$ then $\pi^{\mathscr{V}\mathscr{T}} \simeq v^{\mathscr{V}\mathscr{S}}_{\mathscr{V}\mathscr{T}} \pi^{\mathscr{V}\mathscr{S}}$, hence $F(\pi^{\mathscr{V}\mathscr{T}}) = F(\pi^{\mathscr{V}\mathscr{S}})\, F(v^{\mathscr{V}\mathscr{S}}_{\mathscr{V}\mathscr{T}})$, hence $F(v^{\mathscr{V}\mathscr{S}}_{\mathscr{V}\mathscr{T}}) = F(\pi^{\mathscr{V}\mathscr{S}})^{-1}\, F(\pi^{\mathscr{V}\mathscr{T}})$ is an isomorphism. A polyhedron admits arbitrarily fine triangulations (cf. J. H. C. Whitehead 1939, Thm. 35), i.e. the set of coverings $\mathscr{V}\mathscr{T}$, as $\mathscr{T}$ ranges over all triangulations, is cofinal in the set Ω_P of all numerable coverings; hence (VIII, 5.17), $\check{F}P = \lim\{F\, v\mathscr{U}\} \cong \lim\{F\, v\mathscr{V}\mathscr{T}\}$. Since all morphisms $F(v^{\mathscr{V}\mathscr{S}}_{\mathscr{V}\mathscr{T}})$ of the latter direct system are isomorphisms we obtain $u^{\mathscr{V}\mathscr{T}}$: $F(v\mathscr{V}\mathscr{T}) \cong \check{F}P$ for all $\mathscr{T}$, where u denotes universal transformations; combined with $F(\pi^{\mathscr{V}\mathscr{T}})$: $F(v\mathscr{V}\mathscr{T}) \cong FP$ we get $\rho^{\mathscr{T}} = u^{\mathscr{V}\mathscr{T}}(F\pi^{\mathscr{V}\mathscr{T}})^{-1}$: $FP \cong \check{F}P$. If $\mathscr{U}$ is any numerable covering of P, and $\mathscr{S}$ is a triangulation such that $\mathscr{V}\mathscr{S} \geq \mathscr{U}$ then $u^{\mathscr{U}} = u^{\mathscr{V}\mathscr{S}}\, F(v^{\mathscr{V}\mathscr{S}}_{\mathscr{U}})$, and $\pi^{\mathscr{U}} \simeq v^{\mathscr{V}\mathscr{S}}_{\mathscr{U}}\, \pi^{\mathscr{V}\mathscr{S}}$, hence $F(v^{\mathscr{V}\mathscr{S}}_{\mathscr{U}}) = F(\pi^{\mathscr{V}\mathscr{S}})^{-1} F(\pi^{\mathscr{U}})$, and $u^{\mathscr{U}} = u^{\mathscr{V}\mathscr{S}}\, F(\pi^{\mathscr{V}\mathscr{S}})^{-1} F(\pi^{\mathscr{U}}) = \rho^{\mathscr{S}} F(\pi^{\mathscr{U}})$. If $\mathscr{U}$ is itself of the form $\mathscr{U} = \mathscr{V}\mathscr{T}$ this shows $\rho^{\mathscr{S}} = \rho^{\mathscr{T}}$, hence $\rho = \rho^{\mathscr{T}}$ does not depend on $\mathscr{T}$. For general $\mathscr{U}$ again, we conclude that $\{F(\pi^{\mathscr{U}})\} = \rho^{-1} \circ \{u^{\mathscr{U}}\}$ is indeed universal.

Now let α: $Q \to P$ be a continuous map between polyhedrons. For any numerable covering $\mathscr{U}$ of P we have

$$(F\alpha)(F\pi^{\mathscr{U}}) = F(\pi^{\mathscr{U}}\alpha) \overset{2.16}{=} F(v^{\mathscr{U}}_{\alpha}\, \pi^{\alpha^{-1}\mathscr{U}}) = F(\pi^{\alpha^{-1}\mathscr{U}})\, F(v^{\mathscr{U}}_{\alpha}) \overset{3.5}{=} (\check{F}\alpha)(F\pi^{\mathscr{U}}),$$

hence $F\alpha = \check{F}\alpha$. Thus $\check{F}|\mathscr{P}ol = F$.

Finally, let A be an arbitrary space, and $\mathscr{U}$ a numerable covering of A, with numeration $\pi = \pi^{\mathscr{U}}$: $A \to v\mathscr{U}$, and universal map $u^{\mathscr{U}}_A$: $F(v\mathscr{U}) \to \check{F}A$. By 3.5, we have $(\check{F}\pi^{\mathscr{U}})\, u^{\mathscr{V}}_{v\mathscr{U}} = u^{\pi^{-1}\mathscr{V}}_A\, F(v^{\mathscr{V}}_{\pi})$, for any numerable covering $\mathscr{V}$ of $v\mathscr{U}$. In particular, we take $\mathscr{V} = \mathscr{V}\mathscr{T}$, where $\mathscr{T}$ is the given triangulation of $v\mathscr{U}$ (having $\mathscr{U}$ as the set of vertices). The corresponding numeration $\pi^{\mathscr{V}\mathscr{T}}: v\mathscr{U} \to v\mathscr{V}\mathscr{T}$ is then a (simplicial) homeomorphism, and $u^{\mathscr{V}}_{v\mathscr{U}} = F(\pi^{\mathscr{V}\mathscr{T}})$ by the first part of 3.7. Hence

$$\check{F}\pi^{\mathscr{U}} = u^{\pi^{-1}\mathscr{V}}_A (F\, v^{\mathscr{V}}_{\pi})\, F(\pi^{\mathscr{V}\mathscr{T}})^{-1} = u^{\pi^{-1}\mathscr{V}}_A\, F((\pi^{\mathscr{V}\mathscr{T}})^{-1} v^{\mathscr{V}}_{\pi}).$$

But the map $(\pi^{\mathscr{V}\mathscr{T}})^{-1} v^{\mathscr{V}}_{\pi}$: $v\pi^{-1}\mathscr{V} \to v\mathscr{U}$ agrees with $v^{\pi^{-1}\mathscr{V}}_{\mathscr{U}}$, by the very definitions. Hence $\check{F}\pi^{\mathscr{U}} = u^{\pi^{-1}\mathscr{V}}_A\, F(v^{\pi^{-1}\mathscr{V}}_{\mathscr{U}}) = u^{\mathscr{U}}_A$, the latter because u_A is a transformation. ∎

3.8 Proposition. *The transformation* $\{\check{F}\xi: FR_{\xi} \to \check{F}A\}$, $[\xi] \in \Lambda_A$, *is universal, hence* $\check{F}A = \lim(F \circ R: \Lambda_A \to \mathscr{K}) = F^K A$. *This isomorphism is natural, i.e.* $\check{F} = F^K$.

Proof. Given any transformation $\{\varphi_\xi: FR_\xi \to Y\}$ into an object Y of $\mathscr{K}$ we have to show that there is a unique $\varphi: \check{F}A \to Y$ such that $\varphi \circ (\check{F}\xi) = \varphi_\xi$. Such a φ must satisfy $\varphi \circ (\check{F}\pi) = \varphi_\pi$ for any numeration $\pi: A \to v\mathscr{U}$; since $\{\check{F}\pi\}_{\mathscr{U}\in\Omega_A}$ is universal (by 3.7), this shows that φ is unique, and in fact, we can define $\varphi: \check{F}A \to Y$ by these equations. We must then prove $\varphi \circ (\check{F}\xi) = \varphi_\xi$. But ξ factors through some numeration,

$$\xi: A \xrightarrow{\pi} v\mathscr{U} \xrightarrow{\alpha} R_\xi,$$

by 2.12 (the numeration functor is weakly cofinal); hence

$$\varphi_\xi = \varphi_\pi \circ (F\alpha) = \varphi \circ (\check{F}\pi) \circ (F\alpha) = \varphi \circ \check{F}(\alpha\pi) = \varphi \circ (\check{F}\xi).$$

It remains to prove naturality; let $\beta: B \to A$ a continuous map. Then $(F^K\beta)\, u_A^\xi = u_B^{\xi\beta}$, by definition of $F^K\beta$, where u denotes universal transformations. But $u^\xi = \check{F}\xi$, hence $(F^K\beta)(\check{F}\xi) = \check{F}(\xi\beta) = (\check{F}\beta)(\check{F}\xi)$, hence $F^K\beta = \check{F}\beta$ by universality of $\{\check{F}\xi\}$. ∎

The Kan- resp. Čech-extension of $F: \mathscr{P}ol \to \mathscr{K}$ admits the following abstract characterization.

3.9 Proposition (Universal Property). *If $G: \mathscr{T}op \to \mathscr{K}$ is any homotopy-invariant cofunctor, and $\psi: F \to G|\mathscr{P}ol$ is any natural transformation then there is a unique natural transformation $\Psi: F^K \to G$ (resp. $\check{F} \to G$) such that $\Psi|\mathscr{P}ol = \psi$.*

Proof. Let $\Psi: F^K \to G$ be such that $\Psi|\mathscr{P}ol = \psi$. We can apply naturality of Ψ to the map $\xi: A \to R_\xi$, where $[\xi] \in \Lambda_A$, and we find that $\Psi_A \circ (F^K\xi) = (G\xi) \circ \psi_{R_\xi}$. Since $\{F^K\xi = \check{F}\xi\}$ is universal (3.8) this shows that Ψ_A is uniquely determined, and in fact, we can define $\Psi_A: F^K A \to GA$ by these equations. We must then prove naturality; let $\beta: B \to A$ a continuous map. We get

$$\begin{aligned}(\Psi_B \circ F^K\beta) \circ F^K\xi = \Psi_B \circ F^K(\xi\beta) = G(\xi\beta) \circ \psi_{R_\xi} &= (G\beta) \circ (G\xi) \circ \psi_{R_\xi} \\ &= (G\beta) \circ \Psi_A \circ (F^K\xi),\end{aligned}$$

hence $\Psi_B \circ (F^K\beta) = (G\beta) \circ \Psi_A$ by universality of $\{F^K\xi\}$. ∎

3.10 Corollary. *The numeration functor $\Theta: \Omega_A \to \Lambda_A$ (cf. 2.11) induces an isomorphism $\Theta_\infty: \check{F}A \cong F^K A$.*

This is really a restatement of 3.8; it also follows from 3.9 because Θ_∞: $\check{F}A \cong F^K$ is a natural transformation such that $\Theta_\infty|\mathscr{P}ol = \mathrm{id}$. ∎

In VIII, 6 the name of Čech and the notation $\check{H}$ was used for a construction which differs from the present one. This is justified by the following

3.11 Proposition. *If A is a locally closed subset of a polyhedron P then $\check{F}A = F^K A \cong \lim\{FV\}$, where the limit is taken over the directed set of open (or polyhedral) neighborhoods of A in P.*—This follows immediately from 2.4 and 1.9(ii). ∎

3.12 Example (compare Lee-Raymond, Thm. 2). Let P denote a fixed polyhedron, and let $H_P\colon \mathcal{Top} \to \mathcal{Sets}$ the cofunctor which assigns to every space A the set of homotopy classes of continuous map $f\colon A \to P$, and which assigns to $\beta\colon B \to A$ the map $H_P\beta\colon H_P A \to H_P B$, $(H_P\beta)[f] = [f\beta]$. Let $h_P = H_P|\mathcal{Pol}$. We assert,

$$\check{h}_P = H_P, \tag{3.13}$$

i.e. H_P is the Čech-(or Kan-)extension of its restriction to $\mathcal{Pol}$. This will be clear if we verify the

3.14 Universal Property. *If $G\colon \mathcal{Top} \to \mathcal{Sets}$ is any homotopy-invariant cofunctor, and $\psi\colon h_P \to G|\mathcal{Pol}$ is any natural transformation then there is a unique natural transformation $\Psi\colon H_P \to G$ such that $\Psi|\mathcal{Pol} = \psi$.*

The proof follows by applying the Yoneda-Lemma I, 1.12 both to h_P and H_P; we write out the necessary details: Given any $\Psi\colon H_P \to G$ such that $\Psi|\mathcal{Pol} = \psi$, we apply naturality of Ψ to maps $f\colon A \to P$ and we get $\Psi_A[f] = \Psi_A \circ (H_P f)[\mathrm{id}_P] = (Gf) \circ \psi_P[\mathrm{id}_P]$. This shows that Ψ_A is uniquely determined by ψ (even by $\psi_P[\mathrm{id}_P]$), and in fact, we can define Ψ_A: $H_P A \to GA$, by $\Psi_A[f] = (Gf) \circ \psi_P[\mathrm{id}_P]$. We must prove naturality. Let $\beta\colon B \to A$ a continuous map; then

$$\begin{aligned}(G\beta) \circ \Psi_A[f] &= (G\beta) \circ (Gf) \circ \psi_P[\mathrm{id}_P] = G(f\beta) \circ \psi_P[\mathrm{id}_P] \\ &= \Psi_B[f\beta] = \Psi_B \circ (H_P\beta)[f]. \quad ∎\end{aligned}$$

For instance, if $P = K(G, n)$ is an Eilenberg-MacLane space then h_P agrees with (singular) cohomology, $h_P = H^n(-, G)$; cf. Spanier 8.1.8. It follows that H_P agrees with Čech-cohomology $\check{H}(-, G)$; compare Huber.

3.15 Remarks. It is clear that the Kan-procedure 3.3, but also the Čech-procedure, for extending functors applies to many other situations. For instance, we can replace $\mathcal{Pol}$ by a category of special polyhedrons (say finite, finite-dimensional ...) and/or $\mathcal{Top}$ by a category of special spaces (say compact, finite-dimensional ...). Or we can consider the category $\mathcal{Top}^{(2)}$ of pairs of topological spaces and in it the category $\mathcal{Pol}^{(2)}$ of pairs of polyhedrons. The reader is encouraged to think about these generalizations and modifications; he may consult Lee-Raymond for details.

3.16 Exercises. *1.* Let F_1, F_2: $\mathscr{K} \to \mathscr{L}$ be functors, and let i: $X \to Y$, r: $Y \to X$ be morphisms in $\mathscr{K}$ such that $r\,i = \mathrm{id}$ (X is a *retract* of Y). Show that if $F_1 Y = F_2 Y$, and $F_1(ir) = F_2(ir)$, then $F_1 X \cong F_2 X$; in fact, there is a unique morphism φ: $F_1 X \to F_2 X$ such that $(F_2\, i)\,\varphi = F_1\, i$, or $\varphi(F_1\, r) = F_2\, r$, and φ is isomorphic. Roughly, this says that functors which agree on Y also agree on retracts of Y. If a functor is only defined on some subcategory containing Y and (ir), how would you (try to) extend it to X?

2. For any topological space A define the category $\mathscr{P}_A$ *of polyhedrons over* A dually to 2.1 (objects are homotopy classes ξ: $D_\xi \to A$ whose domain D_ξ is a polyhedron, etc.). Show that $\mathscr{P}_A$ is directed (cf. 1.12 Exerc. 3).

For any covariant homotopy-invariant functor F: $\mathscr{P}ol \to \mathscr{K}$ define $F_K A = \lim(FD$: $\mathscr{P}_A \to \mathscr{K})$, and turn F_K into a (homotopy-invariant) functor F_K: $\mathscr{T}op \to \mathscr{K}$. This is the *Kan-extension* of a covariant functor. Show, by cofinality, that $F_K A = F\,|sA|$, where sA is the singular semi-simplicial set of A, and $|sA|$ is its geometric realization (compare Milnor 1957, Thm. 4). Use this, or (simpler!) show directly that singular homology is (equivalent to) the Kan-extension of simplicial homology. In this case (but not for arbitrary F) one can also replace $\mathscr{P}_A$ by the subcategory of *compact* polyhedrons over A.

3. Let X be a normal space for which $X \times [0,1]$ is also normal, and let $A \subset X$ be a closed subspace. Let $\mathscr{B}$ be a directed (by inverse inclusion) set of closed subsets of X such that (i) $A \subset B$ for every $B \in \mathscr{B}$, and (ii) every neighborhood of A contains at least one $B \in \mathscr{B}$. For instance, $\mathscr{B}$ might be the set of all closed neighborhoods of A. Every continuous map ξ: $A \to R_\xi$ into a polyhedron admits an extension to some B, say e_ξ: $B_\xi \to R_\xi$, $B_\xi \in \mathscr{B}$.

Now let F: $\mathscr{P}ol \to \mathscr{K}$ a homotopy-invariant cofunctor and F^K: $\mathscr{T}op \to \mathscr{K}$ its Kan-extension. Consider the direct system $\{F^K B\}_{B \in \mathscr{B}}$, with arrows induced by inclusion. For every $[\xi$: $A \to R_\xi]$, pick e_ξ: $B_\xi \to R_\xi$ as above, and put $\gamma[\xi] = B_\xi$, $d_{[\xi]} = F^K(e_\xi)$: $FR_\xi \to F^K B_\xi$. Show that (γ, d) passes to the limit (in the sense of A.1.3), hence a limit morphism $\lim(d)$: $F^K A \to \lim\{F^K B\}_{B \in \mathscr{B}}$. The inclusion maps i_B: $A \to B$, on the other hand, induce $F^K i_B$: $F^K B \to F^K A$, hence $\{F^K i_B\}$: $\lim\{F^K B\} \to F^K A$. Show that these are reciprocal isomorphisms, $F^K A \cong \lim\{F^K B\}_{B \in \mathscr{B}}$ (*weak continuity of Čech-extensions*).

Bibliography

ADAMS, J. F.: On the cobar construction. Coll. de Topol. Algébr. Louvain 1956, pp. 81–87.

ADAMS, J. F.: On the nonexistence of elements of Hopf invariant one. Ann. of Math. **72**, 20–104 (1960).

ADAMS, J. F., ATIYAH, M.: *K*-theory and the Hopf invariant. Quart. J. Math. Oxford (2) **17**, 31–38 (1966).

ALEXANDROFF, P., HOPF, H.: Topologie. Berlin: Springer 1935.

ARTIN, E., FOX, R.: Some wild cells and spheres in threedimensional space. Ann. of Math. **49**, 979–990 (1948).

BOREL, A.: Sur la cohomologie des espaces fibrés principaux et des espaces homogènes de groupes de Lie compacts. Ann. of Math. **57**, 115–207 (1953).

BOREL, A.: Sur l'homologie et la cohomologie des groupes de Lie compacts connexes. Amer. J. Math. **76**, 273–342 (1954).

BORSUK, K.: Über eine Klasse von lokal zusammenhängenden Räumen. Fund. Math. **19**, 221–242 (1932).

BOS, W.: Zur Einbettung einer differenzierbaren Mannigfaltigkeit in einen euklidischen Raum. Arch. d. Math. **16**, 232–234 (1965).

BOURBAKI, N.: Topologie générale, Chap. I/II. Paris: Hermann 1947, 3e éd. 1961.

BOURBAKI, N.: Algèbre, Chap. III. Paris: Hermann 1948, 2e éd. 1951.

BROWN, R.: Elements of modern topology. London: McGraw-Hill 1968.

CARTAN, H., EILENBERG, S.: Homological algebra. Princeton: Univ. Press 1956.

CHERN, S. S., HIRZEBRUCH, F., SERRE, J. P.: On the index of a fibered manifold. Proc. Amer. Math. Soc. **8**, 587–596 (1957).

CHEVALLEY, C.: Fundamental concepts of algebra. New York: Acad. Press 1956.

COHN, P. M.: Free ideal rings. J. Algebra **1**, 47–69 (1964).

tom DIECK, T., KAMPS, K. H., PUPPE, D.: Homotopietheorie. Lecture Notes in Math. 157. Berlin-Heidelberg-New York: Springer 1970.

DIEUDONNÉ, J.: Foundations of modern analysis. New York: Acad. Press 1960.

DOLD, A., MAC LANE, S., OBERST, U.: Projective classes and acyclic models. Reports of the Midwest Category Seminar. Lecture Notes in Mathematics 47. Heidelberg: Springer 1967.

DOWKER, C. H.: Topology of metric complexes. Amer. J. Math. **74**, 555–577 (1952).

EILENBERG, S.: Abstract description of some basic functors. J. Indian Math. Soc. **24**, 231–234 (1960).

EILENBERG, S., MAC LANE, S.: Acyclic models. Amer. J. Math. **75**, 189–199 (1953).

EILENBERG, S., STEENROD, N.: Foundations of algebraic topology. Princeton: Univ. Press 1952.

FADELL, E.: On a coincidence theorem of F. B. Fuller. Pacific J. Math. **15**, 825–834 (1965).

GODBILLON, C.: Éléments de topologie algébrique. Paris: Hermann 1971.

GODEMENT, R.: Topologie algébrique et théorie des faisceaux. Paris: Hermann 1958.

GREENBERG, M. J.: Lectures on algebraic topology. New York: Benjamin 1967.

GRIFFITH, H. B.: The fundamental group of two spaces with a common point. Quart. J. Math. Oxford (2) **5**, 175–190 (1954). (Correction) Quart. J. Math. Oxford (2) **6**, 154–155 (1955).
HANNER, O.: Some theorems on absolute neighborhood retracts. Arkiv Mat. **1**, 389–408 (1952).
HILTON, P. J.: An introduction to homotopy theory. Cambridge: Univ. Press 1953.
HU, S. T.: Homotopy theory. New York: Acad. Press 1959.
HU, S. T.: Theory of retracts. Detroit: Wayne State Univ. Press 1965.
HUBER, P.: Homotopical cohomology and Čech cohomology. Math. Annalen **144**, 73–76 (1961).
HUREWICZ, W., WALLMAN, H.: Dimension theory. Princeton: Univ. Press 1948.
KAN, D.: Adjoint functors. Trans. Amer. Math. Soc. **87**, 294–329 (1958).
KAPLANSKY, I.: Infinite abelian groups. Ann Arbor: Univ. of Mich. Press 1954, rev. ed. 1969.
KELLEY, J. L.: General topology. Princeton: Van Nostrand 1955.
KERVAIRE, M. A.: A manifold which does not admit any differentiable structure. Comment. Math. Helv. **34**, 257–270 (1960).
KNESER, H., KNESER, M.: Reell-analytische Strukturen der Alexandroff-Halbgeraden und der Alexandroff-Geraden. Arch. d. Math. **9**, 104–106 (1960).
KOCH, W., PUPPE, D.: Differenzierbare Strukturen auf Mannigfaltigkeiten ohne abzählbare Basis. Arch. d. Math. **19**, 95–102 (1968).
KURATOWSKI, C.: Topologie, 2 vols. Warschau 1948–1950.
KURATOWSKI, C.: Transl. Topology, 2 vols. New York: Acad. Press 1966–1968.
KUROSH, A. G.: The theory of groups. I. New York: Chelsea Publ. Co. 1955.
LEE, C. N., RAYMOND, F.: Čech extensions of contravariant functors. Trans. Amer. Math. Soc. **133**, 415–434 (1968).
MAC LANE, S.: Homology. Berlin-Göttingen-Heidelberg: Springer 1963.
MAC LANE, S.: Categories. For the working mathematician. Berlin-Heidelberg-New York: Springer 1972.
MASSEY, W. S.: Algebraic topology: an introduction. New York: Harcourt, Brace and World 1967.
MATHER, M. R.: Paracompactness and partitions of unity. Mimeographed Notes 1964. Cf. also Ph. D. thesis, Cambridge University 1965.
MILNOR, J.: Construction of universal bundles I. Ann. of Math. **63**, 272–284 (1956).
MILNOR, J.: The geometric realization of a semi-simplicial complex. Ann. of Math. **65**, 357–362 (1957).
MILNOR, J.: On spaces having the homotopy type of a *CW*-complex. Trans. Amer. Math. Soc. **90**, 272–280 (1959).
MILNOR, J.: A survey of cobordism theory. Enseignement Math. (2) **8**, 16–23 (1962).
MILNOR, J.: Morse theory. Ann. of Math. Studies 51. Princeton: Univ. Press 1963.
MILNOR, J., MOORE, J.: On the structure of Hopf algebras. Ann. of Math. **81**, 211–264 (1965).
MITCHELL, B.: Theory of categories. New York: Acad. Press 1965.
MIYAZAKI, H.: The paracompactness of *CW*-complexes. Tohoku Math. J. (2) **4**, 309–313 (1952).
NAGAMI, K.: Dimension theory. New York: Acad. Press 1970.
PONTRJAGIN, L. S.: Topologische Gruppen, I/II. Leipzig: Teubner 1957/58.
SCHUBERT, H.: Topologie. Stuttgart: Teubner 1964.
SEIFERT, H., THRELFALL, W.: Lehrbuch der Topologie. Leipzig: Teubner 1934; Chelsea: Publ. Co. 1947.
SPANIER, E.: Function spaces and duality. Ann. of Math. **70**, 338–378 (1959).
SPANIER, E.: Algebraic topology. New York: McGraw-Hill 1966.
SPECKER, E.: Additive Gruppen von Folgen ganzer Zahlen. Portugaliae Math. **9**, 131–140 (1950).
STEENROD, N.: The topology of fibre bundles. Princeton: Univ. Press 1951, 2nd ed. 1957.

STEENROD, N.: Homology groups of symmetric groups and reduced power operations. Proc. Nat. Acad. Sci. USA **39**, 213–217 (1953).

STEENROD, N., EPSTEIN, D.: Cohomology operations. Ann. of Math. Studies 50. Princeton: Univ. Press 1962.

SWAN, R.: Algebraic *K*-theory. Lecture Notes in Math. 76. Berlin-Heidelberg-New York: Springer 1968.

THOM, R.: Quelques propriétés globales des variétés différentiables. Comment. Math. Helv. **28**, 17–86 (1954).

VERDIER, J. L.: Topologie et faisceaux. Sém. IHES, Bures-sur-Yvette, 1964.

WATTS, CH.: Intrinsic characterisation of some additive functors. Proc. Amer. Math. Soc. **11**, 5–8 (1960).

WHITEHEAD, J. H. C.: Simplicial spaces, nuclei and *m*-groups. Proc. London Math. Soc. (2) **45**, 243–327 (1939).

WHITEHEAD, J. H. C.: Combinatorial homotopy I. Bull. Amer. Math. Soc. **55**, 213–245 (1949).

Index

∞CIM

A. Dold
Lectures on Algebraic Topology
ISBN 3-540-58660-1

F. Hirzebruch
Topological Methods in Algebraic Geometry
ISBN 3-540-58663-6

T. Kato
Perturbation Theory for Linear Operators
ISBN 3-540-58661-X

S. Kobayashi
Transformation Groups in Differential Geometry
ISBN 3-540-58659-8

S. Mac Lane
Homology
ISBN 3-540-58662-8

D. Mumford
Algebraic Geometry I
Complex Projective Varieties
ISBN 3-540-58657-1

C. L. Siegel, J. K. Moser
Lectures on Celestial Mechanics
ISBN 3-540-58656-3

A. Weil
Basic Number Theory
ISBN 3-540-58655-5

K. Yosida
Functional Analysis
ISBN 3-540-58654-7

O. Zariski
Algebraic Surfaces
ISBN 3-540-58658-X